AF598666

Sequence Stratigraphy in British Geology

Geological Society Special Publications
Series Editor A. J. FLEET

GEOLOGICAL SOCIETY SPECIAL PUBLICATION NO. 103

Sequence Stratigraphy in British Geology

EDITED BY

S. P. HESSELBO
Department of Earth Sciences
University of Oxford, UK

AND

D. N. PARKINSON
Western Atlas Logging Services
London Geoscience Centre, UK

1996
Published by
The Geological Society
London

THE GEOLOGICAL SOCIETY

The Society was founded in 1807 as The Geological Society of London and is the oldest geological society in the world. It received its Royal Charter in 1825 for the purpose of 'investigating the mineral structure of the Earth'. The Society is Britain's national society for geology with a membership of around 8000. It has countrywide coverage and approximately 1000 members reside overseas. The Society is responsible for all aspects of the geological sciences including professional matters. The Society has its own publishing house, which produces the Society's international journals, books and maps, and which acts as the European distributor for publications of the American Association of Petroleum Geologists, SEPM and the Geological Society of America.

Fellowship is open to those holding a recognized honours degree in geology or cognate subject and who have at least two years' relevant postgraduate experience, or who have not less than six years' relevant experience in geology or a cognate subject. A Fellow who has not less than five years' relevant postgraduate experience in the practice of geology may apply for validation and, subject to approval, may be able to use the designatory letters C Geol (Chartered Geologist).

Further information about the Society is available from the Membership Manager, The Geological Society, Burlington House, Piccadilly, London W1V 0JU, UK. The Society is a Registered Charity, No. 210161.

Published by the Geological Society from:
The Geological Society Publishing House
Unit 7, Brassmill Enterprise Centre
Brassmill Lane
Bath BA1 3JN
UK
(*Orders*: Tel. 01225 445046
Fax 01225 442836)

First published 1996

British Library Cataloguing in Publication Data
A catalogue record for this book is available from the British Library.

ISBN 1-897799-49-7

Typeset by Type Study, Scarborough, UK

Printed in Great Britain by
The Alden Press, Osney Mead, Oxford, UK

Distributors

USA
AAPG Bookstore
PO Box 979
Tulsa
OK 74101-0979
USA
(*Orders*: Tel. (918) 584-2555
Fax (918) 584-0469)

Australia
Australian Mineral Foundation
63 Conyngham Street
Glenside
South Australia 5065
Australia
(*Orders*: Te! (08) 379-0444
Fax (08) 379-4634)

India
Affiliated East-West Press PVT Ltd
G-1/16 Ansari Road
New Delhi 110 002
India
(*Orders*: Tel. (11) 327-9113
Fax (11) 326-0538)

Japan
Kanda Book Trading Co.
Tanikawa Building
3-2 Kanda Surugadai
Chiyoda-Ku
Tokyo 101
Japan
(*Orders*: Tel. (03) 3255-3497
Fax (03) 3255-3495)

Contents

Sequence stratigraphy in British geology

STEPHEN P. HESSELBO[1] & D. NEIL PARKINSON[2]

[1]*Department of Earth Sciences, University of Oxford, Parks Road, Oxford OX1 3PR, UK*

[2]*Western Atlas Logging Services, 455 London Road, Isleworth, Middlesex TW7 5AB, UK*

In what has now become the standard model for sequence stratigraphy, a *depositional sequence* comprises an unconformity-bounded package of genetically related strata, whose internal geometries are influenced largely by fluctuating sea level (Fig. 1). This model evolved from regional mapping (Sloss 1963), and was later developed from seismic-reflection profiles on passive continental margins (Payton 1977). It was further extended to a higher resolution using data from the surface and sub-surface in many geological settings (Mutti 1985; Wilgus *et al.* 1988; Van Wagoner *et al.* 1990). Here we do not provide a comprehensive summary of the terminology that has grown-up around sequence stratigraphy: judging by its sparse usage within the papers of this volume much of it is redundant anyway. The interested reader should refer to the reviews of Haq (1991), Vail *et al.* (1991), Mitchum & Van Wagoner (1991), Posamentier *et al.* (1992) and Posamentier & James (1993) for further details. In essence, sequence stratigraphy is practised through the recognition of key· surfaces which define the boundaries of packages of genetically related strata (*systems tracts*) and by the recognition

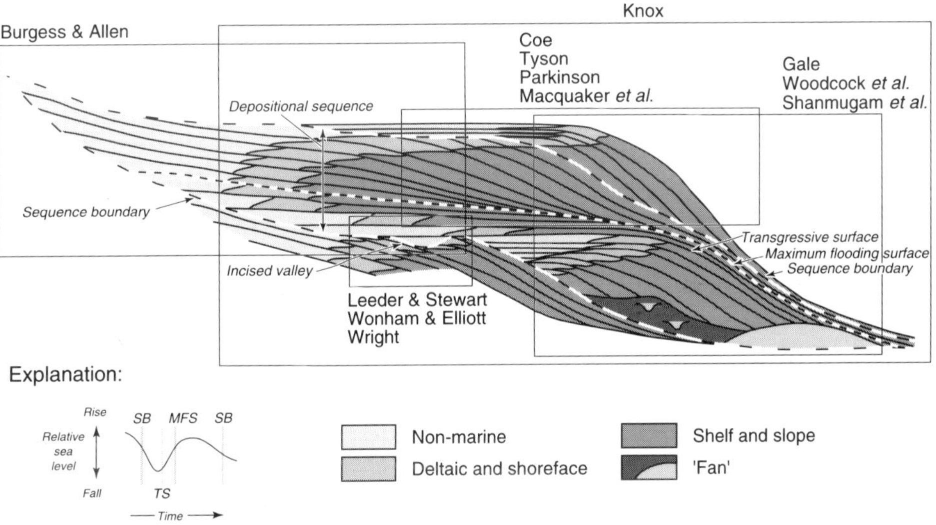

Fig. 1. The standard sequence stratigraphical model for siliciclastic systems (based on Haq *et al.* 1988 and, Christie-Blick & Driscoll 1995) showing the different areas covered by papers in this volume. Wright (this volume) and Coe (this volume) discuss successions that were carbonate ramps, with an implied sequence stratigraphical character not dissimilar to siliciclastic deposits (see Schlager 1992; Tucker *et al.* 1993). A *sequence boundary* (SB) is defined as an unconformity and its basinward correlative conformity. The lower of the sequence boundaries in this figure (*type 1*) is distinguished from the upper one (*type 2*) by its association with 'fan' sedimentation at the toe of slope, which has been related to rapid, high-magnitude, relative sea-level fall. The *maximum flooding surface* (MFS) corresponds to the time of greatest landward extent of the facies belts within a depositional sequence, and is characterized by detrital sediment starvation in basinward regions. The *transgressive surface* (TS) is a widespread horizon across which deep-water sediments abruptly overlie shallow water sediments, and it forms as the underlying progradational sedimentary succession is flooded by rising relative sea level. *Systems tracts* are defined between these key surfaces. For full discussion of this model, definition of systems tracts and their relation to relative and eustatic sea-level change the reader should refer to Haq *et al.* (1988), Posamentier & Vail (1988), Posamentier *et al.* (1988, 1990, 1992), Mitchum & Van Wagoner (1991), Vail *et al.* (1991) Jacquin *et al.* (1992) and Posamentier & James (1993). Useful critiques and alternative models are provided by Miall (1986), Galloway (1989), Carter *et al.* (1991), Fulthorpe (1991), Schlager (1991),Walker (1992) and Christie-Blick & Driscoll (1995), and references therein. As a substitute for depositional sequences as defined above, Galloway (1989) has proposed use of *genetic stratigraphic sequences* bounded by maximum flooding surfaces, and his approach has also been followed by many workers.

From HESSELBO, S. P. & PARKINSON, D. N. (eds), 1996, *Sequence Stratigraphy in British Geology*, Geological Society Special Publication No. 103, pp. 1–7.

of transgressive and regressive facies trends within those packages (Fig. 1). The model is hierarchical and, to some extent, independent of time or physical scale.

The power of the sequence stratigraphical model probably lies in its simplicity (see summary in Christie-Blick & Driscoll 1995) but many of the fundamental questions posed as a result of the early sequence stratigraphical work remain unanswered, for example the relative importance of eustasy versus more localized tectonic effects (e.g. Parkinson & Summerhayes 1985; Hubbard 1988; and many others) or sediment supply (e.g. Galloway 1989). Thus, the topic of sequence stratigraphy provides much fertile ground for further theoretical and empirical work, whether concerned with verification, refutation or modification of the standard sequence stratigraphical model, or whether concerned with the application of the model to help solve wider geological problems.

In addressing the application of sequence stratigraphy to British geology, this book focuses perforce on diverse aspects of sequence stratigraphy that tend to cut across divisions based on depositional environment or age alone. There are two main themes that run through most of the papers in this volume.

(1) Biostratigraphical control is commonly so good in much of the British area, that precise correlations are possible within and between basins, and the geologist need not speculate whether the elements of a depositional sequence in one basin are precisely synchronous with those in another: this can be established as fact or fiction (see for example Hesselbo & Jenkyns 1995, 1996; Coe 1995, this volume; Parkinson, this volume). For this reason, many of the British outcrop sections serve as standards that can be compared to other areas around the world. This was recognized early on by Haq *et al.* (1988) who derived much of their proposed global sea-level history from a sequence stratigraphical analysis of British sections.

(2) Many geologists have to work with the rocks they are given, rather than the rocks they would interpret by choice. Many, perhaps all, of the successions described within this volume may rightly be considered as difficult to interpret using the simple (to some, simplistic) framework of the standard sequence stratigraphical model. However, as is demonstrated by several of the papers in this volume, sequence stratigraphical ideas can cast new light on problematic facies and, in return, problematic facies can offer critical insights into the sequence stratigraphical model (see also, for example, Tucker 1991).

The studies in this volume concern both surface and subsurface geology and cover most parts of the British stratigraphical column (Fig. 2). It was our intention to maximize the stratigraphical coverage in order to gauge the impact that sequence stratigraphy is having upon our understanding of the geology of the British Isles. Nonetheless, the papers fall more evenly into groups based largely on depositional setting, and it is by this criterion that we briefly review the contents below:

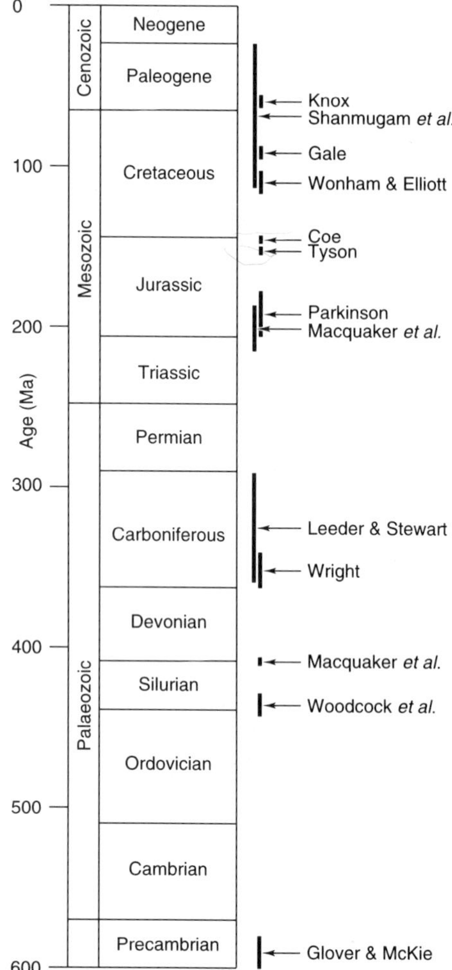

Fig. 2. Stratigraphical column showing the range of geological ages covered by papers in this volume. The time scale is based on Harland *et al.* (1990), Cande & Kent (1992), and Gradstein *et al.* (1994).

Non-marine and paralic sequences

Non-marine settings perhaps represent the most difficult of depositional systems for the application of sequence stratigraphy. Not only are the problems of correlation at their most extreme but also the simple fact of distance from sea means that strata may be totally unaffected by sea-level change. The process by which the standard sequence stratigraphical model has been constructed, a 'distillation' from many ancient and modern case studies by inductive reasoning (Walker 1990, 1992; Christie-Blick & Driscoll 1995), is not of itself adequate for the sedimentary systems with poor chronostratigraphical records such as the fluvial system. Hence, recent attempts to understand the controls of stratal geometries in non-marine settings have concentrated on forward modelling. Two different approaches to modelling fluvial systems are illustrated by papers in this volume. In the first study, **Burgess & Allen** take a geometrical approach, using equilibrium profiles in computer-generated simulations to show that changes in the shape of the fluvial profile may be a first-order control on sequence architecture, irrespective of conditions of sea-level change. In the second study, **Leeder & Stewart** use numerical experiments to model the response of a river to relative sea-level fall and illustrate the rates and magnitudes of channel incision expected from various combinations of river slope, shelf slope and sediment transport coefficients.

In contrast to strictly non-marine systems, paralic settings offer potentially the greatest insights into the sedimentary effects of relative sea-level change, because sea-level fluctuations leave a more direct record in these facies than they do in any other. In paralic environments dominated by siliciclastic sediment, the development of estuarine facies within 'incised valleys' and the influence of relative sea-level change in the organization of these valley fills has become an important issue, not only for documenting sea-level history (e.g. Plint 1988; Ruffell & Wach 1991), but also for predicting the geometries of the numerous hydrocarbon reservoirs found in these settings (Dalrymple *et al.* 1994). In this volume, **Wonham & Elliott** reinterpret the Lower Cretaceous Woburn Sands, a classic example of tidal sedimentation (Dalrymple 1992), as having been deposited largely as an estuarine complex within superimposed incised valleys, rather than either an offshore tidal environment or single transgressive estuary as have been proposed previously (Bridges 1982; Johnson & Levell 1995).

In paralic settings, palaeosols and their attendant diagenesis provide crucial evidence of the subaerial phases in the successional history (e.g. Tucker & Wright 1990; Tucker 1993; Wright 1994). This is of particular interest in peritidal carbonate successions, because vertical thickness variations of metre-scale carbonate cycles have been widely used as a reliable monitor of relative sea-level fluctuations at a number of frequencies (discussed in Drummond & Wilkinson 1993). In this volume, **Wright** reviews the occurrence of palaeosols within peritidal carbonates in Early Carboniferous depositional sequences of southwest Britain, and concludes that their character and spatial distribution do not support the operation of high frequency, moderate amplitude eustatic sea-level fluctuations during that time (cf. Elrick & Read, 1991).

Shallow marine sequences

By virtue of their commonly highly refined biostratigraphy, shallow-marine deposits offer the greatest chance of assessing the synchoneity or otherwise of inferred relative sea-level cycles in widely separated basins (e.g. de Graciansky *et al.* 1996). However, although simpler to correlate, key stratigraphical surfaces from these settings may be no easier to relate to specific relative sea-level changes than they are in any other setting.

In studies using subsurface data, regional correlation of supposed maximum flooding surfaces within shallow marine strata is commonly based largely on correlation of gamma-ray maxima, thought to be generated by relatively argillaceous intervals within sandstones, organic-rich intervals within mudstones, or mineralized surfaces within a condensed section (see for discussion: Milton *et al.* 1990; Partington *et al.* 1993*a,b*; Creany & Passey 1993; Armentrout *et al.* 1993; Hesselbo 1996). Two papers within this volume address different aspects of this topic. **Tyson** reviews the sequence stratigraphical interpretation of organic-rich mudrocks and then applies the principles that he derives to the Upper Jurassic Kimmeridge Clay, the main petroleum source-rock for the North Sea. In contrast to other recent work on the same interval (Herbin *et al.* 1995 and references therein), Tyson concludes that sedimentary dilution, and its affect on preservation, is the principal control on organic-matter enrichment in these settings. Sediment dilution can be related to relative sea-level change by starvation or winnowing, lending some support to the concept that organic-rich mudrocks developed around the time of maximum flooding. **Parkinson** details a regional outcrop-based spectral gamma-ray survey of Lower Jurassic strata, and shows how variations in the relative abundances of Th, K and U (which contain or parent the main gamma-emitting isotopes) can be related to transgressive and regressive facies successions in rocks which are otherwise difficult to interpret in this respect.

Mineralized surfaces are commonly represented by ooidal ironstones or 'bone beds'; two further shallow-marine facies of problematic origin. Their significance in sequence stratigraphy is discussed from a geochemical viewpoint by **Macquaker *et al.***, who conclude that although the phosphate-rich 'bone beds' are most likely to have formed on marine flooding surfaces, oolitic ironstones can have formed over a much wider range of relative sea-level conditions.

A very different approach to the use of sequence stratigraphy as applied to shallow-marine facies is illustrated by **Coe**. After first establishing a sequence stratigraphical template from a relatively complete Upper Jurassic shallow-marine carbonate succession, she then uses this to effect a coherent interpretation on a basin-wide scale, of stratigraphically less complete successions. The interpretation takes correlation to a greater degree of resolution than obtainable through current biostratigraphical methods.

Deep marine sequences

The prediction of deep-marine sedimentary facies and their large-scale depositional geometries is discussed by **Shanmugam *et al.*** who, by reference to subsurface examples from the Paleogene and Cretaceous in and around the North Sea area, challenge the view that 'basin floor fans' as defined by their seismic-reflection and geophysical-log character, are composed predominantly of turbidite facies, preferring instead to consider them as comprising mostly debris flows and slumps.

A combination of detailed section measurement and basin-wide biostratigraphical and lithological correlation has allowed **Gale** to reconstruct the larger scale geometrical relationships of a pelagic succession, the Upper Cretaceous Chalk of southern England. The geometrical relationships of these rocks exhibit strong similarities to those described from outer-shelfal siliciclastic deposits. Importantly, Gale demonstrates that the putative major transgressive hardgrounds correspond to positive carbon-isotope excursions. Although carbon isotopes may be a good proxy for relative sea-level change, these results for Mesozoic pelagic carbonates contrast with those from Cenozoic shallow-marine examples which, according Baum *et al.* (1994), show negative excursions during rising relative sea level.

In a detailed and basin-wide study of predominantly deep-water siliciclastic deposits **Woodcock *et al.*** build on the earlier work of Woodcock (1990) to investigate the probable interactions of tectonics, sediment supply and eustasy as applied to the late Ordovician and early Silurian of the Welsh Basin. In common with the regional studies outlined below, these authors illustrate that the debate concerning tectonics versus eustasy is still very much alive, especially in tectonic settings that are known to have been highly active during deposition. They also highlight the importance of considering multiple sediment sources and the possible distinctly different eustatic, climatic or tectonic influences that may have operated on each source (cf. Galloway 1989).

Regional studies

The interrelationships between eustasy, tectonics, igneous activity and climate are brought out well by regional studies. Drawing on much new work on the Paleocene of northwest Europe summarized in Knox *et al.* (1996), it has been possible for **Knox** to suggest clear relationships in space and time between the uplift of shallow-water areas, deposition in deep-water areas and activity in the British Tertiary Igneous Province. These patterns, if verified as further new data become available, should provide constraints on tectonic and geophysical models for the uplift of the British area during the early phases of sea-floor spreading in the adjacent Atlantic.

The spectral gamma-ray data from the Lower Jurassic of northwest Europe refered to above (**Parkinson**) show regional patterns in elemental abundances that suggest temporal and geographical changes in detrital clay mineral assemblages produced either by means of a large-scale fine-grained sediment-transport path, or changing palaeoclimatic influences on the clay minerals derived locally. This study suggests much future work documenting regional patterns in spectral gamma-ray data and investigating their significance as palaeogeographical or palaeoclimatological indicators via their relationship to clay mineralogy (cf. Hesselbo 1996; Deconinck & Vanderaveroet 1996).

Special problems apply to the sequence stratigraphy of Neoproterozoic successions (Christie-Blick *et al.* 1995). These problems are obviously compounded when the studied succession is sited within an orogenic belt. At first sight, the metasedimentary Dalradian of the Scottish Caledonides would appear to be unpromising ground for the application of sequence stratigraphy, but **Glover & McKie** use the recognition of depositional sequences to provide criteria for the correlation of shallow-marine facies to deep-marine facies and to suggest new locations for the basin depocentres. Further work of this kind may significantly advance our understanding of the early history of the northern margin of Iapetus (see Soper 1994; Soper & England 1995).

Summary

Stratigraphy, and particularly sequence stratigraphy, is an integrating discipline, so we are particularly pleased to have contributions from industry, academia and the British Geological Survey in varying combinations. We hope that this diverse collection of papers may go some way to demonstrating the value of sequence stratigraphy: at several scales of analysis; in a variety of tectonic settings; to a wide range of geological ages; and using many different kinds of geological and geophysical data-set.

Predicting the future direction of any branch of

knowledge will always be a tricky business. However, advances in sequence stratigraphy may come from: ever better chronostratigraphical and chronometrical calibration of ancient successions; technological developments leading to higher resolution imaging of stratigraphical geometries; and a more sophisticated understanding of the relationship between climatic change and the sedimentary record. Whatever the prospect for sequence stratigraphy, we hope that the papers in this volume will provide enough thoughtfully generated stratigraphical data (*sensu lato*) to aid future studies whose aims may be entirely unrelated to our own.

We thank the many referees whose thoroughness has led to significant improvements in the quality of this volume, and Harold Reading for useful comments on a draft of this introduction. We are grateful to BP Exploration for providing the abstract volume at the Burlington House meeting in March 1994 (sponsored by the Stratigraphy Committee and the Petroleum Group of the Geological Society) at which several of these papers were initially presented. We record our particular thanks to Colin Summerhayes for enabling and encouraging much of our early involvement with this topic.

References

ARMENTROUT, J. M., MALECEK, S. J., FEARN, L. B., SHEPPARD, C. E., NAYLOR, P. H., MILES, A. W., DESMARAIS, R. J. & DUNAY, R. E. 1993. Log-motif analysis of Paleogene depositional systems tracts, Central and Northern North Sea: defined by sequence stratigraphic analysis. *In*: PARKER, J. R. (ed.) *Petroleum Geology of Northwest Europe: Proceedings of the 4th Conference*. Geological Society, London, 45–57.

BAUM, J. S., BAUM, G. R., THOMPSON, P. R. & HUMPHREY, J. D. 1994. Stable isotopic evidence for relative and eustatic sea-level changes in Eocene to Oligocene carbonates, Baldwin County, Alabama. *Geological Society of America, Bulletin*, **106**, 824–839.

BRIDGES, P. J. 1982. Sedimentology of a tidal sea: the Lower Greensand of southern England. *In*: STRIDE, A. H. (ed.) *Offshore Tidal Sands*. Chapman and Hall, 183–189.

CANDE, S. C. & KENT, D. V. 1992. A new geomagnetic polarity time scale for the Late Cretaceous and Cenozoic. *Journal of Geophysical Research*, **97**, 13917–13951.

CARTER, R. M., ABBOTT, S. T., FULTHORPE, C. S., HAYWICK, D. W. & HENDERSON, R. A. 1991. Application of global sea-level and sequence-stratigraphic models in Southern Hemisphere Neogene strata from New Zealand. *In*: MACDONALD, D. I. M. (ed.) *Sedimentation, Tectonics and Eustasy: Sea-Level Changes at Active Margins*. Special Publications of the International Association of Sedimentologists, **12**, 41–65.

CHRISTIE-BLICK, N. & DRISCOLL, N. W. 1995. Sequence stratigraphy. *Annual Reviews of Earth and Planetary Science*, **23**, 451–478.

——, Dyson, I. A. & VON DER BORCH, C. C. 1995. Sequence stratigraphy and the interpretation of Neoproterozoic Earth history. *Precambrian Research*, **73**.

COE, A. L. 1995. A comparison of the Oxfordian successions of Dorset, Oxfordshire and Yorkshire. *In*: TAYLOR, P. D. (ed.) *Field Geology of the British Jurassic*. Geological Society, London, 151–172.

CREANEY, S. & PASSEY, Q. R. 1993. Recurring patterns of total organic carbon and source rock quality within a sequence stratigraphic framework. *American Association of Petroleum Geologists, Bulletin*, **77**, 386–401.

DALRYMPLE, R. W. 1992. Tidal depositional systems. *In*: WALKER, R. G. & JAMES, N. P. (eds) *Facies Models: Response to Sea Level Change*. Geological Association of Canada, 195–218.

——, BOYD, R. & ZAITLIN, B. A. (eds) 1994. *Incised-Valley Systems: Origin and Sedimentary Sequences*. Special Publications of the Society for Sedimentary Geology, **51**.

DECONINCK, J. F. & VANDERAVEROET, P. 1996. Eocene to Pleistocene clay mineral sedimentation off New Jersey, western North Atlantic (ODP Leg 150, Sites 903 and 905). *In*: MOUNTAIN, G. S., MILLER, K. G., BLUM, P. & TWITCHELL, D. C. (eds) *Proceedings of the Ocean Drilling Program, Scientific Results, 150, New Jersey continental slope and rise*, in press.

DE GRACIANSKY, P. C., HARDENBOL, J., JACQUIN, T., VAIL, P. R. & FARLEY, M. S.(eds) 1996. *Mesozoic and Cenozoic Sequence Stratigraphy of European Basins*. Special Publications of the SEPM.

DRUMMOND, C. N. & WILKINSON, B. H. 1993. On the use of cycle thickness diagrams as records of long-term sealevel change during accumulation of carbonate sequences. *Journal of Geology*, **101**, 687–702.

ELRICK, M. & READ, J. F. 1991. Cyclic ramp-to-basin carbonate deposits, Lower Mississippian, Wyoming and Montana: a combined field and computer modelling study. *Journal of Sedimentary Petrology*, **61**, 1194–1224.

FULTHORPE, C. S. 1991. Geological controls on seismic sequence resolution. *Geology*, **19**, 61–65.

GALLOWAY, W. E. 1989. Genetic stratigraphic sequences in basin analysis I: architecture and genesis of flooding-surface bounded depositional units. *American Association of Petroleum Geologists, Bulletin*, **73**, 125–142.

GRADSTEIN, F. M., AGTERBERG, F. P., OGG, J. G., HARDENBOL, J. VAN VEEN, P., THIERRY, J. & HUANG, Z. 1994. A Mesozoic time scale. *Journal of Geophysical Research*, **99**, 24051–24074.

HAQ, B. U. 1991. Sequence stratigraphy, sea-level change, and significance for the deep sea. *In*: MACDONALD, D. I. M. (ed.) *Sedimentation, Tectonics and Eustasy: Sea-Level Changes at Active Margins*. Special Publications of the International Association of Sedimentologists, **12**, 3–39.

——, HARDENBOL, J. & VAIL, P. R. 1988. Mesozoic and Cenozoic chronostratigraphy and cycles of sea-level change. *In*: WILGUS, C. K., HASTINGS, B. S., KENDALL, G. ST. C., POSAMENTIER, H. W., ROSS, C. A. & VAN WAGONER, J. C (eds) *Sea-Level Changes: an Integrated Approach*. Special Publications of the

Society of Economic Paleontologists and Mineralogists, **42**, 71–108.

HARLAND, W. B., ARMSTRONG, R. L., COX, A. V., CRAIG, L. E., SMITH, A. G. & SMITH, D. G. 1990. *A Geologic Time Scale 1989*. Cambridge University Press, Cambridge.

HERBIN, J. P., FERNANDEZ-MARTINEZ, J. L., GEYSSANT, J. R, ALBANI, A. EL., DECONINCK, J. F., PROUST, J. N., COLBEAUX, J. P. & VIDIER, J. P. 1995. Sequence stratigraphy of source rocks applied to the study of the Kimmeridgian/Tithonian in the northwest European shelf (Dorset/UK, Yorkshire/UK and Boulonnais/France). *Marine and Petroleum Geology*, **12**, 177–194.

HESSELBO, S. P. 1996. Spectral gamma-ray logs in relation to clay mineralogy and sequence stratigraphy, Cenozoic of the Atlantic Margin, offshore New Jersey. *In*: MOUNTAIN, G. S., MILLER, K. G., BLUM, P. & TWITCHELL, D. C. (eds) *Proceedings of the Ocean Drilling Program, Scientific Results*, **150**, New Jersey continental slope and rise, 000–000.

—— & JENKYNS, H. C. 1995. A comparison of the Hettangian to Bajocian successions of Dorset and Yorkshire. *In*:TAYLOR, P. D. (ed.) *Field Geology of the British Jurassic*. Geological Society of London, 105–150.

—— & —— 1996. Sequence stratigraphy of the British Lower Jurassic. *In*: DE GRACIANSKY, P. C., HARDENBOL, J., JACQUIN, T., VAIL, P. R. & FARLEY, M. S. (eds). *Mesozoic and Cenozoic Sequence Stratigraphy of European Basins*. Special Publications of the SEPM, in press.

HUBBARD, R. J. 1988. Age and significance of sequence boundaries on Jurassic and Early Cretaceous rifted continental margins. *American Association of Petroleum Geologists, Bulletin*, **72**, 49–72.

JACQUIN, T., GARCIA, J-P., PONSOT, C., THIERRY, J. & VAIL, P. R. 1992. Séquences de dépôt et cycles régressif/trangressifs en domaine marin carbonaté : example du Dogger du Bassin de Paris. *Comptes Rendus de l'Académie des Sciences de Paris, Série II*, **315**, 353–362.

JOHNSON, H. D. & LEVELL, B. K. 1995. Sedimentology of a transgressive, estuarine sand complex: the Lower Cretaceous Woburn Sands (Lower Greensand), southern England. *In*: PLINT, A. G. (ed.) *Sedimentary Facies Analysis*. Special Publications of the International Association of Sedimentologists, **22**, 17–46.

KNOX, R. W. O'B., CORFIELD, R. M. & DUNAY, R. E. (eds) 1996. *Correlation of the Early Paleogene in Northwest Europe*. Geological Society, London, Special Publications, **101**.

MIALL, A. D. 1986. Eustatic sea level changes interpreted from seismic stratigraphy: a critique of the methodology with particular reference to the North Sea Jurassic. *American Association of Petroleum Geologists, Bulletin*, **70**, 131–137.

MILTON, N. J., BERTRAM, G. T. & VANN, I. R. 1990, Early Palaeogene tectonics and sedimentation in the Central North Sea. *In*: HARDMAN, R. F. P. & BROOKS, J. (eds) *Tectonic Events Responsible for Britain's Oil and Gas Reserves*. Geological Society, London, Special Publications, **55**, 339–351.

MITCHUM, R. M. JR & VAN WAGONER, J. C. 1991. High frequency sequences and their stacking patterns: sequence-stratigraphic evidence of high-frequency eustatic cycles. *Sedimentary Geology*, **70**, 131–160.

MUTTI, E. 1985. Turbidite systems and their relations to depositional sequences. *In*: ZUFFA, G. G. (ed.) *Provenance of Arenites*. NATO-ASI series, Reidel, Dordrecht, 65–93.

PARKINSON, D. N. & SUMMERHAYES, C. 1985. Synchonous global sequence boundaries. *American Association of Petroleum Geologists, Bulletin*, **69**, 685–687.

PARTINGTON, M. A., COPESTAKE, P., MITCHENER, B. C. & UNDERHILL, J. R. 1993*a*. Biostratigraphic calibration of genetic stratigraphic sequences in the Jurassic of the North Sea and adjacent areas. *In*: PARKER, J. R. (ed.) *Petroleum Geology of Northwest Europe: Proceedings of the 4th Conference*. Geological Society, London, 371–386.

——, MITCHENER, B. C., MILTON, N. J. & FRASER, A. J. 1993*b*. A genetic stratigraphic sequence stratigraphy for the North Sea Late Jurassic and Early Cretaceous: stratigraphic distribution and prediction of Kimmeridgian-Late Ryazanian reservoirs in the Viking Graben and adjacent areas. *In*: PARKER, J. R. (ed.) *Petroleum Geology of Northwest Europe: Proceedings of the 4th Conference*. Geological Society, London, 347–370.

PAYTON, C. E. (ed.) 1977. *Seismic Stratigraphy – Applications to Hydrocarbon Exploration*. American Association of Petroleum Geologists, Memoirs, **26**, Tulsa, Oklahoma.

PLINT, A. G. 1988. Global eustasy and the Eocene sequence in the Hampshire Basin, England. *Basin Research*, **1**, 11–22.

POSAMENTIER, H. W. & JAMES, D. P. 1993. An overview of sequence stratigraphic concepts: uses and abuses. *In*: POSAMENTIER, H. W., SUMMERHAYES, C. P., HAQ, B. U. & ALLEN, G. P. (eds) *Sequence Stratigraphy and Facies Associations*. International Association of Sedimentologists, Special Publications, **18**, 3–18.

—— & VAIL, P. R. 1988. Eustatic controls on clastic deposition II – sequence and systems tract models. *In*: WILGUS, C. K. *et al.* (eds) *Sea-Level Changes: an Integrated Approach*. Society of Economic Paleontologists and Mineralogists, Special Publications, **42**, 125–154.

——, ERSKINE, R. D. & MITCHUM, R. M. JR. 1990. Models for submarine-fan deposition within a sequence stratigraphic framework. *In*: WEIMER, P. & LINK, M. H. (eds) *Seismic Facies and Sedimentary Processes of Submarine Fans and Turbidite Systems*. Frontiers in Sedimentary Geology. Springer-Verlag, London, 127–136.

——, ALLEN, G. P., JAMES, D. P. & TESSON, M. 1992. Forced regressions in a sequence stratigraphic framework: concepts, examples and exploration significance. *American Association of Petroleum Geologists, Bulletin*, **76**, 1687–1709.

——, JERVEY, M. T. & VAIL, P. R. 1988. Eustatic controls on clastic deposition I – conceptual framework. *In*: WILGUS, C. K. *et al.* (eds) *Sea-Level Changes: an Integrated Approach*. Society of Economic Paleontologists and Mineralogists, Special Publications, **42**, 109–124.

RUFFELL, A. H. & WACH, G. D. 1991. Sequence stratigraphic analysis of the Aptian-Albian Lower Greensand in southern England. *Marine and Petroleum Geology*, **8**, 341–353.

SCHLAGER, W. 1991. Depositional bias and environmental change – important factors in sequence stratigraphy. *Sedimentary Geology*, **70**, 109–130.

—— 1992. *Sedimentology and Sequence Stratigraphy of Reefs and Carbonate Platforms*. Continuing Education Course Note Series, **34**. American Association of Petroleum Geologists.

SLOSS, L. L. 1963. Sequences in the cratonic interior of North America. *Geological Society of America, Bulletin*, **74**, 93–114.

SOPER, N. J. 1994. Was Scotland a Vendian RRR junction? *Journal of the Geological Society, London*, **151**, 579–582.

—— & ENGLAND, R. W. 1995. Vendian and Riphean rifting in NW Scotland. *Journal of the Geological Society, London*, **152**, 11–14.

TUCKER, M. E. 1991. Sequence stratigraphy of carbonate-evaporite basins: models and application to the Upper Permian (Zechstein) of northeast England and adjoining North Sea. *Journal of the Geological Society, London*, **148**, 1019–1036.

—— 1993. Carbonate diagenesis and sequence stratigraphy. *In*: WRIGHT V. P. (ed.) *Sedimentology Review*, **1**, 51–72.

—— & WRIGHT, V. P. 1990. *Carbonate Sedimentology*. Blackwell Scientific Publications, Oxford.

——, CALVERT, F. & HUNT, D. 1993. Sequence stratigraphy of carbonate ramps: systems tracts, models and application to the Muschelkalk carbonate platforms of eastern Spain. *In*: POSAMENTIER, H. W., SUMMERHAYES, C. P., HAQ, B. U. & ALLEN, G. P. (eds) *Sequence Stratigraphy and Facies Associations*. International Association of Sedimentologists, Special Publications, **18**, 397–415.

VAIL, P. R., AUDEMARD, F., BOWMAN, S. A., EISNER, P. N. & PEREZ-CRUZ, C. 1991, The stratigraphic signatures of tectonics, eustasy and sedimentology – an overview. *In*: EINSELE, G., RICKEN, W. & SEILACHER, A. (eds) *Cycles and Events in Stratigraphy*. Springer-Verlag, Berlin, 617–659.

VAN WAGONER, J. C., MITCHUM, R. M. JR., CAMPION, K. M. & RAHMANIAN, V. D. 1990. Siliciclastic sequence stratigraphy in well logs, cores, and outcrops. *American Association of Petroleum Geologists, Methods in Exploration Series,* **7**, Tulsa, Oklahoma.

WALKER, R. G. 1990. Facies modelling and sequence stratigraphy. *Journal of Sedimentary Petrology*, **60**, 777–86.

—— 1992. Facies, facies models and modern stratigraphic concepts. *In*: WALKER, R. G. & JAMES, N. P. (eds) *Facies Models: Response to Sea Level Change*. Geological Association of Canada, 1–14.

WILGUS, C. K., HASTINGS, B. S., KENDALL, G. ST. C., POSAMENTIER, H. W., ROSS, C. A. & VAN WAGONER, J. C. (eds) 1988. *Sea-Level Changes: an Integrated Approach*. Special Publications of the Society of Economic Paleontologists and Mineralogists, **42**.

WOODCOCK, N. H. 1990. Sequence stratigraphy of the Palaeozoic Welsh Basin. *Journal of the Geological Society, London*, **147**, 537–547.

WRIGHT, V. P. 1994. Paleosols in shallow marine carbonate sequences. *Earth Science Reviews*, **35**, 367–395.

A forward-modelling analysis of the controls on sequence stratigraphical geometries

P. M. BURGESS[1] & P. A. ALLEN

Department of Earth Sciences, University of Oxford, Parks Road, Oxford OX1 3PR, UK

[1]*Present address: Seismological Laboratory, California Institute of Technology, Pasadena, CA 91125, USA*

Abstract: The sequence stratigraphical depositional model has concentrated primarily on eustasy and tectonic subsidence as the dominant controls on sequence development, with variations in other factors such as sediment supply playing only a modifying part. Little consideration has been given to the significance of the uniqueness problem to the model predictions. Thus it is often assumed that if the sequence stratigraphical model can provide a simple fit with the available data, then it must be the correct and unique solution. This is rarely likely to be the case. Using a quantitative forward model it is possible to reproduce the basic geometries of type 1 and type 2 sequences and then to assess, via a series of sensitivity tests, the significance of some other controls. Use of this forward-modelling approach suggests that fluvial profile behaviour is a first-order control on sequence geometries. It also suggests that variations in the magnitude of sediment supply can significantly alter the development of transgressive ravinement surfaces. These two examples highlight the importance of the uniqueness problem to the sequence stratigraphical model.

Studies of the controls on the deposition and erosion of passive margin stratigraphy have been dominated in recent years by the sequence stratigraphical method. This method is based on the division of stratigraphy into discrete units of genetic significance, or sequences, separated by unconformities and their correlative conformities (Sloss 1963; Mitchum *et al.* 1977). An apparent correlation of the supposed ages of unconformity surfaces in widely separated basins led to eustasy being thought of as the overriding control on sequence development. The model has grown in detail and sophistication since its initial development (Vail *et al.* 1977*a*,*b*), particularly in the attribution of distinct facies packages to particular segments of the eustatic or relative sea-level curve (e.g. Posamentier *et al.* 1988; Van Wagoner *et al.* 1990).

The concept of eustasy as the overriding control on sequence geometry has been criticized on a number of fronts and, as a result, modifications have been made to sequence stratigraphical models. Firstly, the importance of the interplay of tectonically driven subsidence and uplift with eustasy to produce relative sea-level change is now generally recognized and has been incorporated into more recent developments of the sequence stratigraphical model (Vail *et al.* 1984; Jervey 1988; Posamentier *et al.* 1988; Posamentier & Vail 1988). For example, the model has been adapted from the passive-margin environment, where subsidence is dominated by thermal contraction, to the case of foreland basins where the driving mechanism for subsidence is lithospheric loading (Posamentier & Allen 1993*a*). Secondly, it has been suggested that variations in sediment supply, ignored in the prototype models, are important in determining sequence geometry (Galloway 1989; Thorne & Swift 1991). Sediment supply has been incorporated in recent sequence stratigraphical models, though eustasy continues to be regarded as the dominant control, particularly on sequence boundary timing (Posamentier & Allen 1993*b*). Thirdly, the global synchroneity of sequence boundaries has been questioned (Christie-Blick 1991; Miall 1992), primarily on the basis of the lack of sufficient biostratigraphical resolution.

One of the most fundamental problems with sequence stratigraphical interpretations is that of uniqueness. This derives from the fact that a large number of factors may contribute in some way to the stratigraphical product, many of which are linked. This uniqueness problem, and the increasing complexity of the sequence stratigraphical model, makes the area suitable for investigation using quantitative stratigraphical models. Such modelling studies have the advantage over their qualitative equivalents in that they may be more tightly constrained in terms of the definition of their elements, and they may be more rigorously tested (Cross & Harbaugh 1989). A number of quantitative studies has been carried out (e.g. Lawrence *et al.* 1990;

From HESSELBO, S. P. & PARKINSON, D. N. (eds), 1996, *Sequence Stratigraphy in British Geology*, Geological Society Special Publication No. 103, pp. 9–24.

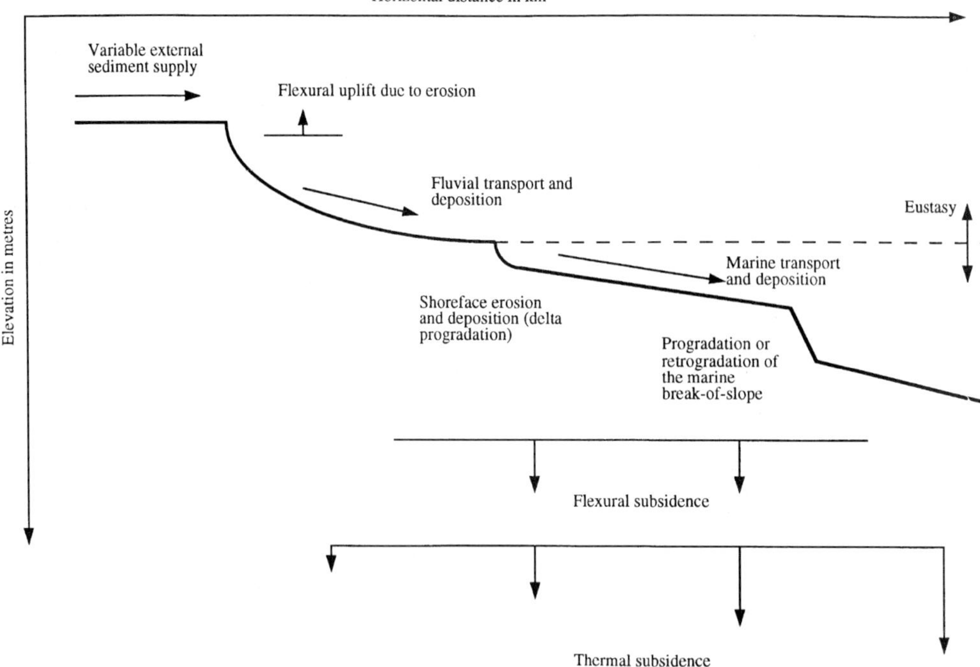

Fig. 1. Components of the quantitative model.

Reynolds *et al.* 1991; Steckler *et al.* 1993) and many of these have already helped to elucidate the problem of the controls on the formation of stratigraphy. Quantitative models generally involve either a physical, process-based approach to the evolution of erosional, transportational and depositional surfaces (e.g. Rivenaes 1992) or, as we have used here, a geometrical approach with equilibrium profiles (e.g. Nummedal *et al.* 1993).

The present study aims to investigate aspects of the controls on stratigraphical patterns using a quantitative stratigraphical model designed to examine, via a series of model sensitivity tests, the influence of factors other than eustasy and tectonic subsidence on sequence architecture. In particular, we wished to explore the role that different variables such as sediment supply and fluvial profile geometry may have in determining stratigraphical patterns. Although the model is necessarily simplistic when compared with the complexities of actual stratigraphical systems, careful use of sensitivity tests and an appreciation of the applicability and weaknesses of the model assumptions allows the model output to be used to illustrate the potential complexity and non-unique nature of stratigraphical patterns.

The components of the quantitative model

The model used here is described more fully in Burgess (1994). It is based on the use of chronostratigraphical lines (chrons) to represent stratigraphy. New chrons are generated at 100 ka intervals using a series of algorithms that describe the various tectonic and geomorphological processes at passive margins. Each chron is subdivided along its length into various environments of deposition such as fluvial, shoreface and offshore shallow marine, based on the water depth at each point.

The model can be summarized as

$$H(x,t+\Delta t) = H(x,t) - S(x,t) + \Delta H(x) + F(x)$$

where H is the elevation of a chron at distance from the origin x at elapsed model time (EMT) t, S is thermal subsidence, ΔH is the change in elevation due to deposition or erosion, F is the change in elevation due to flexure generated by deposition and erosion and Δt is the length of the model time step. The main components of the model are shown in schematic form in Fig. 1. For each time step the contribution to the elevation of the new chron of the different modelled processes is calculated in the sequential order shown in Fig. 2.

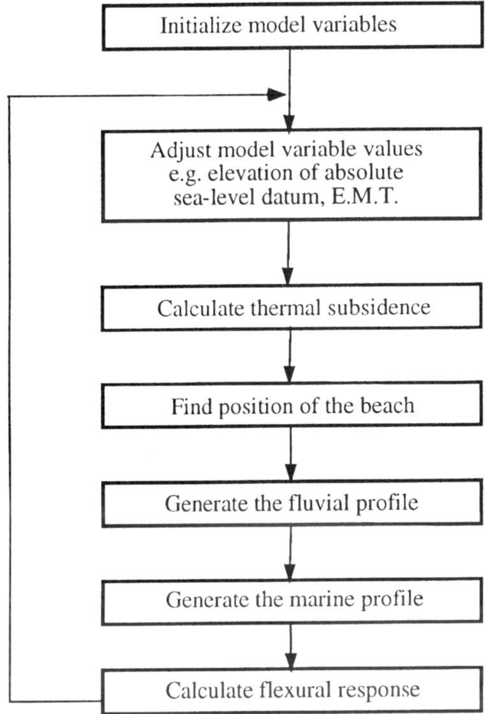

Fig. 2. Flow chart showing the structure of the basic model algorithm.

Thermal subsidence

Thermal subsidence is the fundamental tectonic mechanism by which accommodation space is created in post-rift passive margin basins. It is modelled here using the simple one-layer uniform stretching model of McKenzie (1978) . The stretching factor is varied across the model section from a relatively high value in the distal marine portion of the section to a lower value in the region of the coastal plain. Given the stretching factor, the magnitude of thermal subsidence can be calculated for each time step according to the uniform stretching model and the elevation of the chronostratigraphical surfaces adjusted accordingly. Thus each point on the model profile is treated as a one-dimensional column, which has an initial stretching value, and then subsides through time according to the one-layer stretching model. This model does not include the effects of lateral heat flow (e.g. Steckler & Watts 1982) or two-layer stretching (e.g. Royden & Keen 1980). Though these effects may be important in some basins, they add complexity to the model that is unnecessary to illustrate the uniqueness problem, and so have not been incorporated in the model runs.

Equilibrium profiles

Deposition and erosion of stratigraphy, in both the subaerial and the submarine portions of the basin, are modelled using geometrically defined equilibrium profiles. These are used to calculate a value for ΔH. For values of x where ΔH is positive, deposition occurs; space below the new chron, and above the previous chron surface, is treated as accommodation space and filled with available sediment. At values of x where ΔH is negative, the elevation of previous chrons is greater than the elevation of the new chron, and the elevation of the previous chrons is reduced to that of the new chron to produce erosion (Fig. 3).

The fluvial profile is modelled using a complementary error function curve treated as a purely geometrical entity and fitted between the beach and a point landward of the beach which represents the edge of the coastal plain (see Appendix). This uppermost point is taken to be the highest point on the fluvial profile at which significant deposition of sediment can occur. For example, on the North American Atlantic margin, this point would be at the fall line, which represents the edge of the coastal plain. The landward limit of the fluvial profile is also positioned according to the distance upstream that the effects of relative sea-level change are likely to be felt. It is assumed that any changes in profile geometry due to base-level change can be completed within the duration of the model time step of 100 ka and so the specifics of knickpoint migration and other more detailed profile behaviour are not included in the model. The position of the beach is calculated according to the position of the intersection between the sea-level datum and the previous chron, the area of sediment available for deposition and a value which determines how available sediment is partitioned between fluvial and marine deposition (Fig. 3).

Studies of sediment transport through present day large drainage systems indicate that a significant proportion of the sediment eroded from the catchment fails to reach the ocean, at least in the short term (Milliman & Meade 1983). That a significant proportion remains stored in continental reservoirs in the long term is demonstrated by the abundance of continental rocks preserved in the geological record. The partitioning of the sediment supply into fluvial and marine deposition is accounted for here by the use of a coefficient which is given values in the range 0 to 1. A value of 0.2 means that of the total sediment available for deposition, not including the area of sediment that may be eroded, 20% is deposited as fluvio-deltaic sediment, whereas the remaining 80% is by-passed through the terrestrial system to be deposited as marine sediment, either in the shoreface, on the shelf or on the marine slope (see later).

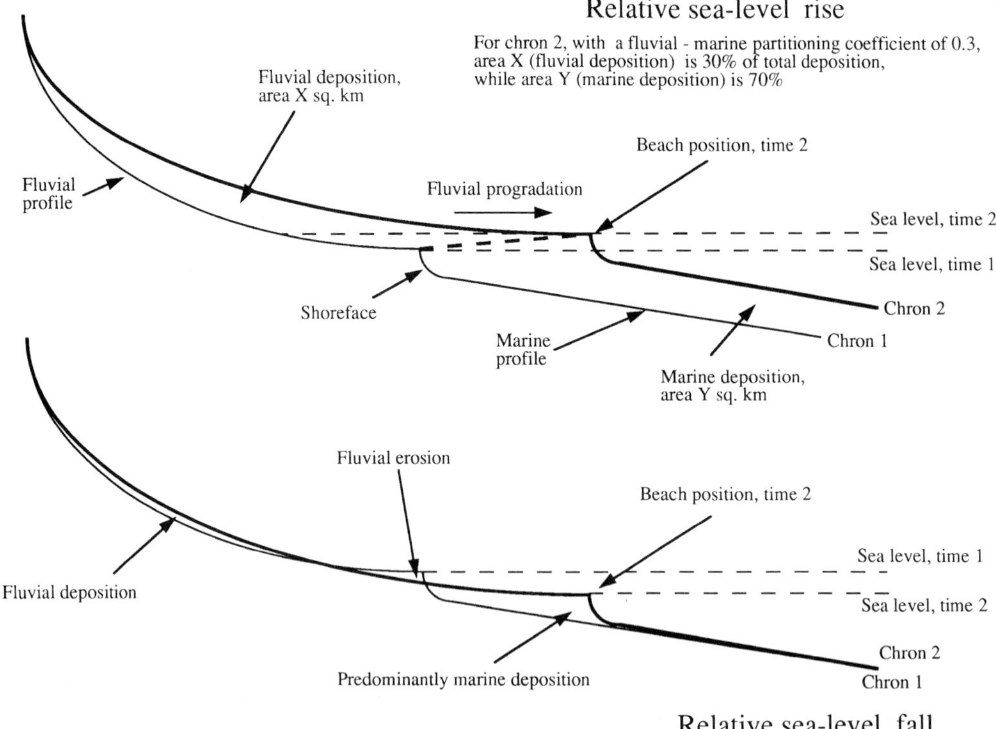

Fig. 3. Geometry of the fluvial profiles and their response to relative sea-level change. The diagram also illustrates the use of the profiles to determine the distribution of deposition and erosion and shows how the terrestrial–marine partitioning coefficient is used to determine the position of the beach and the relative areas of terrestrial and marine deposition.

Snow & Slingerland (1987) showed that exponential, logarithmic and power function curves can provide fits to a more rigorously modelled fluvial profile with only a small error. Burgess (1994) showed that complementary error function curves can be fitted to modern river profiles in a similar manner. Much more problematical is determining how to adjust such curves to reproduce the behaviour of a fluvial profile through time. Such behaviour is currently very poorly constrained, with only small-scale studies over limited time periods (e.g. Leopold & Bull 1979) or experimental flume tank studies (e.g. Begin *et al.* 1981 ; Wood *et al.* 1993). Some workers have adopted a diffusional approach using a combination of the sediment continuity equation and a sediment transport law (Begin *et al.* 1981; Rivenaes 1992). This method allows the change in bed elevation over time to be related to the rate of change of the topographic slope through a coefficient. Although dynamic models such as these have the advantage of a closer link with the observed process, grouping a set of physical processes and behaviour into one geometrical profile has the advantage of simplicity and allows the modeller to maintain a tight control on the profile behaviour through time to test specific possibilities. In addition, Burgess (1994) showed that for a certain range of parameter values, the diffusional and geometrical approaches produce comparable results in terms of river-profile shape and sediment flux leaving the fluvial system. Consequently, in recognition of the problems, and the lack of data available, this model chooses a simple geometrical approach to modelling the fluvial profile through time, with all the inherent, but transparent, weaknesses this imposes on the model results.

The marine profile is composed of two parts. An exponential curve is used to define the shoreface, the width and the depth of which are input as initial model parameters. This then passes seaward into a straight line of a given gradient, which represents the offshore graded shelf. This is obviously a very coarse approximation of a shoreface and shelf geometry. However, the same argument used to justify the choice of fluvial equilibrium profile applies here, perhaps even more strongly, as there is very little published about generally applicable, rigorous, process-based definitions of shoreface

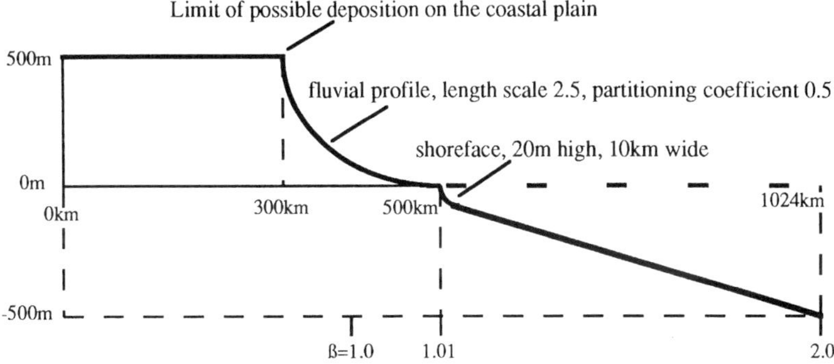

Fig. 4. Initial conditions used to generate the type-1 model example used a reference for the subsequent sensitivity tests.

geometries. Thorne & Swift (1991) gave a quantitative application of their regime model, but this does not appear to include the shoreface. Various other workers (Swift 1970; Everts 1987; Cant 1991; Swift *et al.* 1991) showed similar geometries, but only in schematic form.

The distal end of the marine equilibrium profile consists of a submarine break of slope from the shallow graded shelf to the steeper slope front. The position of the break of slope is determined by sediment supply. Sediment from the fluvial profile and sediment from erosion on the shoreface is fed onto the shelf. As the model is purely two-dimensional, no allowance is made for effects such as submarine canyon erosion, delta lobe switching nor for sediment by-passing the shelf directly onto the slope and into the deep marine environment. The submarine break of slope is thus positioned to use up all the available area of sediment transported onto the shelf.

No grain size information is included in the model. The deposition and erosion of stratigraphy is controlled using purely geometrical algorithms. This precludes the inclusion of a kinetic compaction model as this requires grain size information. Although a more dynamic approach to sediment transport and compaction may reveal interesting features of the stratigraphical system, the purpose here is simply to illustrate, via a model comparable with qualitative sequence stratigraphical equivalents, the importance of the uniqueness problem.

Flexural response

The flexural response of the lithosphere to sediment loading and unloading is calculated using a thin plate elastic model which assumes that the deflections are small compared with the plate thickness. The fourth-order differential flexure equation is solved using a finite difference method (Bodine 1981). This allows for lateral changes in the restoring force, which are caused in this case by changes in the density of the material, either air or water, displaced by flexural uplift. It is assumed that all the uplift or subsidence due to flexure is completed within the duration of a model time step, in this case 100 ka. The contribution of flexure to stratigraphical patterns is assessed in Burgess (1994), who found that in this model the effects of flexure due to deposition and erosion are negligible compared with other model components. For this reason flexure is not included in the sensitivity tests described in the following.

Type-1 and type-2 sequences

The quantitative model can reproduce many of the salient features of type-1 and type-2 depositional sequences described by, for example, Van Wagoner *et al.* (1988, 1990). The initial conditions for the model runs are partly derived by reference to data on actual passive margins, the remainder being essentially arbitrary, chosen to replicate the features described in the sequence stratigraphical model, or to illustrate points about specific controls such as sediment supply or fluvial profile behaviour.

To generate stratigraphy analogous to that described as a type 1 sequence, the quantitative model is run over a period of 2.5 Ma, with 25 time steps, each of 100 ka. Thermal subsidence is calculated using stretching factors ranging from 1.0 at 400 km from the origin of the model profile, to 1.2 at 500 km and 2.0 at 1024 km. These values are based on studies of stretching on the North American Atlantic margin (e.g. LASE Study Group 1986). Absolute sea level is varied sinusoidally with a period of 2.0 Ma and an amplitude of 20 m to reproduce a typical third-order cycle (Haq *et al.* 1988). Sediment supply is held constant at an arbitrary value of 1.0 km^2 per time step. This is a rather high value for the North American Atlantic margin (e.g. Judson 1975), but serves to illustrate sequence development well.

The geometry of the fluvial profile used in the model is based on the rivers on the North American Atlantic Coastal Plain. It is defined using a complementary error function with a parameter value of 2.0 (see Appendix; Burgess 1994), a length of 200 km and fitted between a landward limit at 200 m elevation and the beach position, initially at 0 m elevation. The marine profile consists of a shoreface 10 m high and 2 km wide, a shelf with a gradient of 0.5 m km^{-1}, and a marine slope with a gradient of 43.66 m km^{-1}. These are typical passive margin shelf and slope values (Nummedal *et al.* 1993). The initial conditions are summarized in Fig. 4.

Various systems tracts and sets of chrons showing geometries analogous to parasequence stacking patterns can be identified in the model section (Fig. 5) and compared with the position on the absolute sea-level curve (Fig. 6). The formation of the systems tracts under the influence of absolute sea level and thermal subsidence is clearly demonstrated. Chrons 1–5 form during a time of slow rise and highstand of sea level, thus forming the highstand systems tract. They are subsequently truncated by erosion on the fluvial profile. Chrons 6–15 form during falling relative and absolute sea level, and hence form the lowstand systems tract. During this time, deposition is limited to the marine slope and the proximal fluvial profile. On the lower portion of the fluvial profile and across the previous shelf surface, erosion occurs between chrons 6 and 15, leading to by-passing of sediment directly onto the marine slope.

The type-1 sequence boundary is placed at chron 15. Figure 6 shows the sequence boundary on a chronostratigraphical diagram. It is positioned on the basis of the end of erosion and the beginning of significant deposition on the fluvial profile. Importantly, the sequence boundary is of long duration, cutting as it does 14 chrons over a 1.0 Ma period (Fig. 6). Erosion on the unconformity continues throughout the duration of absolute and relative sea-level fall. This contrasts strongly with the sequence stratigraphical depositional model, which implies that sequence boundaries form over a shorter period and that their formation is directly related to the rate of relative sea-level change (e.g. Posamentier *et al.* 1988).

Chrons 16–25 form the overlying transgressive and highstand systems tracts, formed during an increasing and then decreasing rate of relative sea-level rise. The transgression produces a well-developed marine ravinement surface, which during its formation erodes parts of chrons 15–22. Shoreface progradation begins once more at chron 24 when the rate of relative sea-level rise has slowed sufficiently. The pattern of marine erosion is clearly shown in Fig. 6. Also shown in Fig. 6 is the pattern of stratal onlap formed during rising absolute sea level. It should be noted that this onlap is not coastal onlap, as defined by Vail *et al.* (1977*a*), but instead occurs entirely within fluvial stratigraphy and at elevations up to 200 m. It should be noted that the amplitude of the onlap is controlled predominantly by the geometry of the fluvial profile, rather than by the amplitude of the relative or absolute sea-level rise.

To generate a type-2 sequence and sequence boundary it is necessary to alter some of the model parameters. The magnitude of thermal subsidence is increased by changing the stretching factor on the profile to 2.0 at 500 km and 8.0 at 1024 km. The amplitude of the absolute sea-level curve is also reduced, from 20 to 10 m. Figure 7 shows the section from the model run with these changed parameter values. The most notable difference from the type-1 sequence is the lack of a subaerial erosion surface. Chrons 1–10, which were truncated by the sequence boundary in the type 1 example, are preserved.

The sequence stratigraphical depositional model suggests that a shelf margin systems tract consisting of a slightly progradational to aggradational parasequence set should be developed in a type-2 sequence (e.g. Van Wagoner *et al.* 1988, 1990). Such a pattern is suggested to form as a result of a balance between accommodation space created by subsidence slightly outpacing absolute sea-level fall and sediment supply which is just sufficient to fill this accommodation space. In the model section (Fig. 7) chrons 6–15 show a clearly progradational pattern. Using a quantitative model it is very difficult to achieve the exact balance between subsidence, absolute sea-level fall and sediment supply that is necessary to produce a predominantly aggradational shelf margin systems tract. Hence if this model is realistic with respect to its representation of accommodation development, then either the development of shelf margin systems tracts due to a balance between sediment input and accommodation space generation should be rare, or other mechanisms are involved which bring about such a balance. Such aggradation will certainly require higher levels of tectonic subsidence than are likely to be seen on a post-rift passive margin.

The development of the transgressive ravinement surface in chrons 16–26 is reduced due to the smaller relative sea-level rise. This is visible on both the model section in Fig. 7 and the chronostratigraphical diagram in Fig. 8. The chronostratigraphical diagram also shows the reduction in

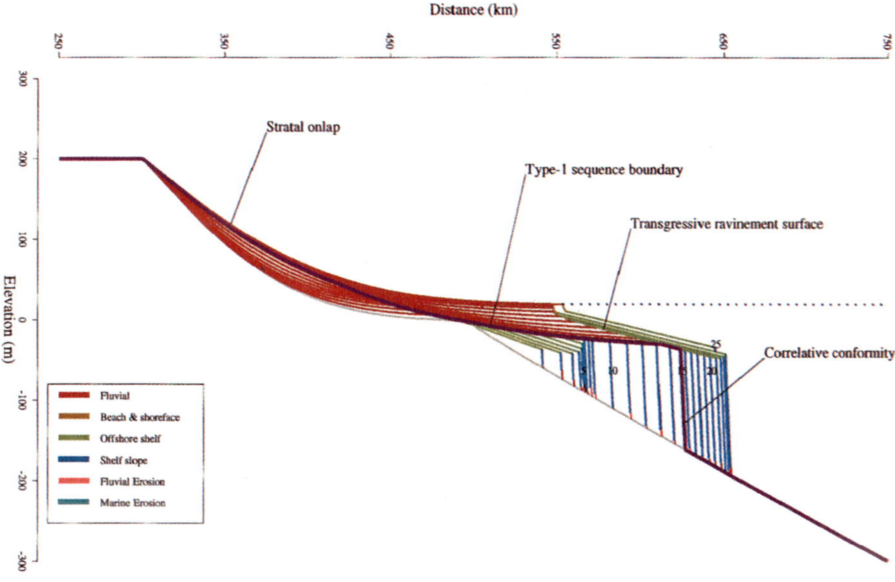

Fig. 5. Stratigraphical cross-section from the type-1 model run with the initial conditions and parameter values shown in Fig. 4 and described in the text. The cross-section demonstrates that the quantitative model is capable of reproducing many of the salient features of the type-1 sequences described in the sequence stratigraphical model (e.g. Van Wagoner *et al.* 1988).

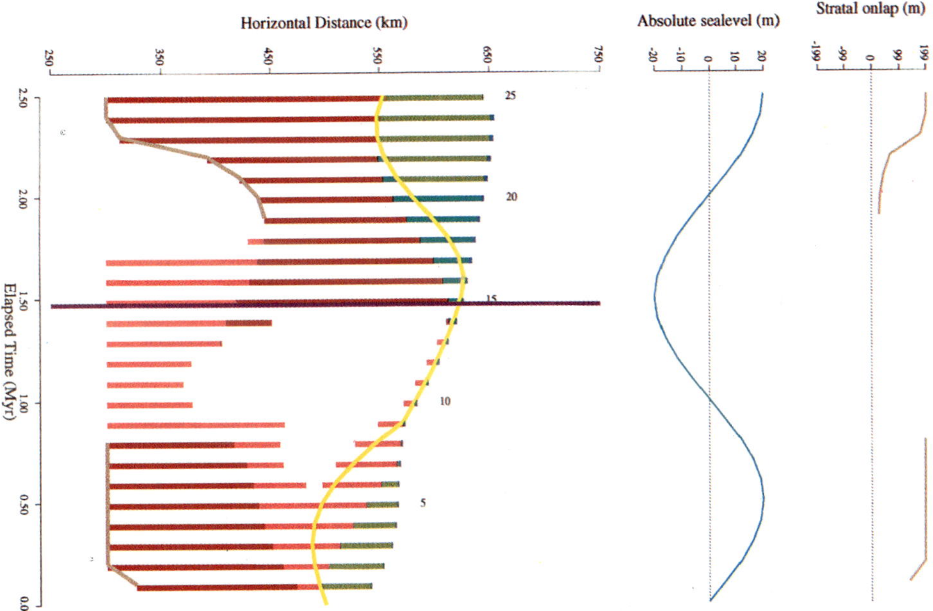

Fig. 6. Chronostratigraphical diagram, an absolute sea- level curve and a stratal onlap curve from the type-1 model run. The control of the absolute sea-level curve on the stratigraphical pattern is apparent. Note particularly the prolonged period of non-deposition (white) and fluvial erosion (pink) on the lower half of the fluvial profile below the type-1 sequence boundary at chron 15, and also the elevation of the stratal onlap on the fluvial profile from chrons 19 to 25. Key as Fig. 5.

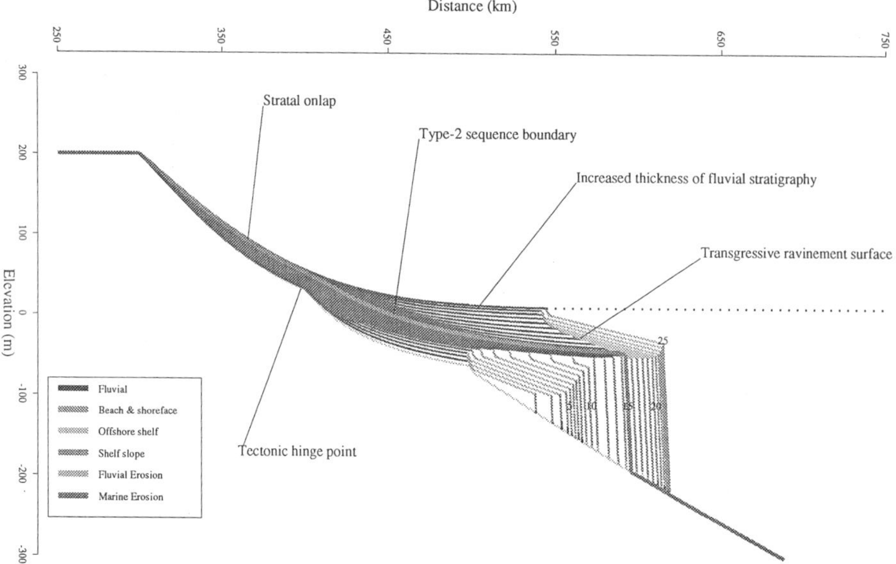

Fig. 7. Stratigraphical cross-section from the model run with the same initial conditions and parameters as the type-1 example, except that the magnitude of the thermal subsidence has been increased and the amplitude of the absolute sea-level curve has been reduced from 20 to 10 m to produce a type-2 sequence. The increased subsidence and reduced magnitude of sea-level change prevents the development of the erosive sequence boundary seen in the type-1 example and leads to increased preservation of fluvial and proximal marine stratigraphy.

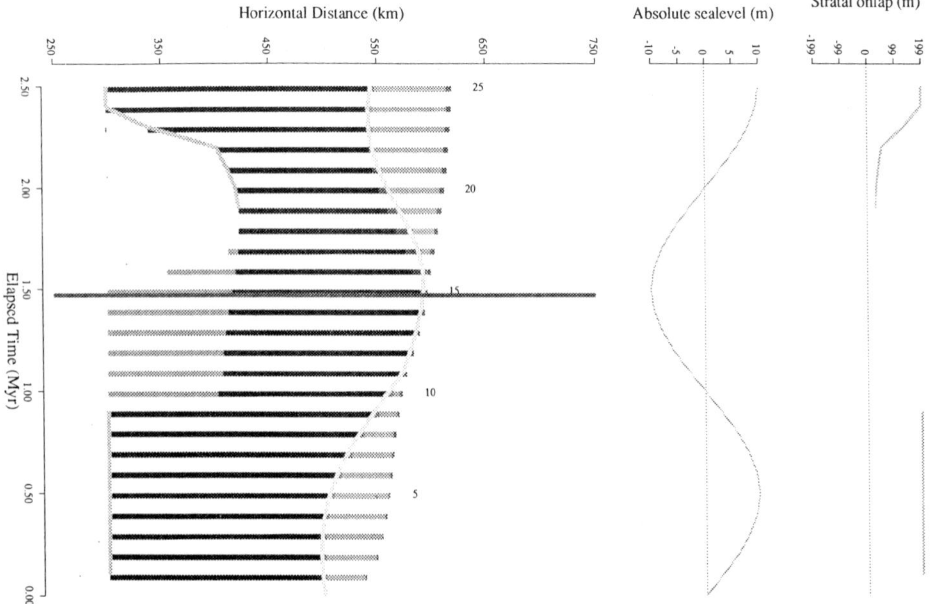

Fig. 8. Chronostratigraphical diagram, an absolute sea-level curve and a stratal onlap curve from the type-2 model example. The chronostratigraphical diagram shows the lack of erosion on the fluvial profile leading to continuous deposition of fluvial stratigraphy on the lower fluvial profile.

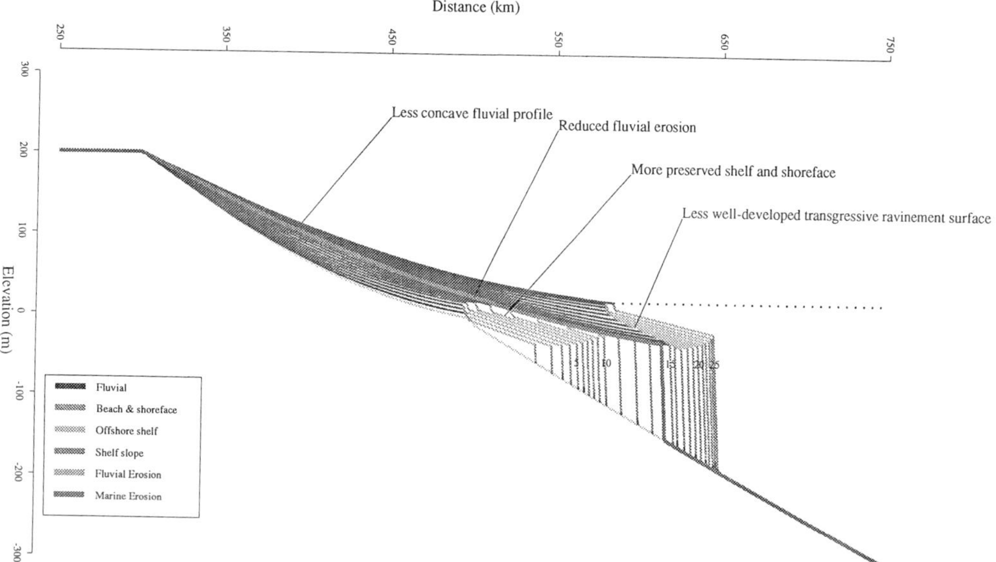

Fig. 9. Stratigraphical cross-section from the model run with a less concave fluvial profile than used in the type-1 example. All other initial conditions and model parameters are the same as the type-1 example. The increased gradients on the lower fluvial profile have reduced the vertical and lateral extent of fluvial erosion, leading to increased preservation of marine and fluvial strata below the sequence boundary.

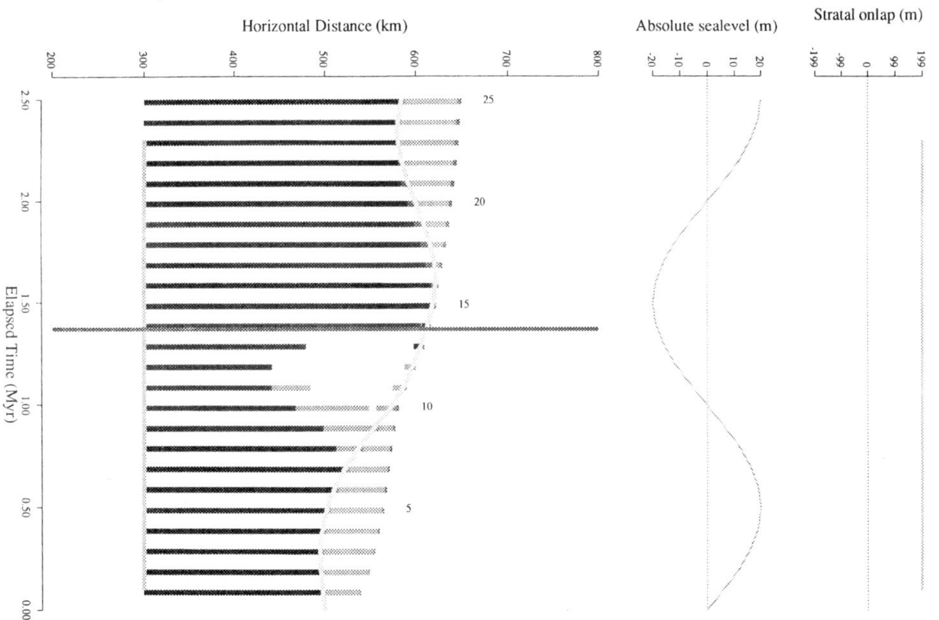

Fig. 10. Chronostratigraphical diagram, an absolute sea-level curve and a stratal onlap curve from the model run with a less concave fluvial profile (complementary error function parameter reduced from 2.0 to 1.2). The reduction in fluvial erosion and non-deposition relative to the type-1 example is clear. The stratal onlap curve shows that the changed fluvial geometry has significantly altered the pattern of onlap in response to the absolute sea-level changes.

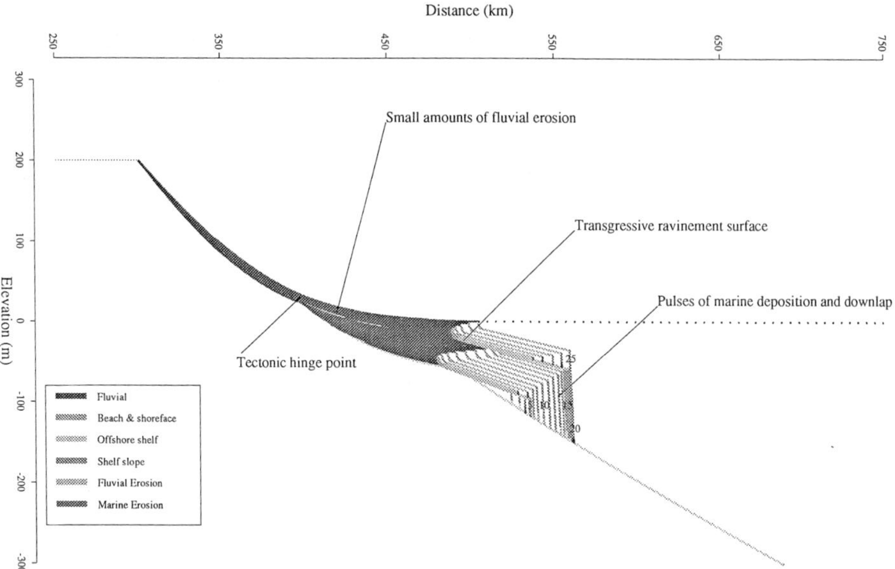

Fig. 11. Stratigraphical cross-section from the model run with constant absolute sea level, high thermal subsidence and variable sediment supply (see Fig. 12). All other model parameters are the same as those for the type-1 example. The section illustrates the development of progradational and retrogradational stacking patterns due to relative sea level rise and variable sediment supply without the need for absolute sea level fluctuations.

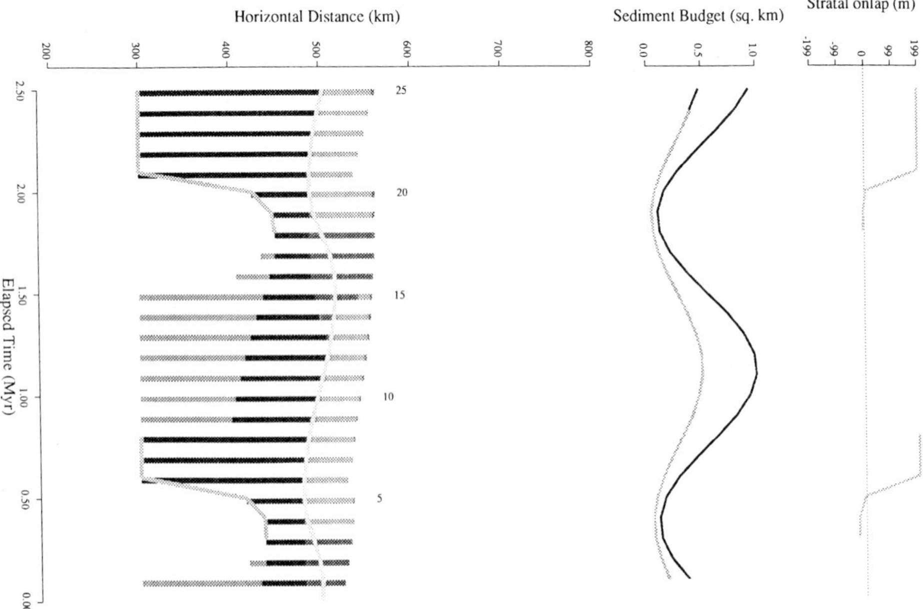

Fig. 12. Chronostratigraphical diagram, a sediment budget curve and a stratal onlap curve from the model run with constant absolute sea level, high thermal subsidence and variable sediment supply. All other model parameters are the same as those for the type-1 example. The plot illustrates the movement of the beach and the pattern of deposition and erosion in response to the rising relative sea level and the sinusoidally varying sediment supply. The higher-amplitude curve is the total supply to the basin and the lower-amplitude curve represents the areas of marine and fluvial deposition. The pattern of stratal onlap is similar to that shown in the type-1 example (Fig. 6).

fluvial erosion, which is now restricted to the upper profile during rising absolute sea level. This can probably be disregarded as an unrealistic model artifact, although evidence from studies of fluvial response to base-level change (e.g. Leopold & Bull 1979) is still inconclusive on the details of profile behaviour above an aggradational wedge. The pattern of stratal onlap in the type-2 sequence example is not significantly changed from that seen in the type-1 example.

Alternative controls on sequence geometry: sensitivity tests

Posamentier & Allen (1993*b*) state that eustasy and tectonic subsidence are the primary controls on sequence geometry, whereas sediment supply and basin physiography play a modifying role; this is qualified with the statement that 'the number of variations on the general sequence stratigraphic-model is limited only by the imagination'. A quantitative model has an advantage over a qualitative model in that it can be rigourously tested via a series of sensitivity tests to determine the relative importance of the various parameters within the model. Hence, depending on the nature of the model components, a quantitative model can provide a useful method to constrain some of the possible variations of factors controlling sequence development. The following section demonstrates the importance in this model of fluvial profile geometry, and of variable sediment supply, and hence the importance of a thorough consideration of controls other than eustasy and subsidence on sequence geometries. A fuller range of sensitivity tests is given in Burgess (1994).

Behaviour of the fluvial profile

Making the fluvial profile in the model less concave increases the mean area of accommodation below the profile during a sea-level cycle, changes the distribution of this accommodation space and reduces the likelihood of fluvial incision and erosion occurring in response to falling absolute sea level. Figure 9 shows a section from the model run with a fluvial profile less concave than that used in the type-1 and type-2 examples. This model run uses an error function parameter value of 1.2. Comparing this section with those from the type-1 and type-2 examples in Figs 5 and 7 shows that the model run with the less concave fluvial profile has produced a stratigraphical pattern intermediate between a type-1 and a type-2 sequence. For example, there is some fluvial erosion on the lower portion of the profile during chrons 10–13 (Fig. 10), but this is much reduced in duration and lateral and vertical extent with respect to the type-1 example. Most of the shoreface stratigraphy deposited in chrons 1–6 has been preserved and fluvial deposition has been more continuous (Fig. 9) than in the type-1 sequence (Fig. 5).

Thus the fluvial profile in this model is a first-order control on the stratigraphical pattern produced. This is a very significant result with respect to controls on sequence geometry, but to interpret it correctly, it is necessary to examine the assumptions in the quantitative model. Aspects of the profile behaviour, such as the small amounts of erosion on the upper portions of the profile in response to base-level rise, are possibly unrealistic, so this should be considered when determining the implications of the results of Figs 9 and 10.

However, even bearing in mind the possibly unrealistic aspects of the modelled fluvial profile, it seems reasonable to suggest that if a stream feeding sediment into a basin can adjust its geometry in response to controls such as climate, sediment load and tectonics, then a change in profile geometry will be a first-order control on the stratigraphical patterns produced. For example, if the profile does not incise in response to a relative sea-level fall, but rather continues to create stratigraphy by delta progradation, then as shown in Figs 9 and 10, the resulting stratigraphical pattern may look more like a type-2 sequence than a type-1 sequence.

Variable sediment supply and sediment partitioning

A model run with variable sediment supply and absolute sea level held constant at 0 m elevation is shown in Figs 11 and 12. The sediment supply curve consists of a sinusoidal curve with a 2.5 Ma period and an amplitude of 0.45 km^2. The influence of the changing sediment supply on the position of the beach is shown on the chronostratigraphical diagram. During times of low sediment supply (e.g. chrons 1–4) the beach is forced landward by a relative sea-level rise driven purely by thermal subsidence. At times of high sediment supply (e.g. chrons 5–15) the beach progrades basinward, producing an unforced regression (cf. Posamentier *et al.* 1992).

Other elements of the stratigraphical pattern seen in the type-1 sequence model run are also visible in this example controlled by sediment supply. Fluvial erosion on the upper portion of the fluvial profile occurs in chrons 6–16, stratal onlap is present in chrons 2–5 and 18–21 and a transgressive ravinement surface is present in chrons 10–18. In the section in Fig. 11 the change from progradation to transgression and retrogradation is visible. Once again, all these effects are produced not by changing absolute sea level, but by variable sediment supply and rising relative sea level due to thermal subsidence. This illustrates the importance of

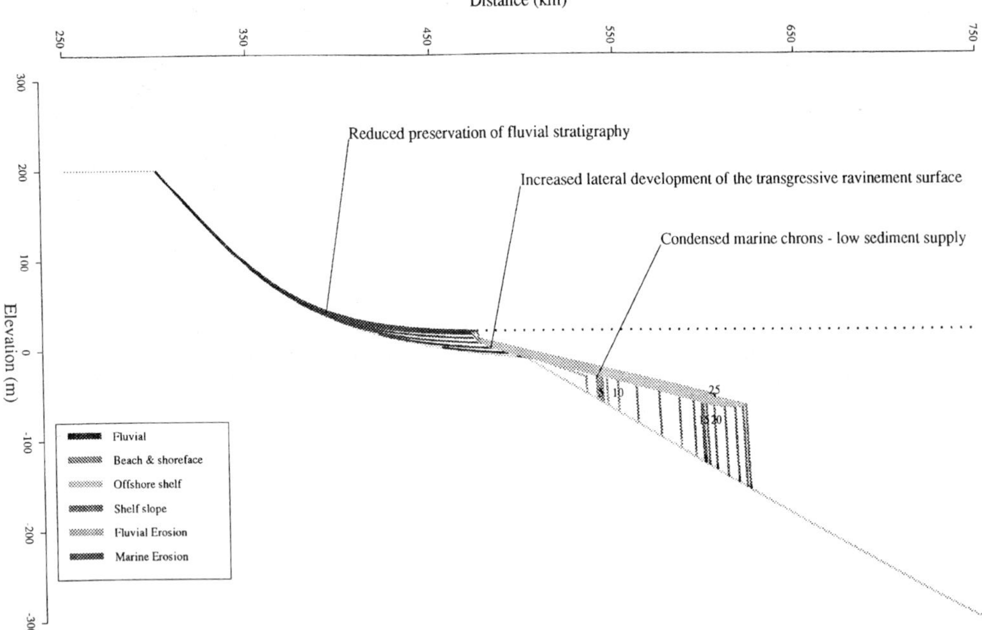

Fig. 13. Stratigraphical cross-section from the model run with variable sediment supply and variable sediment partitioning coefficient values (sinusoidal between 0.2 and 0.8) (see Fig. 14). All other model parameters are the same as those for the type-1 example. The influence of the variable sediment supply is reflected in the increased development of the transgressive ravinement surface relative to the type-1 example (Fig. 5).

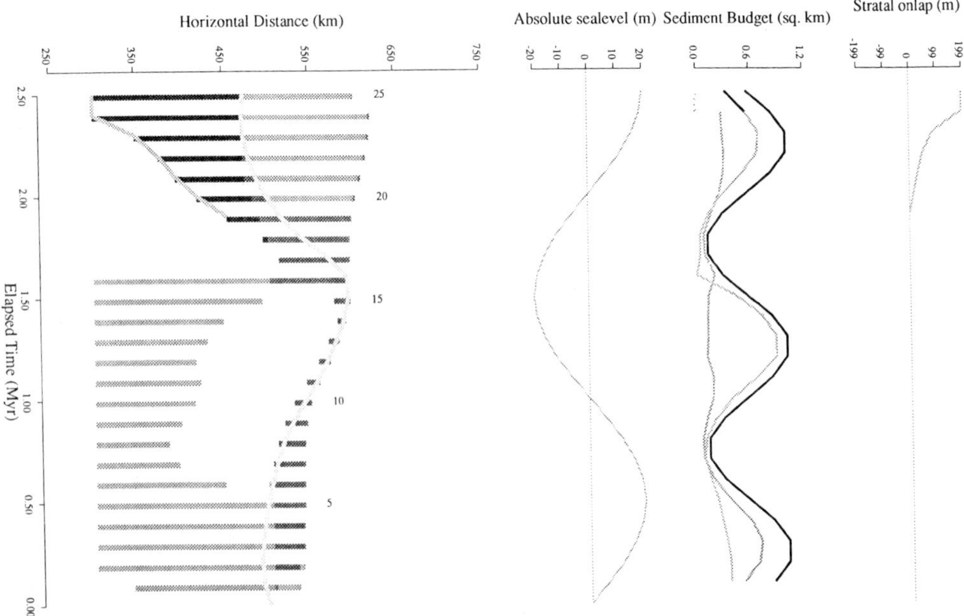

Fig. 14. Chronostratigraphical diagram, an absolute sea- level curve, a sediment budget curve and a stratal onlap curve from the model run with variable sediment supply and variable sediment partitioning coefficient values. All other model parameters are the same as those for the type-1 example. On the sediment supply curve, the highest-amplitude curve is the total sediment supply. The sediment partioning coefficient varies through time, so that fluvial deposition peaks at 0.3 and 1.6 Ma, and marine deposition peaks at 1.3 and 2.3 Ma. The enhanced development of the transgressive ravinement surface is demonstrated by the increased landward movement of the beach from chrons 16 to 23, and the increase in marine erosion of portions of chrons 2–22.

adequate consideration of the uniqueness problem in the sequence stratigraphical model. If such consideration is not given, it seems possible that the causal mechanism of many stratigraphical patterns could be misidentified.

The final sensitivity test illustrates the effects of varying the value of the sediment supply and the sediment partitioning coefficient through a model run. The coefficient is varied according to a sinusoidal curve with a period of 2.5 Ma and a range of values from 0.2 to 0.8. All the other parameters have the same values used for the type-1 sequence example, so absolute sea level is varied in this instance. The most striking feature of the section from the model run (Fig. 13) is the very well-developed transgressive ravinement surface which has removed much of the earlier deposited stratigraphy. This is also visible on the chronostratigraphical diagram (Fig. 14). The erosion on the extensive transgressive ravinement surface can be clearly seen in chrons 1–20.

The ravinement surface was formed at a time of rising relative and absolute sea level when the fluvial–marine partitioning coefficient was low. The low value for the coefficient means that sediment by-passed the fluvial system rather than being deposited in delta progradation and thus the transgression was enhanced. The significance of this effect for actual stratigraphical systems is difficult to assess. Intuitively, it seems obvious that some factor or factors must control the relative volumes that are deposited as terrestrial and marine sediment. However, this is not simply a case of control by accommodation space, because during times of both transgression and highstand there is space available for deltas to prograde, and yet they do not always do so. Other factors such as the grain size of the sediment available, wave power, coastal geomorphology, frequency of storms and longshore drift must play a part. Whatever the reality, from the model results it seems likely that the fluvial–marine partitioning of sediment is another important control on stratigraphical patterns.

Conclusions

The sequence stratigraphical model undoubtedly provides a useful method for the analysis of stratigraphy. However, as Posamentier & James (1993) hinted, there are problems with the predictive elements of the model because similar stratigraphical patterns are non-unique in that they may be produced by a variety of different controls. The quantitative model results presented here illustrate some of the problems in this regard. Using quantitative models it is possible to test the concepts behind the sequence stratigraphical model more rigorously than can be done with qualitative models alone.

More specifically, these model results show that a simple quantitative model can successfully reproduce many of the aspects of stratigraphical patterns used in the sequence stratigraphical model. The results highlight the problems of treating a sequence bounding unconformity as a simple chronostratigraphical surface; the model output suggests that such surfaces may represent prolonged periods of geological time related to the duration of relative sea-level fall and display complex patterns of erosion. Using coastal onlap as an indicator of relative and absolute sea-level change is also problematic. In this model the amplitude of stratal onlap is controlled not by the magnitude of sea-level change alone, but also by the geometry and behaviour of the fluvial profile.

The series of sensitivity test results presented here illustrates the problem of uniqueness with regard to the sequence stratigraphical model. These examples demonstrate the importance of fluvial-profile behaviour and sediment-supply variations. Changing the geometry and hence the erosive behaviour of the fluvial profile is shown to strongly influence the pattern of stratigraphy produced in the model in response to relative and absolute sea-level change. Variable sediment supply and sediment partitioning are demonstrated to be of similar importance, especially during times of relative sea-level rise when reduced sediment supply or increased sediment by-pass into the sea can significantly enhance the formation of a transgressive ravinement surface.

This work demonstrates the importance of a careful consideration of the impact of the uniqueness problem on sequence stratigraphical predictive models. Explanations for particular stratigraphical patterns should be treated with caution. Quantitative forward-modelling can provide a way of dealing with this problem by allowing a more rigorous analysis of the various possible mechanisms that may account for the observed stratigraphical patterns.

Full-colour copies of Figs 5–14 are available in postscript format from P. Burgess.
(e-mail: pete@seismo.gps.caltech.edu)

We acknowledge helpful reviews from M. Leeder and an anonymous referee, and a multitude of useful and stimulating conversations with A. Watts, S. Hesselbo, N. Parkinson, S. Gupta, N. Hovius and P. England. P. Burgess acknowledges the finanical support of a NERC studentship.

References

BEGIN, Z. B., MEYER, D. F. & SCHUMM, S. A. 1981. Development of longitudinal profiles of alluvial channels in response to base-level lowering. *Earth Surface Processes and Landforms*, **6**, 49–68.

BODINE, J. H. 1981. *Numerical Computation of Plate Flexure in Marine Geophysics*. Lamont-Doherty Geological Observatory, Technical Report No. 1, CU 1–80, 36 pp.

BURGESS, P. M. 1994. *A quantitative forward modelling analysis of the controls on passive rift-margin stratigraphy*. DPhil Thesis, University of Oxford.

CANT, D. J. 1991. Geometric modelling of facies migration: theoretical development of facies successions and local unconformities. *Basin Research*, **3**, 51–62.

CHRISTIE-BLICK, N. 1991. Onlap, offlap, and the origin of unconformity bounded depositional sequences. *Marine Geology*, **97**, 35–56.

CROSS, T. A. & HARBAUGH, J. W. 1989. Quantitative dynamic stratigraphy: a workshop, a philosophy, a methodology. *In*: CROSS T. A. (ed.) *Quantitative Dynamic Stratigraphy*. Prentice Hall, Englewood Cliffs, 3–20.

EVERTS, C. H. 1987. Continental shelf evolution in response to a rise in sea level. *In*: NUMMEDAL, D., PILKEY, O. H. & HOWARD, J. D. (eds) *Sea Level Fluctuation and Coastal Evolution*. Society of Economic Palaeontologists and Mineralogists, Special Publications, **41**, 49–57.

GALLOWAY, W. E. 1989. Genetic stratigraphic sequences in basin analysis I : architecture and genesis of flooding-surface bounded depositional units. *Bulletin of the American Association of Petroleum Geologist*, **73**, 125–142.

HAQ, B. U., HARDENBOL, J. & VAIL, P. R. 1988. Mesozoic and Cenozoic chronostratigraphy and cycles of sea-level change. *In*: WILGUS, C. K., HASTINGS, B. S., KENDALL, C. G. ST. C., POSAMENTIER, H. W., ROSS, C. A. & VAN WAGONER, J. C. (eds) *Sea-Level Changes: an Integrated Approach*. Society of Economic Palaeontologists and Mineralogists, Special Publications, **42**, 71–108.

JERVEY, M. T. 1988. Quantitative geologic modelling of siliciclastic rock sequences and their seismic expression. *In*: WILGUS, C. K., HASTINGS, B. S., KENDALL, C. G. ST. C., POSAMENTIER, H. W., ROSS, C. A. & VAN WAGONER, J. C. (eds) *Sea-Level Changes: an Integrated Approach*. Special Publications Society of Economic Palaeontologists and Mineralogists, **42**, 47–69.

JUDSON, S. 1975. Evolution of the Appalachian topography. *In*: MELHORN, W. N. & FLEMAL, R. C. (eds) *Theories of Landform Development*. George Allen & Unwin, 29–44.

LASE STUDY GROUP 1986. Deep structure of the US East Coast passive margin from large aperture seismic experiments. *Marine and Petroleum Geology*, **3**, 234–242.

LAWRENCE, D. T., DOYLE, M. & AIGNER, T. 1990. Stratigraphic simulation of sedimentary basins: concepts and calibration. *Bulletin of the American Association of Petroleum Geologists*, **74**, 273–295.

LEOPOLD, L. B. & BULL, W. B. 1979. Base level, aggradation, and grade. *Proceedings of the American Philosophical Society*, **123**, 168–202.

MCKENZIE, D. P. 1978. Some remarks on the development of sedimentary basins. *Earth and Planetary Science Letters*, **40**, 25–32.

MIALL, A. D. 1992. The Exxon global cycle chart: an event for every occasion. *Geology*, **20**, 787–790.

MILLIMAN, J. D. & MEADE, R. H. 1983. Worldwide delivery of river sediment to the oceans. *Journal of Geology*, **91**, 1–21.

MITCHUM, R. M., VAIL, P. R. & THOMPSON, S. 1977. Seismic stratigraphy and global changes of sea level, part 2: the depositional sequence as a basic unit for stratigraphic analysis. *In*: PAYTON C. E. (ed.) *Seismic Stratigraphy-Applications to Hydrocarbon Exploration*. American Association of Petroleum Geologists, Memoirs, **26**, 53–62.

NUMMEDAL, D., RILEY, G. W. & TEMPLET, P. L. 1993. High-resolution sequence architecture: a chronostratigraphic model based on equilibrium profile studies. *In*: POSAMENTIER, H. W., SUMMERHAYES, C. P., HAQ, B. U., & ALLEN, G. P. (eds) *Sequence Stratigraphy and Facies Associations*. International Association of Sedimentologists, Special Publications, **18**, 55–63.

POSAMENTIER, H. W. & ALLEN, G. P. 1993*a*. Siliciclastic sequence stratigraphic patterns in foreland ramp-type basins. *Geology*, **21**, 455–458.

— & — 1993*b*. Variability of the sequence stratigraphic model: effects of local basin factors. *Sedimentary Geology*, **86**, 91–109.

— & JAMES, D. P. 1993. An overview of sequence-stratigraphic concepts: uses and abuses. *In*: POSAMENTIER, H. W., SUMMERHAYES, C. P., HAQ, B. U., & ALLEN, G. P. (eds) *Sequence Stratigraphy and Facies Associations*. International Association of Sedimentologists, Special Publications, **18**, 3–18.

— & VAIL, P .R. 1988. Eustatic controls on clastic deposition II – sequence and systems tract models. *In*: WILGUS, C. K., HASTINGS, B. S., KENDALL, C. G. ST. C., POSAMENTIER, H. W., ROSS, C. A. & VAN WAGONER, J. C. (eds) *Sea-level Changes: an Integrated Approach*. Society of Economic and Palaeontologists and Mineralogists, Special Publications, **42**, 125–154.

—, ALLEN, G. P., JAMES, D. P. & TESSON, M. 1992. Forced regressions in a sequence stratigraphic framework: concepts, examples, and exploration significance. *Bulletin of the American Association of Petroleum Geologists*, **76**, 1687–1709.

—, JERVEY, M. T. & VAIL, P. R. 1988. Eustatic controls on clastic deposition I – conceptual framework. *In*: WILGUS, C. K., HASTINGS, B. S., KENDALL, C. G. ST. C., POSAMENTIER, H. W., ROSS, C. A. & VAN WAGONER, J. C. (eds) *Sea-level Changes: an Integrated Approach*. Society of Economic Paleontologists and Mineralogists, Special Publications, **42**, 109–124.

REYNOLDS, D. J., STECKLER, M. S. & COAKLEY, B. J. 1991. The role of the sediment load in sequence stratigraphy: the influence of flexural isostasy and compaction. *Journal of Geophysical Research*, **96**, 6931–6949.

RIVENAES, J. C. 1992. Application of a dual lithology, depth-dependent diffusion equation in stratigraphic simulation. *Basin Research*, **4**, 133–146.

ROYDEN, L. & KEEN, C. E. 1980. Rifting processes and thermal evolution of the continental margin of eastern Canada determined from subsidence curves. *Earth and Planetary Science Letters*, **51**, 343–361.

SLOSS, L. L. 1963. Sequences in the cratonic interior of North America. *Bulletin of the Geological Society of America*, **74**, 93–114.

SNOW, R. S. & SLINGERLAND, R. L. 1987. Mathematical modelling of graded river profiles. *Journal of Geology*, **95**, 15–33.

STECKLER, M. S. & WATTS, A. B. 1982. *Subsidence History and Tectonic Evolution of Atlantic-type Continental Margins*. American Geophysical Union Geodynamics Series, **8**, 184–196.

—, REYNOLDS, D. J., COAKLEY, B. J., SWIFT, B. A. & JARRARD, R. 1993. Modelling passive margin sequence stratigraphy. *In*: POSAMENTIER, H. W., SUMMERHAYES, C. P., HAQ, B. U. & ALLEN, G. P. (eds) *Sequence Stratigraphy and Facies Associations*. International Association of Sedimentologists, Special Publications, **18**, 19–41.

SWIFT, D. J. P. 1970. Quaternary shelves and the return to grade. *Marine Geology*, **8**, 5–30.

—, PHILIPS, S. & THORNE, J. A. 1991. Sedimentation on continental margins, V: parasequences. *In*: SWIFT, D. J. P., OERTEL, G. F., TILMAN, R. W. & THORNE, J. A. (eds) *Shelf Sand and Sandstone Bodies, Geometry, Facies and Sequence Stratigraphy*. International Association of Sedimentologists, Special Publications, **14**, 153–187.

THORNE, J. A. & SWIFT, D. J. P. 1991. Sedimentation on continental margins, VI: a regime model for depositional sequences, their component systems tracts, and bounding surfaces. *In*: SWIFT D. J. P., OERTEL, G. F., TILMAN, R. W. & THORNE, J. A. (eds) *Shelf Sand and Sandstone Bodies, Geometry, Facies and Sequence Stratigraphy*. International Association of Sedimentologists, Special Publications, **14**, 189–255.

VAIL, P. R., HARDENBOL, J., & TODD, R. G. 1984. Jurassic unconformities, chronostratigraphy, and sea-level changes from seismic stratigraphy and biostratigraphy. *In*: SCHLEE, J. S. (eds) *Inter-regional Unconformities and Hydrocarbon Accumulation*. American Association of Petroleum Geologists, Memoirs, **36**, 129–144.

—, MITCHUM, R. M. & THOMPSON, S. III. 1977*a*. Seismic stratigraphy and global changes of sea level, part 3: relative changes of sea level from coastal onlap. *In*: PAYTON C. E. (ed.) *Seismic Stratigraphy – Applications to Hydrocarbon Exploration*. American Association of Petroleum Geologists, Memoirs, **26**, 63–81.

—, —, & —1977*b*. Seismic stratigraphy and global changes of sea level, part 4: global cycles of relative changes of sea level. *In*: PAYTON C. E. (ed.) *Seismic Stratigraphy – Applications to Hydrocarbon Exploration*. American Association of Petroleum Geologists, Memoirs, **26**, 83–97.

VAN WAGONER, J. C., MITCHUM, R. M., CAMPION, K. M. & RAHMANIAN, V. D. 1990. *Siliciclastic Sequence Stratigraphy in Well Logs, Cores and Outcrops: Concepts for High Resolution Correlations of time and Facies*. American Association of Petroleum Geologists, Methods in Exploration Series, **7**, 55pp.

——, POSAMENTIER, H. W., MITCHUM, R. M., VAIL, P. R., SARG, J. F., LOUTIT, T. S. & HARDENBOL, J. 1988. An overview of the fundamentals of sequence stratigraphy and key definitions. *In*: WILGUS, C. K., HASTINGS, B. S., KENDALL, C. G. ST. C., POSAMENTIER, H. W., ROSS, C. A. & VAN WAGONER, J. C. (eds) *Sea-level Changes: an Integrated Approach*. Society of Economic Palaeontologists and Mineralogists, Special Publications, **42**, 39–45.

WOOD, L. J., ETHRIDGE, F. G. & SCHUMM, S. A. 1993. The effects of rate of base-level fluctuation on coastal plain, shelf and slope depositional systems: an experimental approach. *In*: POSAMENTIER, H. W., SUMMERHAYES, C. P., HAQ, B. U., & ALLEN, G. P. (eds) *Sequence Stratigraphy and Facies Associations*. International Association of Sedimentologists, Special Publications, **18**, 43–53.

Appendix

The details of the geometry of the fluvial profile are shown in Fig. A1. The curve has the form

$$y(x) = \frac{erfc[(x - x_1) \cdot \bar{\omega} - \theta]}{\phi} + d(t)$$

where

$$\bar{\omega} = \frac{\theta}{(x_1 - x_2)}$$

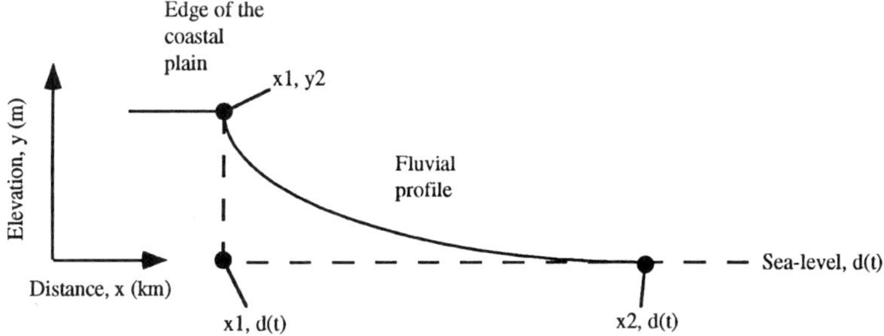

Fig. A1. Coordinate system and the variables used to define the fluvial profile.

and

$$\phi = \frac{1-\theta}{y^1 - d(t)^2}$$

The variables x_1, x_2, y_1 and d(t) are defined in Fig. A1. These equations are essentially scaling functions for the complementary error function used to fit the curve between the desired points at the beach and the head of the coastal plain. The variable θ controls the range of values from the error function used to define the profile, such that the error function parameter value ranges from 1.0 to θ. As the gradient of the error function decreases with higher parameter values, the value of θ determines the portion of the curve used, and hence the concavity of the fluvial profile.

Fluvial incision and sequence stratigraphy: alluvial responses to relative sea-level fall and their detection in the geological record

M. R. LEEDER & M. D. STEWART[1]

Department of Earth Sciences, University of Leeds, Leeds LS2 9JT, UK

[1]Also at School of Geography, University of Leeds, Leeds LS2 9JT, UK

Abstract: A critical question in sequence stratigraphy is whether or not a falling or lowered sea level will cause a river or delta distributary channel to incise and thus help to initiate an Exxon type-1 unconformable surface. River adjustment to relative sea-level fall was quantitatively modelled by linear diffusion. Although not a new approach to the general problem of river incision, these numerical experiments illustrate the rates and magnitudes of channel incision expected for various combinations of river slope, shelf slope and sediment transport coefficient. The results of the modelling broadly agree with, admittedly uncertain, field observations from the Gulf Coastal Plain of the USA. Incised valleys have limited lateral extent because of the slowness of valley side slope retreat. Successive nodal avulsions upstream of the highest point of channel incision may progressively neutralize the effects of sea-level fall in younger channel courses. It is also possible that incision is reduced or eliminated by rivers with high sediment transfer coefficients: it is shown that rivers may 'keep up' with sea-level fall under certain circumstances, constructing a prograding alluvial plain over a previous ramp-like shelf with little slope differential. These conclusions apply to the recognition of geologically ancient incision and thus to the existence or otherwise of local and regional sequence stratigraphical boundaries caused by relative sea-level fall. We briefly discuss examples from the stratigraphical record and conclude that much exciting work remains to be done in the field and on the computer.

A critical question in sequence stratigraphy and alluvial architecture is the extent to which falling or lowered sea level will cause a river or delta distributary channel to incise. Widespread incision is predicted by the conceptual models and thought experiments of sequence stratigraphy (Jervey 1988; Posamentier *et al.* 1988; Posamentier & Vail 1988; Van Wagoner *et al.* 1990). The qualitative effects of sea-level changes on alluvial stratigraphy were originally investigated by Fisk (1944), who proposed an entrenchment/alluviation cycle for the late Pleistocene Mississippi valley. Many sedimentologists have used Fisk's work, often uncritically, to interpret features of ancient alluvial architecture (e.g. McCave 1969; Allen 1974; Leeder 1975; Allen 1976; Nami & Leeder 1978). Existing quantitative stratigraphical models (Allen 1978; Leeder 1978; Bridge & Leeder 1979; Bridge & Mackey 1993) completely ignore the effects of base-level change. This is despite well-documented examples of ancient channel incision and consequent floodplain sediment starvation and soil formation (e.g. Leeder 1976; Allen & Williams 1982; Weimer 1984; Marzo *et al.* 1988).

Major impetus for more research into the effects of sea-level change on river channels comes from two sources. The first is the advent of sequence stratigraphy. As noted earlier, this essentially restated Fisk's Mississippi-based model of channel incision due to relative sea-level fall followed by alluviation during sea-level rise and highstand (Jervey 1988; Posamentier *et al.* 1988; Posamentier & Vail 1988; Van Wagoner *et al.* 1990). Sequence stratigraphical concepts as applied to rivers are geometrical and qualitative and some of these have been criticized (successfully in the present authors' opinion) by several workers, notably Miall (1991) and Schumm (1993). Much attention is currently being paid to the possible sequence stratigraphical implications of river behaviour during periods of sea-level change, particularly the study of incised valley fills (see Dalrymple *et al.* 1994 for a review). This is because the river system, as the major transport route of sediment to coastlines, is the prime control on other depositional responses to sea-level change and because of potential chronostratigraphical significance of erosion surfaces. Laboratory experiments (Koss *et al.* 1994; Wood *et al.* 1994) have shed some valuable qualitative light on the general behaviour of stream channels during base-level fall, although the results suffer from a lack of attention to scaling in the experiments.

The second source is the recognition that Fisk's conclusions regarding incision cannot be applied to the whole length (>1000 km) of the Mississippi valley. Detailed investigations in the last 15 years have established that Pleistocene incision extends only about 350 km up-river from the present day shoreline of the Gulf of Mexico (Saucier 1981), although this estimate is uncertain and may have to be increased by 50% (M. Blum pers. comm. 1995, citing unpublished work by Autin and Aslan). This

From HESSELBO, S. P. & PARKINSON, D. N. (eds), 1996, *Sequence Stratigraphy in British Geology*, Geological Society Special Publication No. 103, pp. 25–39.

begs the vital question of exactly what controls the magnitude and extent of incision upstream from any coastline. Also, many rivers show evidence for Holocene incision/aggradation cycles that are unrelated to sea-level change (Saucier & Fleetwood 1970; Blum 1990; Autin *et al.* 1991; Blum & Valastre 1994).

Finally, it has long been noted (e.g. Harms 1966) that the economic consequences of river responses to sea-level change are considerable, as hydrocarbon reservoir correlation, continuity, magnitude and distribution will all depend on the degree and rate of channel incision and the subsequent infill history of the incised valley.

Incision: general remarks

The sediment continuity equation states generally whether erosion or deposition will occur in any sedimentary environment (Carson & Kirkby 1972; Wilson & Kirkby 1975). It states that changes in the elevation, y, of any established sedimentary surface can occur if there are gradients in space and/or time in the sediment transport rate, i. In symbols and ignoring dimensional considerations of mass

$$-\frac{\partial y}{\partial t} \propto \nabla \cdot i + \frac{\partial i}{\partial t} \qquad (1)$$

This is a surprisingly complicated expression which can be simplified by assuming a steady transport rate with time – that is, $\partial i/\partial t = 0$. Erosion or deposition may thus be due to downstream changes in sediment transport rate caused by changes in water and/or sediment delivery, flow channel geometry and slope. Complications arise because the causes of these changes may be numerous and interdependent.

Climatic and vegetational changes are important controls on erosion or deposition. Thus regional cycles involving stream incision followed by aggradation have occurred in the southwestern USA over the past 1000+ years (Burkham 1972; Hereford 1984; Graf *et al.* 1991). In the Colorado River of Texas there is major late Pleistocene to Holocene (14–12 ka) incision upstream of the subsiding alluvial plain during a period of rising sea level (Blum 1990; Blum & Valastre 1994). As noted earlier, the maximum phase of incision of the Mississippi may only affect the river up to Baton Rouge, about 350 km upstream from the present coastline. In the alluvial valleys to the north aggradation occurred at the glacial maximum of lowest sea level and the early waning stage of rising sea level. This aggradation was caused by massive amounts of sediment and meltwater coming down the river from the north (Autin *et al.* 1991). It is also important to note that large tributaries may control the equilibrium of the main stream by supplying sediment and water from a different runoff regime. Erosion to cause incision cannot be linked automatically to the effects of base-level fall, particularly during past periods when 'hothouse' conditions prevailed. Another complicating factor concerns a river's ability (not yet quantified) to alter its 'wiggledeness' rather than incise or aggrade in response to imposed gradient changes (Schumm 1993).

We refer to the base-level controlled incision which we attempt to model in this paper as *potential incision*. This is because a change in potential (sea level) has caused the incision. We will show that the wave of incision passes upstream from the point at which relative sea-level change occurs (e.g. sea or lake shoreline, upwarp or fault trace) only if there is a slope discontinuity left after the river reacts to the sea-level fall. Terrace development, the incision erosion surface and the initiation of palaeosol horizons will all be time-transgressive, younging upstream. From the evidence of our numerical experiments the distance of upstream migration depends on the magnitudes of the slope discontinuity and the sediment transport coefficient. It also depends on whether avulsion occurs upstream of the migrating knickpoint.

We term discharge-controlled incision (including both sediment and water discharge) of the kind just discussed briefly for Quaternary examples *kinetic incision*. This occurs because the river's ability to transport sediment has changed due to a change in hinterland conditions such as climate. The wave of incision may be expected to young slightly downstream or to be more-or-less instantaneous along the river's course.

With the great complexity of the fluvial system in mind we next seek some simple quantitative solutions concerning the magnitude and extent of potential incision produced by base-level fall.

Diffusional modelling of river–shelf interfaces

A sea-level fall over a simple river–shelf system will result in areas of the shelf being subaerially exposed. If the river has a gradient α and the shelf a gradient β then for $\beta > \alpha$ incision will be initiated at the slope discontinuity at the river–shelf break (Fig. 1).

Over time this incision will express itself as a knickpoint of height, y, travelling up the river. A simple approach to modelling the sediment transport in a river is to consider that the sediment transport at a point is proportional to the discharge of the river times its gradient (Carson & Kirkby 1972). The experimental work of Koss *et al.* (1994) and Wood *et al.* (1994) shows that the rate of upstream migration of a knickpoint is dominantly controlled

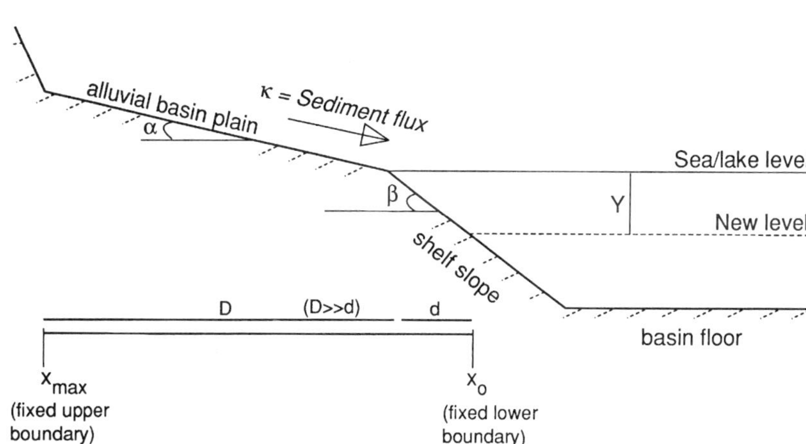

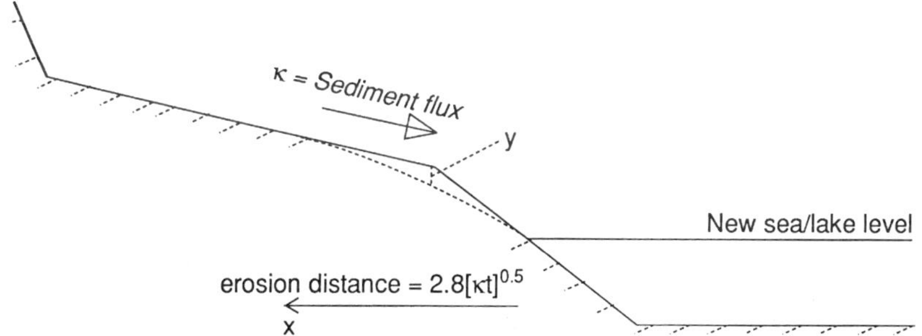

Fig. 1. Definition diagrams for the analysis of fluvial incision due to sea or lake level fall.

by the relative gradients of the river and shelf system. This is to be expected if we consider the erosion of a knickpoint at the local scale. Across the knickpoint the discharge will remain constant, while there will be a steep gradient change. Hence it is probably not unreasonable to model the erosion of a knickpoint as being due to the effect of slope change alone.

By combining the sedimentary continuity equation with a transport law that is dependent on slope only, the linear diffusion equation for sediment erosion or deposition may be derived.
This yields

$$-\frac{\partial y}{\delta t} = \kappa \frac{\partial^2 y}{\partial x^2} \qquad (2)$$

where x and y are the horizontal and vertical components respectively, t is time and κ is the sediment transport coefficient with units L^2T^{-1}. It can be seen that the elevation of a stream bed, y, is time and space dependent and that the actual rate of erosion or deposition is positively dependent on the magnitude of the sediment transport coefficient κ and the rate of change of the local gradient. Channel incision will thus be favoured by large rivers with high transporting power, acting on coastlines with large gradient contrasts between the river and shelf.

Such a linear diffusion approach has been tested for knickpoint retreat in both experimental and field studies by Begin *et al.* (1981) and Begin (1982, 1988), as shown in Fig. 2.

It may be seen that linear diffusion gives a good first-order approximation of the rates of knickpoint migration. In reality, discharge will be important as the knickpoint migrates upstream over time and the size of the discharge crossing it decreases. This will affect the rate of incision propagation. However, as noted by Schumm (1993) the changes in river discharge that occur during periods of sea-level change as a result of climatic variations are poorly constrained. A consideration of these and other complicated (though real) effects is beyond the scope of

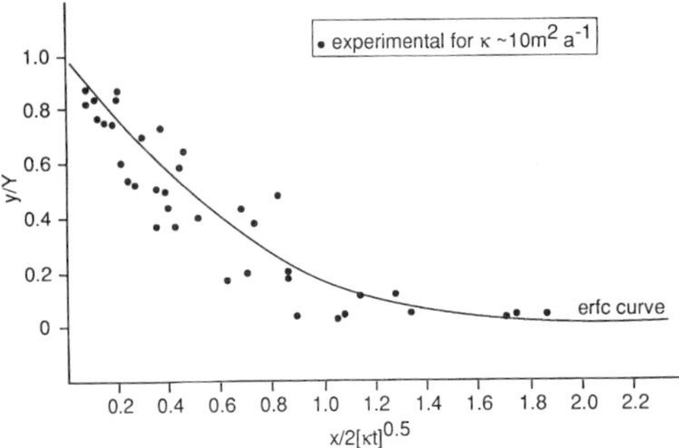

Fig. 2. Experimental results of incision (*y/Y*) compared with the complementary error function (*erfc*) curve as stated in Equation (4) (after Begin *et al.* 1981).

the present study. In contrast linear diffusion gives a simple, well-defined approximation to the processes involved.

Considering the simple situation of a vertical perturbation, Y, of a sloping surface (which approximates to a vertical sea-level fall of magnitude Y), Carslaw & Jaeger (1959) give the general solution to Equation (2) in the form of the complementary error function,

$$y = Y erfc\left(\frac{x}{2\sqrt{\kappa t}}\right) + \beta x \qquad (3)$$

where β is the slope of the initial surface. In the simplest situation $\beta = 0$ and the solution shows that the response to a base-level fall depends on the magnitude of the fall times the ratio between the distance upstream, x, travelled by the incision wave and the inverse of the quantity $\sqrt{\kappa t}$, a diffusion distance. The expression

$$\frac{y}{Y} = erfc\left(\frac{x}{2\sqrt{\kappa t}}\right) \qquad (4)$$

is one which may guide us in our initial attempts to investigate the likely effect of base-level fall on the alluvial system. The solution indicates that elevation changes created by base-level fall migrate rapidly upslope at first, but that the rate of propagation decreases over time, depending on $\sqrt{\kappa t}$. For our purposes we seek a solution for the propagation of a 5 m incision upslope from a 100 m base-level fall. We consider this value of 5% of the initial fall to be the limit of practical detection of incision in the field, as the 'normal' processes of intrinsic scour (Salter 1993) may cause effects of similar magnitude. As 5% incision occurs at $y/Y = erfc\ 0.05$, then $x/2\sqrt{\kappa t} = 1.4$ and $x = 2.8\ \kappa t$. This latter quantity is one that we introduce as a characteristic *erosion* or *incision distance*. Paola *et al.* (1991) has previously modelled knick point retreat by linear diffusion and defined a *response distance* similar to this concept. We define it as the upstream limit to erosion induced by any base-level fall after an elapsed time. In this simple case the erosional distance is dependent both on the elapsed time and the transporting capacity of the river channel in question. Figure 3 gives a plot of incision distance against time for various values of κ.

Application of the model to examples of river–shelf systems

In this section the diffusion equation is solved by finite differencing (see Hanks & Andrews 1989 for analytical solutions) for initial conditions that are based on natural examples of river slope, shelf slope and diffusivity. We consider the erosion of a river profile of gradient, α, after an instantaneous 100 m sea-level fall (i.e. $Y = 100$ m) has exposed a region of shelf of slope β. In all examples $\beta > \alpha$ so that the sea-level fall results in a perturbation of the existing river profile. By considering a sea-level fall across a sloping shelf the model has an appreciation of the two-dimensional effects of sea-level fall deemed essential by Koss *et al.* (1994). The diffusional application of Begin (1988) considered only a vertical perturbation along a profile of constant gradient. A sea-level fall affects the river length-wise as well as vertically as the river must occupy an exposed shelf of differing slope. Hence the models presented here are more directly comparable with the flume studies of Wood *et al* (1994) and Koss *et al.* (1994). Our model also allows for no deposition

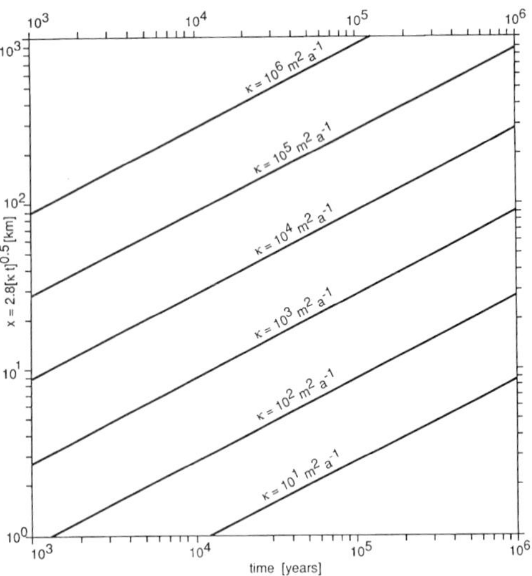

Fig. 3. Plots of incision distance x against time for various values of κ, the sediment transfer coefficient, as computed from Equation (3).

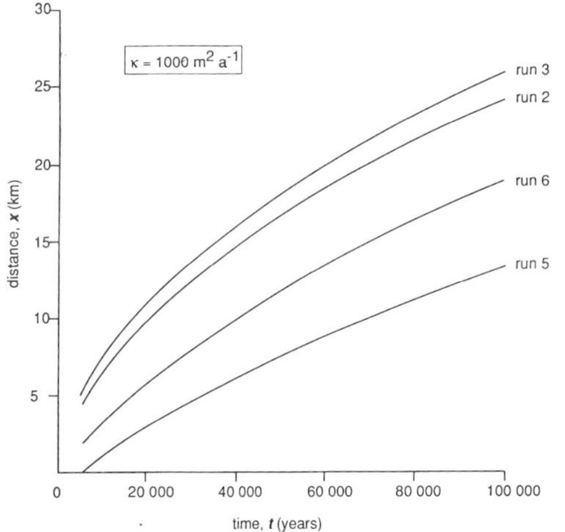

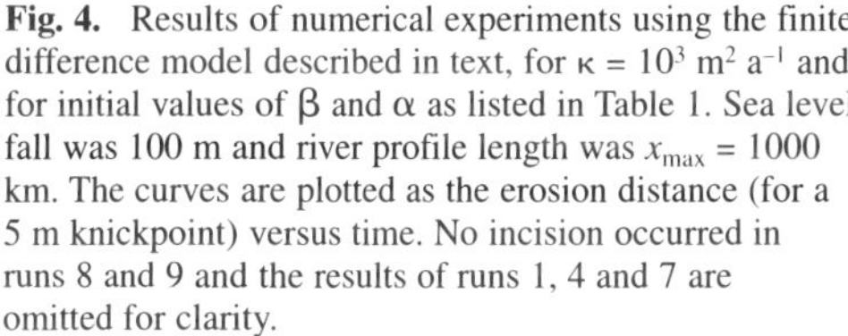
Fig. 4. Results of numerical experiments using the finite difference model described in text, for $\kappa = 10^3$ m^2 a^{-1} and for initial values of β and α as listed in Table 1. Sea level fall was 100 m and river profile length was x_{max} = 1000 km. The curves are plotted as the erosion distance (for a 5 m knickpoint) versus time. No incision occurred in runs 8 and 9 and the results of runs 1, 4 and 7 are omitted for clarity.

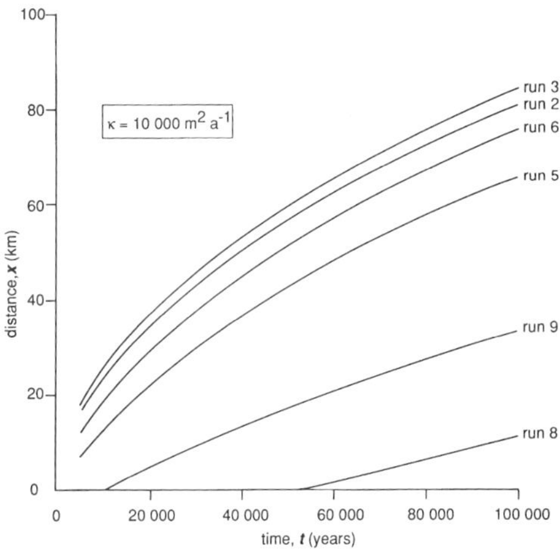

Fig. 5. Results of numerical experiments using the finite difference model described in text, for for κ = 104 m^2 a^{-1} and for initial values of β and α as listed in Table 1. Sea level fall was 100 m and river profile length was x_{max} = 1000 km. The curves are plotted as the erosion distance (for a 5 m knickpoint) versus time.

of sediment on the shelf as the period of sea-level fall is held to be instantaneous. We address this problem semiquantitatively in a later section.

The boundary conditions of the model are such that the elevation of the profile is fixed at both ends, so that x_0 and x_{max} (see Fig. 1) are constant through time. This models the situation where there is no change in the position of the new shoreline (after instantaneous sea-level fall) over time and no lowering of the most upstream part of the river basin. Without a fixed upper boundary the model tends to a final state where the initial base-level fall has been projected right back to the basin divide upstream from the alluvial plain itself. This is the case with the models of Begin *et al.* (1981) and Begin (1982). Field and flume studies (Brush & Wolman 1960; Leopold & Bull 1979; Koss *et al.* 1994) have observed that such a rejuvenation does not occur. Fixing the upper boundary prevents the incision propagating to the upstream limit of the basin drainage divide. The model evolves to a final condition where the river achieves a new grade from the upstream limit of the alluvial plain to the shoreline (Fig. 1). Hence the fixed upper boundary models the effect of the upstream dissipation of the initial sea-level fall.

It is important to note that the fixed upper boundary condition has an effect on the upslope propagation of any perturbation. As the perturbation approaches the upper boundary its rate of propagation will decrease due to the constraints of the boundary condition. Hence consideration of length scales appropriate to rivers is essential to take account of this boundary condition. In his model Begin (1988) placed a fixed upper boundary 3 km away from a 10 m base-level fall. For a 100 m sea-level fall this would place the boundary only 30 km away from the river mouth. A boundary so close to

Table 1. *Experimental data*

River slope α	Shelf slope β	Shelf slope β	Shelf slope β
0.01	Run 1, 0.03	Run 2, 0.05	Run 3, 0.1
0.001	Run 4, 0.003	Run 5, 0.005	Run 6, 0.01
0.0001	Run 7, 0.0003	Run 8, 0.0005	Run 9, 0.001

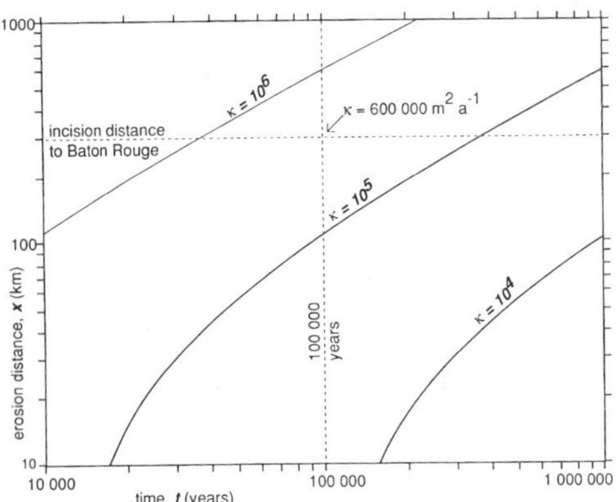

Fig. 6. Erosion distance versus time using the finite difference model described in text for a range of κ values and with initial conditions appropriate to the Mississippi coastal plain and nearshore Gulf of Mexico shelf. The river profile length was 1000 km. The horizontal broken line marks the incision distance observed for the latest Pleistocene sea level fall for the Mississippi (350 km at Baton Rouge; after Saucier 1981; Autin *et al.* 1991). The κ value at 100 000 years is close to Kenyon & Turcotte's (1985) estimate of 560 000 m^2 a^{-1}.

the river mouth would greatly affect knickpoint migration. In the examples which follow, river length >> shelf length (*D>>d* in Fig. 1), such that the migration of the knickpoint is not affected by the upper boundary. The effect of shortening the river length will be discussed explicitly through an experimental example.

Figures 4 and 5 are graphs showing the upstream migration of a 5 m knickpoint through time for a series of initial conditions given in Table 1. The values of κ used are 10^3 and 10^4 m^2 a^{-1}, which fall into the range of values described for rivers by Flemings & Jordan (1989). Each model was run for 100 000 years, approximately equal to the duration of a Milankovitch eccentricity cycle. Alluvial plain length was 1000 km in all runs. As discussed previously, the model predicts that the rate of knickpoint migration is not constant through time, but decreases from an initially high value. Also obvious is the effect of the river and shelf gradients and the κ value. Larger κ values and larger differences between the river and shelf gradient produce markedly increased rates of incision propagation. Differences of up to 80 km in the incision distance due to the effect of the difference in slope alone is predicted from Fig. 5. The effect of using a shelf angle of $\alpha \neq \beta$ is obvious from the results obtained by Begin (1982) when considering only a vertical perturbation on a slope of uniform gradient. These gave a propagation rate at least an order of magnitude higher than those presented here.

In Fig. 6 we consider a series of model runs for the Mississippi River using river and shelf slope values taken from Miall (1991). Figure 6 summarizes the erosion distance over time for a range of κ values. The model is calibrated using the observed (though uncertain) erosion distance at Baton Rouge, about 350 km upstream of the present river mouth (Saucier 1981; Autin *et al.* 1991; M. Blum pers. comm. 1995). For incision commencing 100 000 years ago the value for κ is 600 000 m^2 a^{-1}, remarkably close to the value of κ calculated by Kenyon & Turcotte (1985) for sediment transport onto the Mississippi delta. The value is also in the mid-range of values calculated for κ by Begin (1988) from field studies of knickpoint migration in streams.

In the previous example the length of the river section taken for the Mississippi River was 1000 km. The point of 5 m incision was not able to propagate anywhere near the upper boundary of the model in the time period allowed. In Fig. 7 we now consider shorter river lengths with the same slope values for the Mississippi and a κ value of 600 000 m^2 a^{-1}. The difference in river length dramatically changes the propagation of the knickpoint. In these examples, once the knickpoint has travelled 40% of the way up the river the rate of propagation decreases.

It is important to consider the limitations of our model. We do not consider gradual sea-level change or variations in the magnitude of the diffusion constant in space and time. In natural systems it is also possible that changes in upstream erodibility, such as the exposure of bedrock, may also act as a fixed boundary to incision migration. This could occur at any point along the river, thus dramatically affecting the propagation rate of the knickpoint. The effect of the length scale is also thought to be important for

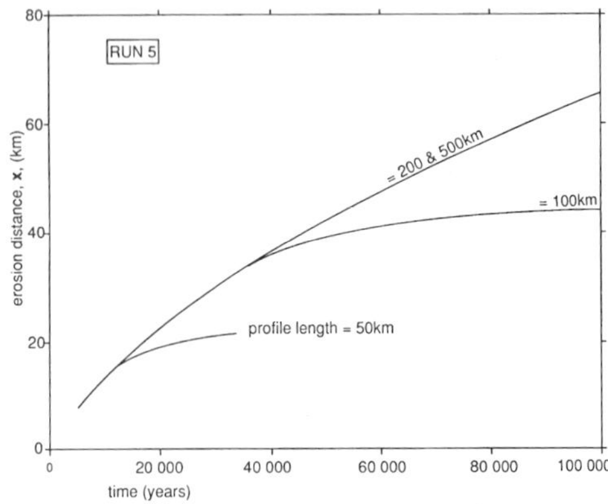

Fig. 7. Results of numerical experiments for the slope conditions of run 5 (see Table 1) and κ = 10^4 m^2 a^{-1} for a range of river lengths (x_{max} of Fig. 1), illustrating the dramatic effect of river length on the erosion distance.

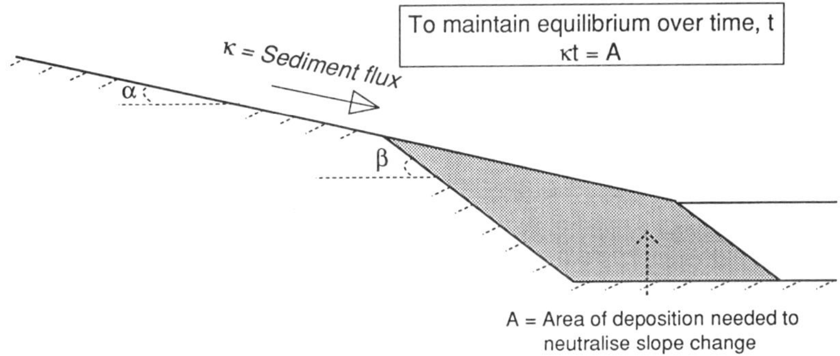

Fig. 8. Definition diagram for the case where deposition on the exposed shelf during gradual sea-level fall neutralizes the inequality $\beta > \alpha$. See text for explanation.

short rivers in climates with high discharge and also for tributaries of a main channel, where knickpoint propagation could be far lower than in the main channel. Despite these limitations we regard our approach as robust as the model predicts many of the features visible in field and laboratory experimental studies, namely the dependence of the incision distance on slope changes, the size of the river (through κ) and the fact that rivers do not incise the entire basin.

Lateral development of incised valleys and significance of avulsions from upstream of incised reaches

Once the wave of incision associated with a particular slope discontinuity passes upstream of any point, then valley widening will begin. The rate of valley widening is expected to be significantly less than the rate of upstream migration of the slope discontinuity because it will depend strongly on hillslope diffusional processes of soil creep and rain splash, both slow compared with rates of fluviatile incision (Kirkby 1971; Carson & Kirkby 1972). In addition to valley widening, gully formation will lead to the lateral propagation of new stream systems and drainage catchments incised into the terrace plains adjacent to the new channel floor.

Implicit in our characterization of alluvial basins affected by base-level change is that there is a moving division between a downstream incised reach and an upstream unincised reach. Although the occurrence of channel avulsions is less likely in the incised downstream part, they may still occur upstream, unaffected by the process of downstream incision. As the frequency of upstream avulsions is expected to be of the order of 10^2–10^3 years (Bridge & Leeder 1979), then it is likely that one or more will occur before an upstream-migrating knickpoint has reached the upstream limit. After avulsion the course of the new channel will now run down to the point where the slope discontinuity began (i.e. the old shoreline) and the process of incision must begin afresh. It thus becomes possible for avulsions to significantly affect the rates of upstream propagation of knickpoints, as they must begin their journeys anew with each new course. In this way the length of an incised valley is further restricted, reinforcing our earlier comments that modelling estimates from linear diffusion are to be regarded as maxima.

The case where deposition neutralizes slope change

It is possible for the delivery of sediment to be of sufficient magnitude that the effects of gradual sea-level fall in exposing a steeper sloping shelf are completely neutralized. This is partly the scenario envisaged by Exxon workers as helping to produce a type-2 sequence boundary and by Soreghan and Dickinson (1994) in the production of their 'keep-up' cycles. Referring to Fig. 8, if A is the area of deposition needed to neutralize slope change, to maintain equilibrium over time t, then $A = \kappa t$.

The value of A will vary directly with the slope of the offshore area and with the magnitude of the sea-level fall. For values of $\alpha = 0.0001°$ and $\beta = 0.003°$, appropriate values for several large modern rivers and their offshore slopes on passive margins (e.g. Rio Grande, Nile, Indus), sedimentary input can neutralize the effects of a 100 m sea-level fall over 20 ka if κ exceeds approximately $10^3 m^2 a^{-1}$. This is not a large figure and we conclude that such conditions could have applied frequently in the stratigraphical record.

This is particularly so for times of 'hothouse' climates when eustatic sea level would have varied slowly over long periods (e.g. 1–5 Ma, or '3rd order' in the Exxon terminology). During such

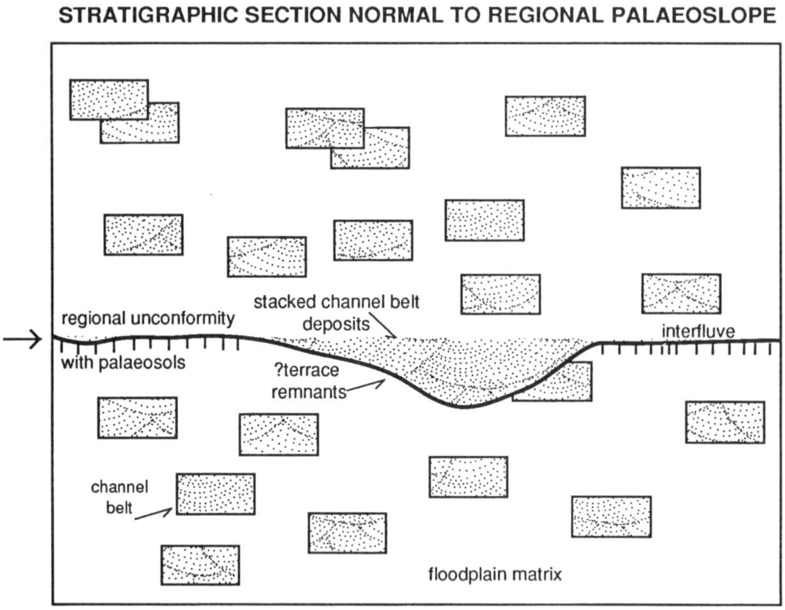

STRATIGRAPHIC SECTION NORMAL TO REGIONAL PALAEOSLOPE

Fig. 9. Sketch section normal to regional palaeoflow to illustrate the major alluvial architectural effects resulting from a major episode of incision as sectioned relatively close to the position of initial base-level fall.

times, constructional conditions would have been the norm, with κ parameters of only between 10^1 and 10^2 required to maintain the slopes noted here. Constructional conditions may also have been attained during 'icehouse' times by very large rivers on passive continental margins. We tentatively conclude that the incidence of riverine incision in the geological record has been overestimated. Further investigation seems warranted, perhaps taking into account the non-trivial problem of spatially and time-varying deposition rates that are to be expected from river mouth to offshore shelf (see Kenyon & Turcotte 1985) during the cycle of sea-level change.

Recognition of ancient incision

To recognize and discuss the possible causes of alluvial plain incision in the stratigraphical record it is necessary to: (l) correctly identify incised fluviatile channel sand bodies and their valley margins; (2) make correlations with other time-equivalent stratigraphical sequences in an effort to see whether the cause of incision was local, regional or global; and (3) determine the causes of the local, regional or global relative sea-level changes.

Only point (l) concerns us here and the most important criteria which are required from field observations are discussed in the following (see also Gibling 1991; Wright & Marriott 1993; Dalrymple *et al.* 1994; Zaitlin *et al.* 1994). Some of the more obvious of these are summarized in Figs 9 and 10.

Physical recognition and mapping of the incised valley

This criterion is difficult to achieve without very good exposures of regional extent or from three-dimensional seismic and well data. Given these conditions we must positively identify and map out the incised valley margin in three dimensions. For example, Shanley & McCabe (1993) attempt to document a sequence stratigraphical model for alluvial architecture, but notably fail to identify a mappable incised valley morphology (see also later comments). Also, erosion represented by the depth of valley-fill margins must be demonstrated to be significantly greater and more widespread in extent than the more localized erosive effects of intrinsic channel scour (Salter 1993).

Loss of section: biostratigraphical and chronostratigraphical considerations

Incision leads to the removal of a time sequence of sediment from below the valley margin erosion surface (Fig. 11). It is a requirement for the logical application of sequence stratigraphical models that the alluvial sequences of the incised valley-fill erosively overlie *entirely unrelated* deposits. It is thus

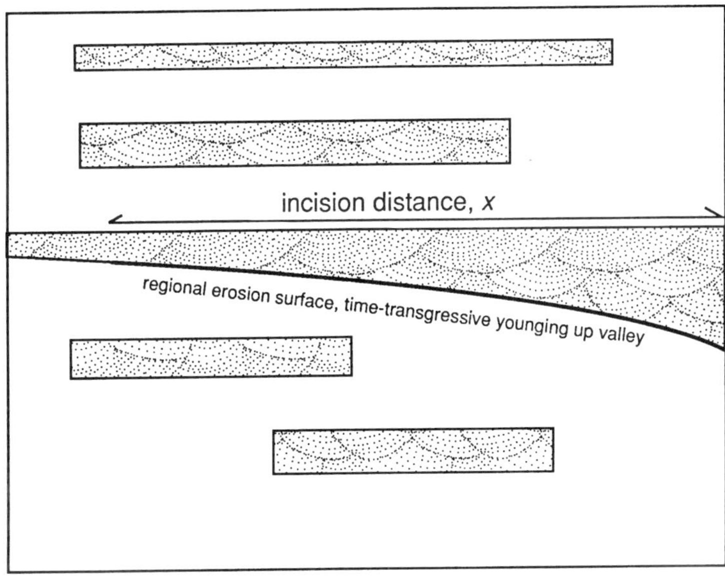

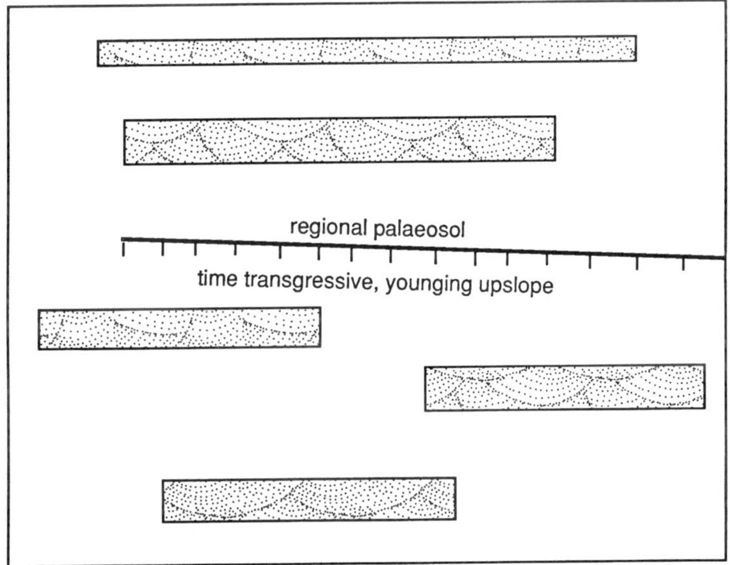

Fig. 10. Sketch sections parallel to regional palaeoflow to illustrate the major stratigraphical and pedogenic effects resulting from a major episode of incision.

vital that some effort be put into substantiating the loss of stratigraphical section using biostratigraphical or chronostratigraphical dating. Otherwise, it is equally possible that time lines run through the alluvial sequences and into the underlying shoreline facies with little (but not zero) displacement, as shown in Fig. 11A.

A good example of the dilemmas raised by

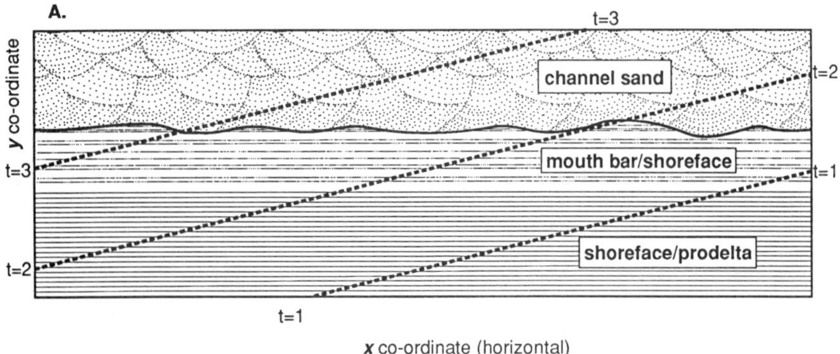

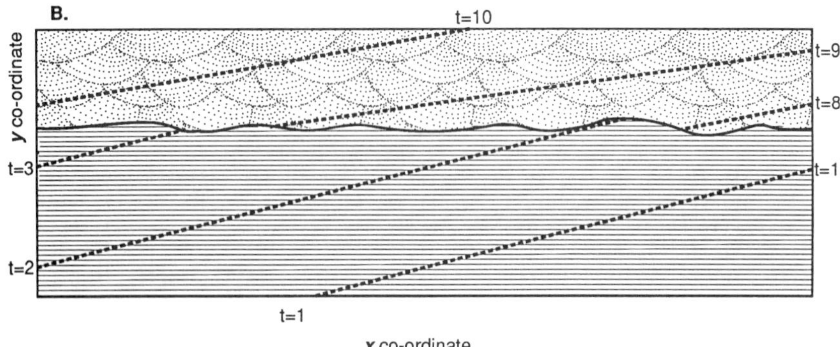

Fig. 11. Sketch sections to illustrate the two cases of non-incision and incision in geological situations where fluviatile channel or fluvio-deltaic channel sand bodies rest with erosion on coastal, delta front and pro-delta deposits. In section A 'normal' coastal progradation from left to right results in diachronous facies boundaries. The channel facies are 'genetically related' to the coastal delta front facies and are not incised, although local erosion due to scour will create some stratigraphical omission. In section B a significant time gap between the two facies (together with evidence such as that discussed in the text and summarized in Figs 9 and 10) allows a sequence stratigraphical interpretation to be made that involves no genetic relationship between the channel and coastal facies.

insufficient time control is given by the example of the fluviatile Moor Grit member of the Middle Jurassic Scalby Formation of the Yorkshire basin, UK. This rests with marked local erosion on coastal and shelf facies of the underlying Scarborough Formation. Alternative models for progradational or incisional scenarios were proposed by Leeder & Nami (1979), the latter ruled out by subsequent biostratigraphical studies (Riding & Wright 1989). New and very detailed biostratigraphical studies (Hogg in press) have enabled the alternative sedimentary models to be properly assessed. A substantial time gap of several million years between the fluviatile and underlying facies has been proposed. This points to, but does not prove, that a prolonged period of erosion occurred before and during valley occupation by the Moor Grit river channel. Poor inland exposures rule out attempts to define the presence and extent of any incised valley.

Relevance of multistorey character or channel stacking pattern

Stacked channel belt deposits with relatively little preserved fine-grained alluvium may fill incised alluvial valleys well upstream of any effects of the highstand shoreline (Shanley & McCabe 1993; Levy & Christie-Blick 1994). This is because the proportion of channel deposits depends (most critically) on the ratio between total floodplain width and channel-belt width (Leeder 1978; Bridge & Leeder 1979; Bridge & Mackey 1993). In an incised valley system this ratio is obviously lower than 'normal'. However, channel stacking and a high channel deposit proportion are *not diagnostic on their own* as criteria for identifying valley-fills in the absence of other confirmatory evidence. Thus an increase in channel deposit proportion may equally well be due to increases in avulsion frequency or reductions in deposition rate due to other allocyclic or autocyclic causes entirely unrelated to base-level changes. These alternative explanations have commonly not been considered by previous workers (e.g. Shanley & McCabe 1993). Conceptual models for the vertical sequences of alluvium expected to be deposited under rising sea level conditions in incised valleys are given by Gibling (1991), Wright & Marriott (1993), Gibling & Wightman (1994) and Zaitlin *et al.* (1994).

Palaeosols

A wave of incision moving up an alluvial channel in response to a major relative base-level fall may be associated with a soil development front. Older floodplain alluvium 'stranded' in terraces by the incision will show a change from wetland to low water-table conditions, with desiccation, oxisol/calcisol development and associated gullying and tributary incision. The absence of such features along supposed incised valley margins and, more importantly, on interfluves, needs careful explanation.

Perhaps the best documented example in the Pleistocene record is that of the ancestral Rio Grande, which began incising its floodplain and earlier alluvium at about 700 ka (see discussion of Mack *et al.* 1993), with about 100 m of incision developed to present. Calcisols progressively developed on the successively younger and lower abandoned ancestral terraced floodplains (summary in Gile *et al.* 1981). Well-documented examples of incision-related palaeosols in the ancient geological record come from the excellent exposures of the Castissent Sandstone analysed by Marzo *et al.* (1988) and the late Carboniferous cyclothems of the Sydney Basin, Nova Scotia (Gibling & Bird 1994; Gibling & Wightman 1994; Tandon & Gibling 1994).

Fluvial incision and base-level change in the Carboniferous 'icehouse' stratigraphical record

Fluviatile and fluvio-deltaic deposits commonly occur interbedded with marine deposits in the Upper Carboniferous (Pennsylvanian) of several continents. A sequence stratigraphical approach seems justified *a priori* as icehouse conditions prevailed world-wide. Previous studies have shown the existence of marine/non-marine cyclicity with a periodicity of around 100-400 ka and with possible amplitudes of sea-level change of >50 m (Heckel 1986; Klein 1990; Crowley & Baum 1991; Maynard & Leeder 1992). A glacio-eustatic origin for Carboniferous sedimentary cycles involving fluvial incision during lowstands is not new, having been proposed first by Wanless & Shepard (1936) in a classic paper. However, the interpretation of Carboniferous cycles in sequence stratigraphical terms is not a simple matter as the rates of tectonic subsidence across foreland basin, extensional synrift and post-rift basin margins were highly variable. The ability to disentangle eustatic, tectonic and autocyclic effects is hampered by the lack of detailed intra-cycle biostratigraphical correlation at time-scales less than a goniatite subzone (< *c* 400 ka), part of a more general problem raised and clearly discussed by Miall (1994). Also, some classic European areas are cursed by the relative poorness of exposure, e.g. the verdant countryside of the northern English Pennines.

Examples of fluvial incision are apparently rare in British Lower to Middle Carboniferous (Mississippian to early Pennsylvanian) clastic/carbonate 'Yoredale-type' cycles (Leeder & Strudwick 1987). The majority of published studies indicate that fluviatile and fluvio-deltaic sand bodies were genetically linked to the underlying delta-front and coastline deposits, although firm evidence for this in the light of sequence stratigraphical theory has yet to be advanced. A possible exception to this general rule is provided by the Great Limestone cyclothem of the northern Pennines. Previous studies have revealed a cyclic sequence caused by autocyclic (delta-switching) mechanisms (Elliott 1975). A regional study, taking into account surface and subsurface data, has led to the identification of narrow zones where the thickness of erosive based sand bodies significantly increased at the expense of clastic delta-front and coastline facies *and* underlying marine carbonate sequences (Hodge & Dunham 1988). These geographically restricted channel sandstones have been identified as marking the main delta distributary channel responsible for the second of three successive 'minor' clastic members that make up the complex cyclothem. We may ask whether they are in fact incised valley-fills. The cyclothem is clearly more complex than a simple sequence stratigraphical scheme would indicate, as the thick marine limestone member is directly overlain in places by a coal and in other places by marine pro-delta facies. Also, the thickened sand bodies, if indeed marking an incised valley-fill, lack significant correlative palaeosols. The diagnosis of valley-fill is clearly not proved and further analysis needed.

The status of fluviatile and fluvio-deltaic sand bodies in the application of sequence stratigraphy to deposits of the Upper Carboniferous (Pennsylvanian) Pennine River of northern Britain is currently under hot dispute (Read & Forsyth 1991; Read 1991; Collinson *et al.* 1992; Martinsen 1993; Maynard 1992*a,b*). There is, to the authors' knowledge, no single study that demonstrates the combination of regional and local evidence for incision that we have discussed above. Thus the evidence for incision cited by Maynard (1992*a*) comprises just one over-thickened section in the fluviatile late Namurian Rough Rock, whereas other workers have noted the great uniformity of thickness of this sheet-like deposit (Read 1991; Collinson *et al.* 1992; Bristow 1993). We attribute the apparent absence of incision to the ramp-like nature of the contemporary shallow marine shelf that the Rough Rock rivers flowed into and over during periods of lowstand. We agree with Collinson *et al.* (1992) that at this time, with marked bathymetric gradients largely eliminated from the previously active Dinantian–early Namurian syn-rift tilt-blocks, significant fluviatile incision would not be expected. In the terminology of the quantitative models introduced in this paper, κ was very high for these large Pennine River channels, but $\beta \approx \alpha$, hence incision was minor or absent. It is possible, however, that earlier Namurian fluviatile deposits such as the Pendle and Kinderscout Grits might exhibit valley-fill features (see Martinsen 1993 for the most important study). The fluviatile facies in these early to mid-Namurian sequences were deposited at a time when marked gradient changes probably existed from coastal plain to offshore shelf due to active extensional tectonics. Further work is clearly needed on this interesting problem.

In the New World a convincing case has been made for the occurrence of incised palaeovalleys and interfluve palaeosols in the late Westpalian–Stephanian of the Sydney Basin, Nova Scotia (Gibling & Bird 1994; Gibling & Wightman 1994; Tandon & Gibling 1994). Studies of the Pennsylvanian Breathitt Group cyclothems of the Kentucky Appalachians (Aitken & Flint 1994, 1995) have also proposed that most major fluviatile sand bodies there were deposited in lowstand incised valleys. Although it would seem churlish to disagree that relative sea level was fluctuating during deposition of the Breathitt Group, some of the criteria noted here for the recognition of incised valleys are not fulfilled. In particular, the two-dimensional nature of the outcrops prevents mapping of the incised valley margins and significant interfluve palaeosols are seemingly absent. We feel in the light of these points that alternative autocyclic and tectonic models for Breathitt Group cyclothems should at least be tested and explored.

Conclusions

We have shown by simple quantitative and conceptual modelling that river channel incision depends on sediment transport capacity and the slope differential between river and shelf. We define two types of incision, potential and kinetic, with each giving rise to distinctive features. From a discussion of linear diffusional modelling we introduce the concept of incision distance and investigate the magnitude of this for various starting conditions appropriate to natural examples. The total starting length of the river profile is a neglected control on the rate of upstream knickpoint propagation. We show that rivers may ‘keep up’ with sea-level fall under certain circumstances, constructing a prograding alluvial plain over a previous ramp-like shelf with little slope differential. The recognition of ancient incision should be accomplished by a thorough review of all the field evidence (three-dimensional incision, palaeosols, channel stacking, biostratigraphy). There is clearly a need for much more critical study of the importance of regional incision features in ancient fluviatile deposits.

We thank M. Kirkby for helping us with our diffusional analysis. J. Bridge gave a typically thorough and much appreciated review of an earlier typescript. Referees M. Blum and P. Burgess were critically encouraging and very helpful. J. Aitken kindly provided preprints of his recent papers. M.S. gratefully acknowledge receipt of a NERC research studentship held jointly between the Departments of Geography and Earth Sciences at Leeds.

References

AITKEN, J. F. & FLINT, S. S. 1994. High-frequency sequences and the nature of incised-valley fills in fluvial systems of the Breathitt Group (Pennsylvanian), Appalachian foreland basin, eastern Kentucky. *In*: DALRYMPLE, R. W., BOYD, R. & ZAITLIN, B. A. (eds) *Incised-valley Systems: Origin and Sedimentary Sequences*. Society for Sedimentary Geology, Special Publications, **51**, 353–368.

—— & —— 1995. The application of high-resolution sequence stratigraphy to fluvial systems: a case study from the Upper Carboniferous Breathitt Group, eastern Kentucky, USA. *Sedimentology*, **42**, 3–30.

ALLEN, J. R. L. 1974. Studies in fluviatile sedimentation: implications of pedogenic carbonate units, Lower Old Red Sandstone, Anglo–Welsh outcrop. *Geological Journal*, **9**, 181–208.

—— 1978. Studies in fluviatile sedimentation: an exploratory quantitative model for the architecture of avulsion-controlled alluvial suites. *Sedimentary Geology*, **21**, 129–147.

—— & WILLIAMS, B. P. J. 1982. The architecture of an alluvial suite: rocks between the Townsend Tuff and Pickard Bay Tuff Beds (Early Devonian), Southwest Wales. *Philosophical Transactions of the Royal Society of London*, **B297**, 51–89.

ALLEN, P. 1976. Wealden of the Weald: a new model. *Proceedings of the Geologists' Association*, **86**, 389–437.

AUTIN, W. J., BURNS, S. F., MILLER, R. T., SAUCIER, R. T. & SNEAD, J. I. 1991. Quaternary geology of the lower Mississippi valley. *In*: MORRISON, R. B. (ed.)

Quaternary Nonglacial Geology: Conterminous US. The Geology of North America, Vol. K-2. Geological Society of America, Boulder, 547–582.

BEGIN, Z. B. 1982. Application of 'diffusion' degradation to some aspects of drainage net development. *In*: BRYAN, R. & YAIR, A. (eds) *Badland Geomorphology and Piping*. GeoBooks, Norwich, 169–179.

—— 1988. Application of a diffusion-erosion model to alluvial channels which degrade due to base-level lowering. *Earth Surface Processes and Landforms*, **13**, 487–500.

——, MEYER, D. F. & SCHUMM, S. A. 1981. Development of longitudinal profiles of alluvial channels in response to base-level lowering. *Earth Surface Processes and Landforms*, **6**, 49–68.

BLUM, M. D. 1990. Climatic and eustatic controls on Gulf Coastal Plain fluvial sedimentation: an example from the late Quaternary of the Colorado River, Texas. *In*: *GCSSEPM Foundation 11th Annual Research Conference*, 71–83.

—— & VALASTRE, S. 1994. Late Quaternary sedimentation, lower Colorado River, Gulf Coastal Plain of Texas. *Bulletin of the Geological Society of America*, **106**, 1002–1016.

BRIDGE, J. S. & LEEDER, M. R. 1979. A simulation model of alluvial stratigraphy. *Sedimentology*, **26**, 617–644.

—— & MACKEY, S. D. 1993. A revised alluvial stratigraphy model. *In*: MARZO, M. & PUIGDEFABREGAS, C. (eds). *Alluvial Sedimentation*. International Association of Sedimentologists, Special Publications, **17**, 319–336.

BRISTOW, C. S. 1993. Sedimentology of the Rough Rock: a Carboniferous braided river sheet sandstone in northern England. *In*: BEST, J. L. & BRISTOW, C. S. (eds) *Braided Rivers*. Geological Society, London, Special Publications, **75**, 291–304.

BRUSH, L. M. & WOLMAN, M. G. 1960. Knickpoint behaviour in noncohesive material for a laboratory study. *Bulletin of the Geological Society of America*, **71**, 59–74.

BURKHAM, D. E. 1972. *Channel Changes of the Gila River in Safford Valley, Arizona, 1846-1970*. US Geological Survey Professional Paper, **655-G**, 24pp.

CARLSLAW, M. S. & JAEGER, J. C. 1959. *Conduction of Heat in Solids*. Clarendon Press, Oxford.

CARSON, M. A. & KIRKBY, M. J. 1972. *Hillslope Form and Process*. Cambridge University Press, Cambridge.

COLLINSON, J. D., HOLDSWORTH, B. K., JONES, C. M. & MARTINSEN, O. J. 1992. Discussion of: 'The Millstone Grit (Namurian) of the southern Pennines viewed in the light of eustatically controlled sequence stratigraphy' by W.A. Read. *Geological Journal*, **27**, 173–180.

CROWLEY, T. J. & BAUM, S. K. 1991. Estimating Carboniferous sea-level fluctuations from Gondwanan ice extent. *Geology*, **19**, 975–977.

DALRYMPLE, R. W., BOYD, R. & ZAITLIN, B. A. 1994. History of research, types and internal organisation of incised-valley systems: introduction to the volume. *In*: DALRYMPLE, R. W., BOYD, R. & ZAITLIN B. A. (eds) *Incised-valley Systems: Origin and Sedimentary Sequences*. SEPM, Special Publications, **51**, 3–10.

ELLIOTT, T. 1975. The sedimentary history of a delta lobe from a Yoredale (Carboniferous) cyclothem. *Proceedings of the Yorkshire Geological Society*, **40**, 505–536.

FISK, H. N. 1944. *Geological Investigation of the Alluvial Valley of the Lower Mississippi River*. Mississippi River Commission, Vicksburg, 78pp.

FLEMINGS, P. B. & JORDAN, T. E. 1989. A synthetic stratigraphic model of foreland basin development. *Journal of Geophysical Research*, **94**, 3851–3866.

GIBLING, M. R. 1991. Sequence analysis of alluvial-dominated cyclothems in the Sydney Basin, Nova Scotia. *In*: LECKIE, D. A., POSAMENTIER, H. W. & LOVELL, R. W. W. (eds) *1991 NUNA Conference on High Resolution Sequence Stratigraphy, Program, Proceedings, Guidebook*. Geological Association of Canada, 15–19.

—— & BIRD, D. J. 1994. Late Carboniferous cyclothems and alluvial palaeovalleys in the Sydney Basin, Nova Scotia. *Bulletin of the Geological Society of America*, **106**, 105–117.

—— & WIGHTMAN, W. G. 1994. Palaeovalleys and protozoan assemblages in a Late Carboniferous cyclothem, Sydney Basin, Nova Scotia. *Sedimentology*, **41**, 699–719.

GILE, L. H., HAWLEY, J. W. & GROSSMAN, R. B. 1981. *Soils and Geomorphology in the Basin and Range Area of Southern New Mexico – Guidebook to the Desert Project*. New Mexico Bureau of Mines and Mineral Resources, Memoirs, **39**, 222pp.

GRAF, J. B., WEBB, R. H. & HEREFORD, R. 1991. Relation of sediment load and floodplain formation to climatic variability, Paria River drainage basin, Utah and Arizona. *Bulletin of the Geological Society of America*, **103**, 1405–1415.

HANKS, T. C. & ANDREWS, M. 1989. Effect of far-field slope on morphologic dating of scarplike landforms. *Journal of Geophysical Research*, **94B**, 565–573.

HARMS, J. C. 1966. Stratigraphic traps in a valleyfill, western Nebraska. *Bulletin of the American Association of Petroleum Geologists*, **50**, 2119–2149.

HECKEL, P. H. 1986. Sea-level curve for Pennsylvanian eustatic marine transgressive depositional cycles along Midcontinent outcrop belt, North America. *Geology*, **14**, 330–334.

HEREFORD, R. 1984. Climate and ephemeral-stream processes: twentieth century geomorphology and alluvial stratigraphy of the Little Colorado River, Arizona. *Geological Society of America, Bulletin*, **95**, 654–668.

HODGE, B. L. & DUNHAM, K. C. 1988. Clastics, coals, palaeodistributaries and mineralisation in the Namurian Great Cyclothem, Northern Pennines and Northumberland Trough. *Proceedings of the Yorkshire Geological Society*, **48**, 323–337.

HOGG, N. M. Dinoflagellate cyst evidence for the age of the Scalby Formation (Middle Jurassic) of the Cleveland Basin, North Yorkshire and regional implications. *Journal of Micropalaeontology*, in press.

JERVEY, M. T. 1988. Quantitative geological modelling of siliciclastic rock sequences and their seismic expression. *In*: WILGUS, C. K., HASTINGS, B. S., KENDALL, C. G. ST. C., POSAMENTIER, H. W., ROSS, C. A. & VAN WAGONER, J. C. (eds). *Sea-level Changes: an Integrated Approach*. Society of Economic Paleontologists and Mineralogists, Special Publications, **42**, 47–69.

KENYON, P. M. & TURCOTTE, D. L. 1985. Morphology of a delta prograding by bulk sediment transport. *Bulletin of the Geological Society of America*, **96**, 1457–1465.

KIRKBY, M. J. 1971. Hillslope process-response models based on the continuity equation. *In*: BRUNSDEN, D. (ed.) *Slopes: Form and Process*. Institute of British Geographers, Special Publications, **3**, 16–30.

KLEIN, G. DE V. 1990. Pennsylvanian time scales and cycle periods. *Geology*, **18**, 455–457.

KOSS, J. E., ETHRIDGE, F. G. & SCHUMM, S. A. 1994. An experimental study of the effects of base-level change on fluvial, coastal plain and shelf systems. *Journal of Sedimentary Research*, **B64**, 90–98.

LEEDER, M. R. 1975. Pedogenic carbonates and flood sediment accretion rates: a quantitative model for alluvial arid-zone lithofacies. *Geological Magazine*, **112**, 257–270.

—— 1976. Palaeogeographic significance of pedogenic carbonates in the topmost Upper Old Red Sandstone of the Scottish Border Basin. *Geological Journal*, **11**, 21–28.

—— 1978. A quantitative stratigraphic model for alluvium, with special reference to channel deposit density and interconnectedness. *In*: MIALL, A. D. (ed.) *Fluvial Sedimentology*. Canadian Society of Petroleum Geologists, Memoirs, **5**, 587–596.

—— & NAMI, M. 1979. Sedimentary models for the non-marine Scalby Formation (Middle Jurassic) and evidence for Late Bajocian/Bathonian uplift of the Yorkshire Basin. *Proceedings of the Yorkshire Geological Society*, **42**, 461–482.

—— & STRUDWICK, A. E. 1987. Delta–marine interactions: a discussion of sedimentary models for Yoredale-type cyclicity in the Dinantian of Northern England. *In*: ADAMS, A., MILLER, J. & WRIGHT, V. P. (eds) *European Dinantian Environments*. Wiley, London, 115–130.

LEOPOLD, L. B. & BULL, W. B. 1979. Base level, aggradation and grade. *Proceedings of the American Philosophical Society*, **123**, 168–202.

LEVY, M. & CHRISTIE-BLICK, N. 1994. NeoProterozoic incised valleys of the eastern Great Basin, Utah and Idaho: fluvial response to changes in depositional base level. *In*: DALRYMPLE, R. W., BOYD, R. & ZAITLIN, B. A. (eds) *Incised-valley Systems: Origin and Sedimentary Sequences*. Society for Sedimentary Geology, Special Publications, **51**, 360–384.

MACK, G. H., SALYARDS, S. L. & JAMES, W. C. 1993. Magnetostratigraphy of the Plio-Pleistocene Camp Rice and Palomas Formations in the Rio Grande Rift of southern New Mexico. *American Journal of Science*, **293**, 49-77.

MARTINSEN, O. J. 1993. Namurian (late Carboniferous) depositional systems of the Craven–Askrigg area, northern England: implications for sequence stratigraphic models. *In*: POSAMENTIER, H. W., SUMMERHAYES, C. P., HAQ, B. U. & ALLEN, G. P. (eds) *Sequence Stratigraphy and Facies Associations*, International Association of Sedimentologists, Special Publications, **18**, 247–282.

MARZO, M., NIJMAN, W. & PUIGDEFABREGAS, C. 1988. Architecture of the Castissent fluvial sheet sandstones; South Pyrenees, Spain. *Sedimentology*, **35**, 719–738.

MAYNARD, J. R. 1992*a*. Sequence stratigraphy of the Upper Yeadonian of northern England. *Marine and Petroleum Geology*, **9**, 197–207.

—— 1992*b*. Discussion of: 'The Millstone Grit (Namurian) of the southern Pennines viewed in the light of eustatically controlled sequence stratigraphy by W.A. Read. *Geological Journal*, **27**, 295–297.

—— & LEEDER, M. R. 1992. On the periodicity and magnitude of late-Carboniferous glacio-eustatic sea level changes. *Journal of the Geological Society, London*, **149**, 303–311.

MCCAVE, I. N. 1969. Correlation of marine and nonmarine strata with example from Devonian of New York State. *Bulletin of the American Association of Petroleum Geologists*, **53**, 155–162.

MIALL, A. D. 1991. Stratigraphic sequences and their chronostratigraphic correlation. *Journal of Sedimentary Petrology*, **61**, 497–505.

—— 1994. Sequence stratigraphy and chronostratigraphy: problems of definition and precision in correlation, and their implications for global eustasy. *Geosciences Canada*, **21**, 1–26.

NAMI, M. & LEEDER, M. R. 1978. Changing channel morphology and magnitude in the Scalby Formation (M. Jurassic) of Yorkshire, England. *In*: MIALL, A. D. (ed.) *Fluvial Sedimentology*. Canadian Society of Petroleum Geologists, Memoirs, **5**, 431–440.

PAOLA, C., HELLER, P. L. & ANGEVINE, C. L. 1991. The response distance of river systems to variations in sea level. GSA Annual Meeting Abstracts with Programs, A170–A171.

POSAMENTIER, H. W. & VAIL, P. R. 1988. Eustatic controls on clastic deposition II – sequences and systems tracts models. *In*: WILGUS, C. K., HASTINGS, B. S., KENDALL, C. G. ST. C., POSAMENTIER, H. W., ROSS, C. A. & VAN WAGONER, J. C. (eds). *Sea-level Changes: an Integrated Approach*. Society of Economic Paleontologists and Mineralogists, Special Publications, **42**, 125–154.

——, JERVEY, M. T. & VAIL, P. R. 1988. Eustatic controls on clastic deposition I – conceptual framework. *In*: WILGUS, C. K., HASTINGS, B. S., KENDALL, C. G. ST. C., POSAMENTIER, H. W., ROSS, C. A. & VAN WAGONER, J. C. (eds). *Sea-level Changes: an Integrated Approach*. Society of Economic Paleontologists and Mineralogists, Special Publications, **42**, 109–124.

READ, W. A. 1991. The Millstone Grit (Namurian) of the southern Pennines viewed in the light of eustatically controlled sequence stratigraphy. *Geological Journal*, **26**, 167–165.

—— & FORSYTH, D. W. 1991. Limestone Coal Group (Pendleian E1) in the Glasgow-Stirling region viewed in the light of sequence stratigraphy. *Geological Journal*, **25**, 85-89.

RIDING, J. B. & WRIGHT, J. K. 1989. Palynostratigraphy of the Scalby Formation (Middle Jurassic) of the Cleveland Basin, NE Yorkshire. *Proceedings of the Yorkshire Geological Society*, **47**, 349–354.

SALTER, T. 1993. Fluvial scour and incision: models for their influence on the development of realistic reservoir geometries. *In*: NORTH, C. P. & PROSSER, D. J. (eds) *Characterisation of Fluvial and Aeolian Reservoirs*. Geological Society, London, Special Publications, **73**, 33–52.

SAUCIER, R. T. 1981. Current thinking on riverine processes and geological history as related to human settlement in the southeast. *Geoscience and Man*, **22**, 7–18.

—— & FLEETWOOD, A. R. 1970. Origin and chronological significance of Late Quaternary terraces, Oachita River, Arkansas and Louisiana. *Bulletin of the Geological Society of America*, **81**, 869–890.

SCHUMM, S. A. 1993. River response to baselevel change: implications for sequence stratigraphy. *Journal of Geology*, **101**, 279–294.

SHANLEY, K. W. & MCCABE, P. J. 1993. Alluvial architecture in a sequence stratigraphic framework: a case history from the Upper Cretaceous of southern Utah, USA. *In*: FLINT, S. S. & BRYANT, I. D. (eds) *The Geological Modelling of Hydrocarbon Reservoirs and Outcrop Analogues*. International Association of Sedimentologists, Special Publications, **15**, 21–56.

SOREGHAN, G. S. & DICKINSON, W. R. 1994. Generic types of stratigraphic cycles controlled by eustasy. *Geology*, **22**, 759–761.

TANDON, S. K. & GIBLING, M. R. 1994. Calcrete and coal in late Carboniferous cyclothems of Nova Scotia, Canada: climate and sea-level changes linked. *Geology*, **22**, 755–758.

VAN WAGONER, J. C., MITCHUM, R. M. JR., CAMPION, K. M. & RAHMANIAN, V. D. 1990. *Siliciclastic Sequence Stratigraphy in Well Logs, Cores and Outcrops: Concepts of High-resolution Time and Facies*. American Association of Petroleum Geologists, Methods in Exploration Series, **7**, 55pp.

WANLESS, A. R. & SHEPARD, F. P. 1936. Sea level and climatic change related to late Paleozoic cycles. *Geological Society of America, Bulletin*, **47**, 1177–1206.

WEIMER, R. J. 1984. Relation of unconformities, tectonics, and sea level changes, Cretaceous of Western Interior, USA. *In*: SCHLEE, J. S. (ed.) *Interregional Unconformities and Hydrocarbon Accumulation*. American Association of Petroleum Geologists, Memoirs, **36**, 7–35.

WILSON, A. G. & KIRKBY, M. J. 1975. *Mathematics for Geographers and Planners*. Clarendon Press, Oxford.

WOOD, L. J., ETHRIDGE, F. G. & SCHUMM, S. A. 1994. An experimental study of the influence of subaqueous shelf angles on coastal plain and shelf deposits. *In*: WEIMER, P. & POSAMENTIER, H. W. (eds) *Siliciclastic Sequence Stratigraphy: Recent Developments and Applications*. American Association of Petroleum Geologists, Memoirs, **58**, 381–391.

WRIGHT, V. P. & MARRIOTT, S. B. 1993. The sequence stratigraphy of fluvial depositional systems: the role of floodplain sediment storage. *Sedimentary Geology*, **86**, 203–210.

ZAITLIN, B. A., DALRYMPLE, R. W. & BOYD, R. 1994. The stratigraphic organisation of incised-valley systems associated with relative sea level changes. *In*: DALRYMPLE, R. W., BOYD, R., & ZAITLIN, B. A. (eds) *Incised-valley Systems: Origin and Sedimentary Sequences*. SEPM, Special Publications, **51**, 45–62.

High-resolution sequence stratigraphy of a mid-Cretaceous estuarine complex: the Woburn Sands of the Leighton Buzzard area, southern England

J. P. WONHAM[1] & T. ELLIOTT

Department of Earth Sciences, University of Liverpool, P.O. Box 147, Liverpool L69 3BX, UK

[1]*Present address: Department of Geology, Imperial College of Science, Technology and Medicine, South Kensington, London SW7 2BP, UK*

Abstract: The Woburn Sands Formation (Aptian–Albian) is a sand-rich, tide-dominated estuarine deposit up to 120 m thick, which is extensively exposed in the Leighton Buzzard area, southern England. Previous studies have interpreted the depositional processes and environments of these strata, but have not satisfactorily resolved their lithofacies complexity. Application of sequence stratigraphical methods and concepts has identified key surfaces (sequence boundaries, tidal and wave ravinement surfaces) which provide a useful framework for correlation of the succession and for understanding the complex facies architecture.

Two regional-scale erosional surfaces identified in the succession are interpreted as erosional unconformities or sequence boundaries. These surfaces resulted from fluvial erosion during periods of relative sea-level fall, followed by extensive erosional modification by tidal currents during the early phases of relative sea-level rise. The depositional sequences defined by the erosional unconformities are dominated by the deposits of incised valleys. The valley-fills mainly comprise sand-dominated tidal current facies which occur as multistorey estuarine channel complexes. In the upper parts of the sequences thin, locally preserved units of shelf facies overlie wave ravinement surfaces. Over part of the study area the upper erosional unconformity erodes directly into the underlying incised valley-fill producing a composite sequence.

The Woburn Sands Formation of southern England is of Aptian–Albian age (Fig. 1) and consists largely of quartz arenites with minor, laterally extensive, heterolithic intervals. Previous studies of the stratigraphy and sedimentology of the Woburn Sands in the vicinity of Leighton Buzzard (Figs 2 and 3) have demonstrated its shallow marine origin (Narayan 1971; Bridges 1982), the development of varying cross-bed types, often of very large scale (Buck 1987; Bristow 1994), and the dominantly tidal character of the depositional system in which the Woburn Sands accumulated (de Raaf & Boersma 1971; Walker 1985). Recent studies have suggested that all or part of the Woburn Sands is estuarine in character (Buck 1987; Wonham 1993; Johnson & Levell 1995).

The paucity of fossils through much of the succession has previously made detailed stratigraphical subdivision of the Woburn Sands difficult. Biostratigraphical studies have been confined to the base (Casey 1961) and to the complex upper parts of the formation (Owen 1972; Dean 1994). The deposits have been subdivided and correlated by various lithostratigraphical schemes (Eyers 1991; Shephard-Thorn *et al.* 1994; Johnson & Levell 1995), but have not so far been investigated using sequence stratigraphical methods and concepts (e.g. Vail *et al.* 1977; Posamentier *et al.* 1988; Posamentier & Vail 1988; Van Wagoner *et al.* 1988; 1990).

The main purpose of this paper is to outline the characteristics of key stratigraphical surfaces within the Woburn Sands and the character of stratal packages defined by these surfaces. A variety of key surfaces is recognized (sequence boundaries, tidal ravinement surfaces and wave ravinement surfaces), which formed in response to fluctuations of relative sea level in a macrotidal, sand-dominated estuary.

Regional setting

The Woburn Sands in the Leighton Buzzard–Woburn area form a unit up to 120 m thick between Upper Jurassic claystones, which they unconformably overlie, and the Middle Albian Gault Clay Formation which oversteps northwestwards (Fig. 1). The basal lag of the Woburn Sands contains nodules which yield a rich fauna of derived Jurassic ammonites, together with a *remanié* of Early Aptian fauna including abundant brachiopods and rare ammonites of the *nutfieldiensis* Zone (Casey 1961).

The overstepping Gault Clay records a regional transgression which terminated deposition of the

From HESSELBO, S. P. & PARKINSON, D. N. (eds), 1996, *Sequence Stratigraphy in British Geology*, Geological Society Special Publication No. 103, pp. 41–62.

S N

STAGE	AMMONITE ZONE	CHANNEL BASIN	WEALD BASIN	BEDFORDSHIRE SHELF	E. MIDLAND SHELF	YORKSHIRE BASIN
96 Ma						
Upper Albian	*dispar*	Upper Greensand	U.G.			Hunstanton
	inflatum		Gault Clay	Gault Clay	Hunstanton (Red Chalk)	Speeton Clay
Middle Albian	*lautus*	Gault Clay				
	loricatus					
	dentatus			Shenley Lst. and Carstone		
102.5 Ma						
Lower Albian	*mammillatum*	Carstone			Carstone	
	tardefurcata	Sandrock	Folkestone	Woburn Sands	Sand/clay	
108 Ma						
Upper Aptian	*jacobi*		Sandgate			
	nutfieldiensis	Ferruginous Sands			Sutterby Marl	
	martinioides		Hythe			
110.5 Ma						
Lower Aptian	*bowerbanki*					
	deshayesi					
	forbesi	Atherfield Clay	Atherfield Clay			
	fissicostatus				Skegness Clay	
113 Ma						
Barremian		Vectis	Weald Clay	Roach		
Hauterivian		Wessex			Tealby Clay	
Valanginian			Hastings Gp.		Claxby Ironstone	
Ryazanian		Purbeck Gp.	Durlston		Spilsby Sst.	
131 Ma						

Fig. 1. Chronostratigraphic correlation of Lower Cretaceous formations between the Channel Basin and the Yorkshire Basin (modified from Casey 1961; Haq *et al.* 1988; Rawson 1992).

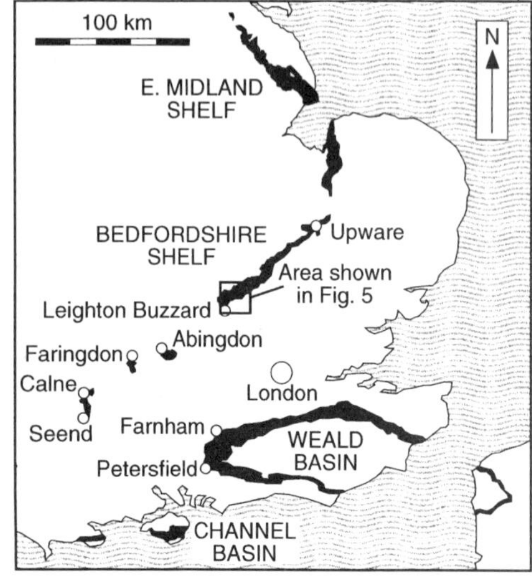

Fig. 2. Outcrops of the Lower Greensand in southern England (modified from Casey 1961).

Woburn Sands. At this level, in the Leighton Buzzard area, two distinctive localized units occur: the Shenley Limestone and the Junction Beds. The earliest of these is the Shenley Limestone, a patchily developed micritic limestone containing abundant brachiopods and rare ammonites of the *regularis* Subzone of the uppermost *tardefurcata* Zone (Casey 1961), which indicate an early Early Albian age. Reworked clasts of Shenley Limestone together with indigenous ammonites of the *kitchini* Subzone of the basal *mammillatum* Superzone (Owen 1988) are found at the base of the overlying Junction Beds (Casey 1961; Owen 1972; Dean 1994), an interval of pebbly claystone deposits with several bands of distinctive black and grey phosphatic nodules. This unit reaches a thickness of 2.5 m at Chamberlain's Barn Pit to the north of Leighton Buzzard (for location, see Appendix) and contains a succession of *mammillatum* Superzone ammonites indicating a late Early Albian age (Wright & Wright 1947; Owen 1972; Dean 1994).

In some parts of the study area, the Woburn

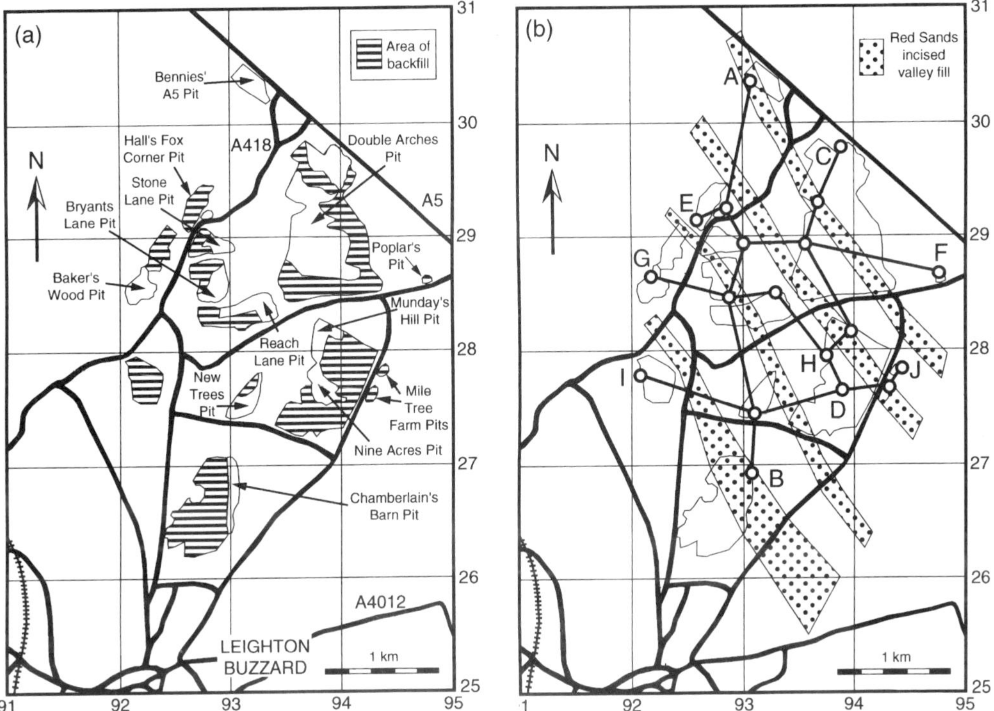

Fig. 3. (**a**) Map showing currently active sand pits in the Leighton Buzzard area. Cross-hatching denotes areas of backfill (modified from Wyatt *et al.* 1988). (**b**) Map showing the position of the main sections (A–B, C–D, E–F, G–H, I–J) and the distribution of incised channels at the base of the Red Sands sequence around Heath and Reach (stippled area).This figure is a reinterpretation of a map created by Bristow (1963; shown in Eyers 1991) together with information from Lamplugh (1922) and new field observations of channel position and palaeocurrents. Location of area is shown in Fig. 5.

Sands is immediately overlain by dark grey claystone of the Middle Albian Gault Clay. This interval is characterized by the presence of bands of pale grey phosphatic nodules and *spathi* Subzone (uppermost *dentatus* Zone) ammonites (Owen 1972). The Woburn Sands (including the Junction Beds) therefore extends from the *nutfieldiensis* Zone to the *mammillatum* Superzone within the Upper Aptian to Lower Albian, a period of approximately 6.5 Ma from 109 to 102.5 Ma BP (based on Haq *et al.* 1988).

At the present day, the Woburn Sands dips gently (1–2°) towards the southeast and crops out to form the largest of a number of Lower Greensand outliers (Ballance 1963; Hesselbo *et al.* 1990) which stretch from Cambridgeshire to Wiltshire (Fig. 2). The distribution of these outliers is thought by Hesselbo *et al.* (1990) to reflect the position of numerous palaeovalleys which may have drained from the northwest towards the southeast. It has been suggested that Jurassic and early Cretaceous faulting may have had some control on the orientation of these palaeovalleys (Hesselbo *et al.* 1990; Ruffell & Wignall 1990). Borehole records show that the outliers connect with the more extensive Lower Greensand deposits of the Weald and English Channel basins to the south.

Previous studies of the Woburn Sands

The Woburn Sands has been subdivided into lithostratigraphical units (Fig. 3) which, in order of deposition, are referred to as: the Orange Sands (or Brown Sands or Lower Woburn Sands), the Heterolithic Sands (or Silver Sands or Lower Woburn Sands or 'Compo'), the Silver Sands (or Upper Woburn Sands), the Silty Beds and the Red Sands (Lamplugh 1922; Shephard-Thorn *et al.* 1994; Eyers 1991; Johnson & Levell 1995). Other units, the thin and discontinuous Shenley Limestone and the Junction Beds, are usually considered as equivalents of the Red Sands. The lithostratigraphical nomenclature of the sandstone beneath the Silty Beds is confusing and the entire unit

(Orange Sands, Heterolithic Sands and Silver Sands) will henceforth be referred to as the Silver Sands notwithstanding the work of Shephard-Thorn *et al.* (1994). Apart from the Shenley Limestone, which is a single bed, the units which make up the Woburn Sands deserve the status of 'members' and will henceforth be referred to as such.

Sedimentological studies of the Woburn Sands have demonstrated complex styles of cross-bedding in which erosive discontinuity surfaces (reactivation surfaces) truncate and divide foresets into bundles overlain by reverse-facing intrasets (de Raaf & Boersma 1971). These structures are interpreted to be the product of periodic flow reversals associated with tidal currents. A range of cross-bed structures produced by tidal currents with pronounced time–velocity asymmetry were recognized in the Woburn Sands by Walker (1985) and Buck (1987), who also made studies of the sedimentary facies, ichnofacies and stratigraphy of the Woburn Sands. Palaeocurrents of the Woburn Sands have previously been found to be dominantly southerly (Schwarzacher 1953) or distributed in a bipolar fashion towards the northeast and southwest, with a dominance of southwest-directed indicators (Buck 1987).

Three depositional models have been proposed for the origin of the Woburn Sands. The first model suggests that deposition occurred in a relatively narrow seaway connecting larger marine basins to the north and south (Casey 1961). It has been suggested that this narrow seaway was of a similar scale to the present day English Channel and consequently the Woburn Sands has been considered as a transgressive marine sand-wave complex in a shelf setting (Narayan 1971; Allen 1982; Bridges 1982; Walker 1985; Schiavon 1988) deposited by a suite of bedforms similar to those described from the English Channel and southern North Sea (Houbolt 1968; McCave 1971). The second model suggests that deposition of the Woburn Sands occurred within an estuary comparable with a number of modern estuaries developed through the Holocene transgression (Johnson & Levell 1980; 1995). A third, hybrid model was proposed by Buck (1987), who recognized both estuarine tidal sand flat and transgressive shelf sand-wave facies.

In this study, an estuarine depositional model is favoured for the Woburn Sands on several grounds: (i) the presence of numerous channels of different scale, some of which contain a basal lag of extra-formational clasts up to cobble grade; (ii) the presence of abundant foreset clay drapes, commonly showing the development of spring-neap tidal bundles; (iii) the presence of abundant reactivation surfaces suggesting deposition by rectilinear tidal currents within the confines of an estuarine channel; (iv) considerable variation in palaeocurrent direction; and (v) rapid bedform migration rates. Estimates of bedform migration rate vary for different facies of the Woburn Sands, but estimates range from 0.06 to 0.2 m per day based on analysis of colonization by *Ophiomorpha*-producing organisms (Pollard *et al.* 1993) to *c.* 2 m per day based on analysis of tidal rhythmicity (Bridges 1982). These rates exceed the measured migration rates of bedforms in shallow tidal shelf settings, e.g. dunes in the Southern Bight of the North Sea, which migrate at 0.04 m per day (McCave 1971). Types of tide-formed structures mentioned in points (ii) and (iii) are reviewed in more detail by Nio & Yang (1991).

Method

Representative sections have been measured in numerous actively quarried sand pits situated to the north of Leighton Buzzard (see Fig. 3a and Appendix for locations). The succession exposed in individual pits varies laterally and the measured sections have therefore been augmented by line drawing interpretations of photographs and photomontages in many of the studied sites. A total of 290 palaeocurrent readings was made. For units dominated by small-scale cross-beds a large number of readings was taken, but where large-scale cross-beds were present and palaeocurrent directions were clearly recognizable, only a few representative measurements were made. Description of transverse bedforms follows the scheme suggested by Ashley (1990).

This study places an equal emphasis on the recognition and characterization of stratigraphically significant surfaces and the description and interpretation of facies. To show lateral variations in the geometry of the subdivisions described, data have been gathered from a number of sources which include details of exposures that no longer exist (Lamplugh 1922; Bristow 1963; Shephard-Thorn *et al.* 1986; Wyatt *et al.* 1988).

Key stratigraphical surfaces of the Woburn Sands

The Woburn Sands includes two regional-scale erosional surfaces, each of which is overlain by a lag of intra- and extra-formational pebble- and cobble-grade material. The lower erosional surface lies at the base of the Woburn Sands, whereas the upper erosional surface lies at the base of the Red Sands Member (Fig. 4). The lower portion of the Woburn Sands that is bounded by these erosive surfaces consists of the Silver Sands Member and the Silty Beds Member whereas the upper portion overlying the upper erosive surface consists of the Red Sands, the Shenley Limestone and the Junction Beds Member.

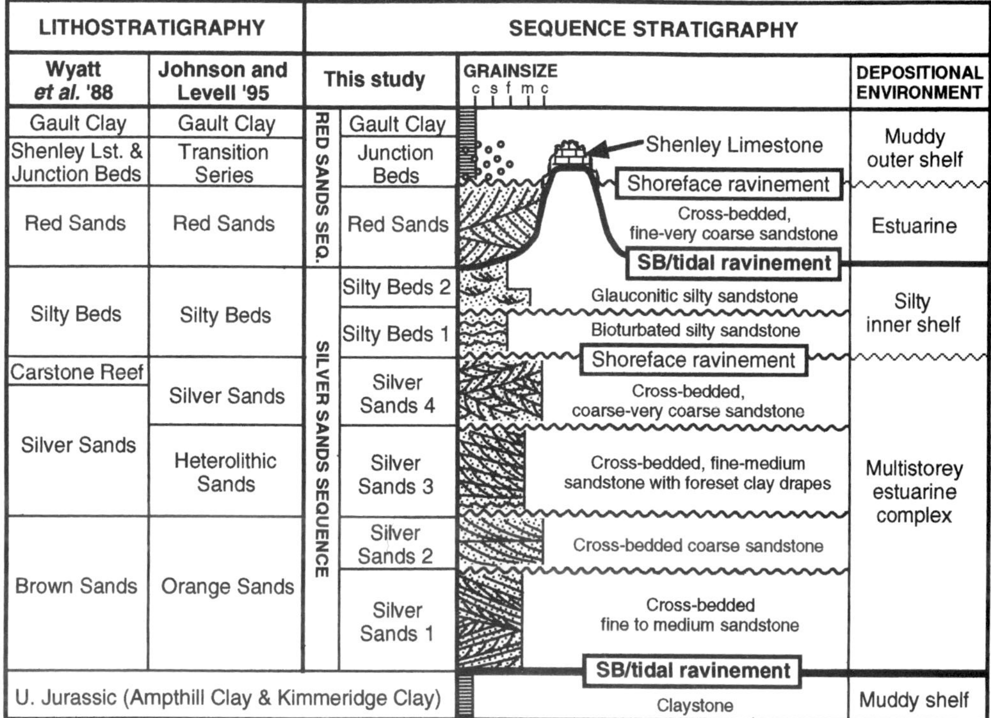

Fig. 4. Summary log of the Woburn Sands showing recent lithostratigraphical classifications together with the sequence stratigraphical subdivision of this study.

Silver Sands basal erosion surface

In a regional sense, the Silver Sands rests with marked unconformity on gently folded Upper Jurassic strata of the Oxford Clay, West Walton, Ampthill Clay and Kimmeridge Clay formations (Shephard-Thorn *et al.* 1994). There is, therefore, a considerable depositional hiatus corresponding to much of the Early Cretaceous. Structural contours on the base of the formation show an irregular, deeply eroded surface (Fig. 5). The base of the Woburn Sands is rarely exposed in the Leighton Buzzard area, but has been observed at Baker's Wood Pit. At this locality a basal lag (15 cm thick) composed of phosphatic casts of ammonites and bivalves in a sandy ferruginous matrix rests on the Ampthill Clay (Wyatt *et al.* 1988).

Phosphate workings in the last century exposed more extensive sections of the basal Woburn Sands at various positions along its broadly depositional strike outcrop length of *c.* 90 km. At Upware (Fig. 2), exposures of the basal conglomeratic bed (30 cm thick) of the Lower Greensand lie with marked unconformity on Upware Limestone and Kimmeridge Clay of Late Jurassic age (Keeping 1883). This basal bed contains phosphatized casts of worn Jurassic ammonites (up to 10 cm) with bivalve borings, together with well rounded extraformational pebbles, ironstone, iron-cemented sandstone and a rich, well-preserved molluscan fauna. At Brickhill (Fig. 5), the base of the Woburn Sands is sharp and larger clasts are scattered throughout the lower few metres of the succession (Teall 1875; Keeping 1883). Exposures also occur showing lag material to be completely absent from the base of the Woburn Sands, as at Little Brickhill, where coarse sand rests directly on Oxford Clay (Bristow 1963). At Brickhill, shelly carbonate-cemented horizons are found near the base (Teall 1875; Bristow 1963), which have a high proportion of gravel, limonite-cemented sandstone clasts and phosphatic nodules, as well as an indigenous fauna indicating a *nutfieldiensis* Zone age for the sands (Casey 1961, p. 497).

Red Sands basal erosion surface

The Red Sands are distinguished by the occurrence of abundant goethite ooids (Schiavon 1988). The erosional surface at the base of the Red Sands is widely exposed and has been described in detail in

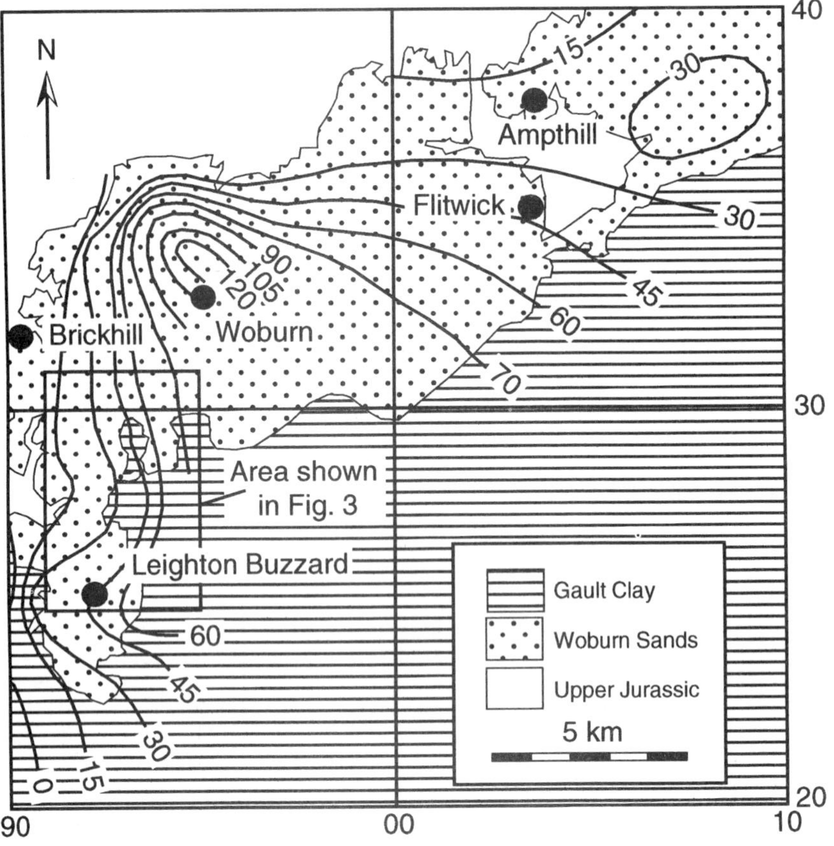

Fig. 5. Isopach map of the Lower Greensand in the Leighton Buzzard area. This map is a composite of two maps covering overlapping areas (Shephard-Thorn *et al.* 1986; Wyatt *et al.* 1988). Reproduced by permission of the Director, BGS: NERC copyright reserved.

previous studies (Lamplugh 1922; Bristow 1963; Bentley 1970). These descriptions are of particular interest in areas not now exposed, for example Chance's Pit, Harris' Pit and Garside's Old Pit. The deeply erosional, incised nature of the boundary has been clearly illustrated by the above workers, with the Silty Beds being locally removed by incision at the base of the Red Sands (Fig. 6).

A phase of ferruginous cementation occurred during the formation of the Red Sands basal surface. Ironstone horizons are developed in the sandstones, siltstones and claystones which immediately underlie the erosive surface, partly following preferentially cemented bedding planes and paralleling the relief of the erosive surface. The ironstone is often a ferruginous claystone with a deep brownish red colour in hand specimen. In thin section, the ironstone consists of approximately 50% iron oxide and 50% quartz (Eyers 1992). The majority of the quartz is monocrystalline with an undulose extinction and a large proportion of the quartz grains show iron oxide infilling fractures.

The basal erosional surface of the Red Sands is usually overlain by a lag (20 cm thick) of conglomeratic material. Extraformational clasts of vein quartz and chert are typically 2 cm in diameter, however, occasional well rounded cobbles of quartzite (up to 12 cm) also occur. Intraformational material, derived from the upper part of the Silver Sands sequence, includes ironstone concretionary nodules and large clasts of broken ironstone (up to 10 cm diameter). *Ophiomorpha* commonly occurs at the surface and locally forms very extensive burrow systems in the underlying sands, e.g. Munday's Hill Pit.

The geometry of the Red Sands basal erosional surface was first shown by Lamplugh (1922), but the stratigraphical relationships between the Red Sands, the Shenley Limestone, the Silty Beds and the Silver Sands are more clearly demonstrated by

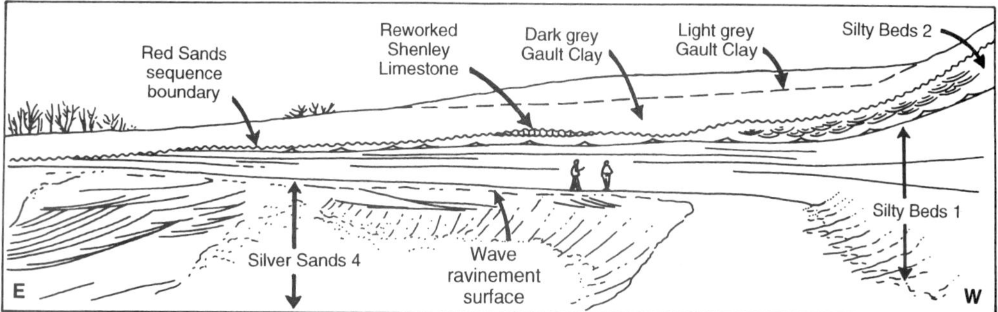

Fig. 6. Erosional base of the Red Sands sequence exposed at Munday's Hill Pit. The Red Sands sequence boundary cuts down through 4 m of strata (Silver Sands 4, Silty Beds 1 and Silty Beds 2) over a distance of roughly 100 m. At this locality (and others) where the sequence boundary comes into contact with coarse-grained sandstone facies (Silver Sands 4), the Silver Sands are strongly cemented by a dark red ferruginous cement. These cemented patches have been referred to previously as 'Carstone Reefs' (human figures sketched for scale).

Bristow (1963), who showed that the Red Sands is incised into the underlying succession. Mapping of the extent of the Red Sands is assisted by the observations of Bristow (1963) and Lamplugh (1922) and allows the geometry of the incised surface to be established (Fig. 3b). North of Leighton Buzzard, the Red Sands occur in at least seven separate channels that are aligned parallel to each other in a NNW–SSE direction. These channels have widths between 20 and 300 m, depths of up to 6 m and are spaced from 50 m to several hundred metres apart. At Double Arches Pit, the incision is *c.* 150 m wide and reaches a depth of 5 m below a flooding surface at the top of the Junction Beds. Passing southwards to Munday's Hill Pit, the incision is *c.* 50 m wide and reaches a depth of 6 m, whereas a number of channels once exposed in the area now referred to as Nine Acre Pit were smaller, mainly less than 50 m wide and up to 4 m deep (Lamplugh 1922). A distinctive feature observed at a number of localities is an undulating ridge and furrow topography at the base of Red Sands channels. The ridges have a relief of up to 1 m and occur at intervals of 10–20 m. This ridge and furrow topography is well exposed at Chamberlain's Barn Pit (Fig. 7) and was also observed by Lamplugh (1922) in other localities.

South of Leighton Buzzard, the Red Sands attains a thickness of at least 12 m and is more laterally extensive than to the north of the town, continuous sections being exposed in a number of large pits over several kilometres (Bristow 1994). However, the basal surface of the Red Sands is not currently exposed in this area.

Formation and significance of erosion surfaces

The Silver Sands and Red Sands lie above regionally extensive erosion surfaces of considerable relief. In the study area, relief at the base of the Silver Sands is inferred from structural contour maps and field mapping (Wyatt *et al.* 1988), whereas relief at the base of the Red Sands, in the form of numerous channel features, can be observed in the field.

These surfaces can also be traced beyond the study area to other areas of southern England. The major hiatus which exists at the base of the Silver

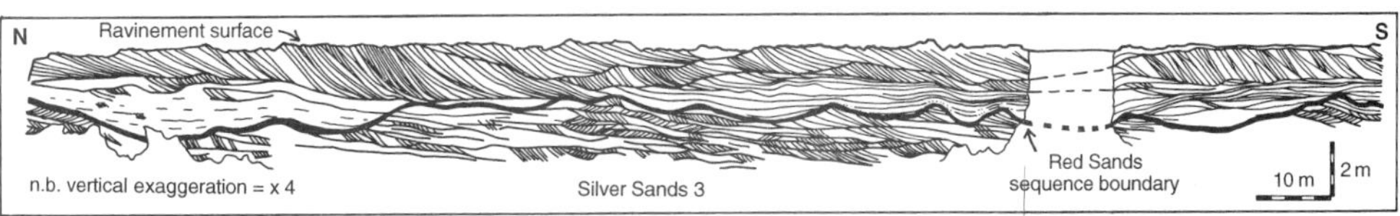

Fig. 7. Red Sands basal unconformity overlying clay-draped cross-bedded sandstone facies (Silver Sands 3) at Chamberlain's Barn Pit. Of particular interest is the undulating nature of the sequence boundary, which shows a number of ridges and furrows with 10–15 m spacing between ridges. This feature may have resulted from intertidal scour during estuary formation within the incised valley.

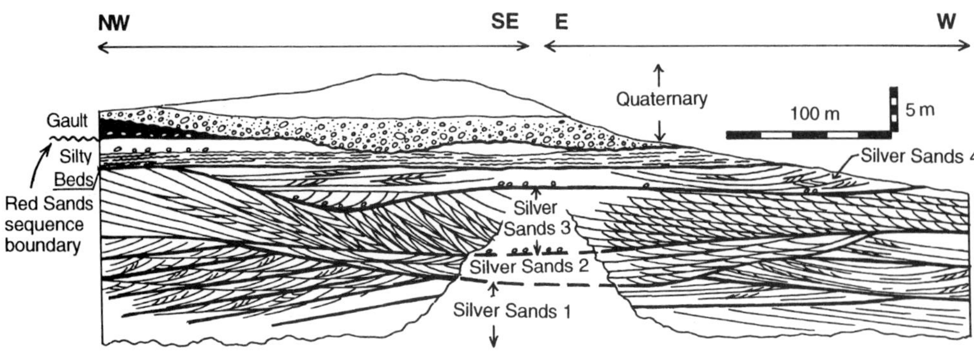

Fig. 8. Sketch illustrating bedform characteristics of successive subdivisions of the Silver Sands sequence in NW–SE and E–W aligned faces exposed at Stone Lane Quarry during the period 1987 to 1994.

Sands coincides with several extensive erosional unconformities in the Weald and Channel basins to the south, the youngest of these being the base *nutfieldiensis* Zone unconformity corresponding to the base of the Sandgate Beds in the Weald Basin (Ruffell & Wach 1991; Ruffell 1992).

The base of the Red Sands may correlate with erosional unconformities elsewhere in southern England, although the lack of indigenous fauna above the surface in most localities makes biostratigraphical correlation imprecise. For example, the base of the Red Sands may correlate with the base of the Carstone Formation in the East Midland Shelf area (Fig. 1); however, ammonites collected from the base of the Carstone are all derived Lower Aptian forms and the only evidence of a later age for the Carstone is its relatively conformable transition with the overlying Albian Red Chalk (Casey 1961). The base of the Red Sands may also correlate with the base of the Carstone in the Channel Basin (Fig. 1) and the base of the condensed Iron Grit at the top of the Folkestone Beds Formation in the Weald Basin (Anderson 1986).

The presence of cobble-grade clasts of quartzite in the basal lag of the Red Sands suggests that fluvial processes supplied this far-travelled material. Importantly, there is no local source from which these cobbles could have been reworked during transgressive erosion. Ironstone horizons which occur immediately below the basal unconformity of the Red Sands are fragmented and transported (as clasts up to 20 cm long), thereby indicating their early origin. These ironstones are interpreted as laterites formed during subaerial exposure under humid tropical conditions (Collinson 1986).

Several pieces of evidence suggest that both of the erosive surfaces discussed here were initially produced by fluvial incision and subsequently modified during relative sea-level rise. This transgressive modification could have been carried out by a number of processes including tidal currents within an estuary, shoreface erosion or erosion by strong tidal currents in a shallow shelf setting. An estuarine origin is favoured, however, based on the character of the facies which immediately overlie these surfaces and on the character of the surface itself. For example, the ridge and furrow topography of the basal erosive surface of the Red Sands is interpreted to reflect tidal current reworking of the fluvially cut surface during transgression. Similar scours are seen at the present day in intertidal areas of estuaries with strong tidal currents such as the Severn estuary (Allen 1990).

The basal erosion surfaces of the Silver Sands and Red Sands members are interpreted as sequence-bounding unconformities developed during periods of relative sea-level fall. Hiatus at the base of the Silver Sands was considerably greater than that at the base of the Red Sands (late Jurassic–Aptian as opposed to Aptian–Albian) and forms a marked angular unconformity. The erosional base of the Silver Sands is the local expression of the 'late Cimmerian unconformity' surface and in a broader regional sense corresponds to the rift to drift transition of the Atlantic spreading centre (Karner *et al.* 1987; Ruffell 1992).

Subdivision of the Silver Sands sequence

Two main lithostratigraphical units have previously been described from the interval referred to in this paper as the Silver Sands sequence. These units, the Silver Sands and the Silty Beds, show markedly different facies characteristics (Fig. 8). A further subdivision of each of these members is also recognized. Four erosively based units (Silver Sands 1–4) are recognized within the Silver Sands exposed in the study area, whereas two erosively based units

(Silty Beds 1–2) are recognized in the Silty Beds. These erosively based units vary in terms of grain size, facies development and palaeocurrents. The amount of relief on the base of each unit also varies. Some basal surfaces show marked relief; others are planar erosional surfaces. Not all the units are present throughout the study area and some have been erosively truncated by overlying deposits. The following sections describe the character of the units which make up the Silver Sands sequence.

Silver Sands 1

This lowermost unit (up to *c*. 25 m thick, base not presently exposed) is bounded by the Silver Sands sequence boundary and the erosional base of overlying Silver Sands 2. It consists of fine- to medium-grained, cross-bedded quartz sands with glauconite and other dark grains. Cross-bedding is large- to very large-scale compound trough cross-bedding (sets up to 5 m thick, troughs 100–200 m wide). Very coarse sand and extraformational pebbles up to 1 cm across lag master bedding surfaces which are inclined at 5–8°. Small superimposed cross-beds (30–50 cm) are seen descending or, less commonly, ascending coset surfaces. Small descending cosets occasionally show thin foreset clay drapes and spring–neap clay-drape bundles (Visser 1980). This facies shows moderate to strong bioturbation throughout, characterized by *Ophiomorpha* and *Skolithos*. Compound cross-beds of this unit are unimodally directed towards the northwest (Fig. 9). This is particularly apparent at Stone Lane Pit, where large troughs have a migration direction out of a large WSW–ENE aligned face.

Silver Sands 2

This unit (up to 8 m thick) overlies an erosive basal surface with slopes of up to 6° and is bounded at the top by the erosive base of Silver Sands 3. The unit consists of fine- to coarse-grained sandstone which is sometimes pebbly and shows medium-scale simple or compound cross-beds (1–2 m thick). Lower foresets occasionally show the development of climbing cosets (10 cm thick). Trough width is up to 30 m and foresets are inclined at 15–20°. Palaeocurrents are west-directed in the northern part of the study area and towards the southwest in the southern part (Fig. 9). *Skolithos* and *Ophiomorpha* are well developed.

Silver Sands 3

This unit (up to 12 m thick) overlies a gently erosive basal surface and is bounded at the top by the erosional base of Silver Sands 4 or Silty Beds 1. Silver Sands 3 consists of fine- to medium-grained, well-sorted pure quartz sandstone which shows small- to medium-scale cross-beds (usually <0.5 m thick) with development of foreset clay drape bundles. Palaeocurrent distribution is bipolar (directed southwest and northeast) with bedforms showing local ebb- or flood-dominance (Fig. 9). The subordinate current is represented either by: (i) climbing sets of current ripples with reverse orientation between foreset clay drapes or (ii) preservation of reversed cross-beds less than 1 m in length, with no clay drapes and strong basal scour of up to 20 cm. In some localities, cross-beds of this unit occur as cosets of very large-scale cross-stratification (9 m thick, dip 10°). At Stone Lane Pit, cosets (directed southwest) are seen to migrate obliquely to the migration direction of the master bedding surfaces (directed south). Towards the base of this unit, clay drapes amalgamate to form a lenticular- to wavy-bedded heterolithic facies which in some areas is only 10 cm thick, e.g. Stone Lane Pit, but which in other areas forms a unit up to 3 m thick, e.g. Bennie's A5 Pit. Clasts of fossil wood (up to 30 cm long) are common throughout this unit. Clay-lined *Skolithos* occurs in cross-bed foresets, whereas *Taenidium* and *Teichichnus* are preserved among amalgamated, clay-draped toesets (Buck 1987).

Silver Sands 4

This unit (up to 10 m thick) consists of medium- to very coarse-grained, well-sorted pure quartz sandstone. The unit has an erosive basal contact lagged by well-rounded extra-formational pebbles of vein quartz, chert and micaceous sandstone (<3 cm) and clasts of limonite-cemented sandstone breccia (<5 cm). The channelized geometry of the sandstone at the base of this subdivision can be clearly demonstrated. Channel margin slopes of up to 5° are observed in the field. Between Nine Acre Pit and Munday's Hill Pit the sandstone thickens from 1 to 10 m over a 200 m interval.

This unit is dominated by distinctive, herringbone style cross-bedding. Cross-beds (10–50 cm thick) are climbing and descending cosets of large-scale, compound cross-beds (<4 m thick; trough width <80 m). Palaeocurrents are bimodal (Fig. 9d) with large-scale cross-bed foresets in adjacent localities directed northwest or southwest. An equal proportion of cosets is seen to climb or descend master-bedding surfaces. Fossil wood showing *Teredolites* borings is abundant and commonly preserved as large clasts (up to 40 cm long). Laminated clay rip-up clasts up to 10 cm long occur, which are in some instances reworked into sand-armoured mud balls. These are probably reworked clay drape material derived from underlying Silver Sands 3.

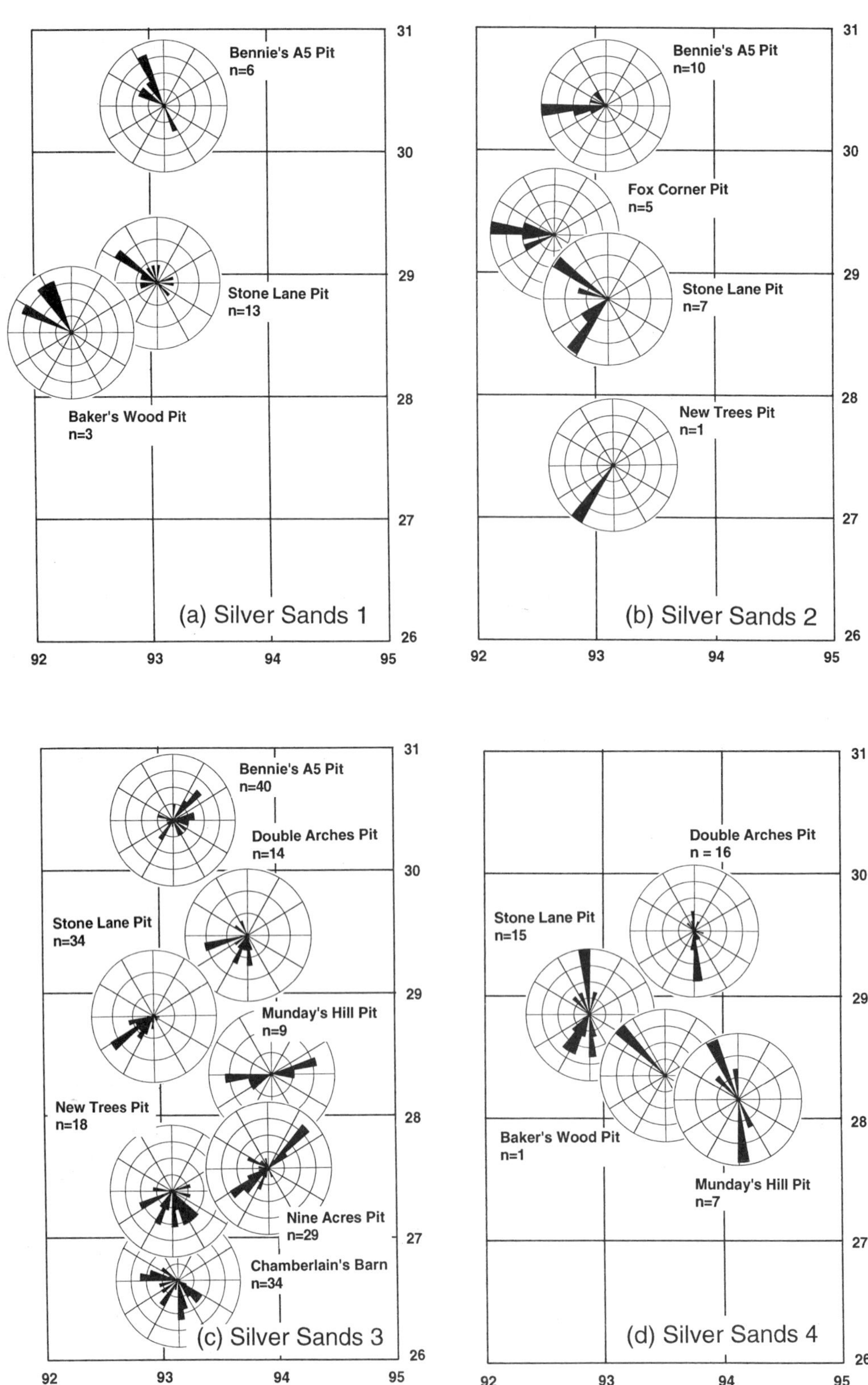
Bennie's A5 Pit
n=6
Stone Lane Pit
n=13
Baker's Wood Pit
n=3
(a) Silver Sands 1
Bennie's A5 Pit
n=10
Fox Corner Pit
n=5
Stone Lane Pit
n=7
New Trees Pit
n=1
(b) Silver Sands 2
Bennie's A5 Pit
n=40
Double Arches Pit
n=14
Stone Lane Pit
n=34
Munday's Hill Pit
n=9
New Trees Pit
n=18
Nine Acres Pit
n=29
Chamberlain's Barn
n=34
(c) Silver Sands 3
Double Arches Pit
n = 16
Stone Lane Pit
n=15
Baker's Wood Pit
n=1
Munday's Hill Pit
n=7
(d) Silver Sands 4
92
93
94
95
26
27
28
29
30
31

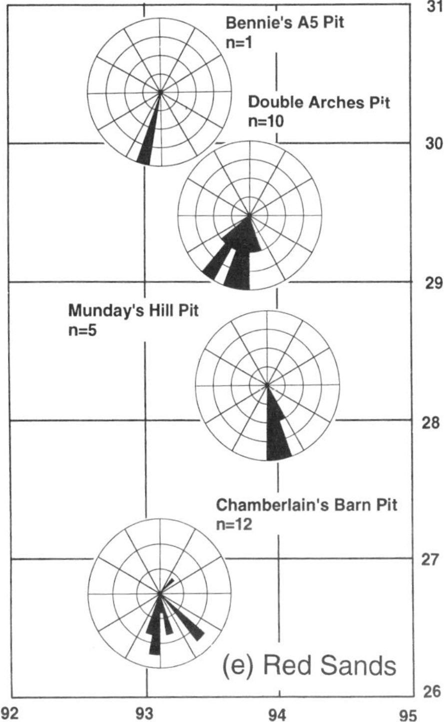

Fig. 9. (a–e) Palaeocurrents of the Silver Sands and Red Sands sequences in the Leighton Buzzard area.

Silty Beds 1

This heterolithic unit (up to 3 m thick) lies above a sharp, planar erosion surface and is bounded above by the erosive base of Silty Beds 2. The sands immediately above the basal surface include abundant small extraformational pebbles of vein quartz and white or grey chert (<2 cm) and contain clay-lined vertical burrows which pass down into underlying Silver Sands 4. Two distinctive facies are recognized within this unit. The lower facies (up to 1 m thick at Stone Lane Pit) consists of two or three discrete sandstone beds (10–30 cm thick) interbedded with carbonaceous siltstone beds (10–20 cm thick) and carbonaceous, laminated claystone beds (up to 4 cm thick). The sandstone units are erosively based and show the development of hummocky bedforms with a wavelength of up to 1 m and a height of up to 25 cm (Fig. 10). The external morphology of these beds is reflected internally by a complex arrangement of convex and concave upwards lamina sets.

The upper heterolithic facies consists of very fine-grained, flaser to wavy bedded sandstone with clay drapes developed over wave ripples (wavelength <20 cm, amplitude <3 cm; Fig. 11). This facies shows the development of low-angle erosion surfaces between decimetre thick sets of rippled strata. Shallow gutters lagged by limonite-cemented sandstone clasts (<2 cm) are also seen. Clay drapes are black due to the high proportion of fine carbonaceous material within them, and wood clasts up to 5 cm also occur. At Munday's Hill Pit this facies is very strongly bioturbated by vertical, clay-lined burrows up to 20 cm long with a meniscate fill. The intensity of bioturbation and carbonaceous content both increase upwards through this unit (Fig. 11).

Silty Beds 2

This unit is bounded by a planar erosion surface at the base and by the Red Sands sequence boundary at the top. Silty Beds 2 shows two distinctive facies types with a similar range of sedimentary features to Silty Beds 1. The basal facies is a strongly glauconitic sandstone interval (20–80 cm thick). At Stone Lane Pit, the glauconitic sandstone interval is 60 cm thick and overlies a sharp surface lagged by extraformational pebbles of vein quartz and black chert up to 3 cm in diameter. Internally, the sandstone shows shallow trough bedforms (50 cm thick, 10 m wide) with hummocky bedded cosets (wavelength 30 cm, amplitude 10 cm). Intraformational clasts are preserved on these surfaces and form the core of nodules with a coat (up to 2 cm thick) of brownish red ferruginous cement. The glauconitic unit passes sharply upwards into a fine-grained, white sandstone, locally with thin clay flasers or small-scale trough cross-bedding. At Stone Lane Pit, the white sands are strongly bioturbated by *Skolithos* and pass gradationally upwards into clayey siltstone (50 cm thick).

Depositional interpretation of the Silver Sands sequence

Multistorey estuarine channel complex (Silver Sands 1-4)

Subdivisions 1–4 of the Silver Sands consist of quartz sandstone facies showing a variety of cross-bed styles interpreted to have been produced by tidal currents of varying ebb–flood strength and asymmetry. Each subdivision overlies an erosional surface interpreted to have been produced by the migration of estuarine channels. Silver Sands 1 and Silver Sands 2 are interpreted as the deposits of large, transverse, three-dimensional dune forms. Abundant *Ophiomorpha* in Silver Sands 1 suggests that dune migration was relatively slow compared with that of overlying units (Buck 1987).

Beds of amalgamated clay drapes of limited extent are developed in Silver Sands 1, but become

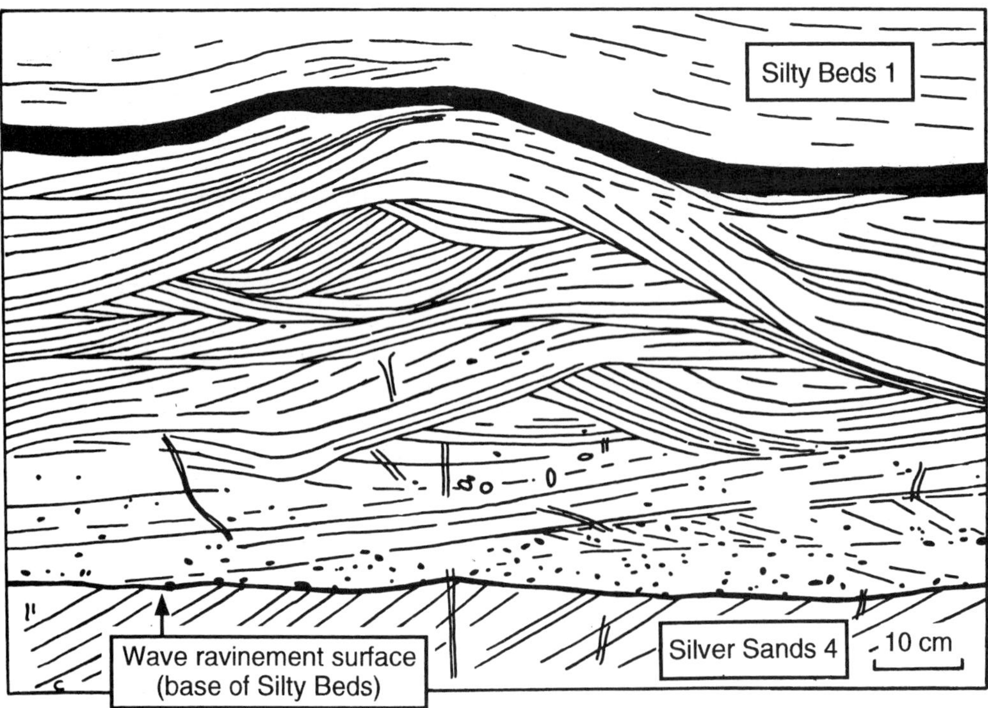

Fig. 10. Wave-modified bedforms of Silty Beds 1 preserved in coarse to very coarse sand above the top of the Silver Sands 4 wave ravinement surface at Stone Lane Pit. Cross-bedded sands of Silver Sands 4 can be seen to underlie the erosive surface at the base of the hummocky interval. Note the pebbly, bioturbated character of the sandstone above the ravinement surface. The hummocky sandstone passes upwards into silty sandstone with extensive layers of laminated claystone seen at the top of the picture.

Fig. 11. Heterolithic facies interpreted as inner shelf deposits (Silty Beds 1 at Munday's Hill Pit). Wave ripples and strong bioturbation can be seen in the lower part of the face (decimetre-interval pole for scale).

common, thickly developed (up to 2.5 m) and laterally extensive (hundreds of metres) in Silver Sands 3. The preservation of foreset clay-drape bundles within Silver Sands 3 is interpreted to reflect deposition during stillstands resulting from strong tidal current asymmetry. The development of paired clay drapes suggests a dominantly subtidal origin, whereas the ordered spacing of clay drapes indicates spring–neap variations in tidal current flow. Similar cross-bed structures have been described from modern estuarine environments, e.g. the Oosterschelde and Haringvliet estuaries in the southwest part of the Netherlands (Terwindt 1981).

Master bedding surfaces observed within Silver Sands 3 are interpreted to have been produced by lateral migration of sand banks (terminology of

Harris 1988; also termed 'sand bars' by Dalrymple *et al.* 1990). These sand banks were at least 9 m high and probably developed in broad channels. Palaeocurrents are bipolar and show local ebb- or flood-dominance, suggesting that the mutually evasive flow of ebb and flood currents may have occurred across sand banks.

Silver Sands 4 overlies a deeply erosive surface lagged by pebbles and large wood or mud clasts. Large-scale cross-bedding with reversed cosets and, in some instances, a herring-bone appearance, reflects the symmetrical nature of the tidal currents at this level and suggests rapid dune migration rates. Silver Sands 4 is interpreted as a high-energy, estuary-mouth inlet deposit in agreement with Johnson & Levell (1995).

Inner shelf deposits (Silty Beds 1 and 2)

The heterolithic facies of Silty Beds 1 and 2 make a strong contrast with the underlying cross-bedded sandstone of Silver Sands 3 and 4. Buck (1987) interpreted the Silty Beds as intertidal sand- and mud-flat deposits of a meso/macrotidal estuary or lagoon. In contrast, Johnson & Levell (1995) interpreted the Silty Beds as a subtidal abandonment facies. Neither of these studies recorded the development of: (i) large, hummocky, wave-influenced structures developed at the bases of both Silty Beds 1 and 2; (ii) erosive planation at the same levels; (iii) the presence of a lag of pebbles at the base of Silty Beds 1 and 2; or (iv) the strongly glauconitic character of the sands at the base of Silty Beds 2.

Planar erosion at the base of Silty Beds 1 is taken as evidence of transgressive wave ravinement at this level. This interpretation is supported by the presence of a pebbly, very coarse-grained sandstone lag above the surface, a local increase in bioturbation and the occurrence of large, hummocky bedforms immediately above the basal surface. These bedforms indicate a marked change in the character of deposition from tide- to wave-dominated. In addition, the large scale of these bedforms may suggest water depths of at least a few metres. These features, together with the strongly bioturbated, laterally extensive character of the Silty Beds and the strongly glauconitic character of Silty Beds 2, is taken as evidence of an inner shelf, tide- and wave-influenced depositional setting characterized by slow sedimentation rates.

Sequence stratigraphy of the Silver Sands sequence

The base of the Silver Sands sequence is rarely exposed for investigation; however, it appears that in this area any lowstand fluvial strata which may have been deposited within the incised valley have been reworked by transgression, leaving only a lag of pebbles, cobble-sized phosphatic nodules and bioclastic material above the basal erosion surface. This surface is immediately overlain by a thick, multistorey estuarine complex (Silver Sands 1–4). The origin of subdivisions 1–4 of the Silver Sands is difficult to prove conclusively. In the following sections, three possible models are considered: (i) a multiphase transgressive origin; (ii) a multiphase base-level downshift origin, and (iii) an autocyclic origin.

Model 1: Multiphase transgressive origin

Many sand-dominated, macrotidal estuaries of the present day obtain their coarse sediment either from offshore areas or from beaches adjacent to the estuary mouth. Consequently, these estuaries commonly fine headward where fine-grained sediment, often fluvially sourced, tends to become trapped (Allen 1994). As a result, progradation of a macrotidal, sand-dominated estuary should theoretically result in the development of a fining-upwards succession.

This simple model of estuary-fill development may need to be modified in many instances, however, to account for the influence of changes in relative sea level. Estuaries form in response to relative sea-level rise which transgressively 'floods' the lower parts of river systems and creates elongate gulfs in which tidal currents are amplified. They are therefore responsive to the influence of changes in relative sea level, e.g. studies of the Gironde Estuary (Allen & Posamentier 1992) and the Salmon River Estuary (Dalrymple & Zaitlin 1994). Both of these studies document the formation of tidal ravinement surfaces in response to transgression. In the case of the Gironde Estuary (a barred or wave-dominated macrotidal estuary) this surface is produced by up-valley migration of the estuary mouth tidal inlet, which may incise as much as 25 m deep. In the case of the Salmon River Estuary (a tide-dominated macrotidal estuary which is funnel-shaped and lacks a wave-influenced barrier mouth) the tidal ravinement surface is not generated by tidal inlet erosion, but by very strong tidal currents throughout most of the estuary which incise down by as much as 7 m.

In both instances, transgression has the potential for introducing coarser grained deposits into parts of the estuary previously dominated by finer grained material. In the case of the Salmon River Estuary, increased tidal current strength assists in the headward transport of coarse sediment from offshore areas, whereas in the Gironde Estuary, coarse sediment is introduced from the coastal zone adjacent to the inlet and deposited on the flood tidal delta.

These findings from modern estuaries may explain the alternation of fine- and coarse-grained sandstone subdivisions in the Silver Sands. Fine-grained sandstone units (Silver Sands 1 and 3) show common clay drape preservation, whereas coarse-grained sandstone units (Silver Sands 2 and 4) do not. Erosive relief is particularly noticeable at the base of the coarser units. Model 1 therefore suggests that two phases of progradation and transgression (Silver Sands 1–2 and Silver Sands 3–4) are developed in the Silver Sands, the coarser units being interpreted to overlie tidal ravinement surfaces formed either by tidal inlet channel erosion or by an increase in tidal current strength throughout the estuary. Note that the development of a tidal ravinement surface does not necessarily imply that the facies beneath the surface were not tidal, only that the tidal current strength was enhanced by transgression.

Model 2: Multiphase base level downshift origin

The presence of four erosively bounded subdivisions in the Silver Sands, each characterized by a basal lag of small (<3 cm diameter) extraformational pebbles and each with a characteristic grain size, palaeocurrent directions and cross-bed style, may provide evidence that successive units of the Silver Sands were deposited following periods of relative fall in sea level. Minor relative falls could have resulted in a basinward shift of estuarine facies (forced regression), whereas more major falls in relative sea level may have brought about subaerial exposure and lowstand fluvial incision.

The evidence for this interpretation is considered to be inconclusive. Extraformational pebbles at the bases of Silver Sands 1–4 are of a size (<3 cm) which could have been transported by tidal currents and consequently no fluvial mechanism need be implicated. In addition, no evidence has so far been found for subaerial exposure during deposition of the Silver Sands, e.g. the development of palaeosols in positions adjacent to channels.

Model 3: Autocyclic origin

The erosively bounded subdivisions of the Silver Sands may simply reflect autocyclic switching of channels within a large macrotidal estuary. However, it is difficult to explain the observed changes in bedform character, grain size and palaeocurrent directions using this model. In addition, Shephard-Thorn *et al.* (1994) record the disappearance of black and glauconitic grains (roughly 5% of the total composition) at the junction of Silver Sands 2 and Silver Sands 3 (their Brown Sands–Silver Sands junction). They take this as evidence of stratigraphical discontinuity at this level. This argument against autocyclic mechanisms is strengthened by the observation of regularly alternating fine and coarse units outlined in Model 1.

From the above arguments it should be clear that, at present, we favour Model 1 (involving two phases of progradation and transgression) to explain the origin of the Silver Sands. Evidence either for or against Model 2 is weak, but may be strengthened by further work on more complete successions at this stratigraphical level, such as those of the Weald or Channel basins.

The hummocky bedded sandstone at the base of the Silty Beds is developed above a widespread, horizontal surface of erosive planation. This surface is interpreted as a wave ravinement surface produced during landward migration of the shoreface across the underlying estuarine complex. The apparent development of two ravinement surfaces (base of Silty Beds 1 and 2) may have resulted from a deepening of wave base following renewed transgression at the start of Silty Beds 2 deposition. The Silty Beds is truncated by the overlying Red Sands basal erosion surface.

Red Sands sequence

Red Sands

The sandstone unit overlying the Red Sands sequence boundary is not as thick as that developed in the Silver Sands sequence and shows a less varied assemblage of facies. The succession is typically dominated by a sandstone facies showing large-scale trough cross-beds. This facies does not vary markedly upwards through the succession, but does vary laterally across the area as documented in the following.

North of Leighton Buzzard, the Red Sands occurs in small erosional valleys (<300 m wide, <5 m deep) and consists of muddy, medium- to coarse-grained sandstone showing large-scale compound cross-beds (2 m thick). The geometry of these cross-beds is shown in flow-parallel sections exposed in Chamberlain's Barn Pit (Fig. 7), which demonstrate strongly swept-out cross-bed toesets. Where foresets are preserved they show two orders of reactivation surface and a maximum angle of repose of 20°. High-order reactivation surfaces typically form at intervals of up to 30 cm, whereas low-order reactivation surfaces form every few metres. Cross-bed palaeocurrents are predominantly unimodal and directed towards the south (Fig. 9e).

South of Leighton Buzzard, the Red Sands forms a thicker unit (>12 m) which extends laterally over several kilometres. The deposit consists of fine- to very coarse-grained sandstone showing very large-scale, unidirectional trough cross-beds (e.g. Pratts Pit and Grovebury Pit). The cross-beds have a mean trough width of 60 m and a height of 3–7 m (Bristow 1994). Cross-beds typically show the development of two orders of reactivation surface: high-order surfaces occurring roughly every 10 cm and low-order surfaces occurring at intervals of 2–3 m. Reversed cosets and clay-draped cross-bed toe-sets are occasionally seen.

Shenley Limestone

The Shenley Limestone is poorly exposed at present, but in the past has been described at Harris's Pit, where it formed lenticular units 3 m long and 60 cm thick (Lamplugh & Walker 1903) and at Reach Lane Pit where similar, smaller units were recorded (Eyers 1992). In all exposures the Shenley Limestone is overlain either by the Junction Beds or the Gault Clay. Although units of Shenley Limestone are laterally impersistant at individual localities, a number of such units of similar character can be correlated across the area and have been described overlying Jurassic claystones 25 km west of the study area at Long Crendon (Lamplugh 1922) and 5 km west at Littleworth (Owen 1972). To the north of Leighton Buzzard, *in situ* Shenley Limestone deposits always overlie the Silver Sands, whereas reworked clasts of Shenley Limestone may occur at the base of the Junction Beds overlying the Red Sands, e.g. Chamberlain's Barn Pit (Owen 1972).

The limestone has been described in detail by Eyers (1992) and is a pale to pinkish yellow micrite with an abundant, well-preserved fauna dominated by brachiopods, but also with bryozoans, echinoids and crinoids (Lamplugh & Walker 1903; Lamplugh 1921, 1922; Casey 1961; Bristow & Kirkaldy 1963; Owen 1972). Rare indigenous ammonites of *regularis* Subzone age have also been reported (Lamplugh 1922; Toombs 1935; Casey 1961). Sandstone is incorporated as lenses within the limestone, together with iron ooids, lime mud clasts and extraformational pebbles. A number of horizons tunnelled by boring bivalves and with a surface coating of glauconite have been observed in some of the Shenley Limestone lenses (Eyers 1992). The faunal character of individual limestone lenses is strongly varied, although certain species remain common to all. Ironstone horizons exposed on the seafloor at the same time as the development of the Shenley Limestone, and intimately associated with it, are encrusted by oysters and serpulids (Lamplugh & Walker 1903; Eyers 1992).

At the southern end of Munday's Hill Pit, blocks of Shenley Limestone, no longer *in situ*, occur as a chaotic unit (20 cm thick, 3 m wide) beneath the Red Sands sequence boundary. The blocks of limestone (up to 50 cm long and 10 cm thick) are supported in a matrix of coarse pebbly sandstone.

Junction Beds

In areas where the Red Sands is not developed, the Junction Beds consists of a lag of pebbles, ironstone clasts, reworked phosphatic nodules and phosphatized ammonites. Phosphatized burrows (up to 15 cm diameter) descend from the basal surface into the sandstone below, e.g. between Double Arches Pit and Garside's Pit. This lag is overlain by the Gault Clay, which has a red colour immediately above the lag and contains crinoid columnals and cirripede valves (Casey 1961).

In areas where the Red Sands is developed, the Junction Beds is generally thicker (up to 2.5 m at Chamberlain's Barn Pit) and consists of interlaminated very fine-grained sandstone, siltstone and claystone with common horizons of black phosphatic nodules with a grey outer coating. These nodules contain gravel and appear to have originally developed around burrows which were later reworked. Matrix supported extraformational pebbles (<2 cm) are common throughout this facies, but tend to be preserved at horizons with reworked phosphatic nodules (<10 cm). Sandstone intervals occasionally show shallow cross-lamination with a wavy form. Within this interval an indigenous fauna of ammonites and belemnites occurs (Wright & Wright 1947; Owen 1972; Dean 1994).

Gault Clay

The Gault Clay consists of dark grey claystone with bands of pale grey phosphatic nodules (<4 cm) and a shelly marine fauna including bivalves, ammonites and belemnites (Owen 1972).

Depositional history and sequence stratigraphy of the Red Sands sequence

The Red Sands is interpreted as an estuarine deposit developed as a result of transgression following a period of lowstand. It is deposited within channels ranging from a few hundred metres to more than a kilometre wide. As the geometry of these channels is known from field-mapping, it can be stated that, in general, smaller channels (towards the north of the study area) are filled by more muddy facies. The large scale and steep foreset angle of the cross-beds probably reflects rapid tidal current flow conditions, the coarse-grained nature of the bedload material and high preservation of individual sets. Abundant tidal reactivation surfaces, clay drape development and occasional preservation of cross-beds produced

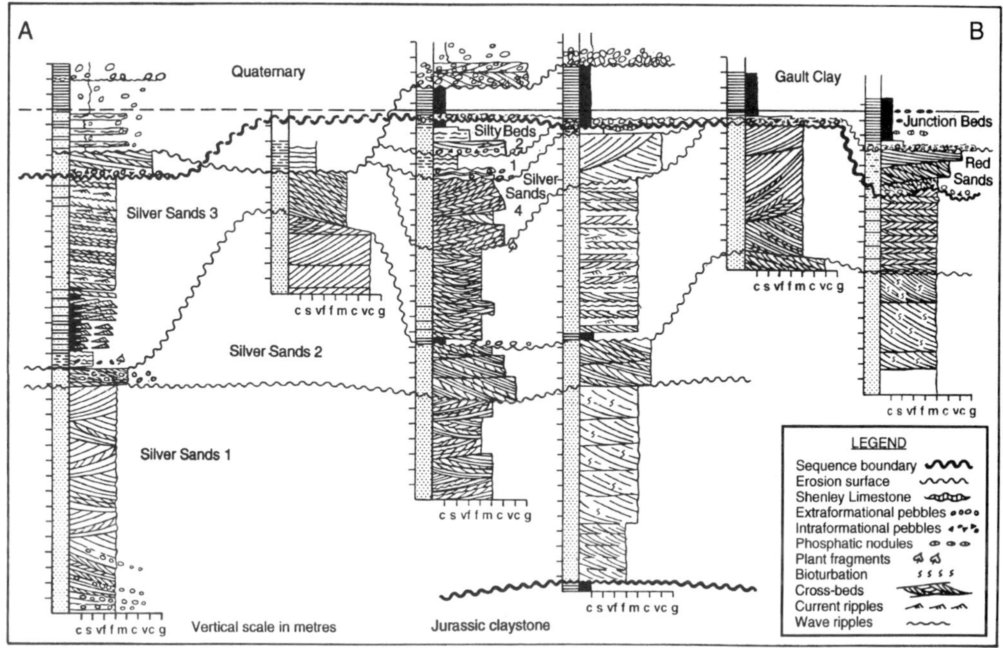

Fig. 12. North–south cross-section (A–B). Location of measured sections shown in Fig. 4.

by the subordinate ebb flow suggest deposition by confined, rectilinear tidal currents typical of estuaries. The presence of goethite ooids (Schiavon 1988) is characteristic of the Red Sands. Goethite ooid formation may have been enhanced by: (i) low sediment supply preceding Red Sands deposition; (ii) continual reworking of substrate by tidal currents; and (iii) lateritic iron-rich cements associated with the basal unconformity of the Red Sands sequence which could provide a supply of iron for the formation of the ooids.

The incised valleys in which the Red Sands was deposited were only partially infilled by estuarine sediments when further transgression occurred and led to the development of a complex succession of facies overlying a shoreface ravinement surface. The earliest deposit to form was the Shenley Limestone, whose development extends beyond the

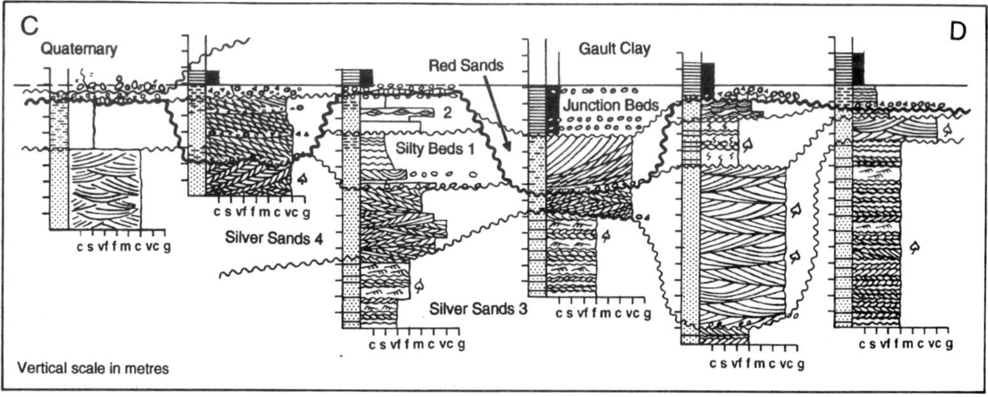

Fig. 13. North–south cross-section (C–D). Location of measured sections shown in Fig. 4. Sketchily drawn section (no longer exposed) is from Bristow (1963). See Fig. 12 for legend.

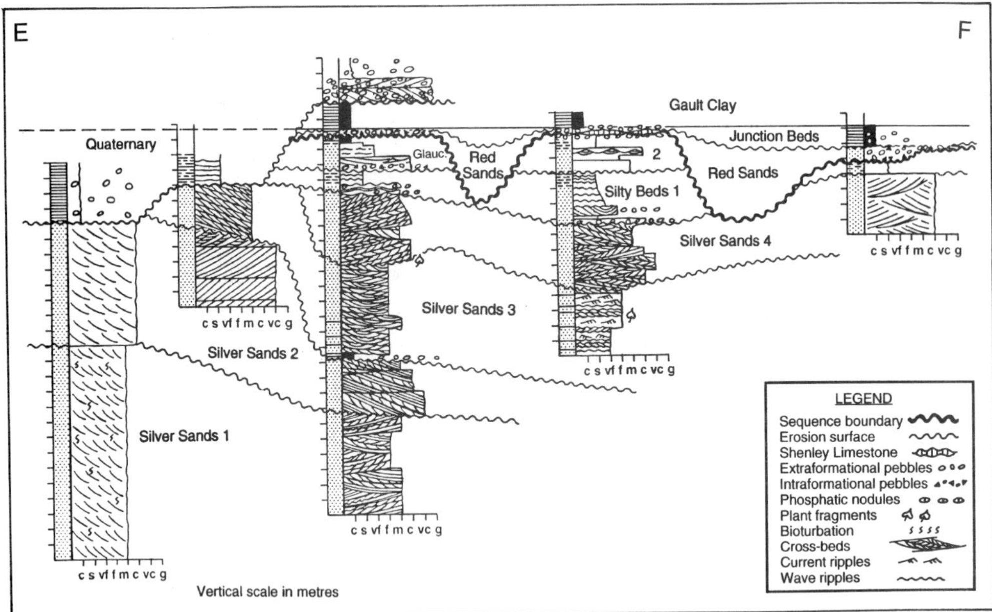

Fig. 14. West–east cross-section (E–F). Location of measured sections shown in Fig. 4. Sketchily drawn sections indicate localities no longer exposed. Hall's Fox Corner Pit is taken from Wyatt *et al.* (1988). Poplars Pit is taken from Bristow (1963: 303).

margins of the underlying estuarine complex onto Jurassic claystones. The fauna of the Shenley Limestone typifies littoral and sublittoral environments (Eyers 1992) and the varying character of this fauna between adjacent units suggests that the limestone underwent multiple stages of development. Early lithification of the limestone appears to have occurred, as indicated by the presence of horizons bored by bivalves and the preservation of reworked blocks of Shenley Limestone at the base of the overlying Junction Beds or Gault Clay. These factors suggest that the Shenley Limestone is a highly condensed deposit associated with very low sedimentation rates in a relatively high energy, littoral to shelf environment. It has also been suggested that the Shenley Limestone was once a laterally extensive unit which was later mainly reworked (Owen 1972).

The Junction Beds is interpreted as the early transgressive deposit of the Gault Clay. Small wave-formed structures within fine-grained clayey sandstone are seen within this interval at Pratt's Pit, together with pebble-grade material derived from adjacent areas of relief. Overall there is a deepening through the Junction Beds into the Gault Clay; however, the preservation of numerous horizons of phosphatic nodules (some consisting of reworked burrows in ferruginous cemented sandstone) and extraformational pebble horizons within the Junction Beds, hints at a more complex depositional history in which phases of erosion (possibly by storms or waves) and the development of transgressive lag deposits occurred.

At Mundays Hill Pit, the Gault Clay onlaps the top of the Silver Sands sequence (Fig. 6). Raised areas at the margins of the Red Sands incised valleys probably continued to exert an influence on sedimentation as knolls on the seafloor (Owen 1972). These knolls may have supplied pebbles to the clay accumulating preferentially within the channels during storms until *spathi* Subzone (uppermost *dentatus* Zone) times and it has been suggested (Owen 1972) that they continued to influence deposition of the Gault Clay until *cristatum* Subzone times. Very slow rates of deposition during this time are indicated by the formation of abundant phosphatic nodules and glauconitic horizons.

The base of the Gault Clay is taken as the datum line for correlation of the succession (Figs 12–16). The Gault Clay may be differentiated from the Junction Beds claystone facies on the basis of: (i) the presence of small pebbles and dark grey phosphatic nodules within the Junction Beds; and (ii) the presence of pale grey phosphatic nodules and the absence of pebbles in the Gault Clay. The sharp surface between these units is interpreted as a regional flooding surface and corresponds to the base of the *spathi* Subzone.

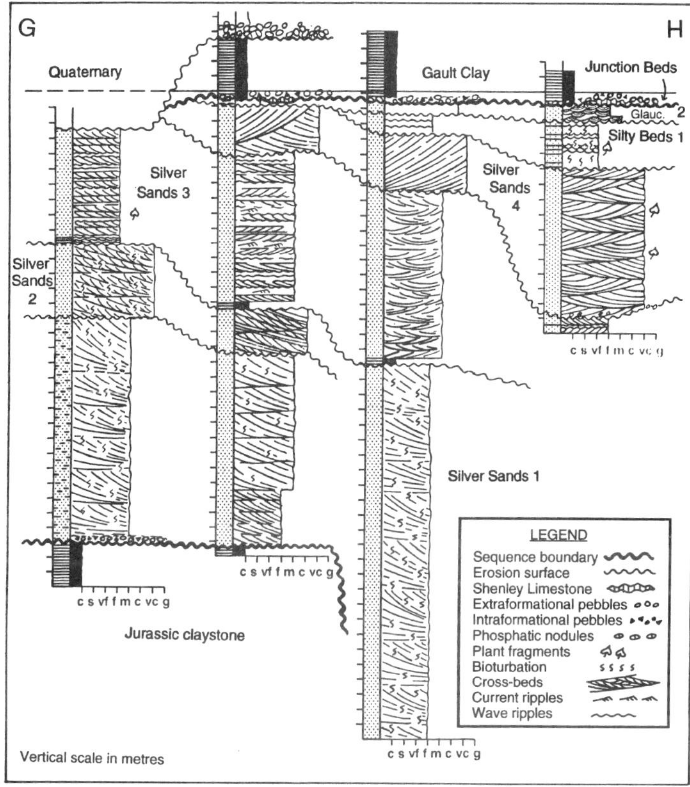

Fig. 15. West–east cross-section (G–H). Location of measured sections shown in Fig. 4. Depth to base of Silver Sands at Bryant's Lane Pit is from Bristow (1963).

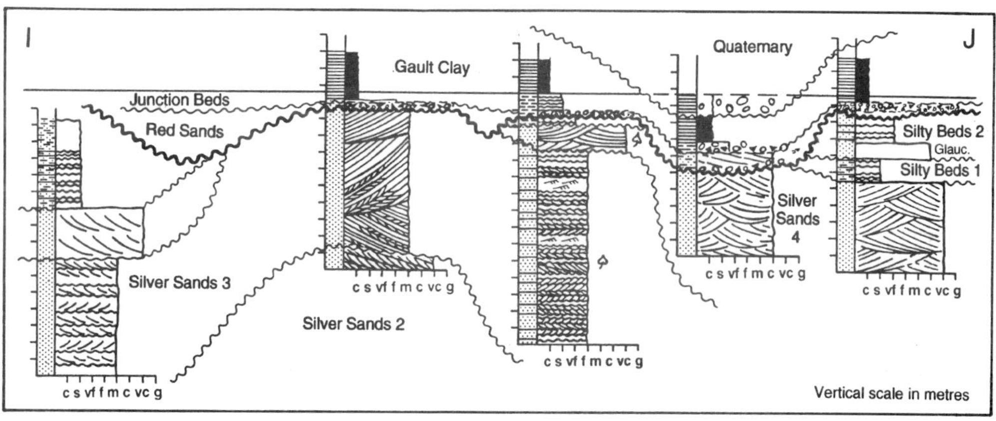

Fig. 16. West–east cross-section (I–J). Location of measured sections shown in Fig. 4. Sketchily drawn sections indicate localities no longer exposed. Section at left [SP921278] is from Bristow (1963: 307). Sections at right [Mile Tree Farm Pits North and South] are from Lamplugh (1922: 19). See Fig. 15 for legend.

Conclusions

Emphasis on the recognition and correlation of surfaces bounding distinctive facies types has allowed the identification of two sequences within the Woburn Sands, namely, the Silver Sands sequence (Upper Aptian) and the Red Sands sequence (Lower Albian). Each sequence is dominated by incised valley-fill deposits interpreted as estuarine in origin.

Three types of sequence stratigraphical surface generated by changes in relative sea level have been documented. These are: (i) fluvially incised sequence boundaries modified by tidal processes during transgression; (ii) tidal ravinement surfaces; and (iii) shoreface ravinement surfaces. Sequence boundaries have been recognized at the base of the Silver Sands sequence and the Red Sands sequence on the basis of several pieces of evidence: (i) the recognition of the unconformable nature of the surfaces from field mapping; (ii) the recognition of considerable erosional relief on the surfaces from field observation and structural contour maps; (iii) the presence of extraformational conglomeratic material above the surfaces; and (iv) the local presence of lateritic ironstones below the surfaces.

Wave ravinement surfaces recognized in both sequences reflect important phases of marine flooding and the passage of the shoreface across the underlying incised valley-fill estuarine deposits. In the case of the Silver Sands sequence, large hummocky bedforms are developed above this surface at the base of the Silty Beds. Marine flooding of the Red Sands incised valley resulted in the development of a highly condensed, micritic Shenley Limestone above which claystone facies with numerous condensed horizons were developed. The development of the Shenley Limestone to the west of Leighton Buzzard, immediately overlying late Jurassic claystones, suggests that this deposit marks a phase of regional transgression following flooding of the incised valley margins. The Red Sands incised valleys remained only partially infilled by estuarine sandstone facies when further transgression occurred which lead, ultimately, to the deposition of the Gault Clay. As a result, the topography of the Red Sands sequence boundary continued to influence deposition on the newly formed shelf. The condensed Shenley Limestone, for example, is preferentially preserved at the margins of the Red Sands incised valleys.

On the basis of the facies identified, the Woburn Sands incised valleys are interpreted as the deposits of wide-mouthed, macrotidal estuaries, similar to the Salmon River Estuary (Dalrymple & Zaitlin 1994). Interpretation of the Woburn Sands as a composite incised valley-fill is in keeping with the Lower Greensand outlier known as the Calne Sands Formation, which crops out well to the west of Leighton Buzzard and has also been interpreted as an incised valley-fill deposit (Hesselbo *et al.* 1990). It is suggested, therefore, that the view of the Woburn Sands being deposited across a tidal shelf (e.g. Rawson 1992; Bristow 1994) is incorrect. Interpretation of the Woburn Sands as a composite incised valley-fill also differs from that of Johnson & Levell (1995), who consider the whole of the Woburn Sands succession (including the Red Sands) as a single phase of transgressive estuarine development.

This study was mainly carried out during tenure of a PhD studentship at Liverpool University (J. P. Wonham) sponsored by Statoil, and was completed at Imperial College, London through the assistance of funding from the European Commission's Joule II Project. Thanks to S. P. Hesselbo and E. R. Shephard-Thorn for their comments on an earlier version of this manuscript. Information given in Fig. 5 and Figs 12–16 is reproduced by permission of the Director, British Geological Survey: NERC copyright reserved.

Appendix: alphabetical listing of localities mentioned in text

Grid references from Ordnance Survey Pathfinder Series map no. 1071 (Leighton Buzzard and Stewkley).

Baker's Wood Pit [SP923285]
Bennie's A5 Pit [SP933302]
Brickhill [SP900318]
Bryant's Lane Quarry [SP929286]
Chamberlain's Barn Pit [SP928267]
Chance's Pit [SP937276]
Double Arches Pit [SP935290]
Fox Corner Pit [SP926291]
Garside's Old Pit [SP937243]
Garside's Pit [SP939295]
Grovebury Pit [SP925237]
Harris' Pit [SP936274]
Little Brickill [SP908323]
Littleworth [SP881233]
Long Crendon [SP695095]
Mile Tree Farm Pit [SP944277]
Munday's Hill Pit [SP937282]
Nine Acre Pit [SP940277]
Poplars Pit [SP947286]
Pratts Pit [SP903240]
Reach Lane Pit [SP933284]
Stone Lane Pit [SP928288]

References

Allen, G. P. & Posamentier, H. W. 1993. Sequence stratigraphy and facies model of an incised valley fill: the Gironde Estuary, France. *Journal of Sedimentary Petrology*, **63**, 378–391.

ALLEN, J. R. L. 1982. Mud drapes in sand wave deposits: a physical model with application to the Folkestone Beds (Early Cretaceous, southeast England). *Philosophical Transactions of the Royal Society of London*, **A306**, 291–345.

—— 1990. The Severn Estuary in southwest Britain: its retreat under marine transgression, and fine-sediment regime. *Sedimentary Geology*, **66**, 13–28.

—— 1994. An introduction to estuarine lithosomes and their controls. *In*: WRIGHT, V. P. (ed.) *Sedimentology Review*, **1**. Blackwell Scientific, Oxford, 123–138.

ANDERSON, I. D. 1986. The Gault Clay – Folkestone Beds junction in West Sussex, southeast England. *Proceedings of the Geologists' Association, London*, **97**, 45–58.

ASHLEY, G. M. 1990. Classification of large-scale subaqueous bedforms: a new look at an old problem. *Journal of Sedimentary Petrology*, **60**, 160–172.

BALLANCE, P. F. 1963. The beds between the Kimmeridge and Gault clays in the Thame–Aylesbury neighbourhood. *Proceedings of the Geologists' Association, London*, **74**, 393–418.

BENTLEY, A. 1970. *Sedimentation studies in the Lower Greensand: a report on the sedimentology of the Aptian and Lower Albian strata near Leighton Buzzard, Bedfordshire*. BSc Thesis, University of Oxford.

BRIDGES, P. J. 1982. Sedimentology of a tidal sea: the Lower Greensand of southern England. *In*: STRIDE, A. H. (ed.) *Offshore Tidal Sands*. Chapman and Hall, London, 183–189.

BRISTOW, C. 1994. A new look at the Lower Greensand using ground-penetrating radar. *Geology Today*, **10**, 24–27.

BRISTOW, C. R. 1963. *The stratigraphy and structure of the Upper Jurassic and Lower Cretaceous Rocks in the area between Aylesbury (Bucks) and Leighton Buzzard (Beds)*. PhD Thesis, University of London.

—— & KIRKALDY, J. F. 1963. Field meeting to the Leighton Buzzard–Aylesbury area. *Proceedings of the Geologists' Association, London*, **73**, 455–459.

BUCK, S. G. 1987. *Facies and sedimentary structures of the Folkestone Beds (Lower Greensand, early Cretaceous) and equivalent strata in southern England*. PhD Thesis, University of Reading.

CASEY, R. 1961. The stratigraphical palaeontology of the Lower Greensand. *Palaeontology*, **3**, 487–621.

COLLINSON, J. D. 1986. Alluvial sediments. *In*: READING, H. G. (ed.) *Sedimentary Environments and Facies*. Blackwell Scientific, Oxford, 20–62.

DALRYMPLE, R. W. & ZAITLIN, B. A. 1994. High-resolution sequence stratigraphy of a complex, incised valley succession, Cobequid Bay – Salmon River estuary, Bay of Fundy, Canada. *Sedimentology*, **41**, 1069–1091.

——, KNIGHT, R. J., ZAITLIN, B. A. & MIDDLETON, G. V. 1990. Dynamics and facies model of a macrotidal sand-bar complex, Cobequid Bay - Salmon River estuary (Bay of Fundy). *Sedimentology*, **37**, 577–612.

DEAN, W. E. 1994. *The relationship between sea-level changes, facies changes and ammonite distributions in the Aptian–Albian of the Anglo–Paris Basin*. PhD Thesis, University College London.

DE RAAF, J. F. M. & BOERSMA, J. R. 1971. Tidal deposits and their sedimentary structures. *Geologie en Mijnbouw*, **50**, 479–504.

EYERS, J. 1991. The influence of tectonics on early Cretaceous sedimentation in Bedfordshire, England. *Journal of the Geological Society, London*, **148**, 405–414.

—— 1992. Sedimentology and palaeoenvironment of the Shenley Limestone (Albian, Lower Cretaceous): an unusual shallow-water carbonate. *Proceedings of the Geologists' Association, London*, **103**, 293–302.

HAQ, B. U., HARDENBOL, J. & VAIL, P. R. 1988. Mesozoic and Cenozoic chronostratigraphy and cycles of sea level change. *In*: WILGUS, C. K., HASTINGS, B. S., KENDALL, C. G. ST. C., POSAMENTIER, H. W., ROSS, C. A. & VAN WAGONER, J. C. (eds) *Sea-level Changes: an Integrated Approach*. Society of Economic Paleontologists and Mineralogists, Special Publications, **42**, 71–108.

HARRIS, P. T. 1988. Large-scale bedforms as indicators of mutually evasive sand transport and the sequential infilling of wide-mouthed estuaries. *Sedimentary Geology*, **57**, 273–298.

HESSELBO, S. P., COE, A. L., BATTEN, D. J. & WACH, G. D. 1990. Stratigraphic relations of the Lower Greensand (Lower Cretaceous) of the Calne area, Wiltshire. *Proceedings of the Geologists' Association, London*, **101**, 265–278.

HOUBOLT, J. J. H. C. 1968. Recent sediments in the southern bight of the North Sea. *Geologie en Mijnbouw*, **47**, 245–273.

JOHNSON, H. D. & LEVELL, B. K. 1980. Sedimentology of Lower Cretaceous subtidal sand complex, Woburn Sands, Southern England. *Bulletin of the American Association of Petroleum Geologists*, **64**, 728–729.

—— & —— 1995. Sedimentology of a transgressive, estuarine sand complex: the Lower Cretaceous Woburn Sands (Lower Greensand), southern England. *In*: PLINT, A. G. (ed.) *Sedimentary Facies Analysis*. International Association of Sedimentologists, Special Publications, **22**, 17–46.

KARNER, G. D., LAKE, S. D., & DEWEY, J. F. 1987. The thermal and mechanical development of the Wessex Basin, southern England. *In*: COWARD, M. P., DEWEY, J. F. & HANCOCK, P. L. (eds) *Continental Extensional Tectonics*. Geological Society, London, Special Publications, **28**, 517–536.

KEEPING, W. 1883. *The fossils and palaeontological affinities of the Neocomian deposits of Upware and Brickhill*. Cambridge University Press (Sedgwick Prize Essay for 1879).

LAMPLUGH, G. W. 1921. The age of the Shenley Limestone. *Geological Magazine*, **58**, 140.

—— 1922. On the Junction of Gault and Lower Greensand near Leighton Buzzard (Bedfordshire). *Quarterly Journal of the Geological Society of London*, **78**, 1–80.

—— & WALKER, J. F. 1903. On a fossiliferous band at the top of the Lower Greensand near Leighton Buzzard (Bedfordshire). *Quarterly Journal of the Geological Society of London*, **59**, 234–265.

MCCAVE, I. N. 1971. Sand waves in the North Sea off the coast of Holland. *Marine Geology*, **10**, 199–225.

NARAYAN, J. 1971. Sedimentary structures in the Lower

Greensand of the Weald, England and Bas-Boulonnais, France. *Sedimentary Geology*, **6**, 73–109.

NIO, S.-D. & YANG, C.-S. 1991. Diagnostic attributes of clastic tidal deposits: a review. *In*: SMITH, D. G., REINSON, G. E., ZAITLIN, B. A. & RAHMANI, R. A. (eds) *Clastic Tidal Sedimentology*. Canadian Society of Petroleum Geologists, Memoirs, **16**, 3–28.

OWEN, H. G. 1972. The Gault and its junction with the Woburn Sands in the Leighton Buzzard Area, Bedfordshire and Buckinghamshire. *Proceedings of the Geologists' Association, London*, **83**, 287–312.

—— 1988. The ammonite zonal sequence and ammonite taxonomy in the *Douvilliceras mammillatum* Superzone (Lower Albian) in Europe. *Bulletin of the British Museum (Natural History) Geology Series*, **44**, 177–231.

POLLARD, J. E., GOLDRING, R. & BUCK, S. G. 1993. Ichnofabrics containing *Ophiomorpha*: significance in shallow-water facies interpretation. *Journal of the Geological Society, London*, **150**, 149–164.

POSAMENTIER, H. W. & VAIL, P. R. 1988. Eustatic controls on clastic sedimentation II -- sequence and systems tract models. *In*: WILGUS, C. K., HASTINGS, B. S., KENDALL, C. G. ST. C., ROSS, C. A., POSAMENTIER, H. W. & VAN WAGONER, J. C. (eds) *Sea-level Changes: an Integrated Approach*. Society of Economic Paleontologists and Mineralogists, Special Publications, **42**, 125–154.

——, JERVEY, M. T. & VAIL, P. R. 1988. Eustatic controls on clastic deposition I – conceptual framework. *In*: WILGUS, C. K., HASTINGS, B. S., KENDALL, C. G. ST. C., ROSS, C. A., POSAMENTIER, H. W. & VAN WAGONER, J. C. (eds) *Sea-level Changes: an Integrated Approach*. Society of Economic Paleontologists and Mineralogists, Special Publications, **42**, 109–124.

RAWSON, P. F. 1992. The Cretaceous. *In*: DUFF, P. MCL. D. & SMITH, A. J. (eds) *Geology of England and Wales*. The Geological Society, London, 355–388.

RUFFELL, A. H. 1992. Early to mid-Cretaceous tectonics and unconformities of the Wessex Basin (southern England). *Journal of the Geological Society, London*, **149**, 443–454.

—— & WACH, G. D. 1991. Sequence stratigraphic analysis of the Aptian–Albian Lower Greensand in southern England. *Marine and Petroleum Geology*, **8**, 341–353.

—— & WIGNALL, P. B. 1990. Depositional trends in the Upper Jurassic–Lower Cretaceous of the northern margin of the Wessex Basin. *Proceedings of the Geologists' Association, London*, **101**, 279–288.

SCHIAVON, N. 1988. Goethite ooids: growth mechanism and sandwave transport in the Lower Greensand (early Cretaceous, southern England). *Geological Magazine*, **125**, 57–62.

SCHWARZACHER, W. 1953. Cross-bedding and grain-size in the Lower Cretaceous sands of East Anglia. *Geological Magazine*, **90**, 322–330.

SHEPHARD-THORN, E. R., HARRIS, P. M., HIGHLEY, D. E. & THORNTON, M. H. 1986. *An Outline Study of the Lower Greensand of Parts of South-east England*. British Geological Survey, Keyworth.

——, MOORLOCK, B. S. P., COX, B. M., ALLSOP, J. M. & WOOD, C. J. 1994. *Geology of the Country Around Leighton Buzzard*. British Geological Survey, Sheet 220 (England and Wales), Memoir. HMSO, London, 127 pp.

TEALL, J. J. H. 1875. *The Potton and Wicken Phosphatic Deposits*. Cambridge University Press (Sedgwick Prize Essay for 1873).

TERWINDT, J. H. J. 1981. Origin and sequences of sedimentary structures in inshore mesotidal deposits of the North Sea. *In*: NIO, S.-D., SHÜTTENHELM, R. T. E. & VAN WEERING, TJ. C. E. (eds) *Holocene Marine Sedimentation in the North Sea Basin*. International Association of Sedimentologists, Special Publications, **5**, 4–26.

TOOMBS, H. A. 1935. Field meeting at Leighton Buzzard, Bedfordshire. *Proceedings of the Geologists' Association, London*, **46**, 432–435.

VAIL, P. R., MITCHUM, R. M. JR., TODD, R. G., WIDMIER, J. M., THOMPSON, S. III, SANGREE, J. B., BUBB, J. N. & HATLELID, W. G. 1977. Seismic stratigraphy and global changes in sea level. *In*: PAYTON, C. E. (ed.) *Seismic Stratigraphy – Applications to Hydrocarbon Exploration*. American Association of Petroleum Geologists, Memoirs, **26**, 49–212.

VAN WAGONER, J. C., MITCHUM, R. M. JR., CAMPION, K. M. & RAHMANIAN, V. D. 1990. *Siliciclastic Sequence Stratigraphy in Well Logs, Cores, and Outcrops: Concepts for High-Resolution Correlation of Time and Facies*. American Association of Petroleum Geologists, Methods in Exploration Series, **7**.

——, POSAMENTIER, H. W., MITCHUM, R. M. JR., VAIL, P. R., SARG, J. F., LOUTIT, T. S. & HARDENBOL, J. 1988. An overview of the fundamentals of sequence stratigraphy and key definitions. *In*: WILGUS, C. K., HASTINGS, B. S., KENDALL, C. G. ST. C., ROSS, C. A., POSAMENTIER, H. W. & VAN WAGONER, J. C. (eds) *Sea-level Changes: an Integrated Approach*. Society of Economic Paleontologists and Mineralogists, Special Publications, **42**, 39–45.

VISSER, M. J. 1980. Neap-spring cycles reflected in Holocene subtidal large-scale bedform deposits: a preliminary note. *Geology*, **8**, 543–546.

WALKER, R. G. 1985. Ancient examples of tidal sand bodies formed in open, shallow seas. *In*: TILLMAN, R. W., SWIFT, D. J. P. & WALKER, R. G. (eds) *Shelf Sands and Sandstone Reservoirs*. Society of Economic Paleontologists and Mineralogists Short Course, **13**, 303–341.

WONHAM, J. P. 1993. *Sedimentology and sequence stratigraphy of tidal sandstone bodies: implications for reservoir characterisation*. PhD Thesis, University of Liverpool.

WRIGHT, C. W. & WRIGHT, E. V. 1947. The stratigraphy of the Albian Beds at Leighton Buzzard. *Geological Magazine*, **84**, 161–168.

WYATT, R. J., MOORLOCK, B. S. P., LAKE, R. D. & SHEPHARD-THORN, E. R. 1988. *Geology of the Leighton Buzzard–Ampthill district*. British Geological Survey, Technical Report, **WA/88/1**.

Use of palaeosols in sequence stratigraphy of peritidal carbonates

V. P. WRIGHT

Postgraduate Research Institute for Sedimentology, The University of Reading, PO Box 227, Whiteknights, Reading RG6 2AB, UK

Abstract: Palaeosols (fossil soils) are common in ancient shallow marine carbonate successions. The degree of development exhibited by the palaeosol reflects the duration of subaerial exposure, with palaeosols occurring at lower order sequence boundaries commonly showing complex, polygenetic histories, and those at higher order boundaries showing less development. Palaeosols in parasequence sets commonly display systematic variations in their degrees of development reflecting the varying lengths of subaerial exposure related to changes in accommodation space creation. These properties of palaeosols can be used to reappraise the sequence stratigraphical positions of peritidal carbonates in the Lower Carboniferous of southwest Britain, serving to illustrate the dangers of assigning facies units to systems tracts based solely on simple facies relationships. The palaeosols and associated sequences do not support the view that high frequency, moderate amplitude eustatic sea-level fluctuations of Milankovitch-type were operating during the Tournaisian or early Viséan.

The ability to recognize subaerial exposure surfaces is critical for identifying sequence boundaries. In shallow marine carbonate successions such surfaces, typically represented by palaeosols, are very common and can be used to significantly refine sequence stratigraphical models. In peritidal units minor cyclothems capped by palaeo-exposure surfaces are commonly stacked into parasequence sets and palaeosols can be used to differentiate various types of sets (e.g. those within highstand versus transgressive systems tracts). In addition, the controls of minor cyclicity can also be assessed using the palaeosols.

A series of examples is presented to demonstrate the principles, followed by an example of their application to reappraise a number of peritidal-shoal carbonate successions in the Lower Carboniferous of southwest Britain. These peritidal units, although all having similar associations with shoal carbonates, all differ significantly in terms of their positions within systems tracts. The example serves to illustrate the dangers of basing sequence stratigraphical interpretations solely on broad facies associations rather than on the detailed analysis of individual units and the importance of a thorough understanding of minor and major boundaries within and between facies units.

Palaeosols, residence time and chronosequences

Palaeosols in carbonate successions represent periods of emersion, weathering and soil formation (Wright 1994). They are common in ancient shallow marine successions and especially in stacked peritidal units (e.g. Read & Horbury 1993).

The degree of development a soil achieves before pedogenesis ceases because of immersion will depend on a number of factors, but the major control is the length of exposure, i.e. the residence time of that exposure surface in a pedogenic environment. Soils exposed for long periods will, under similar climatic conditions, show a greater degree of development than a soil which was active for a shorter time. Related soils, which differ in their properties only because of differences in the duration of pedogenesis, are called chronosequences (Birkeland 1984).

In a sequence stratigraphical framework, residence times are intimately linked to sea-level changes and accommodation space creation and loss. At the simplest level, distinctions can be made between chronosequences at various sequence boundaries of different orders. A major third-order sequence boundary (the orders are here used in the sense of Tucker 1993) will represent a long period of exposure whereas a minor (fifth-order) boundary between metre-scale cyclothems will probably represent a shorter residence time in the soil zone. Such crude patterns in chronosequences have been described from several ancient successions. For example, in the Triassic Latemar platform carbonates in north Italy, Goldhammer *et al.* (1987) noted that the peritidal cyclothems were arranged, on average, in packages of five ('pentacycles'). The palaeosols separating the minor cyclothems within a package were less well developed than those separating the packages. In a detailed study of hierarchies of cyclicity in the Pennsylvanian of the Paradox Basin in the USA, Goldhammer *et al.* (1991) noted that palaeosols separating packages of minor cyclothems were better developed and more

From HESSELBO, S. P. & PARKINSON, D. N. (eds), 1996, *Sequence Stratigraphy in British Geology*, Geological Society Special Publication No. 103, pp. 63–74.

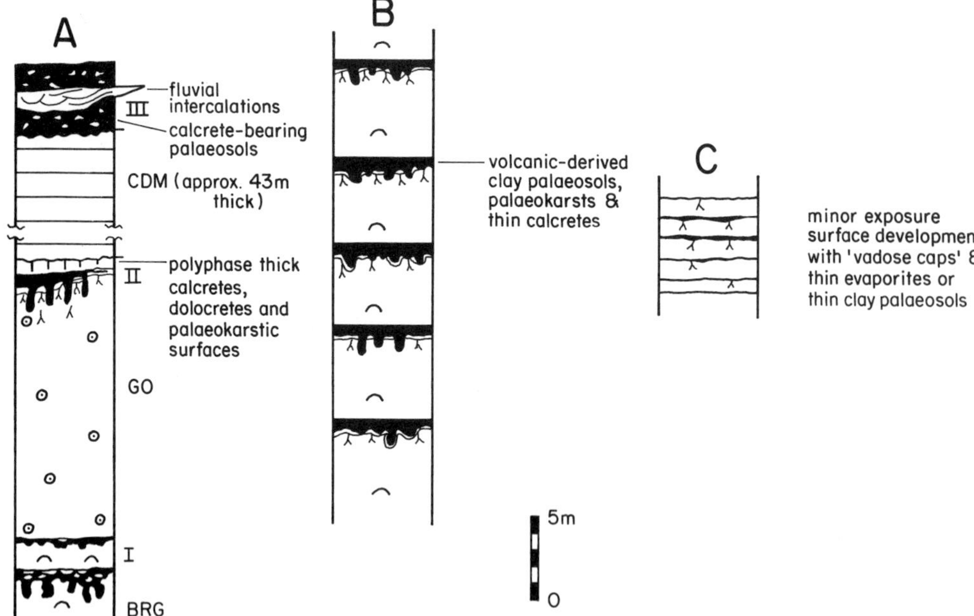

Fig. 1. Palaeosol associations in the Lower Carboniferous sediments of southwest Britain. (**a**). The three major formational boundaries in the Courceyan–Holkerian in the Bristol area. Two complex palaeokarst–speleothem–calcrete horizons separate the Black Rock Group (BRG) (Courceyan) from the Chadian Gully Oolite (GO) (I) corresponding to a major hiatus in deposition (Faulkner *et al.* 1990). The Gully Oolite is capped by a regionally extensive polyphase palaeosol unit (II), the Heatherslade Geosol, which locally contains several separate profiles as well as evidence of karstification phases. The overlying Clifton Down Mudstone (CDM) is capped by a series of fluvial units showing evidence of extensive multiple palaeosol development (III) (Vanstone 1991; Weedon 1987). (**b**) Palaeosol association of Asbian–Brigantian sequences in Britain (Horbury 1989; Vanstone 1993). Each palaeosol is associated with a palaeokarstic surface and thin calcrete profiles (Vanstone 1993). (**c**) Peritidal units containing thin clays associated with supratidal facies (thin evaporites) or minor evidence of vadose diagenesis (Wright 1986); see text.

regionally extensive than the palaeosols capping the smallest cyclothems. Horbury (1989) also noted this relationship in a study of Asbian cyclothems and exposure surfaces in Cumbria, England.

This simple pattern is well seen in the Lower Carboniferous of Britain. Palaeosols developed at major sequence boundaries, such as the Heatherslade Geosol (Riding & Wright 1981; Wright 1987), can be traced from the Bristol area to south Dyfed, a distance of over 150 km along-strike. It is a polygenetic palaeosol representing several distinct phases of soil development over a prolonged period of time, corresponding to the Chadian–Arundian stage boundary in southern Britain (Fig. 1a). A similar situation is found along the northern margin of the South Wales coalfield, where a disconformity between the Oolite Group and the Arundian Llanelly Formation (see later) is represented by up to six palaeosols (Wright 1982). Faulkner *et al.* (1990) have described complex palaeokarst, speleothems and calcretes from the major Tournaisian–Viséan boundary in southwest Britain. Vanstone (1991) has documented highly complex suites of palaeosols and fluvial deposits from the Arundian north of Bristol. These all correspond to major, regional, stratigraphical surfaces at stage boundaries separating decametre-scale formations. The later Asbian–Brigantian successions of Britain are strongly cyclic, with palaeo-exposure surfaces, exhibiting palaeosols and palaeokarsts, occurring every 3–10 m through successions several hundred metres thick (Walkden 1974; Read & Horbury 1993; Vanstone 1993). A strong case has been made that these cyclothems reflect glacio-eustatic sea level changes of approximately 10^5 years duration (Horbury 1989). The palaeosols found at these surfaces are also polygenetic, but are not as complex or as well developed as those separating major formation boundaries (Vanstone 1993) (Fig. 1b). Minor palaeosols also occur capping metre-scale peritidal cycles at several levels in the Lower Carboniferous (e.g. Bridges 1982; see later). In most instances these are poorly developed palaeosols representing only short periods of exposure

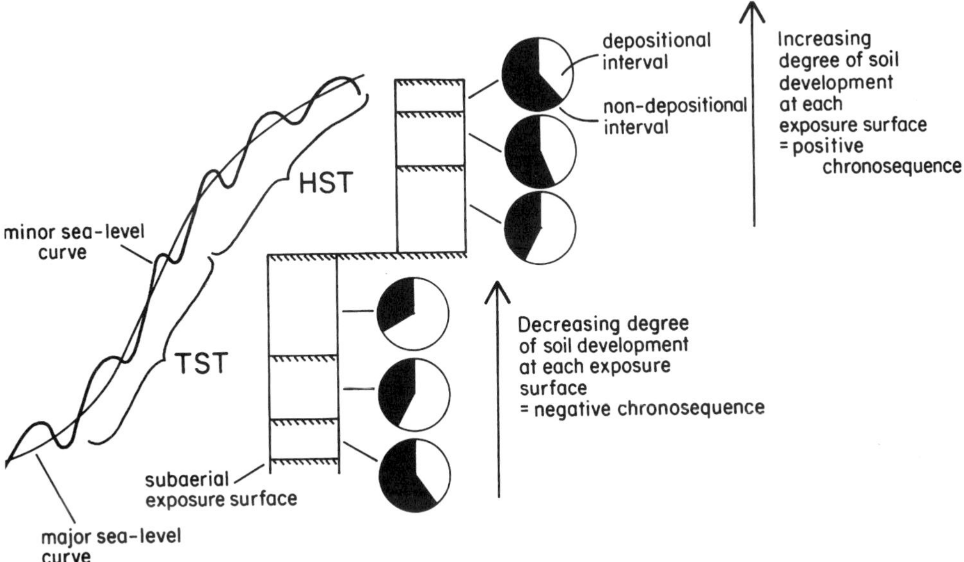

Fig. 2. Schematic representation of origins of positive and negative chronosequences created by the interactions of two orders of sea level change. During construction of the transgressive systems tract (TST) the cyclothems thicken and typically contain deeper water subtidal units, reflecting the higher rate of accommodation space creation. The highstand systems tract (HST) cyclothems are typically thinner and with shallower water lithofacies.

(Fig. 1c). The Asbian–Brigantian exposure surfaces and peritidal palaeosols would correspond to fourth- and fifth-order sequence boundaries in the sense of Tucker (1993).

These simple chronosequence relationships span different stratigraphical units but can also occur within single formations as a result of the interaction of various hierarchies of relative sea-level change. This has been demonstrated by Martin-Chivelet & Gimenez (1992) from the Upper Cretaceous Sierra del Utiel Formation in southeast Spain. The formation contains 179 metre-scale peritidal cycles arranged in three decametre scale major cycles. These workers categorized the palaeosols capping the minor cyclothems into four stages of development. They noticed that within each major cycle, the palaeosols were progressively less or more mature from the base to the top. These changes in palaeosol development corresponded with changes in the thicknesses and compositions of the minor cyclothems. The basal, transgressive systems tract of the major cycle contained less than 10 small cycles which thickened upward (reflecting increased accommodation space creation) and show a decrease in the degree of palaeosol development (reflecting shorter exposure periods as the rate of net sea-level rise increased). The upper part of each major cycle (the highstand systems tract) showed the reverse trend, as the rate of creation of accommodation space decreased. These hierarchical patterns reflect the interaction of more than one order of relative sea-level change (see later). Thus a single formation can display various styles of chronosequence relationships with patterns of increasing or decreasing palaeosol development related to the type of systems tract. The chronosequences which display increasing degrees of development (highstand systems) were referred to as positive chronosequences, and those displaying upwards decreases in the degree of development were termed negative chronosequences (Wright 1994). Not only do such sequences display differences in the degree of soil development and cyclothem type, but also in the degree of exposure-related diagenesis, which can radically affect the porosity and permeability of a limestone (Read & Horbury 1993). These relationships are reviewed in Fig. 2.

Palaeosols in chronosequences do not provide accurate estimates of the time periods of exposure, but can be used in a relative sense if it is possible confidently to determine their relationships to each other as parts of a developmental series. Climatic changes and loss by erosion are two factors which may obscure true developmental relationships. However, the relative stage of development of a palaeosol profile, or the complexity of a polygenetic profile, can be used to distinguish more from less significant subaerial exposure intervals (and hence sequence boundaries) and the presence of positive

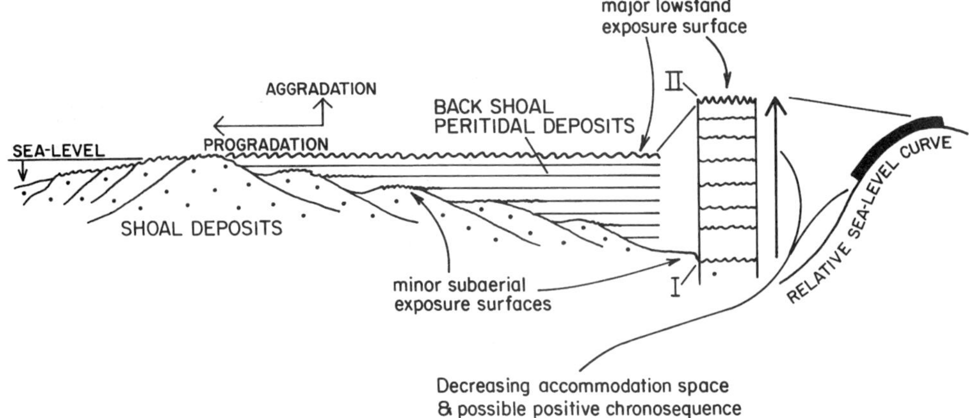

Fig. 3. Hypothetical sequence generated during a highstand progradational phase. The back-barrier deposits aggrade and prograde behind the barrier island. Only a minor exposure surface could form beneath the onlapping back-barrier deposits (I), but a more developed exposure surface should cap the unit (II), corresponding to a lowstand phase. If minor cyclicity developed during deposition of the sequence a positive chronosequence of palaeosols should occur, reflecting the decreasing rate of accommodation space creation.

or negative chronosequences can also be used to distinguish the tract positions of parasequence sets. What follows is an example of how these properties can be used to reappraise relationships of four shoal–peritidal successions in the Lower Carboniferous of southern Britain.

Peritidal carbonates in Lower Carboniferous ramp successions, southwest Britain

Peritidal successions, typically in the range 3–10 m thick, occur at a number of levels in the early Carboniferous ramp successions of southwest Britain. They are also a component of similarly aged successions in, for example, North America (Elrick & Read 1991).

The peritidal units are invariably associated with shoal successions, typically overlying oolitic–bioclastic grainstones, and overlain by bioclastic grainstones. This has led a number of workers to regard the peritidal units as late regressive (i.e. effectively late highstand) units (Ramsbottom 1973; Ramsay 1987; Ross & Ross 1988). Elrick & Read (1991), in their detailed study of early Carboniferous ramp successions (Lodgepole and lower Madison formations) of Wyoming and Montana, recognized peritidal facies development in both highstand and transgressive systems tracts. The latter formed in a narrow inner ramp zone and were not illustrated in their facies sections. The highstand peritidal units are more widespread and major sequence boundaries were placed at the top of these units. These workers note the occurrence of palaeosols, but do not discuss them in detail. The peritidal units are not present in all highstand systems tracts successions and were supposed to have formed behind ooid shoals where the accommodation space was lowest. Elrick & Read (1991) generated computer simulations for the ramp successions, but encountered a number of problems relating to the highstand systems' tracts and peritidal deposits; the synthetic highstand systems tract successions typically generated no peritidal units, but were represented by disconformity capped shallow subtidal units.

Figure 3 illustrates the type of setting and succession which would develop in a highstand systems tract with peritidal deposits forming behind a prograding barrier as envisaged by previous workers, e.g. Elrick & Read (1991). Such a situation could only develop where accommodation space was being created to allow the continued accommodation of back-barrier deposits behind the barrier. These peritidal deposits onlap the shoal deposits. If any higher frequency cyclothems develop, as a result of allocyclic or autocyclic processes, the back-barrier deposits should show a positive chronosequence reflecting decreasing accommodation space as the lowstand phase is approached. The peritidal succession should be capped by the most developed palaeosol (lowstand phase). The barrier deposits might also be capped by an exposure surface, but as the residence time of such a surface in the soil environment would be short, before burial by back-barrier deposits, any soil should not be well developed. If the rate of accommodation space creation is low, no back-barrier deposits should form and the late highstand

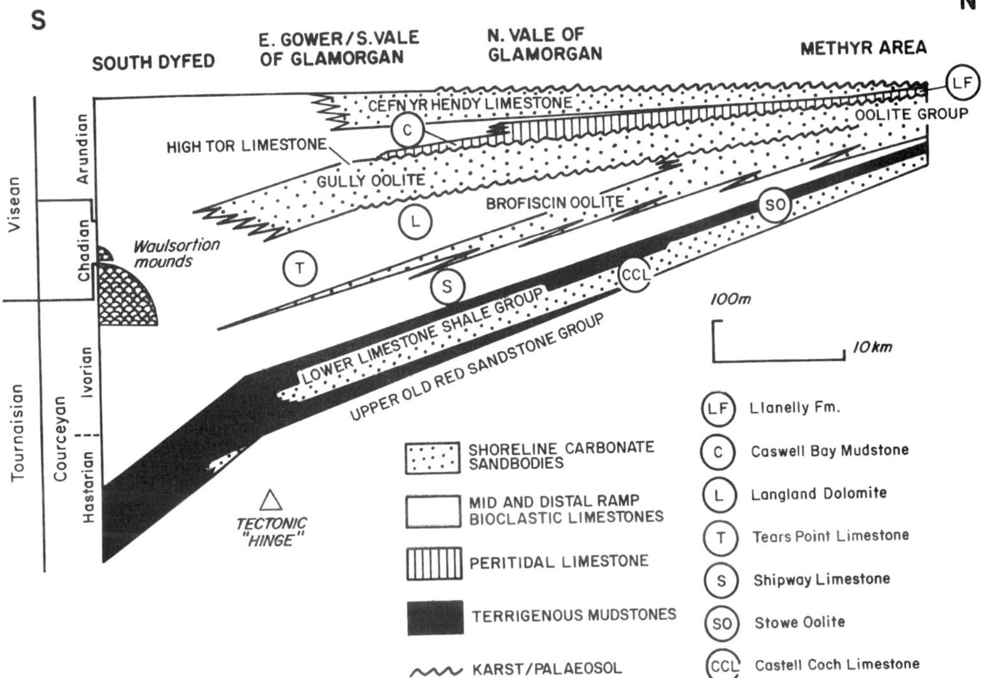

Fig. 4. Sketch cross-section from north to south across part of the Lower Carboniferous ramp stack in South Wales showing the major stratigraphic units discussed in this paper. For details of major facies, see Wright (1986); for details of Lower Limestone Shale Group stratigraphy, see Burchette (1987).

systems tract is likely to be represented by the shoal system prograding as a beach-fronted strandplain capped by a major palaeosol, as in the case of some of the synthetic models generated by Elrick & Read (1991).

In the Lower Carboniferous successions of southwest Britain the positions and chronosequence aspects of palaeosols in peritidal units indicate that the likelihood of such units forming during decreasing accommodation space intervals, such as produced highstand systems tracts, was greatly reduced. To illustrate this point, four peritidal–shoal associations are analysed: the Stowe Oolite and Mitcheldean members of the Castell Coch Formation (Burchette 1987; Burchette *et al.* 1990); the Caswell Bay Mudstone (Riding & Wright 1981): the Flat Holm Member of the Birnbeck Limestone Formation (Weedon 1987) and the Cheltenham Limestone Member of the Llanelly Formation (Wright 1981, 1986). The first is of early Tournaisian age, whereas the latter three are Arundian. Only the Cheltenham Limestone has the attributes of a highstand systems tract and shows clear evidence of minor cyclicity. The other three units comprise transgressive systems tracts, and doubt is cast on earlier interpretations. All these units developed on extensive southerly dipping carbonate ramps which covered part of southwest Britain during the early Carboniferous (Burchette *et al.* 1990; Wright 1986). The landward successions during the Tournaisian and early Viséan (pre-Holkerian) consisted of peritidal and oolitic units, whereas more basinward successions are represented by muddy bioclastic limestones. The base of the ramp stack is a shale-rich unit, the Lower Limestone Shale Group, which contains a complex set of shoal and backshoal deposits (Burchette 1987; Fig. 4).

The Stowe Oolite and Mitcheldean members of the Castell Coch Formation

The Stowe Oolite Member of the Castell Coch Formation is part of the early Tournaisian Lower Limestone Shale Group of southwest Britain (see detailed discussions in Burchette 1987) and is exposed in the Forest of Dean and in southeast Wales. It is a 'shoestring' sand body, up to 5 km across, and can be correlated for at least 40 km along depositional strike. At its maximum it is 15 m thick, but pinches out completely down-dip over 1 km into offshore facies. On its landward margin the Mitcheldean Member was deposited, which consists of fine-grained lagoonal limestones (Burchette

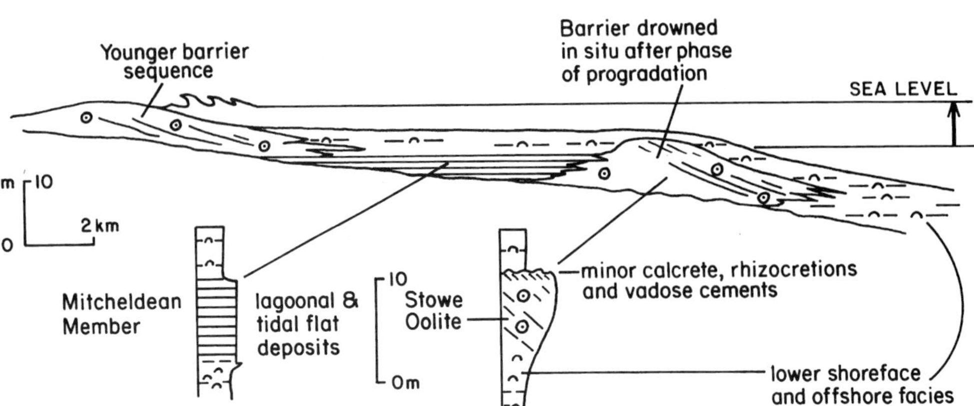

Fig. 5. Relationships between the Stowe Oolite and the Mitcheldean members (from Burchette *et al.* 1990). The peritidal Mitcheldean Member lacks evidence of significant subaerial exposure at the base, top and within the unit. It is underlain and overlain by open marine deposits and formed behind a prograding oolitic barrier during a phase of net sea level rise (i.e. forms a TST).

1987). These limestones, up to 10 m thick with, locally, two shoaling successions, comprise peloidal packstones with oolitic and bioclastic intervals. Calcareous 'algae', such as porostromate-bearing oncoids, are abundant, as are microconchids, indicating restricted conditions.

The Stowe Oolite Member is capped by a minor palaeosol (calcrete) with rhizocretions and vadose cements (Burchette 1987). Both the oolite and the back-barrier sediments are overlain by muddy bioclastic sediments interpreted as offshore ramp deposits.

Unlike the other three examples, the peritidal units are not seen to overlie directly the shoal deposits at outcrop. The likely facies relationships of the Stowe Oolite and Mitcheldean members are shown in Fig. 5. The preservation of the oolite as a near-complete prograding barrier succession was attributed by Burchette *et al.* (1990) to *in situ* drowning and the re-establishment of a new barrier in a more landward position.

The units lack several key features of the proposed highstand systems tract model. The succession is not capped by a major lowstand exposure surface, instead the shoal unit is capped by a minor exposure surface. The back-barrier deposits are predominantly lagoonal in nature and lack any palaeosols with a positive chronosequence. Thus this shoal–peritidal association fails to meet the requirements of a highstand systems tract succession and was interpreted by Burchette *et al.* (1990), from general facies relationships, as a minor progradational unit within a transgressive systems tract.

Caswell Bay Mudstone

The Arundian Caswell Bay Mudstone of South Wales, up to 8 m thick, overlies the Gully (Caswell Bay or Caninia) Oolite (Figs 4 and 6). The succession has been described in detail by Riding & Wright (1981). The shallowing-upwards oolite is capped by a complex palaeosol known as the Heatherslade Geosol (Wright 1987). Although truncated in some sections, in others it exhibits a complex profile with thick calcretized oolitic sands, humus horizons and dolomites (Riding & Wright 1981). In the Bristol district the Caswell Bay Mudstone is known as the Clifton Down Mudstone and also locally exhibits polyphase palaeosol profiles with evidence of as many as four or five phases of soil development and karstification, as at Chipping Sodbury. In the South Wales sections the Caswell Bay Mudstone consists predominantly of laminated micrites, locally fenestral, and restricted subtidal limestones. Detailed descriptions of the intertidal-dominated lithofacies have been given by Riding & Wright (1981) and Ramsay (1987). Ramsay briefly noted some apparent palaeosol profiles in sections in the Caswell Bay Mudstone in Gower, but none of these appears to have reached a significant stage of development and such horizons are absent from sections in the Vale of Glamorgan. The unit lacks any strong cyclicity. Interestingly, the types of laminated limestones are strikingly similar to those described from the Wyoming and Montana sections of Elrick & Read (1991).

The Caswell Bay Mudstone is erosively overlain

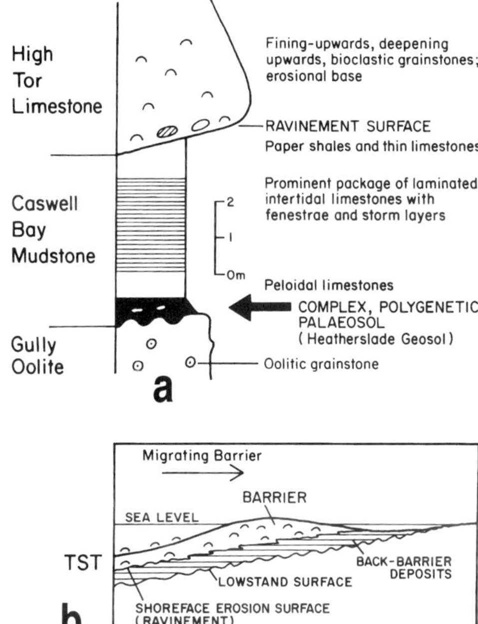

Fig. 6. Caswell Bay Mudstone. (**a**) Simplified log of sequence in Vale of Glamorgan (see Riding & Wright 1981). The unit has a prominent palaeosol at the base, but contains no other well-developed palaeosols or prominent cyclicity. The intertidal component is relatively thick. (**b**) Interpretation of the sequence as a transgressive back-barrier deposit.

by coarse bioclastic grainstones which fine upwards (the High Tor Limestone). The succession was interpreted by Riding & Wright (1981) as an oolitic strandplain (highstand systems tract–lowstand systems tract), capped by a major palaeosol, overlain by back-barrier and barrier deposits (Fig. 6) representing a transgressive succession. Other workers (Ramsbottom 1973; Ramsay 1987; Ross & Ross 1988) have placed the Caswell Bay Mudstone as the uppermost unit in a regressive (i.e. highstand systems tract–lowstand systems tract) succession. The occurrence of a prominent palaeosol beneath the back-barrier unit, the absence of any major palaeosols capping it and the absence of a positive chronosequence within the Caswell Bay Mudstone all point against its origin in a highstand systems tract. The absence of any minor (fifth-order, metre-scale) cyclothems in the peritidal unit might suggest either a lack of any high frequency sea-level oscillations or a relatively high rate of creation of accommodation space. The Gully Oolite–Caswell Bay Mudstone is also not a candidate for a simple, transitional highstand systems tract succession.

Flat Holm Member of the Birnbeck Limestone

In some Arundian sections the High Tor Limestone or its equivalent (the Birnbeck Limestone) contains a cyclic succession known as the Flat Holm Member (Whittaker & Green 1983; Weedon 1987). It is best exposed on the island of Flat Holm in the Bristol Channel, where it consists of 40 m of massive, coarsely bioclastic and oolitic grainstones, mudstones and dolomites (Fig. 7). The grainstones exhibit planar and trough cross-stratification and surfaces covered with small-scale interference and large wave-ripples with wave-lengths of 0.6 m (Weedon 1987). The bioclasts are of a fully marine biota and include rugose and tabulate corals. These grainstones occur interbedded with dolomites and carbonate mudstones, with each lithology in packets up to 0.5–5 m thick, although the lower 20 m of the unit consists solely of grainstones. Contacts between the two lithotypes are generally sharp, but not erosional, with the finer grained tops of the grainstone units, and some of the bases, being dolomitized.

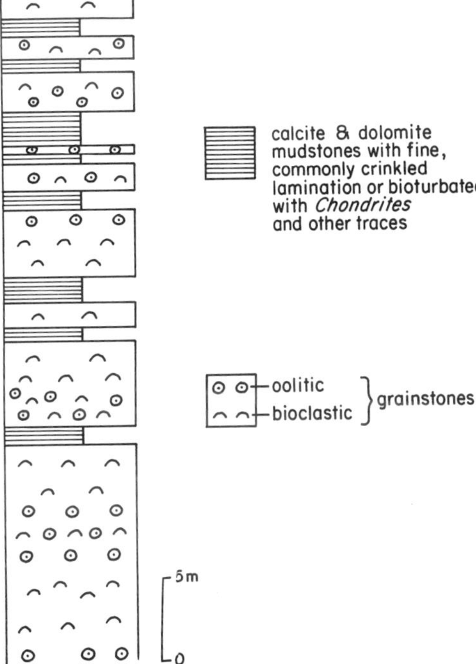

Fig. 7. Simplified log of the Flat Holm Member, Flat Holm. After Weedon (1987). Massively bedded, cross-stratified and water-rippled coarse bioclastic and oolitic grainstones contain diverse fully marine biota, including large numbers of bellerophontid gastropods. The calcitic and dolomitic mudstones either display planar to crinkled (crypt-microbial) lamination or contain locally abundant trace fossils, but lack a diverse skeletal fauna.

The dolomites (dolomudstones) are buff to grey green, non-ferroan or locally ferroan in composition. Some display relict microbial lamination, whereas others are clearly dolomitized bioturbated muddy carbonates with abundant *Chondrites,* but lacking evidence of subaerial exposure.

The Birnbeck Limestone passes down depositional dip into more uniform bioclastic limestones (Whittaker & Green 1983). The Flat Holm Member is overlain by more bioclastic limestones of the offshore facies type and regionally the Birnbeck Limestone is capped by an oolitic unit.

There are similarities between these Flat Holm Member dolomudstones and the dolomite/dolosiltites recorded by Elrick & Read (1991) from Wyoming and Montana, which they interpreted as lagoonal deposits. The restricted biota and microbial laminites support a lagoonal interpretation for the Flat Holm Member dolomites and the interdigitation of shoal and lagoonal deposits suggests a fluctuating shoreline. The absence of any exposure surfaces capping the various dolomites does not support the view that these peritidal units represent a highstand systems tract. No palaeosols occur and there is no suggestion of the development of a positive chronosequence.

Cheltenham Limestone Member of the Llanelly Formation

Along the northeastern rim of the South Wales Coalfield synclinorium the Arundian contains a complex peritidal, oolitic and fluvial unit, the Llanelly Formation (Wright 1981). The formation is underlain by the Oolite Group, which has a prominent palaeokarstic horizon at the top. There is good evidence that this exposure surface was due, at least in part, to local tectonic uplift related to the Neath Disturbance Fault Zone (Wright 1981). The basal member of the Llanelly Formation, the Clydach Halt Member, contains up to six calcrete profiles, each at an advanced stage of development (Wright 1982). The overlying Cheltenham Limestone Member is a medium-bedded package, up to 8 m thick, of peritidal limestones which, although containing some complete shallowing-up cycles, exhibits rather complex changes from restricted subtidal to intertidal and supratidal limestones (Wright 1986). Palaeosols occur at a number of levels, but two prominent and correlatable palaeosols occur near the top of the unit. The top of the member is marked by a widespread, thick calcrete profile named the Cwm Dyar Geosol (Fig. 8) (Wright 1982). The next palaeosol down is distinctive, with a thick calcrete crust with abundant calcified faecal pellets and burrows, the Darrenfelen Geosol (Wright 1983). It is not as well-developed as the Cwm Dyar Geosol. There are no well-developed palaeosols within the remainder of the member, which is overlain by bioclastic and oolitic limestones of the Penllwyn Oolite Member, the lowest unit of which is a coarse bioclastic, intraclastic grainstone with an open marine biota.

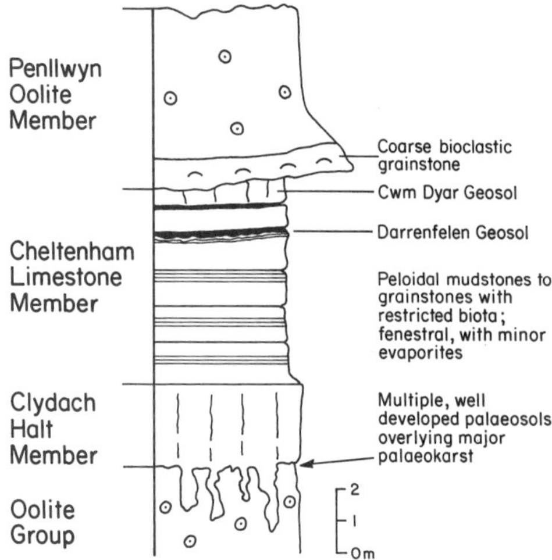

Fig. 8. Cheltenham Limestone Member of the Llanelly Formation. The peritidal unit is underlain by an alluvial sequence, the Clydach Halt Member, which contains up to six well-developed palaeosols above a major palaeokarstic horizon. The Cheltenham Limestone contains a variety of subtidal, intertidal and surpratidal limestones with thin clay horizons. Two of these horizons at the top of the unit exhibit prominent palaeosols. The lower horizon, the Darrenfelen Geosol is less well developed than the upper horizon, the Cwm Dyar Geosol. Both palaeosols can be traced at a number of localities. The upper part of the Cheltenham Limestone represents a positive chronosequence. The Penllwyn Oolite Member is overlain by a fluvial unit with palaeosols.

The Llanelly Formation (and Cheltenham Limestone), although correlated lithostratigraphically and, crudely, biostratigraphically with the transgressive Caswell Bay Mudstone (e.g. Wilson *et al.* 1988), is very different to all the other peritidal units discussed here. In common with the Caswell Bay Mudstone, the peritidal limestones overlie a complex palaeosol unit. However, unlike the Caswell Bay Mudstone, the Cheltenham Limestone contains more varied facies and is cyclic, with the two prominent and highly developed palaeosols towards the top of the unit exhibiting a crude positive chronosequence relationship, the youngest capping the unit. The base of the Cheltenham Limestone must represent a flooding surface, but it bears the major characteristics of a highstand systems tract succession in its upper part.

Discussion

On a general level, all four peritidal units correspond to the simple model of capping shoal successions. However, by assessing the nature of the palaeosols at the unit contacts, and by determining the nature of palaeosols within the peritidal units, marked differences in the relationships of the four units and their likely positions within systems tracts emerge.

Only the Cheltenham Limestone, at least in its upper part, shows any similarity to a highstand systems tract peritidal unit. The Caswell Bay Mudstone is, by association, a transgressive systems tract deposit and is composed of intertidal and subtidal carbonates with minor or no palaeosols. The lack of any minor cyclicity and exposure within it might reflect the absence of any minor sea-level falls or a rapid rate of accommodation space creation overprinting the effects of such falls (see later). The Mitcheldean Member forms a minor progradational succession within a transgressive phase and also lacks evidence of minor sea-level falls and prolonged exposure. The same features apply to the largely subtidal (lagoonal) units of the Flat Holm Member.

Excluding the older Mitcheldean Member, the other three peritidal units are all Arundian in age and although the Caswell Bay Mudstone is older than the overlying Flat Holm Member, the relative age of the Cheltenham Limestone is difficult to assess. Figure 9 shows a model which attempts to place these three Arundian units in a sequence stratigraphical context. The model also includes a progradational unit comparable with the Mitcheldean Member. During a phase early in the formation of a transgressive systems tract, the rate of accommodation space creation appears to have been relatively high, with the result that the barriers migrated over their own back-barrier deposits, preserving them largely complete. The major palaeosol at the base is clear evidence of a prolonged lowstand phase with subaerial exposure. Barrier systems tend to preserve their back-barrier deposits if the rate of sea-level rise or sediment supply to the shoreface is high. If the rate of rise is low, or if the sediment supply is slow, the back-barrier deposits may be completely removed by shoreface erosion. The shoreface erosion event associated with the basal High Tor Limestone could have removed a palaeosol capping the Caswell Bay Mudstone, or any positive chronosequence. However, the complete absence of such features over the outcrop, and the fact that they are well preserved in the Cheltenham Limestone, argues for their original absence. The relatively thick intertidal laminites in the Caswell Bay Mudstone (see Riding & Wright 1981) also favours the presence of a high rate of accommodation space creation to allow the aggradation (rather than progradation) of the tidal flats.

The Flat Holm Member peritidal units reflect short periods when sediment supply allowed the barriers to prograde seawards. The back-barrier deposits do not contain prominent supratidal deposits or palaeosols, indicating relatively high rates of accommodation space creation.

The Cheltenham Limestone may represent a late transgressive to highstand systems tract, when the more landward areas of the ramp became flooded. By this stage the rate of accommodation space creation was reduced so that there was a stronger progradational element to the sedimentation, with the appearance of more supratidal and subaerial lithofacies. As the rate of space creation decreased, a positive chronosequence developed, culminating in the formation of the Darrenfelen Geosol and the more developed Cwm Dyar Geosol. However, the underlying Oolite Group was affected by local uplift related to the Neath Disturbance Fault Zone (Wright 1981). It is likely that during deposition of the Cheltenham Limestone, subsidence was low or uplift occurred so that the exact relationships to the other two Arundian successions in relation to the Arundian transgression is unclear. The pattern within the Cheltenham Limestone corresponds fairly well to that expected of a highstand systems tract, but the decrease in rate of accommodation space creation might reflect only local conditions of reduced subsidence or minor uplift.

The Mitcheldean Member is tentatively added to the model in Fig. 9. The occurrence of a minor palaeosol beneath the unit might reflect a minor sea level fall triggering progradation. Alternatively, the palaeosol might simply cap the backshore deposits of the Stowe Oolite and not actually reflect a fall in sea level.

In summary, only one of the units examined shows, at least in its upper part, some characteristics of a late highstand systems tract peritidal unit formed during a third-order sea level oscillation. In this respect they differ from the highstand systems tract peritidal successions described from Wyoming and Montana by Elrick & Read (1991), but it is only by establishing the nature and positions of palaeosols that the tract positions of the various types of peritidal units can be fully interpreted.

The peritidal successions can also be used to shed light on the causes of minor cyclicity. The causes of such cyclicity in peritidal successions has been the subject of much debate (see review by Drummond & Wilkinson 1993; Wright 1994). Autocyclic models invoke the progradation of tidal flats and the subsequent restriction in the supply of sediment from the reduced subtidal area as being the major cause of shallowing upwards cyclothems. Stacking of such cyclothems must reflect a net rise in sea

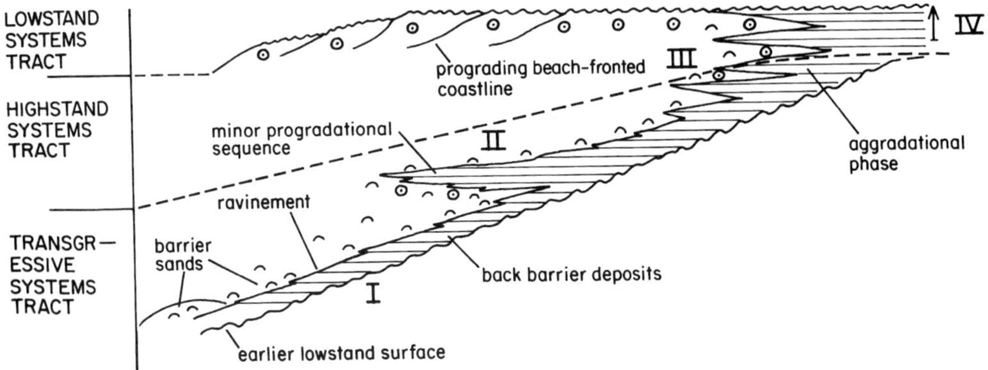

Fig. 9. Hypothetical stratigraphies of barrier and back-barrier sequences developed in different systems tracts. During a transgressive phase back-barrier deposits form behind a migrating barrier. These overlie a lowstand surface and are capped by a shoreface erosion surface (ravinement). The relatively high rate of accommodation space creation may be such that the back-barrier peritidal sequence contains few, if any, well-developed minor exposure surfaces caused by autocyclic or allocyclic processes. This type of sequence (I) corresponds to the Caswell Bay Mudstone of South Wales (Fig. 6). If the rate of accommodation space decreases or if sediment production or supply increases, the barrier may prograde, behind which a back-barrier sequence might develop. Such a sequence will lack a prominent exposure surface at its top and will be overlain by lower shoreface deposits as the transgression continues. This scenario (II) bears some similarities to the Stowe Oolite–Mitcheldean Member sequence (Fig. 5). Of course, this could represent a minor highstand systems tract of a higher order sequence. As the rate of creation of accommodation space decreases later, during the early highstand phase, the barrier system begins to aggrade with interfingering of barrier and back-barrier deposits. This corresponds to the likely setting of the Flat Holm Member (III) (Fig. 7). During the late highstand the behaviour of the shoal belt switches from aggradational to progradational. The back-barrier deposits under such conditions of low accommodation-space creation may be more sensitive to minor sea-level fluctuations or to autocyclic processes. Under such conditions positive chronosequences may be present in the back-barrier units. This corresponds to the upper part of the Cheltenham Limestone Member (IV) (Fig. 8). Eventually, as no new space is created the back-barrier system cannot be maintained and the coastline becomes a beach-fronted strandplain. This is an idealized model and does not explain all the intricacies of the sequences in southwest Britain. Local tectonics have affected the successions in areas such as those where the Llanelly Formation was deposited (Wright 1981).

level rather than simply slow subsidence, which generally occurs at rates too low to generate cyclothems of the types commonly seen (Drummond & Wilkinson 1993). Such autocyclothems do not have prominent exposure surfaces capping the units, and would probably not be well ordered or laterally correlatable as they would have been controlled by local conditions of sediment supply and erosion.

In contrast, allocyclic controls create cyclothems with evidence of subaerial exposure during the sea level fall part of each sea-level oscillation. Not only do palaeosols develop, but overprinting of subtidal facies by such effects can take place (Strasser 1991). The cycles should be more ordered, with both rise and fall components, not simply affected by local variations in sediment supply. Erosional tops are also more common in this type of cyclothem (Strasser 1991). Such metre-scale fifth-order cycles (in the sense of Tucker 1993), are commonly interpreted as resulting from the higher frequency Milankovitch rhythms, such as the precession or obliquity rhythms (e.g. Strasser 1994). If so, the length of exposure represented by most cyclothem tops should be at least half the cycle time (approximately ten or twenty thousand years) (Strasser 1994; Wright 1994). With these long periods, prominent palaeosols should result, unless bioturbation or erosion after or during flooding obliterates the palaeosol.

Such palaeosols are not common in peritidal successions (Wright 1994) and are largely absent from the units described from the Early Carboniferous. The exceptions are those near the top of the Cheltenham Limestone. It is possible that in the Mitcheldean and Flat Holm members evidence of minor, higher frequency sea-level oscillations are absent because the subtidal water depths were too great or because the rates of accommodation space creation, due to third-order effects, were high enough to reduce the effects of *minor* sea-level falls. This might more reasonably explain the absence of exposure surfaces in the Caswell Bay Mudstone. If high frequency falls were small, perhaps only a metre or two, they might not have resulted in significant exposure when the rate of net sea-level rise was high, but only when it was low such as in the late highstand systems tract succession of the upper Cheltenham Limestone.

Elrick & Read (1991) have noted evidence of high frequency (fifth-order type) sea-level oscillations, estimated as having had at least 20–25 m of amplitude, from the early Carboniferous of Montana and Wyoming. If such moderate amplitude and high frequency oscillations had been operating they would have left some evidence in the inner ramp successions, or indeed in the record of offshore deposits. Faulkner (1989), in a detailed study of the offshore ramp successions in southwest Britain, of comparable age to those noted by Elrick & Read (1991), also failed to detect high frequency cyclicity. High frequency sea-level oscillations did develop in the late Viséan, in the Asbian and Brigantian, but probably represent eccentricity rhythm effects with amplitudes of up to 30 m (Horbury 1989). There is no evidence from the inner ramp successions described in this study of moderate amplitude, high frequency sea-level oscillations. If a fifth-order sea-level oscillation were active at all, the evidence favours only a low amplitude.

If a high frequency oscillation did not operate, the absence of both minor cyclothems and positive chronosequences in some of the peritidal units could simply be explained as reflecting no minor sea-level falls. However, the presence of major palaeosols beneath the Caswell Bay Mudstone, the general facies trends of the Mitcheldean Member and Flat Holm members, and the complete absence of lowstand exposure features capping all these units, argues for them having formed during transgressive rather than highstand systems tracts.

The 'picture' is clearly complex and the characteristics and associations of peritidal units need to be assessed very carefully before they can be confidently assigned to a particular systems tract.

I thank J. Banham and A. Cross for help in preparing the text and illustrations. The thorough reviews by A. Horbury and A. Strasser were much appreciated. T. Burchette is thanked for advice on sequence stratigraphical concepts. The generous support of Exxon, Houston for my studies of Mississippian sand bodies and related units is gratefully acknowledged.

References

BIRKELAND, P. W. 1984. *Soils and Geomorphology*. Oxford University Press, Oxford.

BRIDGES, P. H. 1982. The origin of cyclothems in the late Dinantian platform carbonates at Crich, Derbyshire. *Proceedings of the Yorkshire Geological Society*, **44**, 159–180.

BURCHETTE, T. P. 1987. Carbonate-barrier shorelines during the basal Carboniferous transgression: the Lower Limestone Shale Group, South Wales and western England. *In*: MILLER, J., ADAMS, A. E. & WRIGHT, V. P. (eds) *European Dinantian Environments*. Wiley, Chichester, 239–263.

——, WRIGHT, V. P. & FAULKNER, T. J. 1990. Oolitic sand-body depositional models and geometries, Mississippian of southwest Britain: implications for petroleum exploration in carbonate ramp settings. *Sedimentary Geology*, **68**, 87–115.

DRUMMOND, C. N. & WILKINSON, B. H. 1993. Carbonate cycle stacking patterns and hierarchies of orbitally forced eustatic sealevel change. *Journal of Sedimentary Petrology*, **63**, 369–377.

ELRICK, M. & READ, J. F. 1991. Cyclic ramp-to-basin carbonate deposits, Lower Mississippian, Wyoming and Montana: a combined field and computer modeling study. *Journal of Sedimentary Petrology*, **61**, 1194–1224.

FAULKNER, T. J. 1989. *Carbonate facies on a Lower Carboniferous storm-influenced ramp in SW Britain*. PhD Thesis, University of Bristol.

——, WRIGHT, V. P., PEETERS, C. & GARVIE, L. A. J. 1990. Cryptic exposure horizons in the Carboniferous Limestone of Portishead, Avon. *Geological Journal*, **25**, 1–17.

GOLDHAMMER, R. K., DUNN, P. A. & HARDIE, L. A. 1987. High frequency glacio-eustatic sealevel oscillations with Milankovitch characteristics recorded in Middle Triassic platform carbonates in northern Italy. *American Journal of Science*, **187**, 853–892.

——, OSWALD, E. J. & DUNN, P. A. 1991. Hierarchy of stratigraphic forcing: example from Middle Pennsylvanian shelf carbonates of the Paradox Basin. *In*: FRANSEEN, E. K., WATNEY, W. L., KENDALL, C. G. ST. C. & ROSS, W. C. (eds) *Sedimentary Modeling: computer simulations and methods for improved parameter definition. Kansas Geological Survey Bulletin*, **233**, 362–413.

HORBURY, A. D. 1989. The Relative Roles of Tectonism and Eustasy in the Deposition of the Urswick Limestone in South Cumbria and North Lancashire. *Yorkshire Geological Society, Occasional Publications*, **6**, 153–169.

MARTIN-CHIVELET, J. & GIMENEZ, R. 1992. Palaeosols in microtidal carbonate sequences, Sierra de Utiel Formation, Upper Cretaceous, SE Spain. *Sedimentary Geology*, **81**, 125–145.

RAMSAY, A. T. S. 1987. Depositional environments of Dinantian limestones in Gower, South Wales. *In*: MILLER, J., ADAMS, A. E. & WRIGHT, V. P. (eds) *European Dinantian Environments*. Wiley, Chichester, 265–308.

RAMSBOTTOM, W. H. C. 1973. Transgressions and regressions in the Dinantian: a new synthesis of British Dinantian stratigraphy. *Proceedings of the Yorkshire Geological Society*, **39**, 567, 601.

READ, J. F. & HORBURY, A. D. 1993. Eustatic and tectonic controls on porosity evolution beneath sequence-bounding unconformities and parasequence disconformities on carbonate platforms. *In*: HORBURY, A. D. & ROBINSON, A. G. (eds) *Diagenesis and Basin Development*. American Association of Petroleum Geologists, Studies in Geology, **36**, 155–198.

RIDING, R. & WRIGHT, V. P. 1981. Palaeosols and tidal-flat/lagoon sequences on a Carboniferous carbonate shelf. *Journal of Sedimentary Petrology*, **51**, 1323–1339.

ROSS, C. A. & ROSS, J. R. P. 1988. *Late Paleozoic Transgressive – Regressive Deposition*. Society of

Economic Paleontologists and Mineralogists, Special Publications, **42**, 227–247.

STRASSER, A. 1991. Lagoonal–peritidal sequences in carbonate environments: autocyclic and allocyclic processes. *In*: EINSELE, G., RICKEN, W. & SEILACHER, A. (eds) *Cyclic and Event Stratification*. Springer Verlag, Berlin, 709–721.

—— 1994. Milankovitch Cyclicity and High-resolution Sequence Stratigraphy in Lagoonal–Peritidal Carbonates (Upper Tithonian – Lower Berriasian, French Jura Mountains). *In*: DE BOER, P. L. & SMITH, D. G. (eds) *Orbital Forcing and Cyclic Sequences*. Special Publications International Association of Sedimentologists, **19**, 285–301.

TUCKER, M. E. 1993. Carbonate diagenesis and sequence stratigraphy. *Sedimentology Review*, **1**, Blackwell Science, Oxford, 51–72.

VANSTONE, S. D. 1991. Early Carboniferous (Mississippian) paleosols from south west Britain: influence of climatic change on soil development. *Journal of Sedimentary Petrology*, **61**, 445–457.

—— 1993. *Early Carboniferous (Mississippian) palaeosols and their palaeoclimatic significance*. PhD Thesis, University of Reading.

WALKDEN, G. M. 1974. Palaeokarstic surfaces in Upper Viséan (Carboniferous), limestones of the Derbyshire Block, England. *Journal of Sedimentary Petrology*, **44**, 1232–1247.

WEEDON, M. J. 1987. *The analysis of a peritidal lithosome: the Lower Carboniferous Clifton Down Mudstone and correlative Chadian/Arundian Units*. MSc Thesis, University of Bristol.

WHITTAKER, A. & GREEN, G. W. 1983. *The Geology of the Country Around Weston-Super-Mare*. Geological Survey of Great Britain, Memoirs, Sheet 279 and Parts of 263 and 295.

WILSON, D., DAVIES, J. R., SMITH, M. & WATERS, R. A. 1988. Structural controls on Upper Palaeozoic sedimentation in south-east Wales. *Journal of Geological Society, London*, **145**, 901–914.

WRIGHT, V. P. 1981. *The stratigraphy and sedimentology of the Llanelly Formation between Pendergn and Blorenge, South Wales*. PhD Thesis, University of Wales.

—— 1982. Calcrete palaeosols from the Lower Carboniferous Llanelly Formation, South Wales. *Sedimentary Geology*, **33**, 1–33.

—— 1983. A rendzina from the Lower Carboniferous of South Wales. *Sedimentology*, **30**, 159–179.

—— 1986. Facies sequences of a carbonate ramp: the Carboniferous Limestone of South Wales. *Sedimentology*, **33**, 221–241.

—— 1987. The ecology of two early Carboniferous paleosols. *In*: MILLER, J., ADAMS, A. E. & WRIGHT, V. P. (eds) *European Dinantian Environments*. Wiley, Chichester, 345–358.

—— 1994. Paleosols in shallow marine carbonate sequences. *Earth Science Reviews*, **35**, 367–395.

Sequence-stratigraphical interpretation of organic facies variations in marine siliciclastic systems: general principles and application to the onshore Kimmeridge Clay Formation, UK

RICHARD V. TYSON

Newcastle Research Group, Fossil Fuels and Environmental Geochemistry (Postgraduate Institute), Drummond Building, University of Newcastle, Newcastle upon Tyne NE1 7RU, UK

Abstract: The occurrence and stratigraphical distribution of source rock potential in distal fine-grained facies is strongly related to the prevailing palaeo-oxygenation regime, especially during the deposition of the condensed section (CS). The CS will tend to correspond to organic facies B-AB, with good to excellent source-rock potential, only when conditions are dysoxic–anoxic, and organic facies D-CD, with poor source rock potential, when conditions are oxic. The applicability of the Creaney and Passey organic facies model is limited by its assumption that palaeo-oxygenation remains constant within the basinal facies. The CS commonly shows decimetre (Milankovitch)-scale redox cyclicity characterized by total organic carbon (TOC) and organic facies variations whose amplitude increases as sediment accumulation rates decrease. The greater frequency of dysoxic–anoxic conditions in the CS may be related to changes in accommodation space via the relationship between bottom water volume and the rate of deoxygenation, as observed in modern settings.

The onshore Kimmeridge Clay diverges from the classic Exxon basin model as it was deposited in an intra-shelf setting where focusing and ponding of clayey sediment resulted in the basinal sections being both thicker and more complete. However, the best source rock intervals still coincide with the maximum flooding surfaces. The long-term Late Jurassic (shallow to deep to shallow) sea level curve appears to influence the mean (oxic to dysoxic–anoxic to oxic) palaeo-oxygenation regime in the basin and the background trend in sediment accumulation rate; it is expressed in a symmetrical increase in TOC to 4–6 wt% from background values of about 1 wt%. The superimposed decimetre- to metre-scale redox cycles appear to be controlled by Milankovitch climatic controls on water mass stratification. The relative spacing of these cyclic organic-rich interbeds decreases during relatively condensed (maximum flooding surface) intervals associated with the short-term sea-level curve, and the interval TOC values also increase and broaden. Intervals deposited during or shortly after lowstands are less organic-rich and have fewer and more widely spaced organic-rich beds. The individual organic-rich beds also appear to be condensed relative to adjacent mudstones; palynofacies and TOC versus hydrogen index data suggest the origin of these beds is due more to cyclically decreased dilution and redox conditions rather than enhanced palaeoproductivity.

Although sequence stratigraphy is a well established and almost universally adopted approach to basin analysis, relatively few publications have so far discussed the inter-relationship between sequence stratigraphy and either the distribution of potential petroleum source rocks or variations in organic facies or palynofacies characteristics (but see Leckie *et al.* 1990, 1992; Pasley *et al.* 1991, 1993; Curiale *et al.* 1992; Creaney & Passey 1993; Katz & Pratt 1993; Tyson 1995; Frank & Tyson, 1995). This review will attempt to provide an overview of the sequence stratigraphical expression of organic facies variations in marine siliciclastic systems, and to highlight some of the interpretive principles which may allow improved prediction of petroleum source rocks. These principles are then applied to an example from the British stratigraphical record, the onshore Kimmeridge Clay Formation.

Although several of the general sequence stratigraphical texts by the Exxon group briefly reported an association between the condensed section (CS) and the deposition of organic-rich sediments (Loutit *et al.* 1988: 186; Van Wagoner *et al.* 1990: 45; Vail *et al.* 1991: 653), there was no attempt to provide a rationale for this observation until the paper by Creaney & Passey (1993). The latter paper is based on a highly simplified model which assumes that palaeoproductivity was constant and that the basinal facies were continuously 'anoxic', such that the only other major control on total organic carbon (TOC) (whole rock wt% total organic carbon)

From HESSELBO, S. P. & PARKINSON, D. N. (eds), 1996, *Sequence Stratigraphy in British Geology*, Geological Society Special Publication No. 103, pp. 75–96.

ORGANIC FACIES:	A	AB	B	BC	C	CD	D
% 'AOM' of kerogen	dominant			mod	usually low/absent		
'AOM' matrix fluoresence	highest		mod-weak		weak	usually absent	
% prasinophytes of plankton	highest	mod	rare	usually very rare			
% phytoclasts of kerogen	low dilution			mod	usually dominant		
opaque: translucent phytoclasts	often high			usually low		increases	
Geochemical characteristics: (for immature sediments)							
Hydrogen Index	≥850	≥650	≥400	≥250	≥125	50-125	≤50
Kerogen Type	I	I/II	II	II/III	III	III/IV	IV
TOC%	5-20+	3-10+		3-3+	≤3	<0.5	
Environmental factors:							
Proximal-distal trend	distal			proximal			distal
Oxygen regime	anoxic	anoxic-dysoxic			oxic		v.oxic
Sediment accumulation rate	low	varies		high		mod	low
ORGANIC FACIES:	A	AB	B	BC	C	CD	D

Fig. 1. Summary of organic facies scheme of Jones (1987), showing some of the major kerogen and environmental trends (after Tyson 1995).

variations is, inevitably, dilution by siliciclastic sediment. As the extent of dilution is a function of sediment input, TOC variations are then simply related to progradational cycles, with the highest TOC (least dilution) occurring within the CS, especially at the maximum flooding surface (MFS) itself. Also, as progradation in siliciclastic systems is typically associated with a greater input of gas-prone (type III) or inert (type IV) terrestrial phytoclasts, it also results in reduced oil potential as well as lower TOC, such that the abundance and quality of the organic matter tend to be auto-correlated. Consequently, the relative influence of progradation within an anoxic basin provides a reasonable indication of probable variation in source rock quality (TOC and mean oil potential).

The general applicability of the Creaney & Passey (1993) model is limited because basinal palaeo-oxygenation regimes are often very variable and prolonged continuous anoxia is rare. The mean oil potential of the kerogen thus often reflects redox-dependent variations in the preservation of the marine organic matter, not just progradation-related dilution of marine (oil-prone) organic matter by terrestrial (gas-prone or inert) components. Other potential limitations of the predictive capability of the Creaney & Passey (1993) model stem from pragmatic generalizations in the sequence stratigraphical model itself (cf. Thorne 1992), in particular the assumption that basins are filled by progradation, with sigmoidal to oblique clinoform sediment bodies building outward from a single, well-defined sediment source area and accumulating in a single depocentre, with a simple proximal–distal relationship between the two. The latter model predicts that the basin centre is the site of condensed sedimentation relative to the margin, whereas sediment bypassing and focusing often result in intra-shelf basins being the areas of greatest sediment accumulation (cf. Wignall 1991), if not necessarily the areas of most rapid deposition.

Organic facies and palynofacies

The author's preferred definition of an organic facies is: 'a body of sediment containing a distinctive assemblage of organic constituents which can either be recognized by microscopy, or is associated with a characteristic bulk organic geochemical composition' (Tyson 1995; cf. Rogers 1980; Jones 1984, 1987). The key issues are the source of the organic matter, its preservation state and its relative abundance (TOC). In practice, organic facies are defined primarily by elemental analysis, and to a lesser extent by Rock–Eval pyrolysis, TOC and microscopy. Palynofacies represents that aspect of organic facies which may be determined by palynological study of the organic matter.

Jones's (1987) organic facies classification scheme (see Fig. 1) recognizes four major organic facies (A–D) and three transitional organic facies (AB, BC and CD). The four main organic facies

Table 1. *Depositional variables that change through marine depositional sequences and their effect on organic facies and palynofacies variables*

Depositional variables that change through marine sedimentary sequences	Associated organic facies and palynofacies variables
Proximity to sediment input: distance (−) and duration of transport (−)	Absolute (+) and relative (+) phytoclast content Phytoclast size (+), sorting (−) opaque : translucent ratio (−), and preservation (+/−) Absolute sporomorph content (+) Sporomorph : plankton ratio (+) Sporomorph size (+), sorting (−), and preservation (+) Absolute (+) and relative (+) freshwater algal content
Sediment accumulation rate (+)	Dilution of TOC (+) Preservation of marine carbon, burial efficiency (+)
Water depth, accommodation space (−)	Phytoplankton carbon flux (+) Meroplankton : holoplankton ratio (+) Likelyhood of watermass stratification (−) Relative bottom water volume (−) Disposition to dysoxia–anoxia and to Milankovitch redox cycles (−)
Modal grain size (+) and % coarse fraction (+)	TOC content (−) Relative phytoclast content (+) Phytoclast size (+) and sorting (−)
Erosion (+)	Absolute (+) and relative (?) content of recycled kerogen Absolute (+) and relative (?) content of reworked palynomorphs
Downslope redeposition and translocation	Meroplankton : holoplankton ratio (+) Other proximal indicators (+)

The '+' symbol indicates a relative proximal increase; the '−' symbol indicates a relative distal increase (see also Tyson 1993, 1995).

(A–D) have bulk organic geochemical characteristics that are broadly equivalent to the four main kerogen types (I–IV) defined by Tissot *et al.* (1974, 1980). However, the organic facies approach emphasizes the fact that we are dealing with a continuum of kerogen assemblages rather than discrete types, whose character is controlled by preservation as well as by organic matter source(s), and thus on a diversity of environmental and sedimentological factors (Jones & Demaison 1982).

As has already been shown directly or indirectly by Jones (1987), the major controls on organic facies characteristics are: (1) palaeo-oxygenation and redox conditions in the depositional basin; (2) the relative proximity of the depositional site to active siliciclastic sediment sources; and (3) climatic controls on the production, preservation and export of terrestrial organic matter. It is the second of these major controls which provides the most direct link with sequence stratigraphy. Jones's (1987) organic facies classification emphasizes the distinction between dysoxic–anoxic facies (organic facies BC, B, AB and A), oxic proximal deltaic to shelf facies (organic facies C and CD) and highly oxic, distal and condensed facies (organic facies CD to D). Within the dysoxic–anoxic facies, organic facies BC represents sediments with a relatively high phytoclast input, whereas organic facies A, AB and B occur where there is a relatively low phytoclast input due to climate, magnitude of runoff or distance from fluvio-deltaic source areas. Most marine siliciclastic dysoxic–anoxic 'black shale' source rocks correspond to organic facies B, although occasionally AB if they are enriched in prasinophyte algae. Oxic shelf facies whose kerogen assemblage is dominated by partially degraded marine amorphous organic matter (AOM) may correspond to organic facies C, and thus the latter facies do not necessarily equate with sediments enriched in terrestrial woody (vitrinitic) organic matter (Tyson 1995).

Systems tracts and proximal–distal facies shifts

As sequence stratigraphy deals with familiar geological issues such as relative sea level and transgressive–regressive facies changes, we can predict which variables are likely to show significant

	SYSTEM TRACTS				
	Early LST	Late LST	TST	Early HST	Late HST
General trend:	V. Proximal →	Proximal →	Distal	→ V. Distal	→ Proximal
Phytoclasts:					
% of kerogen	Highest	High	Decreases (Minimal in CS)	Low	High
% opaque	Lowest	Low	Increases Maximal in CS	High	Low
equant: lath opaques	Highest	High	Decreases (Minimal in CS)		High
'AOM':					
% of kerogen Oxic CS Anoxic CS	Low	Low	Minimal (Increases ± peak at MFS)		Low
preservation Oxic CS Anoxic CS	Poor	Poor	Poorest Good to excellent		Poor
Organic facies: Oxic CS Anoxic CS	C C	C C	CD → D CB → B[→ AB]	D → CD B → CB	C C
Reworked palynomorphs:	Highest	High	Minimal in CS		Increases
Freshwater algae:	Common	Common	Rare		Common
***In situ* plankton:**					
% of palynomorphs	Lowest	Low	Maximum in CS		Decreases
dino. diversity Oxic CS Anoxic CS	Low	Low	Peak in TST/CS May be low or high		Decreases
% prasinophytes Oxic CS Anoxic CS	Low	Low	Low Maximum in CS		Low

Fig. 2. Summary of idealized organic facies and palynofacies trends through a depositional sequence (after Tyson 1995).

variation through depositional sequences and the general response of those organic facies and palynofacies parameters likely to be affected (Table 1). Many of these organic facies and palynofacies parameters are already focused on proximal–distal variations and the relative degree of terrestrial influence exhibited by total kerogen and palynomorph assemblages (Tyson 1993, 1995). The progression from lowstand (LST), to transgressive (TST) and early and late highstand (eHST and lHST) systems tracts ideally records a relative proximal to distal and back to proximal shift in facies with respect to the siliciclastic sediment source (Fig. 2). For any given depositional site, the most distal organic facies characteristics should correspond to the MFS (boundary between the TST and HST), and the most proximal organic facies characteristics to the LST or its equivalents.

The typical proximal–distal trends shown in Fig. 2 may be modified by tectonic and climatic factors (Table 2). Arid climates will greatly reduce the production and delivery of contemporary terrestrial phytoclasts and result in a relative increase in the proportion of both oxidized refractory and older recycled material. In general, organic facies deposited under arid climatic regimes will appear more distal in character relative to those deposited under humid regimes. The nature, scale and detailed timing of the stratigraphical changes in organic facies characteristics are also strongly determined by the relative proximal–distal position of the locality within the depositional basin (Table 3).

The effects of progradation in siliciclastic systems are primarily related to changes in the sediment accumulation rate (the supply of allochthonous particles: terrigenous sediment and terrestrial organic matter). According to the published work, an increase in sediment accumulation rate can have either a positive effect on TOC values (via increased preservation) or a negative effect (via increased dilution). Increased sediment accumulation rates in modern oxic settings certainly result in greater preservation of the marine organic matter reaching the seafloor – that is, a higher carbon

Table 2. *Important modifiers of organic facies characteristics*

Modifier	Response
Climate	Nature of parent flora:
	Woody versus herbaceous [herbaceous = swamp/marsh (+) or arid grassland (−)]
	C3(+) versus C4 (−) floras (post-Oligocene)
	Absolute production of contemporary phytoclasts (+)
	Preservation of phytoclasts (+)
	Opaque : translucent phytoclast ratio at sediment source (−)
	Absolute production of contemporary sporomorphs (+)
	Spore : pollen ratio of palynoflora (+)
	Magnitude (+) and seasonality of runoff
	Relative percentage of aeolian organic matter flux (−)
	Continental weathering regime and thus sediment supply (+)
	Occurrence (+), origin and stability (+) of watermass stratification
	Frequency and intensity of storm redeposition
Tectonics	Bottom topography (basin morphometry)
	Bottomwater volume (via degree of lateral confinement)
	Sediment bypassing (of shallow shelf, slope or highs)
	Sediment focusing and ponding (intra-shelf/intra-basinal lows)
	Rate of progradation and coastline migration
	Water depth (local relative sea level and accommodation space)
	Number, distribution and magnitude of sediment sources (local base level alteration)
	Sediment source terrane (provenance)
	Abundance, age and nature of recycled kerogen and palynomorphs
	Yield and character of siliciclastic sediment in catchment

Under 'climate' '+' symbol indicates a relative increase under more pluvial conditions and the '−' symbol indicates a relative increase under more arid conditions.

burial efficiency (Henrichs & Reeburgh 1987; Canfield 1989, 1994; Betts & Holland 1991), at least up to sediment accumulation rates of around 60–100 cm/ka (Tyson 1995). However, most of the evidence for a positive effect on TOC values (e.g. Müller & Suess 1979; Ibach 1982; Stein 1986) comes from pelagic or hemipelagic facies where the fluxes of organic matter and of the largely biogenic sediment are auto-correlated. A TOC dilution effect seems to be much more typical of siliciclastic systems, resulting in deltaic sediments rarely having TOC values greater than 1–2 wt% (e.g. Demaison & Moore 1980; Berner 1982). This dilution relationship is seen in the seston : particulate-organic-carbon-ratio of river waters; where seston (suspended matter) concentrations exceed about 50 mg/l (as during high discharge or in major river systems), the particulate organic carbon concentration seldom exceeds 2 wt% (Tyson 1995). Consequently, TOC values often show an inverse correlation with both siliciclastic sediment accumulation rate and progradation rate (e.g. Bustin 1988). The absolute carbon burial (organic carbon accumulation rate – mass per unit area per unit time) will remain high and is usually positively correlated with sediment accumulation rate and progradation. However, because of its relationship with hydrocarbon expulsion efficiency, the TOC is more important in a petroleum source rock context.

As high sediment accumulation rates in siliciclastic facies generally reflect proximity to terrestrial sediment sources, not only the absolute mass, but also the percentage of terrestrial organic matter (of the total organic matter) typically increase with sediment accumulation rate (e.g. Kontovorich *et al.* 1971; Hedges & Mann 1979; Burtner & Wartner 1984; Gough *et al.* 1993). Although the subordinate planktonic TOC may be better preserved at higher sediment accumulation rates, the greater admixture of hydrogen-poor terrestrial material means that sediment hydrogen indices (HI, kerogen hydrocarbon pyrolysis yield normalized to sample TOC, mg/g TOC) are unlikely to increase (Marzi & Rullkötter 1986). Phytoclast particles are also hydrodynamically equivalent to coarse silt and fine sand siliciclastic particles (Tyson 1995) and thus their relative content will usually increase if the grain size increases (through progradation, shallowing or redeposition).

The simultaneous dilution of the total organic matter by sediment and of the marine organic matter by terrestrial phytoclasts means that hydrogen indices tend to fall as the TOC decreases. However, this relationship can also occur in the complete absence of terrestrial organic matter and be solely a function of poorer preservation of the marine organic matter in response to increasingly more oxidizing conditions or increased exposure to oxic conditions (e.g. Dean *et al.* 1986). The relative importance of these processes are best distinguished by using palynofacies or organic petrological data.

Importance of palaeo-oxygenation regime

The effects of sediment accumulation rate on organic matter preservation and TOC are strongly dependent on the prevailing palaeo-oxygenation

Table 3. *Varied expression of some palaeoenvironmental factors and sequence stratigraphical characteristics depending on location within depositional basin*

	Proximal margin	Outer shelf/slope	Distal deep basin
Relative change of *in situ* accommodation space	Large	Modest	Negligible
Relative change in sediment accumulation rate and modal grain size (siliciclastic sediment)	Very large	Modest, except where redeposition (e.g. in fan systems)	Negligible (CS)
Timing of progradational effects	Soonest	Delayed distally	Little or no effect
Continuity of sedimentation	Intermittent	Less intermittent	Quasi-continuous
Stratigraphical completeness	Poor	Good	Excellent
Sediment thickness (starved)	Moderate	High	Minimal
Sediment thickness (ponded)	Minimal	Moderate	Maximal
Parasequence definition	Good	Poor	None
Conspicuous scale of variation	Parasequences Systems tracts Sequences	Parasequences? Systems tracts Sequences	Systems tracts? Sequences
Systems tract preservation	LST may be reduced by bypassing, HST reduced by erosion	LST may be reduced by bypassing on slope	All represented but less distinct
Best correlative features	Sequence boundaries The MFS		The MFS
Relative development of good source rock, if basin remains dysoxic–anoxic	Minimal: MFS only?	Moderate: TST–eHST	Maximal: all; peak at MFS?
Onset and duration of dysoxic–anoxic conditions	Later onset? Minimal duration	Intermediate duration	Earlier onset (if significantly deeper) Maximum duration
TOC cycles, if basin remains dysoxic–anoxic	Sharp base, truncated top?	Asymmetrical HTB	Symmetrical about the MFS
Amplitude of TOC redox cycles (starved basin)	Moderate?	Moderate	Very high

The trends shown are highly generalized to illustrate the potential importance of location and will not apply in all instances (e.g. depending on tectonics and the actual bathymetric profile).

regime. For example, although progradation in a dysoxic–anoxic system may result in lower TOC and HI values, progradation in an oxic system may result in some increase in both TOC and HI values. In anoxic systems the relative preservation of the potentially fossilizing marine organic matter should already approximate to its maximum value (relative to the mean water depth and position of the redoxcline). A progradation-related increase in sediment accumulation rate is thus unlikely to increase the preservation of marine organic matter to any significant additional extent, but it is certain to dilute the TOC and is often associated with the introduction of terrestrial organic matter of a lower HI value, resulting in a downgrading of the organic facies (from AB or B to B or BC). This is confirmed by empirical evidence: most ancient black shales appear to have been deposited relatively slowly and many show their highest source rock quality in the most condensed intervals, accumulated at rates of 0.1–1.0 cm ka^{-1} and seldom at more than 10.0 cm/ka (Tyson 1995). Holocene dysoxic–anoxic facies also show either a largely independent or negative relationship between TOC and sediment accumulation rate (Demaison & Moore 1980; Summerhayes 1983; Stein 1986; Huc 1988).

By contrast, the preservation of marine organic matter is poor or minimal in condensed oxic facies and the kerogen assemblages are typically enriched in small refractory particles of black wood (inertinitic) phytoclasts, corresponding to organic facies D or CD. Progradation and an increase in sediment accumulation rate will both increase the preservation (burial efficiency) of marine organic matter and result in the introduction of fresher and less-oxidized (brown wood or vitrinitic) phytoclasts, together producing a change to organic facies CD or perhaps C. Because of the very low TOC values that characterize organic facies D, the TOC may increase significantly during this transition, but it is unlikely to exceed 3 wt% in marine sediments.

Several studies have provided data supporting the

Table 4. *Theoretical TOC values resulting from 1–8% preservation of palaeoproductivity, a siliciclastic flux of 10g m^{-2} a^{-1} and a range of palaeoproductivity values*

	Primary productivity (PP, g C m^{-2} a^{-1})				
	Ocean	Shelf		Upwelling	
Preservation factor	(50)	(100)	(200)	(400)	(800)
1% of PP	4.6	8.6	15.2	24.4	35.1
3% of PP	12.1	20.3	30.6	41.1	49.6
8% of PP	24.4	35.1	44.9	52.3	56.9

Simplifying assumptions include exclusively organic-walled phytoplankton, no terrestrial carbon input and no positive change in preservation with sediment accumulation rate. The value of 3% preservation is the lower limit for modern laminated dysoxic–anoxic facies and the value of 8% is based on modern outer shelf and upper slope dysoxic–anoxic sites (after Bralower & Thierstein 1987). The siliciclastic flux value is taken as representative of condensed sections; 10 g m^{-2} a^{-1} is the approximate mean for the deep Atlantic for the last 150 Ma (from Ehrmann & Thiede 1985). The calculations also utilize a TOC to organic matter conversion factor of 1.6.

relationship between petroleum source rocks and the CS, although it is clear that the best source rock interval can occur either at, above or below the MFS, and thus in the TST or the early HST (e.g. Leckie *et al.* 1990, 1992; Curiale *et al.* 1992; Rasmussen *et al.* 1992; Creaney & Passey 1993; Katz & Pratt 1993 and references cited therein). Variations in the sequence stratigraphical position of the source rock maximum will at least partly be a function of the location of the depositional site within the basin (Table 3). At relatively proximal sites, where the effects of highstand progradation are likely to occur earlier, the most condensed interval and the source rock maximum are likely to occur either at or just below the MFS (i.e. within the TST); the LST and HST may also, respectively, be attenuated by sediment bypassing and erosion. At more distal or bypassed sites, where highstand progradation is either delayed or has little impact, the source rock maximum may range up into at least the early HST. The timing of the onset of source rock deposition within the TST may be partly a function of the initial water depths at the depositional site (i.e. those occurring at the initiation of the TST), which determine the magnitude of the *in situ* relative change necessary to allow dysoxia–anoxia.

From the preceding discussion it is clear that there should be no general relationship between source rocks and the CS unless the development of anoxia itself is at least partly correlated with relative sea level and thus accommodation space changes (Tyson 1995). Although the combination of low sediment accumulation rates (minimal dilution) and anoxia (maximum preservation) may represent the most favourable case for source rock deposition, the combination of low sediment accumulation rates and oxic conditions represents the worst possible case (prolonged exposure to aerobic degradation). Modern oxic pelagic and hemipelagic facies deposited at sediment accumulation rates of ≤1 cm ka^{-1} (condensed sections *sensu* Loutit *et al.* 1988) generally exhibit the lowest burial efficiencies, have TOC values less than 0.5 wt%, and correspond to organic facies D. It is therefore apparent that the most critical control on petroleum source rock development is the basin redox condition, as this is the main control on the preservation of marine organic matter at low sediment accumulation rates. For example, the study of Curiale *et al.* (1992) shows that the best overall source rock potential is associated with the most consistently dysoxic–anoxic, rather than the most condensed, part of the sequence (cf. Arthur & Sageman 1994, pp. 534–535).

Simple calculations show that at the dilution levels thought to be typical of condensed sections, it only requires the preservation of around 1% of the palaeoproductivity (assuming exclusively organic-walled phytoplankton) to produce TOC values in the 5–15 wt% range, even at low to normal shelf palaeoproductivities (Table 4; Tyson 1995, p.109). By comparison, asymptotic levels of preservation are said to represent at least 3% (≤17%) of the productivity in Quaternary dysoxic–anoxic facies (Bralower & Thierstein 1987; Tyson 1995, p. 140). Such calculations also show that if modern deep-sea CS sediments exhibited near-perfect (80%) preservation of the carbon delivery flux, the resulting TOC values could reach 25–40 wt%, despite great depths (mean 5000 m) and low primary productivity (mean 65 gC m^{-2} a^{-1}), whereas in most other facies this 'maximum potential TOC' is often only around 4–6 wt% due to dilution and autodilution (Tyson 1995). An 80% burial efficiency is, of course, improbable, but even at the 49% mean burial efficiency seen in modern anoxic facies these modern CS sediments could still potentially exhibit TOC values of 15–30 wt%. Contrary to the views of Pedersen & Calvert (1990), this suggests that the TOC of fine-grained distal (CS) sediments is primarily limited by preservation and dilution, rather than by carbon flux and dilution. However, preservation is partly linked to the absolute carbon delivery flux, which always needs to be high enough to at least significantly depress the redox state of the surficial sediment layers, even if it does not result in bottom water dysoxia–anoxia. As shelf CS conditions would be associated with higher mean primary palaeoproductivities (≤ × 3) and much higher relative carbon fluxes ($\leq \times 10$) and also more prone to dysoxia–anoxia than the modern deep sea, there is no need to resort to arguments based on unusually high palaeoproductivity.

Where the CS is not lithologically uniform, but consists of a cyclic decimetre- to metre-scale alternation of dysoxic–anoxic and oxic facies, it will exhibit variations in TOC and organic facies whose amplitude will decrease as the sediment accumulation rate increases and as the magnitude of oxic–anoxic differences in preservation falls. At higher sediment accumulation rates burial efficiency increases in the oxic phase, while the anoxic phase of the cycles is more diluted, together diminishing the oxic–anoxic TOC contrast (Tyson 1995). Although the individual most organic-rich beds may be black shales within the MFS interval, such interbeds are often thin and often only ≤50% of the overall interval, and thus the volume of source rock with these properties may be too small to be commercially significant. Thicker acyclic dysoxic–anoxic shale beds with a lower mean TOC, as may occur above or below the MFS, may then represent the interval with the overall best source rock potential (e.g. Arthur & Sageman 1994, pp. 534–535).

The occurrence of Milankovitch-scale redox cyclicity is actually very common in distal facies of both shelf and ocean basins, e.g. the Blue Lias and onshore Kimmeridge Clay of the British Jurassic (House 1985), the Cenomanian–Turonian Greenhorn and Niobrara Formations of the USA (e.g. Arthur & Sageman 1994) and the mid-Cretaceous Hatteras Formation in the central Atlantic (Waples 1983). Such cyclicity is thus a significant consideration in terms of source rock volumetrics. The occurrence of these cycles is apparently a function of climate, water depth and palaeoproductivity: basins need to be sufficiently deep to allow dysoxic–anoxic conditions to develop, but not so deep or so stable that they are able to remain continuously anoxic. In deep oceanic basins such cyclicity may be largely controlled by climatic regulation of palaeoproductivity, but climatic regulation of watermass stability and relative bottom water volumes will also be important in shallower waters.

Palaeo-oxygenation and sea level

For the CS to be associated with petroleum source rocks requires that it is also associated with dysoxic–anoxic bottom waters. This is clearly not a universal relationship, but why should it so commonly be the case? The condensed section develops within the TST to early HST when water depths are either rapidly increasing or at their peak. This increase in water depth may lead to larger areas of the shelf being below either fair weather or storm wave base. Reduced turbulent mixing may make it possible for watermass stratification to occur where none did before, or for any pre-existing stratification to become more stable. Either of these watermass changes will tend to result in less efficient ventilation of the seafloor on at least a seasonal basis, dysoxia–anoxia and thus a greater potential for the preservation of marine organic matter and the formation of petroleum source rocks. However, although the empirical relationship between the frequency of black shales and periods of marine transgression has been apparent for decades, it is clear that dysoxic–anoxic conditions are not just associated with maximum water depths, as inferred from either the Exxon global sea-level curve, regional or basinal data (e.g. Arthur & Sageman 1994, p.533; Herbin *et al.* 1995).

The occurence of anoxia may be mechanistically related to the magnitude of the changes in accommodation space. As noted by Tyson & Pearson (1991, pp. 13–18), the volume of the shelf bottom water exerts a significant control on rates of seasonal deoxygenation and therefore the occurrence of dysoxia–anoxia via the size of the dissolved oxygen reservoir at the start of the seasonal water mass stratification cycle. There may thus be an indirect correlation between accommodation space and the size of the oxygen reservoir in basinal settings (where depth is the main variable). Seasonal open shelf anoxia is currently most common in water depths of about 15–70 m, where the sub-thermocline volume is comparatively small (≤20 m thick). The increasing rarity of dysoxia–anoxia in modern deeper waters is partly a function of increasing bottom water volume, but also because it reflects the transition to the deeper water of the slope, where the oxygenation of the bottom water is buffered by the open ocean oxygen reservoir. In ancient settings these deeper water areas were often intra-shelf deeps within an epeiric sea, isolated from the open ocean and thus prone to lower oxygenation and even permanent anoxia depending on the stability of the watermass. The modern pattern cannot therefore be extrapolated in a simple fashion. Furthermore, the thickness of the bottom water layer also depends on climate-controlled thermocline thickness and basin morphometry. Strong salinity stratification was probably of limited importance in open shelf settings (Tyson & Pearson 1991).

Although primary productivity is important, whether oxygen depletion develops in shelf waters (under a given mixing regime) is largely a function of the ratio between the oxygen demand (carbon flux) and the size of the oxygen reservoir on which it is acting (the bottom water volume, which in open shelf waters is primarily determined by the

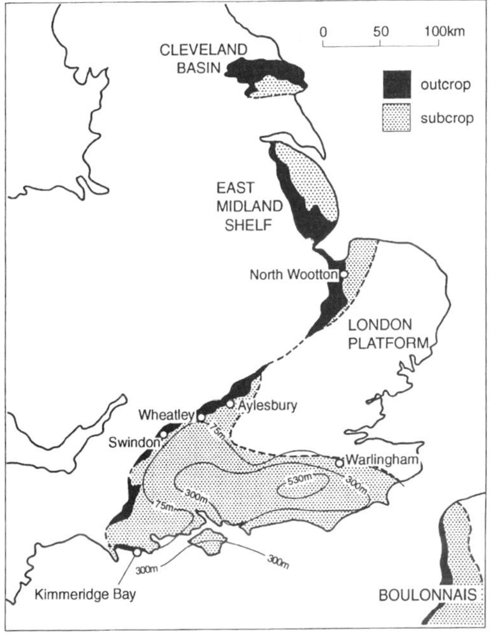

Fig. 3. Location map for onshore Kimmeridge Clay Formation. Total Kimmeridge Clay isopachs in the Wessex Basin after Sellwood *et al.* (1986).

thickness of the sub-pycnocline layer). As the depth of the pycnocline (thermocline) is generally much less variable than the total depth, significant variations in relative bottom water volume are associated primarily with shelf waters; significant bottom water volume variations in deeper waters will relate more to lateral topographic restrictions. The bottom water volume model could partly explain why open shelf dysoxic–anoxic facies can become less common with either shallowing (no stratification possible, or too frequent overturns) or, more significantly, deepening (bottom water volume and dissolved oxygen reservoir too great for recurrent prolonged seasonal anoxia), but are generally more common early in transgressive sequences.

Source rocks and sequence stratigraphy in intra-shelf basins: the onshore Kimmeridge Clay Formation as an example

The onshore Kimmeridgian–Volgian Kimmeridge Clay Formation extends from the Wessex Basin in southern England, through the Eastern England/ Midlands Shelf of Lincolnshire, to the Cleveland Basin of Northern Yorkshire (Fig. 3). In general, it is a rather homogenous dark-coloured mudstone facies with a few apparently isochronous carbonate and organic-rich marker beds that can be correlated over virtually the whole of the onshore area (Cox & Gallois 1981). The predominant mudstone facies are basinal in character and all deposited below normal wave base. Lateral transitions to coarser grained shallow-marine facies are poorly developed or preserved, but they do occur on the northwestern edge of the Wessex Basin and on its eastern edge in the Boulonnais of northeast France; however, there is no indication of major deltaic sediment sources. At its type area in the Wessex Basin the formation is approximately 480 m thick, but in the centre of the subsurface Weald sub-basin of southeast England it reaches a maximum of 556 m (Sellwood *et al.* 1986); in Lincolnshire it is variably truncated, but attains thicknesses of up to 180 m (Penn *et al.* 1986) and in the Cleveland Basin it reaches at least 220 m (Herbin *et al.* 1991, 1993). The thickness of the formation increases towards the centre of the onshore basins and thins towards their flanks, where condensed facies and stratigraphical gaps are common (e.g. Wignall 1991). Locally, at least, this pattern of sediment distribution does not really conform to the progradational basin-filling assumed in the Exxon sequence stratigraphical model (Wignall 1991; cf. Thorne 1992), but rather seems to reflect more aggradational infilling due to sediment focusing and ponding within an intra-shelf depression.

The average sediment accumulation rate of the Kimmeridge Clay type section has been estimated at about 70 m Ma^{-1} (Miller 1990; see also Oschmann 1988, p.242); after corrections for compaction this is probably about ten times greater than that said to typify condensed sections (≤10 m Ma^{-1}; Loutit *et al.* 1988). As the basin centre sedimentation is not strictly condensed according to this widely accepted definition, and there are no obvious progradational wedges, it is interesting to examine the relationship between source rock distribution and the sequence stratigraphical interpretation. A further major complication is that much of the onshore Kimmeridge Clay also exhibits clear decimetre- to metre-scale redox cyclicity and, unlike the coeval graben facies in the North Sea offshore, the organic-rich beds are thus intercalated within oxic to dysoxic mudstone facies.

Stratigraphical distribution of 'oil shale bands'

The frequency of cyclic decimetre-scale organic-rich interbeds is greatest in the middle (late Kimmeridgian to early Volgian, *eudoxus* to *pectinatus* zones) part of the onshore Kimmeridge Clay when water depths were apparently deepest, at the extreme limit of storm influence. This organic-rich part of the sequence is about 200 m thick in the

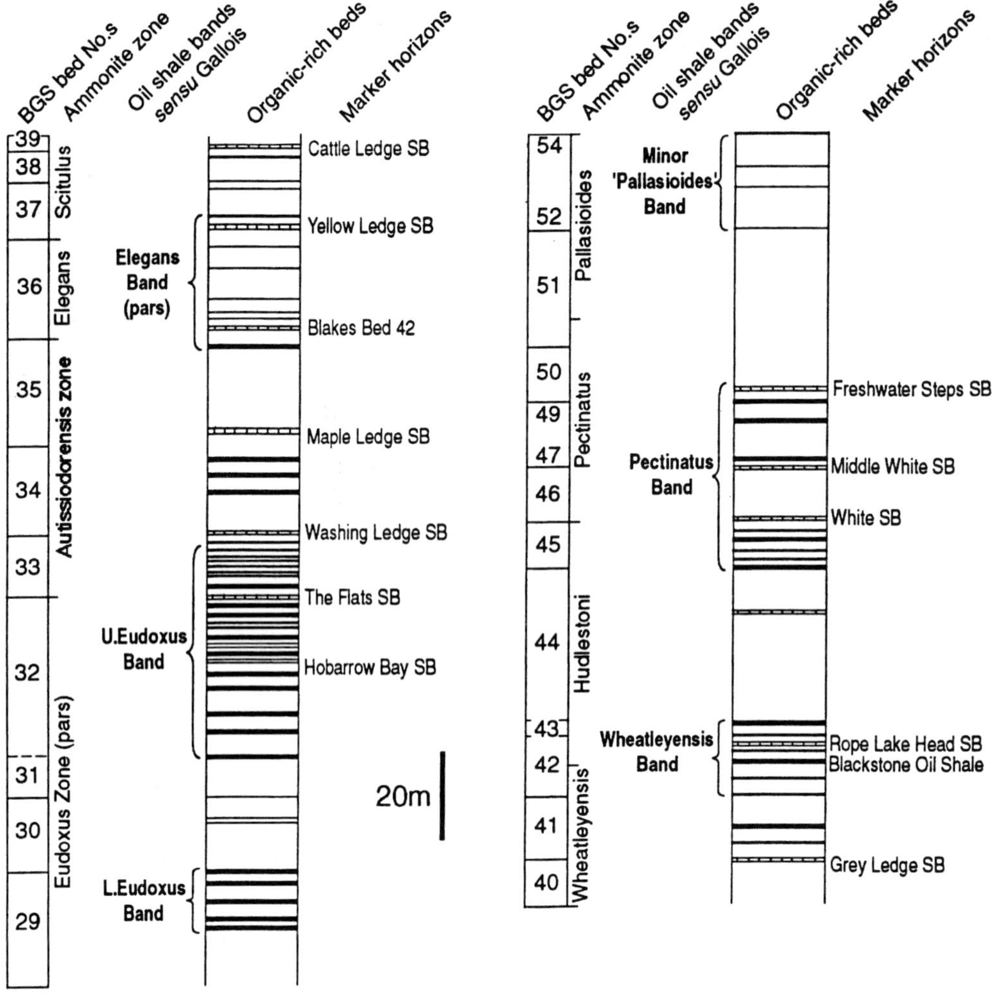

Fig. 4. Generalized stratigraphical column of the middle, organic-rich, part of the type Kimmeridge Clay of the Dorset coast (modifed after Cox & Gallois 1981). The lithological column is simplified to emphasize organic-rich beds and the 'oil shale bands' of Gallois (1978), together with the main marker horizons (SB = Stone Band). For more detailed published lithological logs see Cox & Gallois (1981) and Wignall (1990). The bed numbers for the *pectinatus* and *pallasioides* zones are after Wignall (1990, 1991).

Kimmeridge type section, representing 40% of the total thickness (Fig. 4).

The organic-rich interbeds (the 'oil shales' of Gallois 1978, the 'oil shales' and 'bituminous shales' of Downie 1955 and Tyson *et al.* 1979) show a variable frequency through this interval and are mostly concentrated in five 5–30 m thick bands, separated by more uniform mudstones with less organic-rich layers (Gallois 1978). Each bundle of organic-rich beds is thought to represent about 0.5 Ma (Herbin & Geyssant 1993). The five bundles, evident as peaks on gamma ray logs, are closely correlated with the MFSs identified in this stratigraphical interval (Melnyk *et al.* 1992, 1993; cf. Whittaker *et al.* 1985; Table 5). It should be noted that although the MFS is a planar surface at the seismic scale, it can only be resolved as a stratigraphical interval in these basinal sediments. The MFS intervals of Melnyk *et al.* (1992, 1994) clearly correspond to episodes of onlap on the basin margin (Wignall 1991; see Fig. 5; Table 5). There is also a reasonable correlation with several of the MFSs which have been defined on North Sea offshore geophysical and micropalaeontological data (see Partington *et al.* 1993*a,b*), although the stratigraphical resolution of the latter is inevitably lower.

Table 5. *Correlation of major intervals of organic-rich 'oil shales' in the onshore Kimmeridge Clay Formation (KCF) (Gallois 1978) with the MFS horizons reported for the onshore Wessex Basin (Melnyk* et al. *1992) and the Boulonnais (Proust* et al. *1993; Herbin* et al. *1995). Also shown are phases of onlap during deposition of the onshore Kimmeridge Clasy (Wignall 1991)*

Onshore KCF 'oil shale' intervals	Onshore onlap	Wessex MFS horizons	Boulonnais MFS horizons	North Sea MFS horizons
M. eudoxus (L. eudoxus Band) BGS 'bed' 29		*M. eudoxus*	*M. eudoxus*	Eudoxus TEMFS† ('possibly extends into *L. autissiodorensis*')
	M-U. eudoxus			
U. eudoxus–L. autissiodorensis (U. eudoxus Band) BGS 'beds' 32 & 33		*eudoxus–autissiodorensis* boundary(?)	*L. autissiodorensis**	Autissiodorensis MFS ('intra-*autissiodorensis*')
elegens–basal *scitulus* (Elegans Band) BGS 'beds' 36–37	*L. elegans–M. scitulus*	*elegans–scitulus* boundary(?)	*L. elegans**	Not distinguished
U. wheatleyensis–basal *hudlestoni* (Wheatleyensis Band) BGS 'beds' 42 & 43)	*U. wheatleyensis*	*M. wheatleyensis*	*?wheatleyensis**	Hudlestoni MFS ('approx.' *hudlestoni*)
Uppermost *hudlestoni–L. pectinatus* (Pectinatus Band) BGS 'beds' 45–47	*U. hudlestoni–L. pectinatus*	Top(?) *pectinatus*	*L. pectinatus?*	Pectinatus MFS (*?pectinatus*)
U. pallasioides (unnamed minor band) BGS 'beds' 52–54	*U. pallasioides*	None indicated	*M. pallasioides*	Fittoni TEMFS ('approx.' *fittoni*, post-*pectinatus*)

The correlation with the MFS horizons defined in the subsurface Kimmeridge Clay of the North Sea region (Partington *et al.* 1993*a*) is approximate and uncertain because of the lower biostratigraphical precision. The BGS bed numbering for the onshore Kimmeridge Clay is after Cox & Gallois (1981) and Wignall (1990).

* Only these three MFS horizons are association with organic enrichment in the Boulonnais (cf. Herbin & Geyssant 1993; Herbin *et al.* 1995).

† TEMFS, Tectonically enhanced maximum flooding surface.

Thus although the onshore Kimmeridge Clay may not conform to the Creaney & Passey (1993) model (which is based on normal progradational infilling and continuous anoxia), the intervals of greatest source rock development are still primarily associated with the MFS intervals.

As might be expected, not all the MFS intervals occur in organic-rich facies everywhere in the Wessex Basin. On its most sand-rich and probably shallowest preserved margin in the Boulonnais (northwest France), only three of the regional MFS intervals (*elegans* and *wheatleyensis* and, to a lesser extent, the lower *autissiodorensis* zones) are recorded by peaks in TOC (Dunn 1972; Proust *et al.* 1993; Herbin & Geyssant 1993; Herbin *et al.* 1995; Table 5). It appears that either these intervals were the only ones associated with a relative change in water depth sufficient to allow dysoxic–anoxic conditions to be established in this marginal setting, or that other watermass factors were unfavourable.

Sequence boundaries ('K5–K8' of Wignall 1991) separate the phases of onlap on the margins of the Wessex Basin (Fig. 5); in the basinal areas these sequence boundaries correlate with the more uniform and organic-poor mudstone intervals separating the five bundles of organic-rich beds (Wignall 1991). Cyclostratigraphical evidence suggests correlative minor hiatuses in the basinal section during the upper *wheatleyensis* and upper *hudlestoni* zones (Melnyk *et al.* 1992), but sedimentological expression of the sequence boundaries is generally negligible in the basinal facies above the *elegans* Zone (below which they are marked by thin siltstone layers, as at the top of the *autissiodorensis* and the bases of the *eudoxus* and earlier zones; Bellamy 1979; Hantzpergue 1985; Wignall 1991). The major

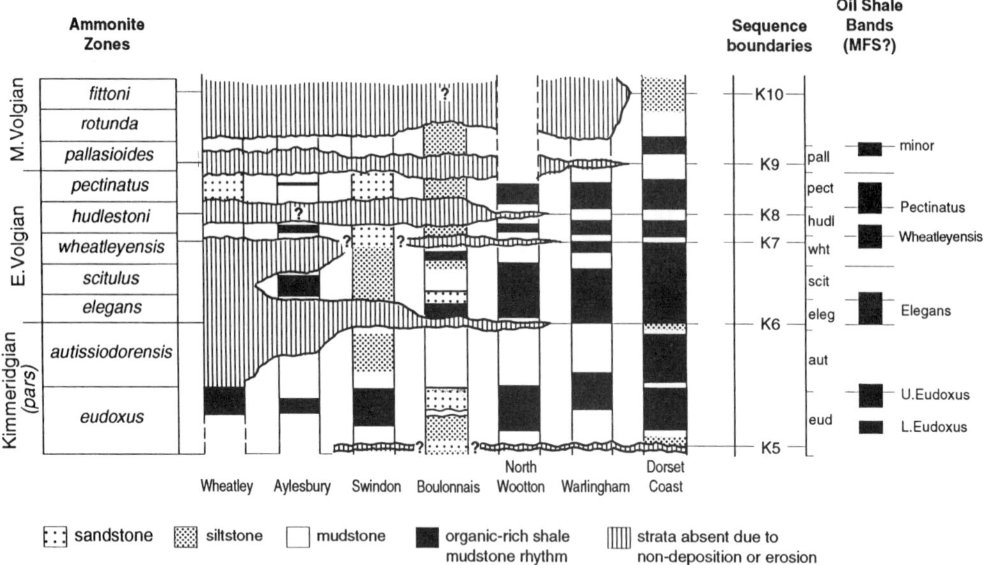

Fig. 5. Chronostratigraphical chart of the *eudoxus–fittoni* zones of the onshore Kimmeridge Clay, based on Wignall (1991), with the Boulonnais log corrected according to Proust *et al.* (1993). The localities are given on Fig. 3. The stratigraphical positions of the 'oil shale bands' of Gallois (1978) have been added to the right-hand side of the diagram.

monotonous mudstone intervals, like the very conspicuous one in the mid-*hudlestoni* Zone (bed 44, Fig. 4; cf. Wignall & Ruffell 1990), probably reflect a higher sediment accumulation rate in the basin; they are associated with erosion on the basin margin (cf. Wignall 1991) and thus probably equate to a basinal expression of the lowermost systems tract, or perhaps the earliest TST. As well as increasing the wavelength of the Milankovitch-scale cycles, the higher sediment accumulation rate may obscure some weaker cycles by diluting the rhythmic variations in biogenic components that define them.

All MFS intervals should be associated with reduced siliciclastic sedimentation in basinal areas and the correlation between the bundles of organic-rich beds and the MFS intervals thus indicates that the bundles should be relatively condensed. Wignall (1991, pp. 439–440), although noting an association between phases of marginal onlap and deposition of organic-rich shales in the basin, asserted that there is no convincing candidate for condensed sections (*sensu stricto*) in the onshore Kimmeridge Clay, and concluded that this seriously invalidates the Exxon model. However, this reasoning is perhaps too influenced by unspecified taphonomic arguments and the view that the occurrence of subordinate, rapidly deposited coccolith limestone interbeds can somehow negate a CS interpretation. Conceptually, the CS should be related primarily to the siliciclastic flux; the relative biogenic component of the sediment should always increase during the CS (Loutit *et al.* 1988), but the absolute pelagic carbonate flux may also increase due to productivity effects (especially in shelf facies), such that the total sediment accumulation rate may also locally increase to some extent, provided carbonate dissolution is not also enhanced. Bellamy (1979, p. 137) and Oschmann (1988, p. 242, 245) estimate that the individual lower *pectinatus*-Zone coccolith limestones were deposited up to ten times faster than adjacent shales; although the palaeoproductivity was probably 'high' during these events, the massive autodilution of the type II kerogen results in TOC values of only a few per cent.

It is difficult to prove whether the five bundles of organic-rich beds were deposited more slowly than the intervening intervals, although it is noteworthy that they are generally thinner and that the relative separation of the individual organic-rich beds (cf. Gallois 1978) apparently decreases within the organic-rich bundles. If the individual organic-rich beds occur with a fixed Milankovitch periodicity (about 30–40 ka per cycle according to House 1985), and all cycles are expressed, then this pattern suggests an up to ten-fold slowing of mean sediment accumulation rate within the bundles. Such spacing patterns can also be produced by the cumulative effect of several superimposed cycles of different wavelengths, coupled with a specific environmental threshold (cf. House 1985, p. 725);

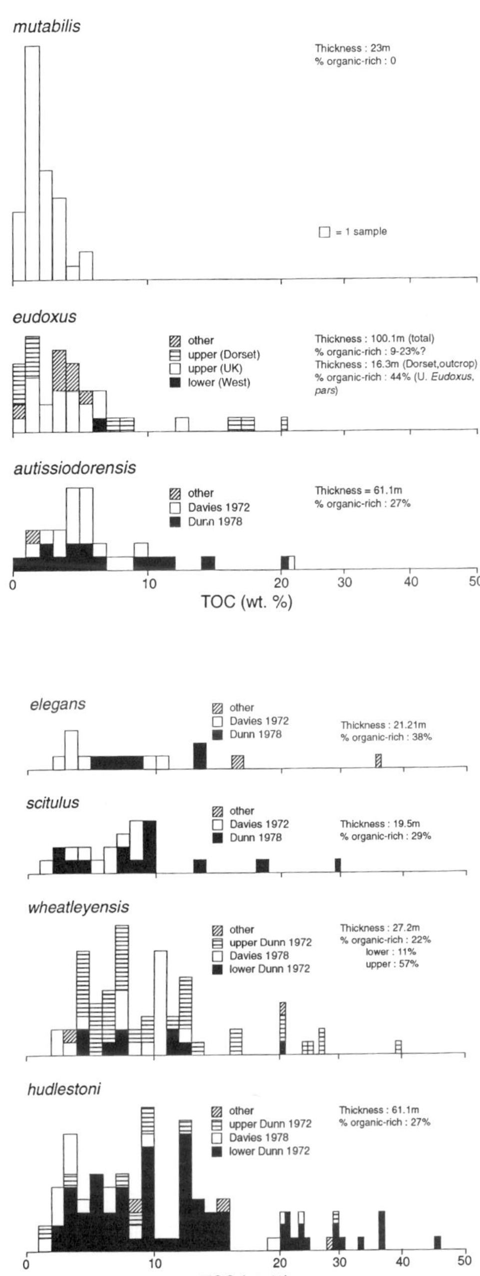

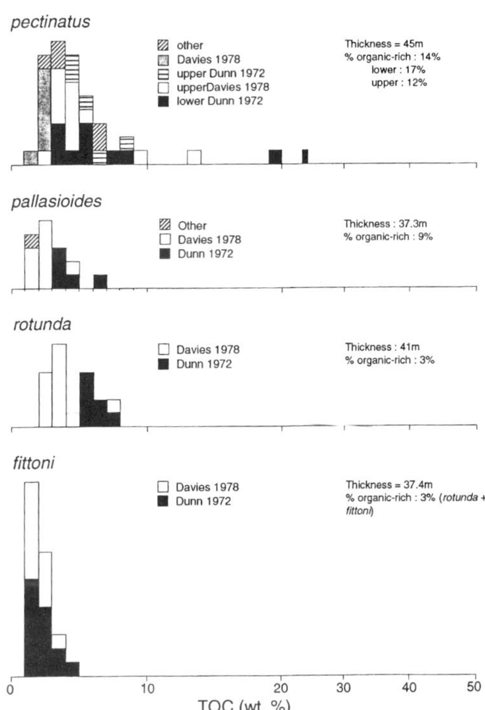

Fig. 6. Frequency histograms of reported TOC values for the *mutabilis* to *fittoni* zones of the Kimmeridge Clay of the Wessex Basin. There are no data for the oxic condensed facies of the *baleyi* (3.8 m) and *cymodoce* (5.0 m) zones, but values are very unlikely to exceed those for the *mutabilis* Zone. As the *mutabilis* Zone is not exposed at Kimmeridge, the *mutabilis* Zone data are from Westbury, Dorset (after Nøhr-Hansen 1989); the remainder are primarily derived from the data of Dunn (1972) and Davies (1978) for the type Kimmeridge Clay section (Kimmeridge Bay to Chapman's Pool, Dorset). The data from the lower and upper parts of the zones are distinguished on some of the diagrams. Although the studies by Dunn (1972) and Davies (1978) use fairly balanced sampling of the different lithofacies, the general tendency to over-sample the organic-rich interbeds means that the supra-*mutabilis* Zone data cannot be used to calculate the overall mean TOC per ammonite zone. The relative proportion of conspicuous organic-rich shale beds within each ammonite zone, as determined from sedimentological logs, is also given next to the histograms. Note the change in the scale of the TOC axis for values above 20%.

the geochemical data of Herbin *et al.* (1991, 1993) certainly suggest a series of 'nested' cycles with periodicities ranging from about 25 to 280 ka (cf. Oschmann 1988, p. 243). Interference effects between various scales and amplitudes of cyclicity are likely to produce occasional anomalies and additional arhythmia will also result from the differing rates, magnitudes and symmetries of relative sea-level or climatic change during any given interval.

It is difficult to determine the distribution of time within the Kimmeridge Clay. Based on geophysical log cyclostratigraphy and assuming a Milankovitch

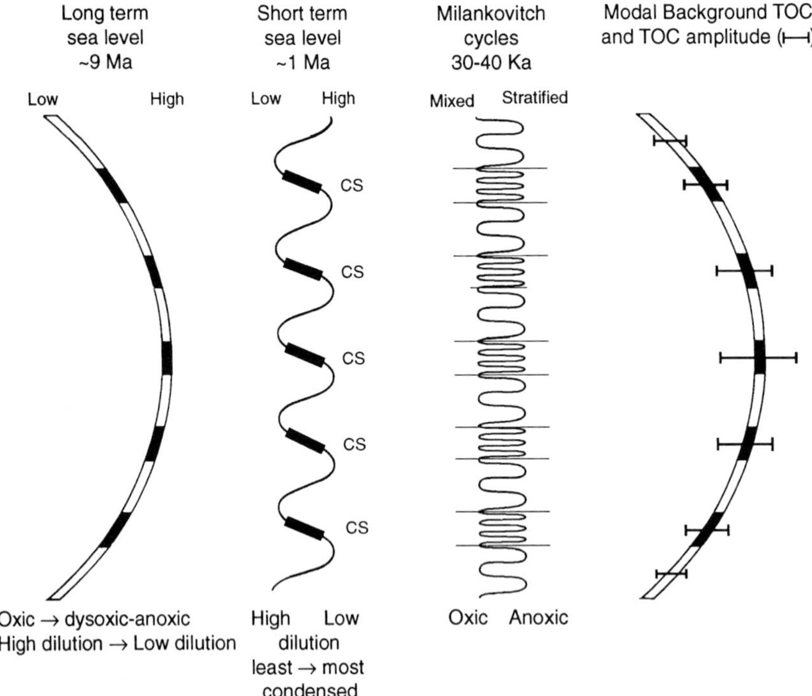

Fig. 7. Highly simplified schematic model for the interaction of the long-term and short-term sea-level curves with Milankovitch redox cycles and their effect on the distribution of organic-rich interbeds, background TOC and amplitude of redox TOC cycles. Note that the vertical scale is time and not thickness; the organic-rich beds should tend to be thinner and closer together towards the middle of the major cycle, producing a higher percentage of source rock in that interval. The Milankovitch column on the schematic diagram is not to scale and differences in the wavelength and numbers of cycles are meant to indicate only the general effect of sediment accumulation rate on the relative *number* of cycles (of fixed duration) that occur per stratigraphical interval. The relative scales for TOC background values and amplitudes are independent of each other.

cycle duration of 40 ka, Melnyk *et al.* (1994, p. 45) have suggested that the *wheatleyensis* through *pectinatus* zones generally represent the interval with the highest sediment accumulation rates within the Kimmeridge Clay of the Wessex Basin. By contrast, plots of total thickness per ammonite zone suggest a regionally consistent pattern of relatively slow sedimentation (*c.* 20–35 m Ma^{-1}) in the *elegans* to *pectinatus* zones (Penn *et al.* 1986: 409), although with a significant increase in the *hudlestoni* Zone (*c.* 70 m Ma^{-1}), probably mainly associated with Bed 44. However, the latter argument assumes an equal duration of the zones, which has yet to be substantiated (cf. House 1985, p. 723; Wignall 1990, p. 63). Wignall & Ruffell (1990) argue for decreased sediment accumulation rates in the mid-*hudlestoni* to mid-*pectinatus* zones due to the influence of more arid climates. A generally (but not uniformly) low sediment accumulation in the middle part of the sequence, upper *wheatleyensis* Zone to lower *hudlestoni*-Zone, would certainly seem to fit better with its higher frequency of organic-rich interbeds and the TOC maxima in the background mudstones.

Long-term TOC trend

The background mudstone facies of the onshore Kimmeridge Clay show a very variable TOC content, from close to the average for shales (about 1 wt%) to very high values (≤10 wt%) that are greater than some of the best values exhibited by many other source rock sequences (Tyson 1989). Even the richest of these mudstones still often contain a benthic macrofauna (see Wignall 1990); their high content of TOC (type II/III to type II kerogen) has thus been interpreted in terms of enhanced preservation related to generally reducing sediments (with the redox boundary near the sediment surface) and in some instances recurrent periodic or seasonal, but not continuous bottom water anoxia (Oschmann 1988, 1991; Tyson 1989).

The TOC distribution by ammonite zone through the Kimmeridge Clay type section shows a change from unimodal to multimodal and back to unimodal (Fig. 6). The peak of the first (lowest) TOC mode, that which is most representative of the least organic-rich and mostly mudstone facies, is seen to increase from values of around 1 wt% in the *mutabilis* Zone to around 4–6 wt% in the middle of the formation (upper *wheatleyensis*–lower *hudlestoni* zones) and then decline back to around 1 wt% above the lower *pectinatus* zone. The decrease in TOC range observed in the lower *pectinatus* Zone relative to the lower *hudlestoni* Zone in the Kimmeridge Clay type section is probably related to the greater autodilution by coccolith carbonate (Tyson 1989, p. 142), which may also affect the mid-*hudlestoni* and upper *scitulus*–lower *wheatleyensis* zones. Herbin *et al.* (1993, pp. 84–85) also observe an overall upward increase on their synthetic TOC log of the Yorkshire Kimmeridge Clay in the *cymodoce* to the lower *pectinatus* zones: the minimum values increase from about 1 to 4 wt% and the mean of the peak values increases from about 2 to 12 wt%. The upper *pectinatus* and younger zones of the Kimmeridge Clay are not preserved in Yorkshire or Lincolnshire.

The long-term TOC trend seen in the Kimmeridge Clay type section corresponds well to the Late Jurassic long-term sea-level curve of Haq *et al.* (1987), apparently reflecting its influence on the mean basin palaeo-oxygenation state and background sediment accumulation rate (Fig. 7). It is improbable that this trend results from a consistent increase in palaeoproductivity, as watermass stratification, and thus nutrient trapping, seems to have increased in importance and the climatic trend suggests decreasing runoff and a lower supply of external nutrients.

The terrestrial kerogen content of the onshore English Kimmeridge Clay remains consistently low and shows little or no relation to thickness; PhytOC values (percentage relative numerical frequency of phytoclasts/100 × TOC) are ≤1 throughout most of the Wessex Basin and Eastern England Shelf (Tyson 1989; Scotchman 1991). This may partly be attributable to climatic factors. In terms of the terrestrial phytoclasts alone, nearly all the basinal mudstones show distal characteristics typically associated with organic facies D or CD (small particles and a high relative proportion of black wood, inertinitic material; see Tyson 1989). However, the overall organic facies of the background mudstones facies is often BC, which is due mainly to a predominance of variably preserved marine AOM (e.g. Nøhr-Hansen 1989), especially in the middle part of the Kimmeridge Clay.

Decimetre- to metre-scale cycles

As noted above, the MFS intervals in the onshore Kimmeridge Clay are not represented by discrete uniform units, but by a bundle of alternating organic-rich and organic-poor beds. This decimetre- to metre-scale alternation has a Milankovitch periodicity (House 1985) and thus appears to reflect climatic control of the basin watermass characteristics. Although some degree of enhanced terrestrial input is implied by their palynofacies characteristics, the relatively organic-poor interbeds of these cycles are not simply the product of dilution, but also represent improved palaeo-oxygenation in the basin related to watermass changes (Tyson *et al.* 1979; Oschmann 1988, 1991). The relative magnitude of these changes in palaeo-oxygenation is not uniform throughout the sequence. Because the long-term relative sea-level cycle apparently influences the mean palaeo-oxygenation state of the 'most-oxic' phase of the Milankovitch cycles, it should also influence the amplitude of the TOC changes. The actual palaeo-oxygenation conditions during these 'most-oxic' phases varies from oxic to dysoxic and thus the scale of palaeo-oxygenation changes that occur during each whole cycle may, respectively, vary from large (oxic to anoxic) to small (dysoxic to anoxic). The preservational consequences will also be modified by sediment accumulation rate (e.g. the slow/dysoxic combination should be less favourable for qualitative preservation than the fast/dysoxic combination, although the TOC may be lower in the latter instance due to higher dilution).

Much of the recent work on the Yorkshire Kimmeridge Clay considers the individual organic-rich beds to be primarily a function of cyclically increased palaeoproductivity in the Kimmeridge Clay sea (e.g. Belin & Brosse 1992; Pradier & Bertrand 1992; Ramanampisoa *et al.* 1992; Bertrand & Lallier-Vergès 1993; Lallier-Vergès *et al.* 1993; Tribovillard *et al.* 1994; Ramanampisoa & Disnar 1994; Herbin *et al.* 1995). The arguments put forward to support this productivity hypothesis are not especially convincing. The 'sulphate reduction index' argument of Bertrand and Lallier-Vergès (1993) and Lallier-Vergès *et al.* (1993) is most likely a function of iron-limitation controls on pyrite formation and thus an unreliable indicator of organic matter flux (Tyson 1995; cf. Tribovillard *et al.* 1994). The argument that the relative compositional uniformity of the mineralic fraction implies that no significant change in sediment accumulation rate occurred through the individual redox cycles also seems tenuous in such a distal facies and at such detailed scales of observation. The organic geochemical criteria used by Ramanampisoa & Disnar (1994) also do not satisfactorily distinguish

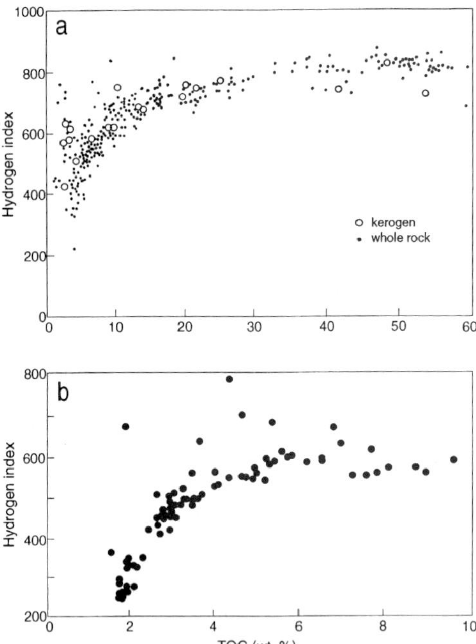

Fig. 8. Hydrogen index (HI, mg HC/g TOC) versus total organic carbon (TOC wt%) for (**a**) the Wheatleyensis 'oil shale band' of the Dorset type section (after Huc *et al.* 1992), and (**b**) an individual cycle from the Upper Eudoxus 'oil shale band' of the Marton 87 'Yorkim'/IFP borehole in Yorkshire (after Bertrand & Pradier 1992). Note the very different TOC values at which HI becomes asymptotic, but the small differences in HI values at this point.

between the effects of productivity plus preservation and those of preservation plus variable sediment dilution.

In the Cleveland Basin (Yorkshire), Herbin *et al.* (1993) have observed that the organic-rich cycles of the Kimmeridge Clay tend to become richer in TOC towards the deeper parts of the basin, where there was also greater sediment accumulation. However, most of the lateral differences in thickness in the onshore Kimmeridge Clay are probably associated with changes in the background mudstone facies, rather than with the organic-rich beds themselves, as is observed between intra-basinal highs and lows within the Wessex Basin (Downie 1955; Bellamy 1979, pp. 137, 141; see also Melnyk *et al.* 1994, p. 42). This pattern implies that the apparently isochronous oil shale and coccolith limestone interbeds are largely pelagic deposits, but that the intervening mudstones were partly deposited from dilute turbid clay-rich clouds of sediment which preferentially gravitated towards the (structurally controlled) intra-basinal lows before depositing much of their sediment load. The basinward focusing of clayey sediment would probably have been much reduced during periods of greater and more stable water mass stratification when wave and current resuspension of fine-grained sediment were limited largely to the very margins of the basin above the pycnocline.

The individual organic-rich interbeds exhibit relatively more distal palynofacies characteristics than do the background mudstone facies, including a higher relative abundance of small black phytoclasts, buoyant bisaccate pollen and prasinophyte algae (Tyson, pers. obs.; Waterhouse 1992). This indicates lower energy and more pelagic deposition (cf. Tyson 1995). The increased concentration of phytoclasts (particles per gram of sediment) in the individual organic-rich beds (Waterhouse 1992), despite a likely decreased or constant, and possibly partly aeolian, background input of these particles, also suggests that these beds are more condensed than the intervening mudstones.

Several studies have shown that the overall aspect of the palynofacies assemblages (especially the AOM : phytoclast ratio) apparently remain relatively constant throughout some of the individual metre thick cycles in the *eudoxus* Zone of the Yorkshire Kimmeridge Clay, although the character of the phytoplankton-derived AOM does show significant changes (e.g. Belin & Brosse 1992, p. 718; Ramanampisoa *et al.* 1992, p. 1496; Boussafir *et al.* 1994, p. 356). Transmission electron microscope studies show that the relative proportion of the ultralamina part of the AOM, believed to represent refractory phytoplankton cell walls, decreases in the most organic-rich part of the cycles (Boussafir *et al.* 1994, p. 360). This pattern suggests that the increase in organic content is due to a greater preservation of more labile phytoplanktic organic matter, rather than a higher flux of refractory ultralaminae-forming algaenans, and is thus at odds with the 'algaenan hypothesis' of kerogen formation (e.g. de Leeuw *et al.* 1991; cf. Tyson 1995).

If the TOC changes through the Yorkshire Kimmeridge Clay cycles were primarily due to changes in productivity, the AOM : phytoclast ratio ought to show major changes as the higher supply of marine AOM input dilutes the background terrestrial component. As it is unlikely that the absolute input of the phytoclasts is positively correlated with palaeoproductivity, the apparent relative uniformity of the AOM : phytoclast ratio in these cycles argues against a major productivity control (or indeed major differences in redox conditions sufficient to affect the quantitative preservation, i.e. the percentage of the AOM). This leaves differential siliciclastic dilution as the probable controlling factor, as changing sediment dilution can produce the observed change in TOC without significantly affecting the AOM : phytoclast ratio (partly because the phytoclast input represents a low level background in a distal setting and is thus largely decoupled from sediment input), while still allowing the concentration of phytoclasts (particles per gram of sediment) to increase in the organic-rich, condensed part of the cycle. Other individual cycles, especially those where the minimum TOC is less than 3 wt% and the background mudstone facies is more oxic than dysoxic, may show more variation in AOM : phytoclast ratios according to organic petrological observations (e.g. Pradier & Bertrand 1992), but such changes could reflect either a greater input or preservation of the marine component (or both together).

Examination of HI versus TOC curves from the onshore Kimmeridge Clay (Fig. 8) indicates that the TOC level at which optimum organic matter preservation is established (i.e. at which HI values more or less stabilize) varies considerably depending on the interval considered, ranging from fairly typical values of about 5–6 wt% TOC (e.g. Pradier & Bertrand 1992; cf. Creaney & Passey 1993, p. 396; Ramanampisoa & Disnar 1994, p. 1162) to values as high as 20–30 wt%, such as occur in the upper *wheatleyensis* Zone–basal *hudlestoni* Zone interval at Kimmeridge (Huc *et al.* 1992, p. 472) and the *eudoxus*-Zone 'upper cycle' of Tribovillard *et al.* (1994). However, the asymptotic HI values are rather similar, 600 versus 750 mg HC/g TOC, suggesting comparable levels of preservation and similar organic matter inputs. Coupling these observations with the preceding arguments suggests that the two to three fold variation that occurs in the TOC at more or less constant HI is probably due to differences in relative siliciclastic sediment accumulation rate (dilution). The higher (seven versus five fold) amplitude of the TOC variation in the more organic-rich cycles and their greater variation in redox conditions and organic matter preservation, as determined by trace element analysis and Fourier transform infra-red spectroscopy (Tribovillard *et al.* 1994), fits well with this hypothesis. As noted earlier, at lower sediment accumulation rates, the peak TOC values will be higher, but redox–palaeo-oxygenation dependent differences in preservation should reach their maximum expression. The relatively high value of the minimum TOC in these specific organic-rich high amplitude cycles presumably reflects the modest scale of change in the palaeo-oxygenation and redox conditions that occur through these particular cycles, together with low background dilution levels.

Gallois (1973, p. 65) and Cox and Gallois (1981, p. 13) indicate that the individual, organic-rich–organic-poor cycles of the onshore Kimmeridge Clay are characterized by a sharp base and a gradational top, with fissile organic-rich beds passing up into more blocky and relatively organic-poor mudstones. This cycle asymmetry is not very evident in the Kimmeridge Clay type section in Dorset: both the base and top of the organic-rich interbeds appear fairly distinct in the field and asymmetry is certainly not a conspicuous feature of the detailed data of Huc *et al.* (1992, p. 471). However, the detailed TOC and HI profiles of some individual *eudoxus* Zone cycles in the Yorkshire Kimmeridge Clay do seem to support an asymmetrical pattern (Herbin *et al.* 1991, 1993; Pradier & Bertrand 1992; Belin & Bross 1992; Tribovillard *et al.* 1994; Ramanampisoa & Disnar 1994; see also the *elegans* Zone data of Herbin *et al.* 1995, p. 190). Asymmetrical patterns are predicted by the Creaney & Passey (1993) model, although on a systems tracts scale, with 'HTB' (high TOC base) units being produced by maximum condensation at the base of the cycle (MFS) and an upward progressive increase in dilution due to progradation. The HTB mechanism might also apply at the parasequence 'scale', presuming that parasequences are expressed in anoxic basinal facies, which is questionable as their expression declines in a distal direction (cf. Van Wagoner *et al.* 1990). However, as the individual minor cycles of the onshore Kimmeridge Clay appear to be the product of Milankovitch controls on basin water mass properties, rather than the relative sea-level changes commonly assumed for parasequences (Van Wagoner *et al.* 1990), the reason for the TOC asymmetry at this scale is not obvious.

The sediment record of palaeo-oxygenation cycles is always likely to be asymmetrical because of the differing scale and preservation potential of the oxic and anoxic phases. Although an anoxic event may only deposit a single, thin organic-rich lamina or bed, an oxic event may potentially be associated with a bioturbated layer several centimetres in thickness, biasing the record, in relative thickness terms, towards the oxic state. The change from oxic to anoxic facies probably occurs in the form of an increasing frequency (and perhaps duration) of transient anoxic events, rather than a uniform shift in mean palaeo-oxygenation. Most of the earlier transient anoxic events will be destroyed (consumed by bioturbation) during the subsequent return to more oxygenated conditions; these bioturbation events will be especially effective if the sediment accumulation rate is decreasing. The final switch to consistently laminated suboxic–anoxic facies will only occur once a critical threshold is reached (the establishment of either continuous suboxia–anoxia, or annually recurrent prolonged seasonal anoxia that prevents recolonization by bioturbating benthos; see Tyson & Pearson 1991). The major fabric change in facies at the base of the cycle may thus be relatively abrupt, although there are likely to be some progressive high-gradient geochemical changes preceding this. By contrast, as palaeo-oxygenation increases again on the other side of the cycle, the re-establishment of oxic conditions, and the progressive recolonization of increasingly larger burrowing taxa (e.g. Savrda & Bottjer 1991), will tend to produce increasing obliteration of the latter stages of the anoxic record, producing a more gradual transition in fabric and geochemical parameters in the upper half of the cycle. This is especially so if the sediment accumulation rate is also simultaneously increasing. Differential compaction of the organic-rich and organic-poor parts of the cycles will further exaggerate this asymmetry. Consequently, small-scale 'HTB-like' units, when associated with palaeo-oxygenation trends, do not necessarily reflect a simple progradation-dilution mechanism.

It is possible that sediment supply was indeed assymetrical across the minor redox cycles. The 'false bottom effect' of pycnoclines (Tyson 1985, p. 70) traps sediments, especially coarser sediments, at the basin margin (e.g. Exon 1972; Sly 1978; Pollehne 1986), leading to primarily fine-grained sedimentation in more offshore areas, except for turbidites and tempestites. The progressive development of water mass stratification, or its increasing intensity, should lead to a progressive reduction in sediment supply to the basin, analogous to the observation that sediment focusing in lakes primarily occurs during the seasonal overturn, not during the stratified period. However, when the importance of stratification wanes, it is possible that the finer sediment that built up at the margins of the basin during the stratified period is then remobilized and focused into the basin, resulting in a temporarily higher sediment supply than during the earlier part of the cycle. Coarser sediments may also be allowed to spread outwards, especially in proximal areas, where the breakdown or weakening of stratification allows currents to re-impinge on the basin floor. This would decrease nearshore accommodation and promote bypass to the basin. Destabilization of the watermass could thus perhaps trigger *minor* progradational cycles at the margin of the basin. Such sedimentological effects of changes in watermass stability may partly explain the association between basinal redox cycles and nearshore marine parasequences reported for the Greenhorn cyclothem by Elder *et al.* (1994).

Conclusions

It is evident that there are logical relationships between sequence stratigraphy and the palaeoenvironmental and sedimentological variables that control organic facies characteristics. Not surprisingly, the best inter-relationships relate to the allochthonous terrestrial component of the kerogen assemblages, whose input to the basin is generally correlated with the input of siliciclastic sediment and thus progradation, although this can be modified by climate. The relationship between marine organic matter and sequence stratigraphy is less direct and redox-dependent, and is probably due to the influence of relative sea level on mixing regimes, bottom water volumes and thus palaeo-oxygenation. The sequence stratigraphical distribution of TOC depends on the interaction of sediment accumulation rate and the palaeo-oxygen regime. In oxic regimes sediment accumulation can have a positive effect on preservation (and thus perhaps TOC), whereas in dysoxic–anoxic regimes it acts mainly as a diluent. The relative importance of the palaeo-oxygen regime increases as sediment accumulation rates fall and is thus greatest in the TST to early HST. Although palaeoproductivity is a significant factor, the generally high mean productivity and carbon flux associated with shelf environments, where most source rocks were deposited, probably mean that it is seldom limiting; most of the observed variation appears to be explainable by redox-related changes in preservation and changes in sediment supply tied to climate and relative sea level.

In accord with the Creaney & Passey (1993) model, the best development of source rocks in the onshore Kimmeridge Clay is clearly associated with the MFS intervals. However, the long-term relative sea level cycle and short-term Milankovitch climate controls are also significant controls on the pattern of source rock development. The long-term sea-level cycle is expressed via its influence on the mean water depth, mean ambient palaeo-oxygenation and background sediment accumulation rate, which result in the highest background and peak TOCs within the middle part of the sequence. The short-term sea-level cycle results in varying sediment accumulation rates and thus the highest TOC values and percentages of source rock facies focused within the MFS intervals. Milankovitch climate controls on watermass characteristics result in varying palaeo-oxygenation and thus decimetre- to metre-scale variation in organic facies, even in the MFS intervals, especially in the middle onshore Kimmeridge Clay. As predicted from the general principles outlined here, it is these interactions between the changing sediment accumulation rates and palaeo-oxygenation regimes that appear to determine the background TOC trend, the peak TOC trend, the amplitude of TOC cyclicity and the occurrence, percentage, spacing and quality of the organic-rich interbeds. The combined use of palynofacies data (AOM : phytoclast ratios, phytoclast type and phytoclast numbers per gram of sediment) and HI versus TOC plots offers a promising means of assessing the relative roles of dilution, preservation and palaeoproductivity through cyclic sequences. Any small-scale cycle asymmetry may be an inherent feature of redox cycles, but could also be produced by minor changes in the pattern of fine-grained sediment distribution caused by variations in watermass stability.

I thank S. Hesselbo for his invitation to contribute a review paper to this book. My thanks also to the anonymous reviewer who gave constructive comments on the first draft of the manuscript, and to B. Brown who draughted the diagrams.
Author's e-mail address: r.v.tyson@newcastle.ac.uk.

References

ARTHUR, M. A. & SAGEMAN, B. B. 1994. Marine black shales: depositional mechanisms and environments of ancient deposits. *Annual Review of Earth and Planetary Sciences*, **22**, 499–551.

BELIN, S. & BROSSE, E. 1992. Petrographical and geochemical study of a Kimmeridgian organic sequence. *Revue de l'Institut Français du Pétrole*, **47**, 711–725.

BELLAMY, J. 1979. *Carbonates within bituminous shales in the British Jurassic – their petrography and diagenesis*. PhD Thesis, University of Southampton.

BERNER, R. A. 1982. Burial of organic carbon and pyrite sulfur in the modern ocean: its geochemical and environmental significance. *American Journal of Science*, **282**, 451–473.

BERTRAND, P. & LALLIER-VERGÈS, E. 1993. Past sedimentary organic matter accumulation and degradation controlled by productivity. *Nature*, **364**, 786–788.

BETTS, J. N. & HOLLAND, H. D. 1991. The oxygen content of ocean bottom waters, the burial efficiency of organic carbon, and the regulation of atmospheric oxygen. *Palaeogeography, Palaeoclimatology, Palaeoecology (Global and Planetary Change Section)*, **97**, 5–18.

BOUSSAFIR, M., LALLIER-VERGÈS, E., BERTRAND, P. & BADAUT-TRAUTH, D. 1994. Structure ultrafine de la matière organique des roches mères du Kimméridgien du Yorkshire (UK). *Bulletin de la Societé Géologique de France*, **165**, 353–361.

BRALOWER, T. J. & THIERSTEIN, H. R. 1987. Organic carbon and metal accumulation rates in Holocene and mid-Cretaceous sediments: palaeoceanographic significance. *In*: BROOKS, J. & FLEET, A. J. (eds) *Marine Petroleum Source Rocks*, Geological Society, London, Special Publications, **26**, 345–369.

BURTNER, R. L. & WARTNER, M. A. 1984. Hydrocarbon generation in Lower Cretaceous Mowry and Skull Creek Shales of the northern Rocky Mountain area. *In*: WOODWARD, J., MEISSNER, F. F., & CLAYTON, J. L. (eds) *Hydrocarbon Source Rocks of the Greater Rocky Mountain Region*. Rocky Mountain Association of Geologists, Denver, 449–468.

BUSTIN, R. M. 1988. Sedimentology and characteristics of dispersed organic matter in Tertiary Niger Delta: origin of source rocks in a deltaic environment. *American Association of Petroleum Geologists Bulletin*, **72**, 277–298.

CANFIELD, D. E. 1989. Sulfate reduction and oxic respiration in marine sediments: implications for organic carbon preservation in euxinic environments. *Deep-Sea Research*, **36**, 121–138.

—— 1994. Factors influencing organic carbon preservation in marine sediments. *Marine Chemistry*, **114**, 315–329.

COX, B. M. & GALLOIS, R. W. 1981. *The Stratigraphy of the Kimmeridge Clay of the Dorset Type Area and its Correlation with Some Other Kimmeridgian Sequences*. Report of the Institute of Geological Sciences, **80/4**.

CREANEY, S. & PASSEY, Q. R. 1993. Recurring patterns of total organic carbon and source rock quality within a sequence stratigraphic framework. *Bulletin of the American Association of Petroleum Geologists*, **77**, 386–401.

CURIALE, J. A., COLE, R. D. & WITMER, R. J. 1992. Application of organic geochemistry to sequence stratigraphic analysis: Four Corners Platform Area, New Mexico, U.S.A. *In*: ECKARDT, C. B., MAXWELL, J. R., LARTER, S. R. & MANNING, D. A. C. (eds) *Advances in Organic Geochemistry 1991*. Pergamon, Oxford. *Organic Geochemistry*, **19**, 53–75.

DAVIES, R. A. 1978. *A petroleum source rock study of the Kimmeridgian section in Dorsetshire, England*. PhD Thesis, University of London.

DEAN, W. E., ARTHUR, M. A. & CLAYPOOL, G. E. 1986. Depletion of ^{13}C in Cretaceous marine organic matter: source, diagenetic, or environmental signal? *Marine Geology*, **70**, 119–157.

DE LEEUW, J. W., VAN BERGEN, P. F., VAN AARSEN, B. G. K., GATELLIER, J.-P. L. A., SINNINGHE DAMSTÉ & COLLINSON, M. E. 1991. Resistant biomacromolecules as major contributors to kerogen. *Philosophical Transactions of the Royal Society of London*, **B333**, 329–337.

DEMAISON, G. J. & MOORE, G. T. 1980. Anoxic environments and oil source bed genesis. *American Association of Petroleum Geologists Bulletin*, **64**, 1179–1209.

DOWNIE, C. 1955. *The nature and origin of the Kimmeridge oil shale*. PhD Thesis, University of Sheffield.

DUNN, C. E. 1972. *Trace element geochemistry of Kimmeridge sediments in Dorset, north west France and northern Spain*. PhD Thesis, University of London.

EHRMANN, W. U. & THIEDE, J. 1985. *History of Mesozoic and Cenozoic Sediment Fluxes to the North Atlantic Ocean*. Contributions to Sedimentology, **15**.

ELDER, W. P., GUSTASON, E. R. & SAGEMAN, B. B. 1994. Correlation of basinal carbonate cycles to nearshore parasequences in the Late Cretaceous Greenhorn Seaway, Western Interior, U.S.A. *Bulletin of the Geological Society of America*, **106**, 892–902.

EXON, N. 1972. Sedimentation in the outer Flensburg Fjord area (Baltic Sea) since the last Glaciation. *Meyniana*, **22**, 5–62.

FRANK, M. C. & TYSON, R. V. 1995. Parasequence-scale organic facies variations through an Early Carboniferous Yoredale cyclothem (Middle Limestone

Group, Scremerston, Northumberland). *Journal of the Geological Society, London*, **52**, 41–50.

GALLOIS, R. W. 1973. Some detailed correlations in the Upper Kimmeridge Clay in Norfolk and Lincolnshire. *Bulletin of the Geological Survey of Great Britain*, **44**, 63–75.

—— 1978. *A Pilot Study of Oil Shale Occurrences in the Kimmeridge Clay*. Report of the Institute of Geological Sciences, **78/13**.

GOUGH, M. A., FAUZI, R., MANTOURA, C. & PRESTON, M. 1993. Terrestrial plant biopolymers in marine sediments. *Geochimica et Cosmochimica Acta*, **57**, 945–964.

HANTZPERGUE, P. 1985. Les discontinuités sédimentaires majeures dans le Kimméridgien Français: chronologie, extension et corrélations dans les bassins ouest-européens. *Geobios*, **18**, 179–194.

HAQ, B. U., HARDENBOL, J. & VAIL, P. R. 1987. Chronology of fluctuating sea levels since the Triassic. *Science*, **235**, 1156–1167.

HEDGES, J. I. & MANN, D. C. 1979. The lignin geochemistry of marine sediments from the southern Washington coast. *Geochimica et Cosmochimica Acta*, **43**, 1809–1818.

HENRICHS, S. M. & REEBURGH, W. S. 1987. Anaerobic mineralization of marine sediment organic matter: rates and the role of anaerobic processes in the oceanic carbon economy. *Geomicrobiology Journal*, **5**, 191–238.

HERBIN, J. -P. & GEYSSANT, J. R. 1993. 'Ceintures organiques' au Kimméridgien/Tithonien en Angleterre (Yorkshire, Dorset) et en France (Boulonnais). *Comptes Rendus Academie des Sciences Paris, Series II*, **317**, 1309–1316.

——, FERNANDEZ-MARTINEZ, J. -L., EL ALBANI, A., DECONINCK, J. -F., PROUST, J. N., COLBEAUX, J. -P., & VIDIER, J. -P. 1995. Sequence stratigraphy of source rocks applied to the study of the Kimmeridgian/Tithonian in the north-west European shelf (Dorset/UK, Yorkshire/UK and Boulonnais, France). *Marine and Petroleum Geology*, **12**, 177–194.

——, MÜLLER, C., GEYSSANT, J. R., MÉLIERES, F. & PENN, I. E. 1991. Hétérogénéité quantitative et qualitative de la matière organique dans les argiles du Kimméridgien du val de Pickering (Yorkshire, UK): cadre sédimentologique et stratigraphique. *Revue de l'Institut Français du Pétrole*, **46**, 675–712.

——, ——, ——, —— & —— 1993. Variation of the distribution of organic matter within a transgressive systems tract: Kimmeridge Clay (Jurassic), England. *In*: KATZ, B. J. & PRATT, L. M. (eds) 1993. *Source Rocks in a Sequence Stratigraphic Framework*. American Association of Petroleum Geologists Studies in Geology, **37**, 67–100.

HOUSE, M. R. 1985. A new approach to an absolute timescale from measurments of orbital cycles and sedimentary microrhythms. *Nature*, **315**, 721–725.

HUC, A. Y. 1988. Sedimentology of organic matter. *In*: FRIMMEL, F. H. & CHRISTMAN, R. F. (eds) *Humic Substances and their Role in the Environment*. Wiley, Chichester, 215–243.

——, LALLIER-VERGÈS, E., BERTRAND, P., CARPENTIER, B. & HOLLANDER, D. J. 1992. Organic matter response to change of depositional environment in Kimmeridgian shales, Dorset, U.K. *In*: WHELAN, J. K. & FARRINGTON, J. W. (eds) *Productivity, Accumulation, and Preservation of Organic Matter in Recent and Ancient Sediments*. Columbia University Press, New York, 469–486.

IBACH, L. E. J. 1982. Relationship between sedimentation rate and total organic carbon content in ancient marine sediments. *American Association of Petroleum Geologists Bulletin*, **66**, 170–188.

JONES, R. W. 1984. Comparison of carbonate and shale source rocks. *In*: PALACAS, J. G. (ed.) *Petroleum Geochemistry and Source Rock Potential of Carbonate Rocks*. American Association of Petroleum Geologists, Studies in Geology, **18**, 163–180.

—— 1987. Organic facies. *In*: BROOKS, J. & WELTE, D. (eds) *Advances in Petroleum Geochemistry*, Vol. 2, Academic Press, London, 1–90.

—— & DEMAISON, G. J. 1982. Organic facies – stratigraphic concept and exploration tool. *In*: SALDIVAR-SALI, A. (ed.) *Proceedings of the Second ASCOPE Conference and Exhibition, Manilla, October 7–11, 1981*. Asian Council on Petroleum, 51–68.

KATZ, B. J. & PRATT, L. M. (eds) 1993. *Source Rocks in a Sequence Stratigraphic Framework*. American Association of Petroleum Geologists, Studies in Geology, **37**.

KONTOROVICH, A. E., POLYAKOVA, I. D. & FOMICHEV, A. S. 1971. Laws characterising the accumulation of organic matter in old sedimentary bodies (using the example of the Mesozoic deposits of Siberia). *Lithology and Mineral Resources*, **6**, 657–667.

LALLIER-VERGES, E., BERTRAND, P., HUC, A. Y., BÜCKEL, D. & TREMBLAY, P. 1993. Control of the preservation of organic matter by productivity and sulphate reduction in Kimmeridgian shales from Dorset, UK. *Marine and Petroleum Geology*, **10**, 600–605.

LECKIE, D. A., SINGH, C., BLOCH, J., WILSON, M. & WALL, J. 1992. An anoxic event at the Albian–Cenomanian boundary: the Fish Scale Marker Bed, northern Alberta, Canada. *Palaeogeography, Palaeoclimatology, Palaeoecology*, **92**, 139–166.

——, ——, GOODARZI, F., & WALL, J. H. 1990. Organic-rich, radioactive marine shale: a case study of a shallow-water condensed section, Cretaceous Shaftesbury Formation, Alberta, Canada. *Journal of Sedimentary Petrology*, **60**, 101–117.

LOUTIT, T. S., HARDENBOL, J. & VAIL, P. R. 1988. Condensed sections: the key to age determination and correlation of continental margin sequences. *In*: WILGUS, C. K., HASTINGS, B. S., KENDALL, C. G. ST.C., POSAMENTIER, H. W., ROSS, C. A. & VAN WAGONER, J. C. (eds) *Sea-level Changes: an Integrated Approach*. Society of Economic Paleontologists and Mineralogists, Special Publications, **42**, 183–215.

MARZI, R. & RULLKÖTTER, J. 1986. Organic matter accumulation and migrated hydrocarbons in deep-sea sediments of the Mississippi Fan and adjacent intraslope basins, northern Gulf of Mexico. *Mitteilungen Geologische Paläontologische Institut Universitat Hamburg*, **60**, 359–379.

MELNYK, D. H., ATHERSUCH, J. & SMITH, D. G. 1992. Estimating the dispersion of biostratigraphic events in the subsurface by graphic correlation: an example from the Late Jurassic of the Wessex Basin, UK. *Marine and Petroleum Geology*, **9**, 602–607.

——, SMITH, D. G. & AMIRI-GARROUSSI, K. 1994. Filtering and frequency mapping as tools in subsurface cyclostratigraphy, with examples from the Wessex Basin, U.K. *In*: DE BOER, P. L. & SMITH, D. G. (eds) *Orbital Forcing and Cyclic Sequences*. International Association of Sedimentologists, Special Publications, **19**, 35–46.

MILLER, R. G. 1990. A paleoceanographic approach to the Kimmeridge Clay Formation. *In*: HUC, A. Y. (ed.) *Deposition of Organic Facies*. American Association of Petroleum Geologists, Studies in Geology, **30**, 13–26.

MÜLLER, P. J. & SUESS, E. 1979. Productivity, sedimentation rate, and sedimentary organic matter in the oceans. 1. Organic carbon preservation. *Deep-Sea Research*, **A26**, 1347–1362.

NØHR-HANSEN, H. 1989. Visual and chemical kerogen analyses of the Lower Kimmeridge Clay, Westbury, England. *In*: BATTEN, D. J. & KEEN, M. C. (eds) *Northwest European Micropalaeontology and Palynology*. British Micropalaeontological Society Series, Ellis Horwood, Chichester, 118–131.

OSCHMANN, W. 1988. Kimmeridge Clay sedimentation – a new cyclic model. *Palaeogeography, Palaeoclimatology, Palaeoecology*, **65**, 217–251.

—— 1991. Distribution, dynamics and palaeoecology of Kimmeridgian (Upper Jurassic) shelf anoxia in western Europe. *In*: TYSON, R. V. & PEARSON, T. H. (eds) *Modern and Ancient Continental Shelf Anoxia*. Geological Society, London, Special Publications, **58**, 381–395.

PARTINGTON, M. A., COPESTAKE, P., MITCHENER, B. C. & UNDERHILL, J. R. 1993*a*. Biostratigraphic calibration of genetic stratigraphic sequences in the Jurassic–lowermost Cretaceous (Hettangian to Ryazanian) of the North Sea and adjacent areas. *In*: PARKER, J. R. (ed.) *Petroleum Geology of Northwest Europe: Proceedings of the 4th Conference*. Geological Society, London, **1**, 371–386.

——, MITCHENER, B. C., MILTON, N. J . & FRASER, A. J. 1993*b*. Genetic sequence stratigraphy for the North Sea Late Jurassic and Early Cretaceous: distribution and prediction of Kimmeridgian–Ryazanian reservoirs in the North Sea and adjacent areas. *In*: PARKER, J. R. (ed.) *Petroleum Geology of Northwest Europe: Proceedings of the 4th Conference*. Geological Society, London, 347–370.

PASLEY, M. A., GREGORY, W. A. & HART, G. F. 1991. Organic matter variations in transgressive and regressive shales. *Organic Geochemistry*, **17**, 483–509.

——, RILEY, G. W. & NUMMEDAL, D. 1993. Sequence stratigraphic significance of organic matter variations: example from the Upper Cretaceous Mancos Shale of the San Juan Basin, New Mexico. *In*: KATZ, B. J. & PRATT, L. M. (eds) 1993. *Source Rocks in a Sequence Stratigraphic Framework*. American Association of Petroleum Geologists, Studies in Geology, **37**, 221–241.

PEDERSEN, T. F. & CALVERT, S. E. 1990. Anoxia vs. productivity: what controls the formation of organic-carbon-rich sediments and sedimentary rocks? *American Association of Petroleum Geologists Bulletin*, **74**, 454–466.

PENN, I. E., COX, B. M. & GALLOIS, R. W. 1986. Towards precision in stratigraphy: geophysical log correlation of Upper Jurassic (including Callovian) strata of the Eastern England Shelf. *Journal of the Geological Society, London*, **143**, 381–410.

POLLEHNE, F. 1986. Benthic nutrient regeneration processes in different sediment types of Kiel Bight. *Ophelia*, **26**, 359–368.

PRADIER, B. & BERTRAND, P. 1992. Étude à résolution d'un cycle du carbon organique de roche-mère du Kimmeridgien du Yorkshire (G.B.): relation entre composition pétrographique du contenu organique observé *in situ* teneur en carbone organique et qualité pétroligène. *Compte Rendus Academie des Sciences Paris, Series II*, **315**, 187–192.

PROUST, J. -N., DECONINCK, J. -F., GEYSSANT, J. R., HERBIN, J. -P., & VIDIER, J. -P. 1993. Nouvelles données sédimentologiques dans le Kimméridgien et le Tithonien du Boulonnais (France). *Comptes Rendus Academie des Sciences Paris, Series* II, **316**, 363–369.

RAMANAMPISOA, L. & DISNAR, J. -R. 1994. Primary control of paleoproduction on organic matter preservation and accumulation in the Kimmeridge rocks of Yorkshire (UK). *Organic Geochemistry*, **21**, 1153-1167.

——, BERTRAND, P., DISNAR, J. -R., LALLIER-VERGES, E., PRADIER, B., & TRIBOVILLARD, N. -P. 1992. Étude à haute résolution d'un cycle de carbone organique des argiles du Kimméridgien du Yorkshire (Grande-Bretagne): résultats preliminaires de géochemie et de pétrographie organique. *Comptes Rendus Academie des Sciences Paris, Series II*, **314**, 1493–1498.

RASMUSSEN, A., KRISTENSEN, S. E., VAN VEEN, P. M., STØLAN, T. & VAIL, P. R. 1992. Use of sequence stratigraphy to define a semi-stratigraphic play in Anisian sequences, southwestern Barents Sea. *In*: VORREN, T. O., BERSAGER, E., DAHL-STAMNES, Ø. A., HOLTER, E., JOHANSEN, B., LIE, E. & LUND, T. B. (eds) *Arctic Geology and Petroleum Potential*. Norsk Petroleumsforening Special Publications, **2**, Elsevier, Amsterdam, 439–455.

ROGERS, M. A. 1980. Application of organic facies concepts to hydrocarbon source rock evaluation. *In*: *Proceedings of the 10th World Petroleum Congress, Bucharest 1979*, Vol. 2, Heyden, London, 23–30.

SAVRDA, C. E. & BOTTJER, D. J. 1991. Oxygen-related biofacies in marine strata: an overview and update. *In*: TYSON, R. V. & PEARSON, T. H. (eds) *Modern and Ancient Continental Shelf Anoxia*. Geological Society, London, Special Publications, **58**, 201–219.

SCOTCHMAN, I. C. 1991. Kerogen facies and maturity of the Kimmeridge Clay Formation in southern and eastern England. *Marine and Petroleum Geology*, **8**, 278–295.

SELLWOOD, B. W., SCOTT, J. & LUNN, G. 1986. Mesozoic basin evolution in southern England. *Proceedings of the Geologists' Association, London*, **97**, 259–289.

SLY, P. G. 1978. Sedimentary processes in lakes. *In*: LERMAN, A. G. (ed.) *Lakes – Chemistry, Geology, and Physics*. Springer, Berlin, 70–89.

STEIN, R. 1986. Organic carbon and sedimentation rate –

further evidence for anoxic deep-water conditions in the Cenomanian/Turonian Atlantic Ocean. *Marine Geology*, **72**, 199–209.

SUMMERHAYES, C. P. 1983. Sedimentation of organic matter in upwelling regimes. *In*: THIEDE, J. & SUESS, E. (eds) *Coastal Upwelling: its Sediment Record Part B: Sedimentary Records of Ancient Coastal Upwelling*. NATO Conference Series IV, 10b. Plenum Press, New York, 29–72.

THORNE, J. A. 1992. An analysis of the implicit assumptions of the methodology of seismic sequence stratigraphy. *In*: WATKINS, J., FENG, Z. & MCMILLEN, K. (eds) *Geology and Geophysics of Continental Margins*. American Association of Petroleum Geologists, Memoirs, **53**, 375–397.

TISSOT, B., DEMAISON, G., MASSON, P., DELTEIL, J. R. & COMBAZ, A. 1980. Paleoenvironment and petroleum potential of Middle Cretaceous black shales in Atlantic Basins. *Bulletin of the American Association of Petroleum Geologists*, **64**, 2051–2063.

——, DURAND, B., ESPITALIE, J. & COMBAZ, A. 1974. Influence of nature and diagenesis of organic matter in formation of petroleum. *Bulletin of the American Association of Petroleum Geologists*, **58**, 499–506.

TRIBOVILLARD, N. -P., DESPRAIRIES, A., LALLIER-VERGÈS, E., BERTRAND, P., MOUREAU, N., RAMDANI, A. & RAMANAMPISOA, L. 1994. Geochemical study of organic-matter rich cycles from the Kimmeridge Clay Formation of Yorkshire (UK): productivity versus anoxia. *Palaeogeography, Palaeoclimatology, Palaeoecology*, **108**, 165–181.

TYSON, R.V. 1985. *Palynofacies and sedimentology of some Late Jurassic sediments from the British Isles and northern North Sea*. PhD Thesis, The Open University, Milton Keynes.

—— 1989. Late Jurassic palynofacies trends, Piper and Kimmeridge Clay Formations, UK onshore and northern North Sea. *In*: BATTEN, D. J. & KEEN, M. C. (eds) *Northwest European Micropalaeontology and Palynology*. British Micropalaeontological Society Series, Ellis Horwood, Chichester, 135–172.

—— 1993. Palynofacies analysis. *In*: JENKINS, D. G. (ed.) *Applied Micropalaeontology*. Kluwer, Dordrecht, 153–191.

—— 1995. *Sedimentary Organic Matter: Organic Facies and Palynofacies*. Chapman & Hall, London.

—— & PEARSON, T. H. 1991. Modern and ancient continental shelf anoxia: an overview. *In*: TYSON, R. V. & PEARSON, T. H. (eds) *Modern and Ancient Continental Shelf Anoxia*. Geological Society, London, Special Publications, **58**, 1–24.

——, WILSON, R. C. L. & DOWNIE, C. 1979. A stratified water column environmental model for the Type Kimmeridge Clay. *Nature*, **277**, 377–380.

VAIL, P. R., AUDEMARD, F., BOWMAN, S. A., EISNER, P. N. & PEREZ-CRUZ, C. 1991. The stratigraphic signatures of tectonics, eustasy and sedimentology – an overview. *In*: EINSELE, G., RICKEN, W. & SEILACHER, A. (eds) *Cycles and Events in Stratigraphy*. Springer-Verlag, Berlin, 617–659.

VAN WAGONER, J. C., MITCHUM, R. M., CAMPION, K. M. & RAHMANIAN, V. D. 1990. *Siliciclastic Sequence Stratigraphy in Well Logs, Cores and Outcrops: Concepts for High-Resolution Correlation of Time and Facies*. American Association of Petroleum Geologists, Methods in Exploration Series, **7**, 55pp.

WAPLES, D. W. 1983. Reappraisal of anoxia and organic richness, with emphasis on Cretaceous of North Atlantic. *American Association of Petroleum Geologists Bulletin*, **67**, 963–978.

WATERHOUSE, H. K. 1992. *Quantitative palynofacies analysis of Jurassic climatic cycles*. PhD Thesis, University of Southampton.

WHITTAKER, A., HOLLIDAY, D. W. & PENN, I. E. 1985. *Geophysical Logs in British Stratigraphy*. Geological Society, London, Special Reports, **18**.

WIGNALL, P. B. 1990. *Benthic Palaeoecology of the Late Jurassic Kimmeridge Clay of England*. Special Papers in Palaeontology, **43**.

—— 1991. Test of the concepts of sequence stratigraphy in the Kimmeridgian (Late Jurassic) of England and northern France. *Marine and Petroleum Geology*, **8**, 430–441.

—— & RUFFELL, A. H. 1990. The influence of a sudden climatic change on marine deposition in the Kimmeridgian of northwest Europe. *Journal of the Geological Society of London*, **147**, 365–371.

Sedimentological and geochemical controls on ooidal ironstone and 'bone-bed' formation and some comments on their sequence-stratigraphical significance

J. H. S. MACQUAKER[1], K. G. TAYLOR[1], T. P. YOUNG[2] & C. D. CURTIS[1]

[1]*Department of Earth Sciences, University of Manchester, Manchester M13 9PL, UK*

[2]*Department of Earth Sciences, University of Wales Cardiff, PO Box 914, Cardiff CF 3YE, UK*

Abstract: Ironstones and 'bone-beds' are both formed in shallow marine settings where the prevailing sediment accumulation rates are very low. A review of published work shows that both sediment types cap coarsening upward successions and are commonly developed in mudstone-dominated settings. Both can be shown to contain similar sedimentary structures (e.g. rip-up clasts, disarticulated shell debris) and a diverse assemblage of ichnogenera. However, the early diagenetic assemblages within each sediment type are both different from each other and from the mudstones within which they are enclosed. In ooidal ironstones, the cement assemblage comprises glaucony, berthierine and siderite; in 'bone-beds' it comprises early apatite, glaucony and later calcite and pyrite; in the mudstones it is predominantly pyrite. It is argued that the physical environment in which the ironstones and 'bone-beds' formed was similar, whereas their diagenetic environments were rather different. In particular, the characteristic cement assemblage of ironstones was produced in suboxic pore waters which were subjected to intense Fe reduction, whereas the characteristic cement assemblage of 'bone-beds' was produced in suboxic pore waters where there was limited Fe reduction. As a consequence of these chemical and stratigraphical constraints it is believed that 'bone-beds' formed on marine flooding surfaces and ironstones formed either at sequence boundaries, major flooding surfaces or maximum flooding surfaces.

Ooidal ironstones and 'bone-beds' form a distinctive, if volumetrically small, component of the sedimentary record. They are easily recognized in the field, in core and using downhole logging techniques and may have economic significance. Both ironstones and 'bone-beds' are most common within shallow marine sediments and often occur together in the same sedimentary succession. Despite these similarities they are usually considered in isolation from each other and from the sediment in which they are present. The objectives of this paper are: (1) to summarize the sedimentological conditions under which these deposits occur and (2) to describe their characteristic mineral assemblages. From these considerations the chemical and sedimentological conditions in which they formed are constrained and models presented for their formation. Finally, we consider the significance of ooidal ironstones and 'bone-beds' as key horizons within a sequence stratigraphical context (*sensu* Van Wagoner *et al.* 1990).

Sedimentological setting

Ironstones

Ooidal ironstones (terminology from Young 1989*a*) generally form stratigraphically condensed intervals within marine mudstone successions (Bayer 1989; Dreesen 1989; Gehring 1989; Fürsich *et al.* 1992; Young 1993). Where lateral relationships can be observed, many ironstones can be seen to correlate with stratigraphically expanded marine mudstone intervals, e.g. the Liassic Frodingham and Pecten Ironstones of eastern England and Cretaceous ironstones of Alberta, Canada (Plint *et al.* 1993; Taylor & Curtis 1995). It has also been recognized that ooidal ironstones commonly cap coarsening upward cycles (e.g. Hemingway 1951; Van Houten & Karasek 1981; Van Houten & Bhattacharyya 1982; Rawson *et al* 1983; Bayer 1989; Teyssen 1989; Young *et al.* 1991) and that they are overlain by marine mudstones. The stacking patterns of the ironstones and their associated sediments have been variously interpreted as indicating ironstone formation either at sea level lowstands (e.g. Hallam & Bradshaw 1979; Bayer et al. 1985; Dreesen 1989; Madon 1992), during transgression (e.g. Embry 1982; Young 1989*b*; Chan 1992) or at times of maximum flooding (Young 1992). More recently, Burkhalter (1995) has argued that ironstones occur at a number of discontinuities associated with breaks in sediment accumulation on a parasequence scale.

Ironstones commonly contain high-angle cross-bedding, storm couplets, abundant sediment reworking, internal erosion surfaces, intraclasts and

From HESSELBO, S. P. & PARKINSON, D. N. (eds), 1996, *Sequence Stratigraphy in British Geology*, Geological Society Special Publication No. 103, pp. 97–107.

abraded thick-walled shell material. The presence of these components is usually interpreted to indicate that the ironstones were deposited in shallow high-energy conditions (e.g. Sellwood 1972; Hallam & Bradshaw 1979; Bayer *et al.* 1985; Bayer 1989; Guerrak 1989; Teyssen 1989; Young *et al.* 1991). In addition, it has been proposed (e.g. Sellwood 1972) that the faunal evidence indicates that ironstone formation occurred within shallower environments than the underlying and overlying mudstones. Although many workers have proposed that ironstone deposition took place in low-oxygen depositional environments (e.g. Cotter & Link 1993), the majority of ironstones contain an abundant and diverse shelly fauna with well-developed, often tiered infaunal communities (e.g. Sellwood 1972), which suggests that at the time of their formation the bottom waters were well oxygenated.

'Bone-beds'

'Bone-beds' are usually developed on widespread transgressive surfaces (e.g. those from the well-known Upper Silurian Downton Group, Pridoli Series succession at Ludlow, Shropshire, England (Bassett *et al.* 1982) and, like ironstones, their formation has often been linked to the development of a condensed section (e.g. Kidwell 1989). Where lateral relationships can be observed, 'bone-beds' can be correlated with either marine mudstones or erosion surfaces (Macquaker 1994). 'Bone-beds' commonly overlie coarsening-upward successions (e.g. those from the Westbury Formation, Penarth Group, Upper Triassic of southwest Britain) and are themselves usually overlain by marine mudstones. Like the ironstones they commonly contain disarticulated shell debris and reworked intraclasts. 'Bone-beds' are commonly intensely bioturbated and may contain a diverse, tiered assemblage of ichnogenera (e.g. *Thalassinoides*, *Diplocraterion*, *Peleycpodichnus* and *Chondrites*, which have been found in the 'bone-beds' from the Westbury Formation by Macquaker 1994). Researchers have suggested that their formation is linked to both storm reworking (e.g. Smith & Ainsworth 1989) and areas of high organic productivity (Macquaker 1994). The origin of the vertebrate-derived skeletal debris (predominantly fish scales, teeth, coprolites and other robust skeletal elements) in these units has been discussed by many workers (summarized in Macquaker 1994). It has variously been suggested that this material is derived from: (a) mass mortality events; (b) the product of winnowing a large body of sediment; (c) shoreline lags; and (d) reduced clastic dilution in upwelling environments. The current consensus seems to be that 'bone-beds' formed in areas of high organic productivity and low sediment accumulation rates in association with a major flooding event and the development of a condensed section (e.g. Macquaker 1994).

Mineralogy and petrography

Ironstones

The predominant iron minerals within oolitic ironstone deposits are berthierine, chamosite, siderite and goethite. Both berthierine and chamosite are phyllosilicates with similar chemical composition,

Fig. 1. (**a**) Low magnification optical micrograph of an ironstone from the Dunlin Formation, Block 210/25, UK continental shelf. The section illustrates calcite- and siderite-cemented ooidal ironstone. The ooid cores comprise glaucony (g); these are initially cemented by tangential berthierine (b) and latterly by a radial calcite cement (1). The matrix comprises detrital quartz and minor clay and late authigenic siderite (s) and calcite (2). In this micrograph the textural relationship between the late rhombic siderite and the matrix calcite is equivocal. Field of view 700 μm. (**b**) Low magnification backscattered electron micrograph of an ironstone (Pecten Ironstone Formation, Lower Jurassic, East Midlands Shelf onshore UK). The micrograph illustrates the edge of an ooid. The core of the ooid is composed of a glaucony grain (g) with a low η coefficient; this is enclosed by a tangential berthierine cement (b) with a slightly higher η coefficient, which in turn is cemented by siderite (s) with a high η coefficient and calcite (1) with an intermediate η coefficient. Siderite authigenesis both pre- and post-dates calcite precipitation. Field of view 494 μm. (**c**) High magnification backscattered electron micrograph of an ironstone (Frodingham Ironstone Formation, Lower Jurassic, East Midlands Shelf, onshore UK). The micrograph illustrates rhombic siderite crystals (s) with a high η coefficient, enclosed in a berthierine matrix (b) intermediate η coefficient. Field of view 140 μm. (**d**) Low magnification optical micrograph of a 'bone-bed' from the Upper Silurian (Downton Group, Pridoli Series) succession from Ludlow, Shropshire, England. The section illustrates phosphatic debris [e.g. fish scales (f), teeth (t), small nodules (n)] and quartz grains (q) which have been enclosed by a calcite cement (1). Field of view 700 μm. (**e**) Low magnification backscattered electron micrograph of the 'bone-bed' from the Upper Silurian (Downton Group, Pridoli Series) succession from Ludlow, Shropshire, England. The image illustrates that the detrital phosphatic debris (d) with intermediate η coefficient, is enclosed by a zoned early apatite cement (z) with an intermediate and variable η coefficient; and that this in turn is enclosed by a calcite cement (1) with a lower η coefficient. Field of view 70 μm. (**f**) High magnification backscattered electron micrograph of a 'bone-bed' from the Upper Triassic (Westbury Formation, southwest England). The image illustrates the complex zonation fabrics (z) picked out by their variable η coefficient in the early apatite cements. Field of view 49 μm.

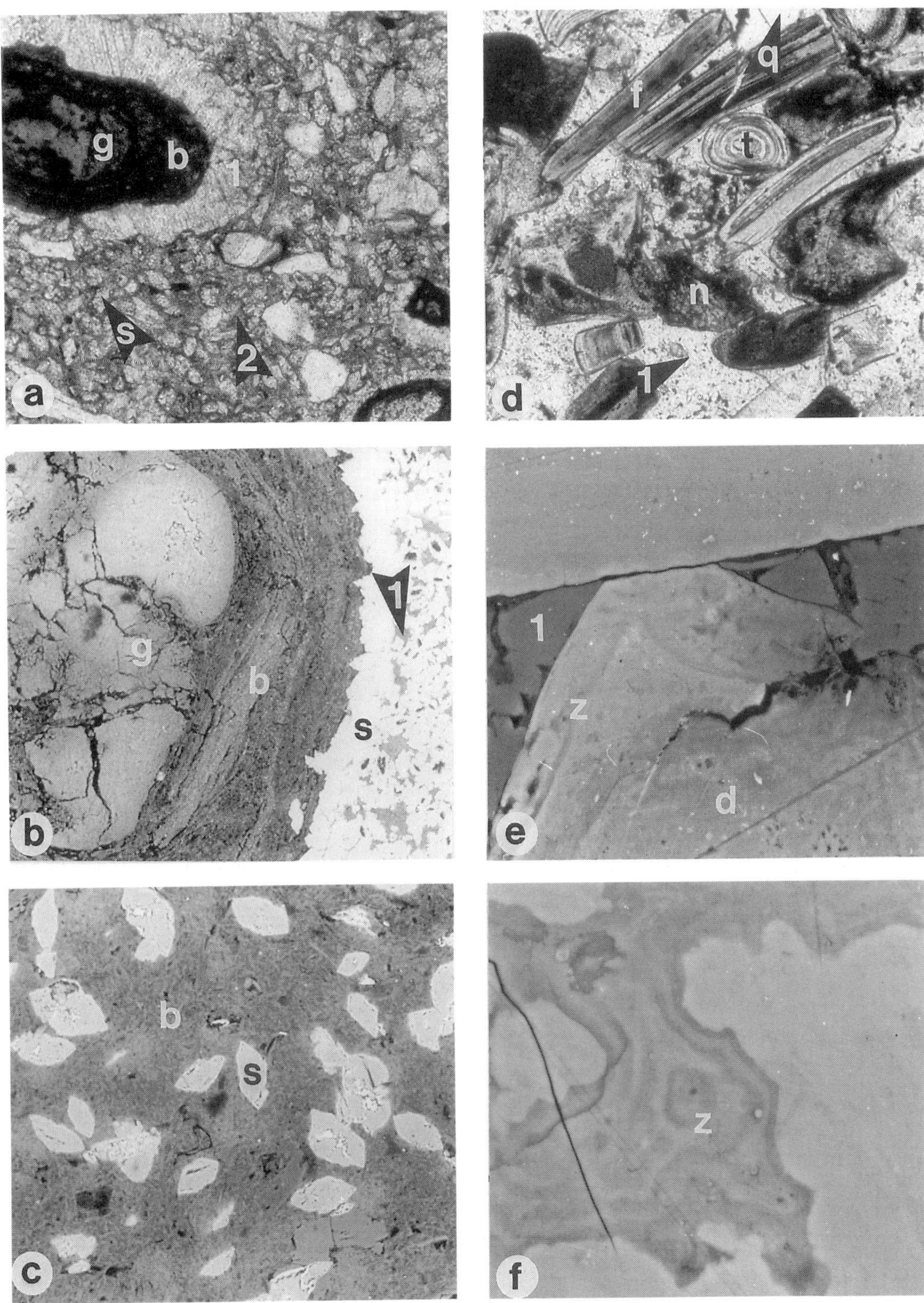
a
g
b
1
s
2
d
f
q
t
n
1
b
g
b
1
s
e
1
z
d
c
b
s
f
z

but with differing structures (0.7 nm and 1.4 nm basal spacings, respectively). It is generally believed that chamosite is a higher temperature alteration product of berthierine (Iijima & Matsumoto 1982; Ahn & Peacor 1985; Jahren & Aagaard 1989). Berthierine may occur as ooids (Fig. 1a), or grain-rimming (Fig. 1b) and pore-filling (Fig. 1c) cements. Berthierine ooids commonly exhibit a tangential fabric of berthierine crystals surrounding a nucleus (Fig. 1b), which may either be a detrital grain, biogenic fragment, glaucony grain (Fig. 1a, and b) or a pellet of berthierine mud. Grain-rimming and pore-filling berthierine cements are commonly observed to enclose earlier berthierine and goethite ooids. Siderite is commonly present as a later cement (Fig. 1a–c) within ironstones. It occurs both as individual rhombohedra set within a berthierine matrix (Fig. 1c) and as a replacement mineral for berthierine ooids (Gehring 1990). In some ironstones, e.g. the Liassic Pecten Ironstone, glaucony pellets are present (Taylor & Curtis 1995; Fig. 1b)

Common to all ironstones is the low pyrite content (Spears 1989; Young 1989*a*). Where pyrite is present it can be seen to post-date berthierine and siderite authigenesis and commonly replaces berthierine ooids (e.g. the 'Snap Band' in the Frodingham Ironstone Formation and the 'Sulphur Band' in the Cleveland Ironstone Formation; Young *et al.* 1991). In many ironstones a range of later carbonate cements is present, which have been variously ascribed to either meteoric water ingress or the redistribution of primary shell material, e.g. the Frodingham Ironstone (Aggett 1990). Light $\delta^{18}O$ data for early carbonate within ironstones have been interpreted as evidence for meteoric ingress (e.g. Aggett 1990; Plint *et al.* 1993), supporting the sedimentological interpretation of shallow water deposition.

'Bone-beds'

The predominant framework grains in 'bone-beds' are composed of quartz and phosphatic debris (Fig. 1d). These components are initially cemented by minor, zoned apatite cements which form both discrete nodules and infill pore space (Fig. 1d–f) and latterly by either pore-filling carbonates (e.g. non-ferroan calcite and ferroan calcite; Fig. 1d & e) and/or pyrite cements. Textural evidence suggests that the pyrite cements, where present, post-date apatite authigenesis. In addition to the major components described above, some 'bone-beds' also contain minor glaucony. In these units the glaucony grains are commonly enclosed by both carbonate and pyrite cements; however, the textural relationship between these two is usually equivocal.

Discussion

The presence of early cements within both ironstones and 'bone-beds' and their stratigraphical relationships with other units suggest that they both formed during periods of low net sediment accumulation in aerated and reworked sediment [note the presence in both of aerobic, tiered infaunal communities (cf. Bromley 1990) and rip-up clasts]. Although the physical conditions governing the formation of each of these units is likely to have been broadly similar, their contrasting diagenetic mineral assemblages suggests that the composition of the pore waters, at the time of their formation, was rather different. The following section discusses the geochemical constraints that control the formation of both sediment types in an attempt to shed light on the specific conditions required for the formation of each.

Geochemical constraints on ooidal ironstone formation

In ironstones the presence of the Fe(II)-rich authigenic minerals – berthierine and siderite – clearly indicates that a supply of reduced iron was available in the early diagenetic pore waters. In addition, carbonate must also have been present for siderite to have precipitated. A major mechanism by which iron oxides are reduced during early burial, following the removal of free oxygen from the pore waters, is by a process of bacterially mediated, organic matter oxidation and iron reduction (Lovely & Phillips 1986, 1988; Canfield 1989). As a result of these reactions, Fe^{2+}, bicarbonate ions and hydroxyl ions are released into the pore waters [summarized in Equation 1; Curtis 1987]

$$212Fe_2O_3 + (CH_2O)_{106}(NH_3)_{16}H_3PO_4 + 318H_2O \rightarrow 424Fe^{2+} + 106HCO_3^- + 16NH_3 + H_3PO_4 + 742OH^- \quad (1)$$

Within most modern shallow shelf sediments, on removal of free oxygen, the predominant mechanism of organic matter oxidation is bacterially mediated sulphate reduction (Claypool & Kaplan 1974; Canfield *et al.* 1993). This process causes sulphide to be released into the pore waters (Berner 1970, 1984). Curtis and Spears (1968) have shown that in the presence of any sulphide, iron in solution will preferentially react to form iron sulphides, limiting the formation of other iron-rich minerals. The presence of abundant early berthierine and siderite cements and the absence of early pyrite suggests that sulphate reduction was not a significant process during the early diagenesis of ironstones. In addition, the fact that both berthierine and siderite pre-date any pyrite authigenesis in ironstones

indicates that the precipitation of these minerals took place before sulphate reduction.

Within hemipelagic environments, Froelich *et al.* (1979) recognized a zone of Mn and Fe reduction on removal of pore water oxygen before sulphate reduction. The reactions which occurred in this zone were termed suboxic by Froelich *et al.* (1979) and non-sulphidic-post-oxic by Berner (1981). Within the iron reduction zone, high Fe^{2+} activity coupled with negligible sulphide activity gives iron minerals other than iron sulphide the potential to precipitate. From this discussion, it is likely that the characteristic ironstone mineral assemblage must have formed within this zone. In addition, this interpretation is supported by the limited stable carbon isotope data available for siderite within ironstones (Gehring 1990).

During the early stages of suboxic diagenesis, iron reduction produces four times as much Fe(II) as HCO_3^- [see equation (1)]. Thus the Fe^{2+} pore water activities are relatively high, whereas carbonate activities would have been relatively low. Under such conditions it has been shown (Curtis 1987; Taylor & Curtis 1995) that berthierine precipitation can take place as a result of the destabilization of detrital clay minerals. During continued suboxic diagenesis, however, carbonate activities increase and siderite precipitation is more likely (Curtis & Spears 1968). This is supported by the common observation within ironstones of siderite replacing berthierine (e.g. Gehring 1990).

In addition to the Fe(II)-rich minerals, ironstones commonly contain small amounts of the predominantly ferric mineral glaucony. Glaucony is commonly present in the cores of the ooids (e.g. the Pecten Ironstone). Fe(III) is only sparingly soluble at normal marine pH (6–8), but transient Fe(III) solubility would be expected at the interface between oxic and anoxic conditions and it is at this interface that the glaucony is thought to have formed (Berner 1981). Thus the presence of glaucony texturally enclosed by berthierine and siderite in the ironstones is unsurprising as the glaucony would have precipitated at the oxic–suboxic interface before berthierine and siderite precipitation.

The predominance of suboxic diagenesis in ironstones can be related to the long residence time of the precursor sediment in the mobile surficial sediment veneer, caused by both very low sediment accumulation rates and extended physical reworking. These conditions are likely to have allowed both prolonged oxic zone diagenesis and frequent re-oxidation of buried anoxic sediment. As a result, the sediment buried would have been relatively low in organic matter and high in iron oxide. It has been shown that both low organic matter and high iron oxide contents should be expected to lead to the inhibition of sulphate reduction by iron reduction (Berner 1981; Coleman 1985; Lovley & Phillips 1987; Lovley 1991; Chapelle & Lovley 1992). A similar mechanism has been argued to explain the predominance of iron reduction within modern Amazon Shelf sediments (Aller *et al.* 1986; Aller 1993). Therefore the physical conditions of ironstone deposition led to the predominance of iron reduction during early diagenesis and the precipitation of berthierine and siderite cements.

There has been much speculation as to whether verdine deposits in recent shelf sediments are a modern analogue for ancient berthierine deposits (Odin 1988). The green grains of the recent verdine deposits are dominated by the poorly known 0.7 nm clay mineral odinite (a ferric, 1:1 clay mineral; Bailey 1988), which some workers have proposed converts to berthierine on burial. In addition, Kimberley (1989) has reported berthierine ooids from modern sediments on the Venezualan continental shelf. Before conclusions can be drawn about possible modern analogues, however, detailed comparisons between ancient and modern examples are required.

Geochemical constraints on 'bone-bed' formation

In 'bone-beds' the presence of early apatite cements, which are enclosed by calcite and pyrite cements, indicates that during early diagenesis the pore water phosphate activities were high and that apatite authigenesis pre-dates sulphate reduction. Given the preceding discussion it is likely that this occurred in either the oxic or suboxic zones. The origin of phosphorus in marine pore waters has been extensively discussed (e.g. Krum & Berner 1982; Cook & McElhinny 1979; Chase & Sayles 1980; Burnett 1990) and it seems most probable that the phosphorus in these 'bone-beds' was derived from the dissolution of phosphatic skeletal debris, and the breakdown of organic matter (particularly phospholipids) [Equation 2]

$$(CH_2O)_{106}(NH_3)_{16}H_3PO_4 + 106O_2 \rightarrow 106HCO_3^- + 16NH_3PO_4 + 106H^+ \quad (2)$$

This reaction supplies both phosphate and bicarbonate to the pore waters and generates weak acids. In normal marine pore waters, where the pH is approximately 7 and there is high carbonate activity, carbonate precipitation is to be expected at the sediment–water interface (Drever 1971). The presence of early apatite cements in these units is perhaps surprising; it implies either that the pore waters were not normal marine or that they were acidic, as it has been shown that apatite precipitation is favoured over carbonate precipitation at

a)

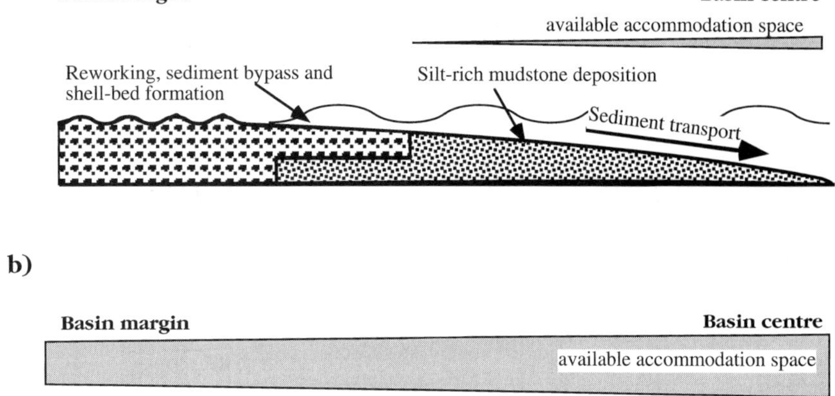

b)

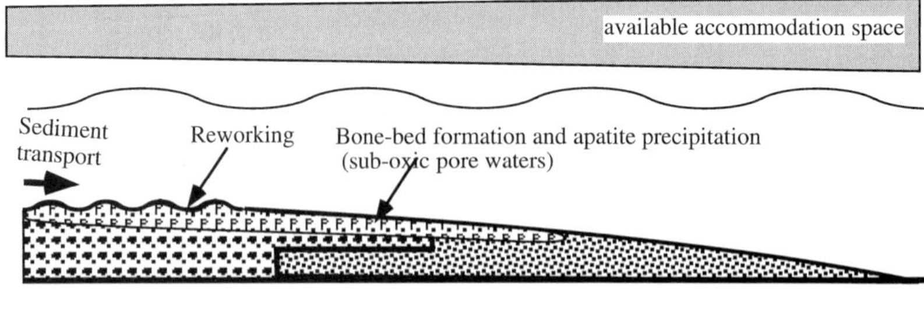

c)

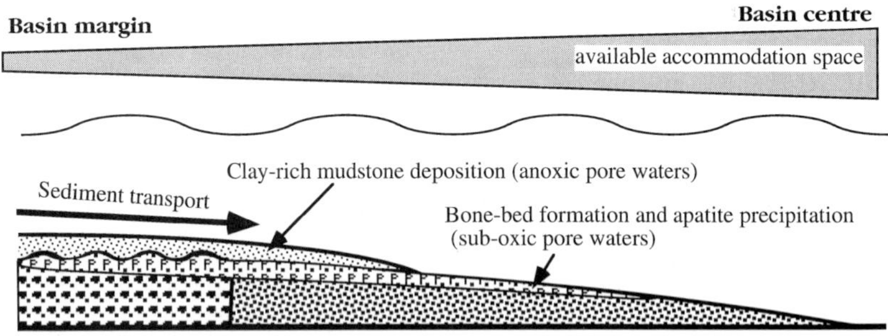

Fig. 2. 'Bone-bed' formation. (**a**) Starting condition; (**b**) rapid flooding, which causes an increase in the length of the sediment transport path, sediment reworking and apatite precipitation on the maximum flooding surface; and (**c**) sediment gradually progrades over maximum flooding surface, re-establishing anoxic – sulphidic pore waters and restricting apatite precipitation.

low pH values (Atlas & Pytkowicz 1977; Nathan & Sass 1981). In marine sediments one of the main contributors to alkalinity is iron reduction [see Equation (1)]. Given this observation, it is potentially possible to generate the relatively acidic pore waters and thereby limit carbonate formation by removing most of the detrital iron [Equation (2); see also Macquaker 1994]. In addition, as the sediments are obviously reworked and accumulated very slowly, it is also possible to generate this acidity by (a) sulphide oxidation [Equation (3); see also Benmore *et al.* (1983)] and (b) Fe(II) oxidation [Equation (4)] at the oxic–anoxic interface.

$$H_2S + 2O_2 \rightarrow H_2SO_4 \quad (3)$$

$$4Fe^{2+} + O_2 + 6H_2O \rightarrow 4FeO \cdot OH + 8H^+ \quad (4)$$

However, it is worth recalling that in recent apatite-precipitating environments, Fe-redox cycling provides a key mechanism by which phosphate is transported from the surface oxic zone into the deeper diagenetic zones (Heggie *et al.* 1990).

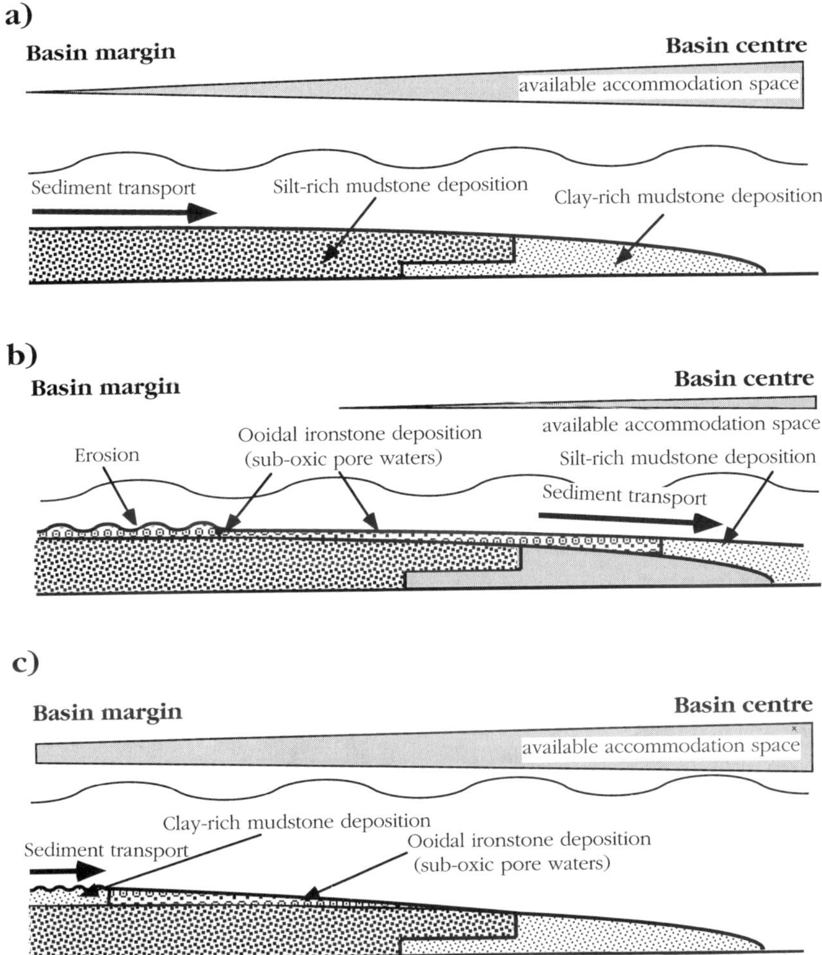

Fig. 3. Ironstone formation. (**a**) Starting condition; (**b**) ooidal ironstone formation at sequence boundaries – in this setting the availability of accommodation space is limited, sediment by-pass occurs and overall sediment accumulation rates are very low; and (**c**) ooidal ironstone formation on maximum flooding surfaces – in this setting unfilled accommodation space is available, the sediment transport path lengths are long and the rates of sediment accumulation are low.

Given the applicability of this analogue it is (a) probable that Fe-redox cycling was contributing to the processes of apatite precipitation and (b) unlikely that the supply of detrital iron was cut off completely.

Given these comments it is proposed that apatite precipitation occurred just above the oxic–sub-oxic/sulphidic interface in conditions where both pore-water phosphate concentrations and carbonate activities were high, and where the pH was low [as a result of the limited acid buffering effect of Fe(II) reduction to offset the acid-generating reactions of organic matter, sulphide and Fe(III) oxidation; Macquaker 1994]. Inevitably, apatite precipitation would also be favoured if there was high primary organic productivity within the environment to produce the necessary increased dissolved pore water phosphate concentrations.

Depositional environments of ironstones and 'bone-beds': implications for sequence stratigraphy

Both ironstones and 'bone-beds' were formed in relatively shallow water, where the sediments were well mixed and the rates of sediment accumulation were either very slow or zero. In addition, as both sediment types were reworked and bioturbated it is likely that, initially at least, the sediment pore waters were oxic. Following initial oxic diagenesis, the diagenetic paths of each sediment type diverged. In ironstones Fe reduction was the dominant suboxic diagenetic process and glaucony, berthierine and siderite cement assemblages were precipitated. In contrast, within 'bone-beds' the suboxic diagenetic processes were iron-limited and the resulting cement assemblage which precipitated was composed of glaucony and apatite. Diagenesis within both ironstones and 'bone-beds' differed to that which occurred within the enclosing mudstones, where sulphate reduction was the predominant process and pyrite and non-ferroan calcite cement formed (e.g. Irwin *et al.* 1977).

In a sequence stratigraphical context, very slow sediment accumulation rates occur: (a) in regions where there is near-source sediment trapping of the coarser clastic sediment; (b) in distal regions where there is very little sediment supply; and (c) in proximal regions where there is abundant sediment supply but no available accommodation space (Jervey 1988; Loutit *et al.* 1988; Kidwell 1989; Van Wagoner *et al.* 1990).

The availability of iron and the relative organic productivity of the environment probably had a significant bearing on which oxic/suboxic cement type precipitated. Macquaker (1994) has postulated that the availability of iron might be controlled by the availability of detrital sediment. During marine flooding events basinal sediments would become iron-poor as a result of the increased availability of accommodation space updip within the basin and, consequently, clastic sediment deposition being restricted to these regions. Assuming that apatite precipitation is indeed linked to a reduced supply of iron, then 'bone-beds' may be interpreted to form on flooding surfaces. These processes are summarized in Fig. 2.

Conversely, ironstones are likely to form wherever net sediment accumulation rates are very low, energy conditions are high and Fe(III) is available. Such conditions can be envisaged to occur either during periods of sediment by-pass, when the water is shallow and there is no available accommodation space, or in shallow conditions where near-source sediment trapping has occurred. If suitable conditions are present, ironstones can be envisaged to form on a number of different stratal surfaces. These alternatives are summarized in Fig. 3. Distinguishing between formation at lowstand, transgression and maximum flooding for any particular deposit will, therefore, involve integrating data from the surrounding sediments and cannot unambiguously be determined from an examination of the ironstone unit in isolation.

Conclusions

Ironstones and 'bone-beds' share many characteristics. Firstly, they both occur in marine sediments close to the top of coarsening upward successions and are usually overlain by mudstones. Secondly, they show evidence of extensive reworking and of having been deposited in regions of very low net sediment accumulation. As a result of these specific sedimentological conditions the diagenesis of both, at least initially, was dominated by oxic and suboxic diagenetic reactions in contrast with the enclosing mudstones, where sulphidic diagenetic reactions were dominant. The nature of the early diagenesis of these two rock types, however, varies significantly, with ironstone formation requiring a supply of iron to form a glaucony, berthierine, siderite assemblage, whereas 'bone-bed' formation requires no iron, low pore water pH, high concentrations of dissolved phosphate and carbonate to precipitate an apatite-dominated assemblage.

The sedimentological and geochemical data show that both ironstones and 'bone-beds' form on major stratal surfaces. The different geochemical requirements have led us to infer that 'bone-beds' formed during periods of marine flooding where the detrital iron supply was restricted and sediment accumulation rates were low. In contrast, ironstones formed in shallow conditions (above storm wave base), where net sediment accumulation rates were low and detrital Fe(III) was available. The conditions apply at sequence boundaries, major flooding surfaces or maximum flooding surfaces.

We are very grateful to S. Burley, J. Aggett and R. Gawthorpe for stimulating discussions during the gestation period of this work. We acknowledge NERC (Grant Number GR3/7887 and the Directorate Controlled PES programme) and Esso/Shell for financial support to undertake this research. We are also very grateful to the Universities of Manchester, Cardiff and Sheffield for logistical support. Finally, we thank P. Manning and Esso for allowing us to illustrate some of the key textures with their samples.

References

AGGETT, J. R. 1990. *The sedimentology, mineralogy, and geochemistry of the Frodingham Ironstone: implications for the genesis of ooidal ironstones*. PhD Thesis, University of Manchester.

AHN, J. H. & PEACOR, D. R. 1985. Transmission electron microscopic study of diagenetic chlorite in Gulf Coast argillaceous sediments. *Clays and Clay Minerals*, **33**, 228–236.

ALLER, R. C. 1993. Influence of terrestrial weathering on early diagenetic reactions in continental shelf sediments. *Chemical Geology*, **107**, 437–438.

——, MACKIN, J. E. & COX, R. T. 1986. Diagenesis of Fe and S in Amazon inner shelf muds: apparent dominance of Fe-reduction and implications for the genesis of ironstones. *Continental Shelf Research*, **6**, 263–289.

ATLAS, E. & PYTKOWICZ, R. M. 1977. Solubility behaviour of apatites in sea-water. *Limnology and Oceanography*, **22**, 290–300.

BAILEY, S. W. 1988. Odinite, a new dioctahedral-trioctahedral Fe^{3+} rich 1:1 clay mineral. *Clay Minerals*, **23**, 237–247.

BASSETT, M. G., LAWSON, J. D. & WHITE, D. E. 1982. The Downton Series as the fourth Series of the Silurian System. *Lethaia*, **15**, 1–24.

BAYER, U. 1989. Stratigraphic and environmental patterns of ironstone deposits. *In*: YOUNG, T. P. & TAYLOR, W. E. G. (eds) *Phanerozoic Ironstones*. Geological Society, London, Special Publications, **46**, 105–117.

——, ALTHEINER, E. & DEUTSCHLE, W. 1985. Environmental evolution in shallow epicontinental seas: sedimentary cycles and bed formation. *In*: BAYER, U. & SEILACHER, A. (eds) *Sedimentary and Evolutionary Cycles*. Lecture Notes in Earth Sciences No. 1, Springer, Heidelberg, 347–381.

BENMORE, R. A., COLEMAN, M. L. & MCARTHUR, J. M. 1983. Origin of sedimentary francolite from its sulphur and carbon isotope composition. *Nature*, **302**, 516–518.

BERNER R. A. 1970. Sedimentary pyrite formation. *American Journal of Science*, **268**, 1–23.

—— 1981. A new geochemical classification of sedimentary environments. *Journal of Sedimentary Petrology*, **51**, 359–365.

—— 1984. Sedimentary pyrite formation: an update. *Geochimica et Cosmochimica Acta*, **48**, 605–617.

BROMLEY, R. G. 1990. *Trace Fossils: Biology and Taphonomy*. Special Topics in Palaeontology. Unwin Hyman, London.

BURKHALTER, R. M. 1995. Ooidal ironstones and ferruginous microbialites: origin and relation to sequence stratigraphy (Aalenian and Bajocian, Swiss Jura Mountains). *Sedimentology*, **42**, 57–74.

BURNETT, W. C. 1990. Phosphorite growth and sediment dynamics in the modern Peru shelf upwelling system. *In*: BURNETT, W. C. & RIGGS, S. R. (eds) *Phosphate Deposits of the World, Vol. 3*. Cambridge University Press, Cambridge, 62–72.

CANFIELD, D. E. 1989. Reactive iron in marine sediments. *Geochimica et Cosmochimica Acta*, **53**, 619–632.

——, THAMDROP, B. & HANSEN, J. W. 1993. The anaerobic degradation of organic matter in Danish coastal sediments: iron reduction, manganese reduction and sulfate reduction. *Geochimica et Cosmochimica Acta*, **57**, 3867–3883.

CHAN, M. A. 1992. Oolitic ironstone in the Cretaceous Western Interior Seaway, east-central Utah. *Journal of Sedimentary Petrology*, **62**, 693–705.

CHAPELLE, F. H. & LOVLEY, D. R. 1992. Competitive exclusion of sulfate reduction by Fe(III)-reducing bacteria: a mechanism for producing discrete zones of high-iron ground water. *Ground Water*, **30**, 29–36.

CHASE, E. M. & SAYLES, F. L. 1980. Phosphorus in suspended sediments of the Amazon River. *Estuarine and Coastal Marine Science*, **11**, 383–391.

CLAYPOOL, G. E. & KAPLAN, I. R. 1974. The origin and distribution of methane in marine sediments. *In*: KAPLAN, I. R. (ed.) *Natural Gases in Marine Sediments*. Plenum Press, New York, 99–139.

COLEMAN, M. L. 1985. Geochemistry of diagenetic non-silicate minerals: kinetic considerations. *Philosophical Transactions of the Royal Society, London*, **A315**, 39–56.

COOK, P. J. & MCELHINNY, M. W. 1979. A re-evaluation of the spatial and temporal distribution of sedimentary phosphate deposits in the light of plate tectonics. *Economic Geology*, **74**, 315–330.

COTTER, E. & LINK, J. E. 1993. Deposition and diagenesis of Clinton ironstones (Silurian) in the Appalachian Foreland Basin of Pennsylvania. *Bulletin of the Geological Society of America*, **105**, 911–922.

CURTIS, C. D. 1987. Données récentes sur le réactions entre matières organiques et substances minérales dans les sédiments et sur leurs conséquences minéralogiques. *Mémoire Société Géologique France*, **151**, 127–141.

—— & SPEARS, D. A. 1968. The formation of sedimentary iron minerals. *Economic Geology*, **63**, 257–270.

DREVER J. I. 1971. *The Geochemistry of Natural Waters*. Prentice Hall, Englewood Cliffs.

DREESEN, R. 1989. Oolitic ironstones as event-stratigraphical marker beds within the Upper Devonian of the Ardenno–Rhenish Massif. *In*: YOUNG, T. P. & TAYLOR, W. E. G. (eds) *Phanerozoic Ironstones*. Geological Society, London, Special Publications, **46**, 65–78.

EMBRY, A. F. 1982. The Upper Triassic – Lower Jurassic Heiberg deltaic complex of the Sverdrup basin. *In*: EMBRY, A. F. & BALKWILL, H. R. (eds) *Arctic Geology and Geophysics*. Canadian Society of Petroleum Geology, Memoirs, **8**, 189–215.

FROELICH, P. N., KLINKHAMMER, G. P. & 8 OTHERS 1979. Early oxidation of organic matter in pelagic sediments of the eastern equatorial Atlantic: suboxic diagenesis. *Geochimica et Cosmochimica Acta*, **43**, 1075–1090.

FÜRSICH, F. T., OSCHMANN, W., SINGH, I. B. & JAITLEY, A. K. 1992. Hardgrounds, reworked concretion levels and condensed horizons in the Jurassic of western India: their significance for basin analysis. *Journal of the Geological Society, London*, **149**, 313–331.

GEHRING, A.U. 1989. The formation of goethitic ooids in condensed deposits in northern Switzerland. *In*: YOUNG, T. P. & TAYLOR, W. E. G. (eds) *Phanerozoic Ironstones*. Geological Society, London, Special Publications, **46**, 133-139.

—— 1990. Diagenesis of ferriferous phases in the Northampton Ironstone in the Cowthick Quarry near Corby (England). *Geological Magazine*, **127**, 169–176.

GUERRAK, S. 1989. Time and space distribution of Palaeozoic oolitic ironstones in the Tindouf Basin, Algerian Sahara. *In*: YOUNG, T. P. & TAYLOR, W. E. G. (eds) *Phanerozoic Ironstones*. Geological Society, London, Special Publications, **46**, 197–212.

HALLAM, A. & BRADSHAW, M. J. 1979. Bituminous shales and oolitic ironstones as indicators of transgressions and regressions. *Journal of the Geological Society, London*, **136**, 157–164.

HEGGIE, D. T., DYRING, G. W. & 6 OTHERS 1990. Organic carbon cycling and modern phosphorite formation on the East Australian continental margin: an overview. *In*: NOTHOLT, A. J. G. & JARVIS, I. (eds) *Phosphorite Research and Development*. Geological Society, London, Special Publications, **52**, 87–117.

HEMINGWAY, J. E. 1951. Cyclic sedimentation and the deposition of ironstones in the Yorkshire Lias. *Proceedings of the Yorkshire Geological Society*, **28**, 67–74.

IIJIMA, A. & MATSUMOTO, R. 1982. Berthierine and chamosite in coal measures of Japan. *Clays and Clay Minerals*, **30**, 264–274.

IRWIN, H., CURTIS, C. D. & COLEMAN, M. L. 1977. Isotopic evidence for source of diagenetic carbonates formed during burial of organic-rich sediments. *Nature*, **269**, 209–213.

JAHREN, J. S. & AAGAARD, P. 1989. Compositional variations in diagenetic chlorites and illites, and relationships with formation-water chemistry. *Clay Minerals*, **24**, 157–170.

JERVEY, M. T. 1988. Quantitative geological modeling of siliciclastic rock sequences and their seismic expression. *In*: WILGUS, C. K., HASTINGS, B. S., KENDALL, C. G. ST. C., POSAMENTIER, H. W., ROSS, C. A. & VAN WAGONER, J. C. (eds) *Sea-level Changes: an Integrated Approach*, Society of Economic Paleontologists and Mineralogists, Special Publications, **42**, 47–70.

KIDWELL, S. M. 1989. Stratigraphic condensation of marine transgressive records: origin of major shell deposits in the Miocene of Maryland. *Journal of Geology*, **97**, 1–24.

KIMBERLEY, M. M. 1989. Exhalative origins of iron formations. *Ore Geology Reviews*, **5**, 13–145.

KRUM, M. D. & BERNER, R. A. 1982. Diagenesis of phosphorus in a near shore marine sediment. *Geochimica et Cosmochimica Acta*, **45**, 207–216.

LOUTIT, T. S., HARDENBOL, J., VAIL, P. R. & BAUM, G. R. 1988. Condensed sections: the key to age determination and correlation of continental margin sequences. *In*: WILGUS, C. K., HASTINGS, B. S., KENDALL, C. G. ST., C. POSAMENTIER, H. W., ROSS, C. A. & VAN WAGONER, J. C. (eds) *Sea-level Changes: an Integrated Approach*. Society of Economic Paleontologists and Mineralogists, Special Publications, **42**, 183–213.

LOVLEY, D. R. 1991. Dissimilatory Fe(III) and Mn(IV) reduction. *Microbiological Reviews*, **55**, 259–287.

—— & PHILLIPS, E. J. P. 1986. Organic matter mineralisation with reduction of ferric iron in anaerobic sediments. *Applied and Environmental Microbiology*, **51**, 683–689.

—— & —— 1987. Competitive mechanisms for inhibition of sulfate reduction and methane production in the zone of ferric iron reduction in sediments. *Applied and Environmental Microbiology*, **53**, 2636–2641.

—— & —— 1988. Novel mode of microbial metabolism: organic carbon oxidation coupled to dissimilatory reduction of iron and manganese. *Applied and Environmental Microbiology*, **54**, 1472–1480.

MACQUAKER, J. H. S. 1994. Palaeoenvironmental significance of 'bone-beds' in organic-rich mudstone successions: an example from the Upper Triassic of south west Britain. *Zoological Journal of the Linnean Society*, **112**, 285–308.

MADON, M. B. H. 1992. Depositional setting and origin of berthierine oolitic ironstones in the Lower Miocene Terengganu Shale, Tenggol Arch, Offshore Peninsular Malaya. *Journal of Sedimentary Petrology*, **62**, 899–916.

NATHAN, Y. & SASS, E. 1981. Stability relations of apatites and calcium carbonates. *Chemical Geology*, **34**, 103–111.

ODIN, G. S. 1988. *Green Marine Clays. Oolitic Ironstone Facies, Verdine Facies, Glaucony Facies and Celadonite-Bearing Facies – a Comparative Study*. Developments in Sedimentology, **45**, Elsevier, Amsterdam, 445pp.

PLINT, A. G., HART, B. S. & DONALDSON, W. S. 1993. Lithospheric flexure as a control on stratal geometry and facies distribution in Upper Cretaceous rocks of the Alberta foreland Basin. *Basin Research*, **5**, 69–77.

RAWSON, P. F., GREENSMITH, J. T. & SHALABY, S. E. 1983. Coarsening-upwards cycles in the uppermost Staithes and Cleveland Ironstone Formations (Lower Jurassic) of the Yorkshire Coast, England. *Proceedings of the Geologists' Association, London,* **94**, 91–93.

SELLWOOD, B.W. 1972. Regional environmental changes across a Lower Jurassic stage boundary in Britain. *Palaeontology*, **15**, 125–157.

SMITH, R. D. A. & AINSWORTH, R. B. 1989. Hummocky cross-stratification in the Downton of the Welsh borderland. *Journal of the Geological Society, London*, **146**, 897–900.

SPEARS, D. A. 1989. Aspects of iron incorporation into sediments with special reference to the Yorkshire Ironstones. *In*: YOUNG, T. P. & TAYLOR, W. E. G. (eds) *Phanerozoic Ironstones*. Geological Society, London, Special Publications, **46**, 19–30.

TAYLOR, K. G. & CURTIS, C. D. 1995. Stability and facies association of early diagenetic mineral assemblages: an example from a Jurassic ironstone–mudstone succession, U.K. *Journal of Sedimentary Research*, **A65**, 358–368.

TEYSSEN, T. 1989. A depositional model for the Liassic Minette ironstones (Luxemburg and France), in comparison with other Phanerozoic oolitic ironstones. *In*: YOUNG, T. P. & TAYLOR, W. E. G. (eds) *Phanerozoic Ironstones*. Geological Society, London, Special Publications, **46**, 79–92.

VAN HOUTEN, F. B. & BHATTACHARYYA, D. P. 1982. Phanerozoic oolitic ironstones – geological record and facies model. *Annual Review of Earth and Planetary Science*, **10**, 441–457.

—— & KARASEK, R. 1981. Sedimentology framework of the Late Devonian oolitic iron formation, Shatti Valley, west-central Libya. *Journal of Sedimentary Petrology*, **51**, 415–427.

VAN WAGONER, J. C., MITCHUM, R. M., CAMPION, K. M. & RAHMANIAN, V. D. 1990. *Siliciclastic Sequence Stratigraphy in Well Logs, Cores and Outcrops: Concepts for High Resolution Correlation of Time and Facies*. American Association for Petroleum Geologists, Methods in Exploration Series, **7**, 55pp.

YOUNG, T. P. 1989*a*. Phanerozoic Ironstones: an introduction and review. *In*: YOUNG, T. P. & TAYLOR, W. E. G. (eds) *Phanerozoic Ironstones*. Geological Society, London, Special Publications, **46**, ix–xxv.

—— 1989*b*. Eustatically controlled ooidal ironstone deposition: facies relationships of the Ordovician open-shelf ironstones of western Europe. *In*: YOUNG, T. P. & TAYLOR, W. E. G. (eds) *Phanerozoic Ironstones*. Geological Society, London, Special Publications, **46**, 51–63.

—— 1992. The Ordovician ooidal ironstones of Gondwana: a review. *Palaeogeography, Palaeoclimatology, Palaeoecology*, **99**, 321–347.

—— 1993. Sedimentary ironstones. *In*: PATTRICK, R. A. D. & POLYA, D. A. (eds) *Mineralization in Britain*. Chapman and Hall, London, 446–489.

——, PARSONS, D. & AGGETT, J. R. 1991. The Frodingham Ironstone. *In*: YOUNG, T. P. (ed.) *Jurassic and Ordovician Ooidal Ironstones*. 13th International Sedimentological Congress Field Guides, 32–50.

Unconformities within the Portlandian Stage of the Wessex Basin and their sequence-stratigraphical significance

ANGELA L. COE
Department of Earth Sciences, University of Oxford, Parks Road, Oxford OX1 3PR, UK
Present address: Department of Earth Sciences, Open University, Walton Hall, Milton Keynes MK7 6AA, UK

Abstract: At outcrop, the Portlandian Stage of the western Wessex Basin, southern England comprises a complex series of marine and non-marine carbonates deposited in moderate to shallow water depth, with the occasional development of siliciclastic deposits near to the basin margins. The strata record part of a marked overall long-term regression on which are superimposed four shorter term relative sea-level cycles. The base of each of these short-term cycles is marked by a major unconformity or its correlative conformity (sequence boundary) formed during the maximum rate of relative sea-level fall; these are denoted P1 to P4. The south Dorset exposures are each in slightly different tectonic settings within the Central Channel Sub-basin. The Isle of Purbeck sections are stratigraphically the most complete, but the sections on the Isle of Portland, previously assumed to be complete, contain at least two significant stratal gaps. The sections at Ringstead Bay and Dungy Head are thin, condensed and in part incomplete. Proximally, in the Vale of Wardour (Mere Sub-basin), Vale of Pewsey (Pewsey Sub-basin) and on the London Platform at Swindon, in Oxfordshire and in Buckinghamshire the succession becomes condensed, the major unconformities are frequently angular and there is at least one incised valley.

The Wessex Basin is an epicontinental basin composed of a series of half-graben sub-basins, each downstepping towards the south (Whittaker 1985; Chadwick 1986; Karner *et al.* 1987; Lake & Karner 1987). This paper concentrates on a line of section across the western part of the basin through the discontinuous Portlandian outcrop from near the basin centre in Dorset, to the basin margin in Buckinghamshire (Figs 1–4). The Portlandian is the terminal Jurassic stage in southern England. It encompasses the *Progalbanites albani* to the *Subcraspedites lamplughi* ammonite zones (Wimbledon & Cope 1978; Wimbledon 1980 and references cited therein). The ammonites provide biostratigraphical control and are used for the standard zonation and for correlation between exposures (Fig. 2). In Dorset, the outcrop can be split into three structural settings: the Isle of Purbeck and Gad Cliff sections lie immediately on the downthrown side of the Purbeck–Isle of Wight fault system and are stratigraphically the most complete; the Ringstead Bay and Dungy Head sections are on the upthrown side of the same fault system; the Isle of Portland sections lie to the south of the main fault system, possibly on a low relief, intrabasinal, hanging-wall high (Fig. 5). Inland, the outcrop is discontinuous and exposures are limited to small quarries and road cuttings. There is an outcrop in the Vale of Wardour (Mere Sub-basin) on the downthrown side of the Mere Fault and also a small area of outcrop in the Vale of Pewsey (Pewsey Sub-basin). Discontinuous outcrop occurs across the basin margin (London Platform) from Swindon, through Oxfordshire and into Buckinghamshire (Fig. 1).

The lithological characteristics are varied. In the basinal areas marine and non-marine carbonates predominate; these include dolomites, cherty packstones and wackestones, shelly limestones, oolitic limestones and non-marine marls, limestones and palaeosols (Fig. 6). The marginal areas are characterized by quartzose limestones, condensed limestones, sandstones and pebble beds. The overall sedimentation rate is low. Lithostratigraphical subdivisions are summarized in Fig. 4.

Sequence stratigraphy (e.g. Van Wagoner *et al.* 1988; Vail *et al.* 1991) emphasizes the importance of unconformities within the stratigraphical record and provides a basic model for their interpretation and that of the intervening sedimentary packages. Unconformities are a fundamental part of the stratigraphical record. Periods of erosion and non-deposition are very commonly recorded in sedimentary successions, but they are not always easily recognized. Assessment of the importance of an unconformity, its lateral continuity, the amount of time not represented in the stratigraphical record, and the location of its laterally equivalent correlative conformity are of great importance if we are to place all the intervening sedimentary rocks into context and

From HESSELBO, S. P. & PARKINSON, D. N. (eds), 1996, *Sequence Stratigraphy in British Geology*, Geological Society Special Publication No. 103, pp. 109–143.

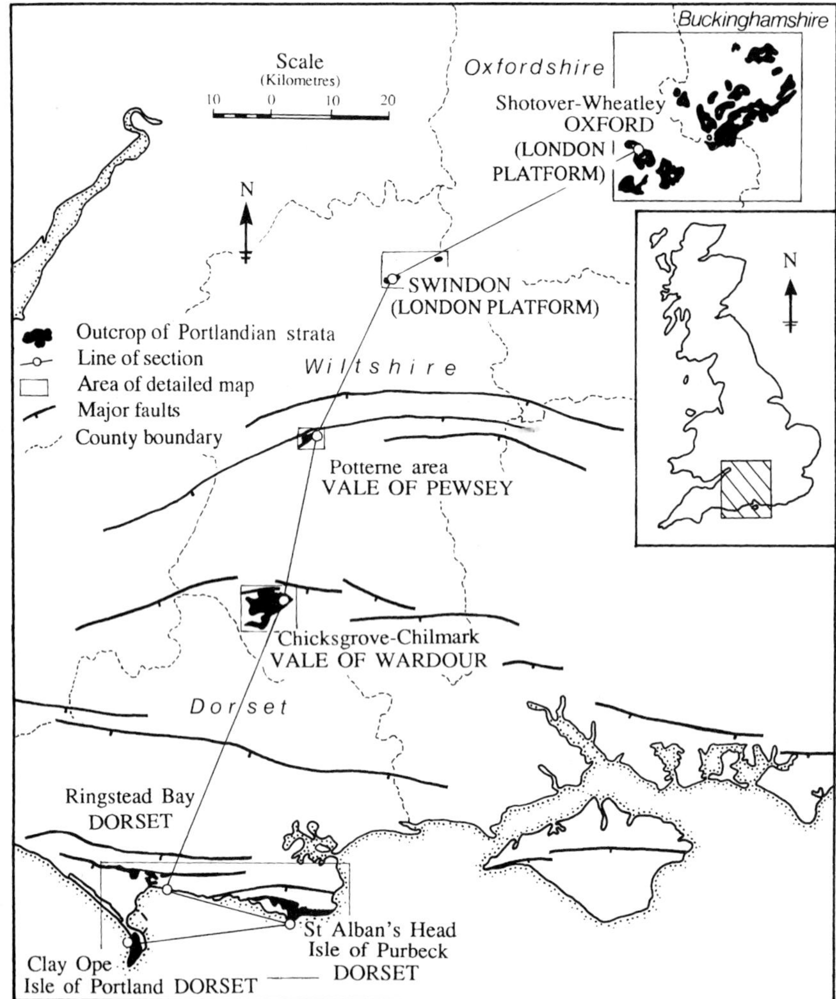

Fig. 1. Map showing the outcrop of the Portlandian strata in the Wessex Basin. Location of major faults after Whittaker (1985).

fully understand the implications of lateral facies changes. This paper aims to : (1) demonstrate that there are four major unconformities, informally denoted P1 to P4, within the Portlandian of the Wessex Basin; (2) show that these major unconformities are sequence boundaries; (3) provide a full sequence stratigraphical interpretation; and (4) suggest a higher resolution correlation of the Portlandian across the Wessex Basin than is currently provided by the ammonite biostratigraphy.

Major Portlandian unconformities and their associated sedimentary packages

Unconformity P1, middle glaucolithus *Zone*

The first major unconformity P1 occurs within the Portland Sand Formation. On the London Platform and in the Vale of Pewsey it is associated with a complex biostratigraphical gap which extends

Fig. 3. Summary graphical lithological logs of the Portlandian age strata to show the thickness changes and the key sequence stratigraphical surfaces. Localities shown in Fig. 1 and summary of lithostratigraphical nomenclature in Fig. 4. Compiled from Arkell (1933, 1935, 1947b), Casey and Bristow (1964), Clements (1969), Townson (1971), Wimbledon (1974*b*, 1976, 1980), Wimbledon & Hunt (1983), Morter (1984) and Coe (1992).

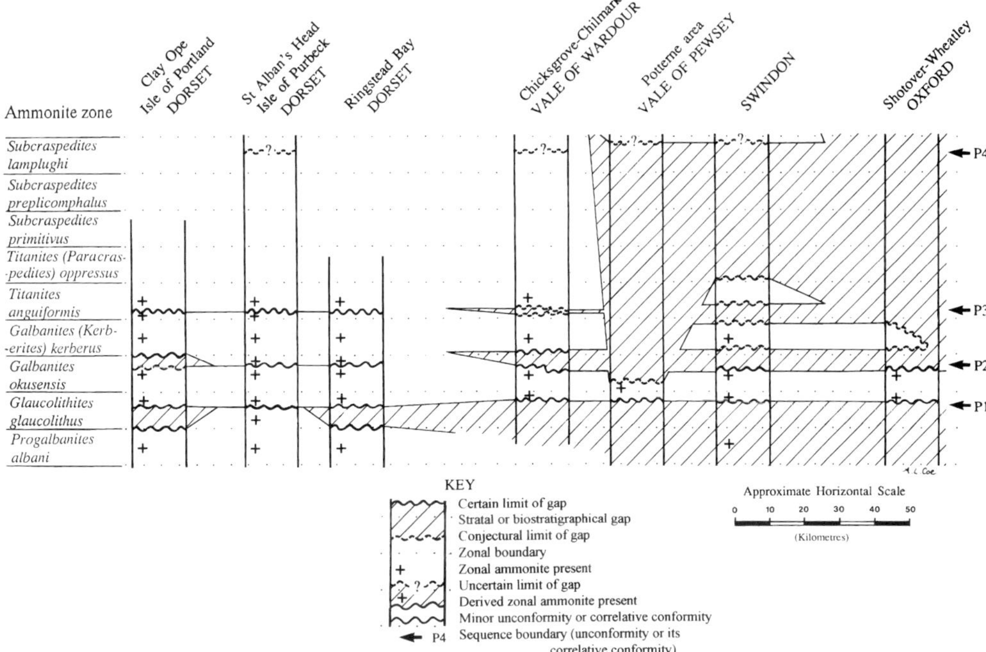

Fig. 2. Summary of the biostratigraphical correlation of the Portlandian across the Wessex Basin. Localities shown in Fig. 1. Compiled from Arkell (1933, 1935, 1947*b*), Casey & Bristow (1964), Coe (1992), Wimbledon & Cope (1978), Wimbledon & Hunt (1983), Anderson (1985).

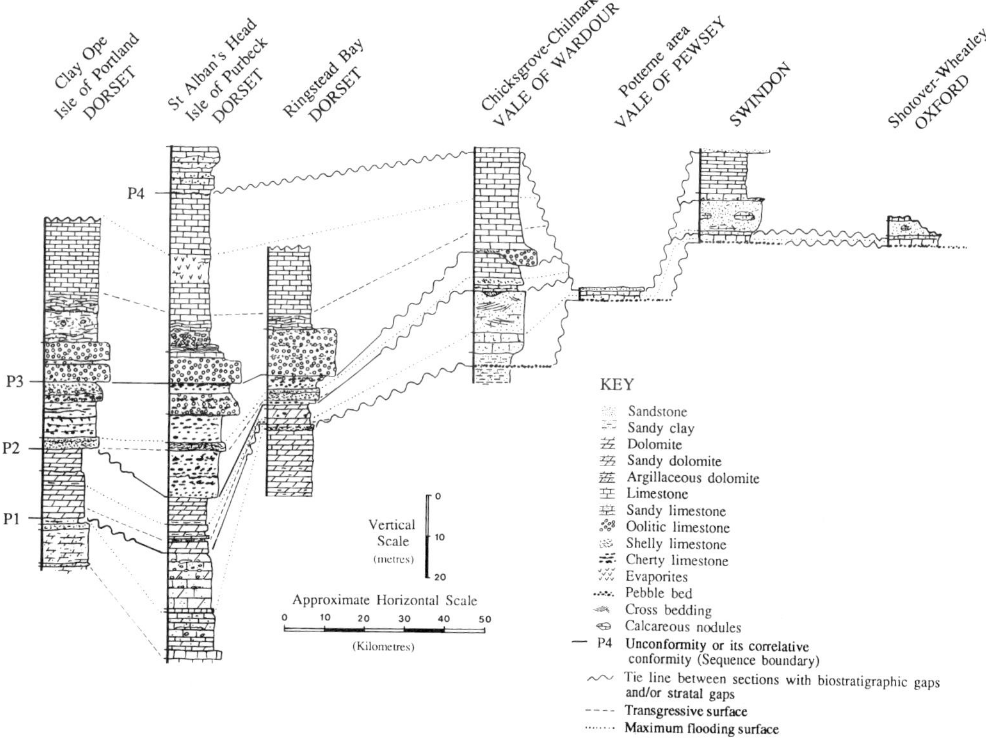

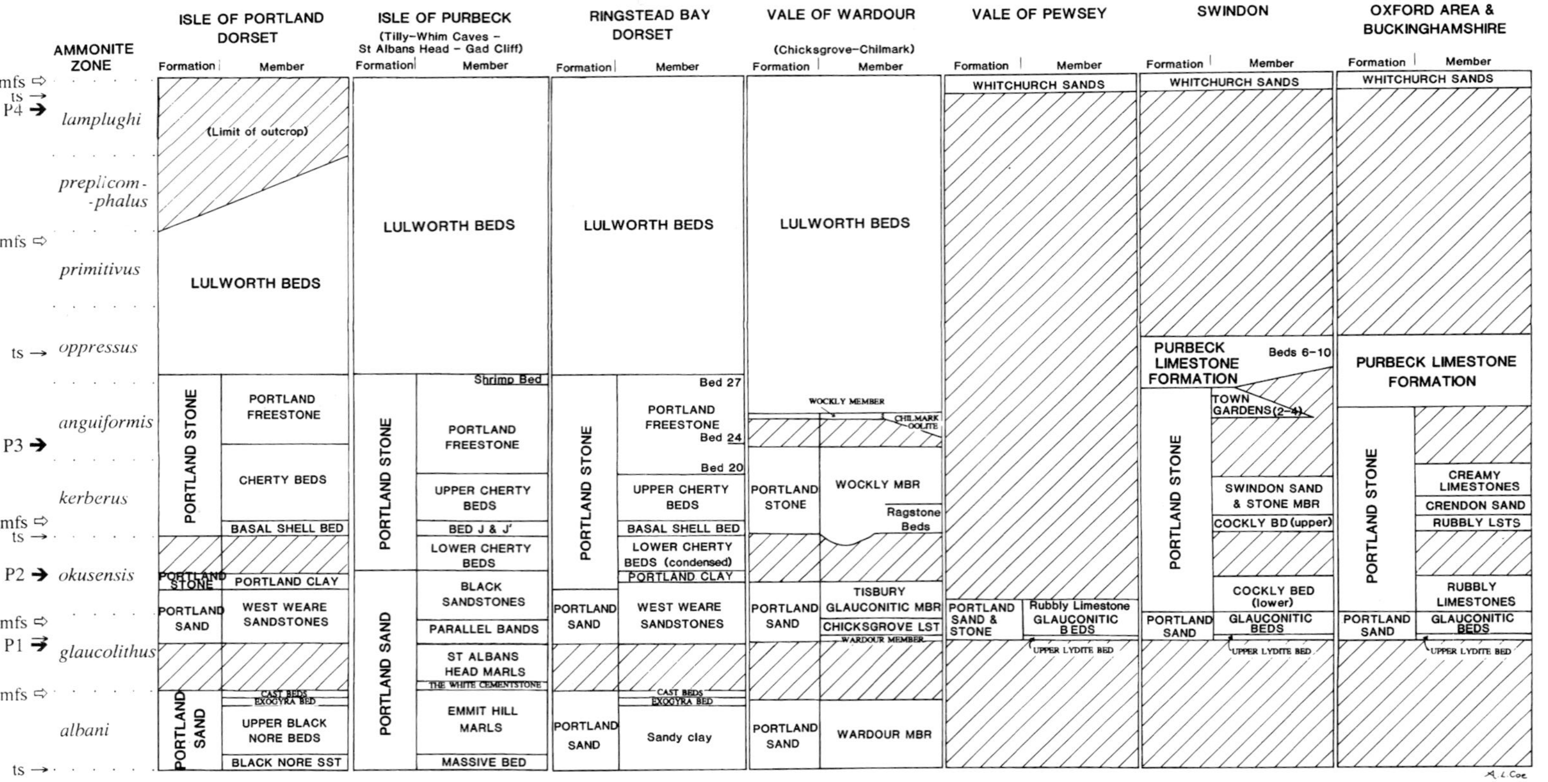

Fig. 4. Lithostratigraphical nomenclature and correlation of Portlandian strata in the Wessex Basin. The lithostratigraphical names for the divisions of the Portland Beds are those widely accepted today (Wimbledon & Cope 1978; Wimbledon 1980, 1986; House 1989) and are mostly those collated or proposed by Arkell (1933, 1935, 1947*a*). The base of the Purbeck Limestone at Swindon and in Oxfordshire and Buckinghamshire may be younger than shown. The Rubbly Limestones are equivalent to the Aylesbury Limestone in the Aylesbury area.

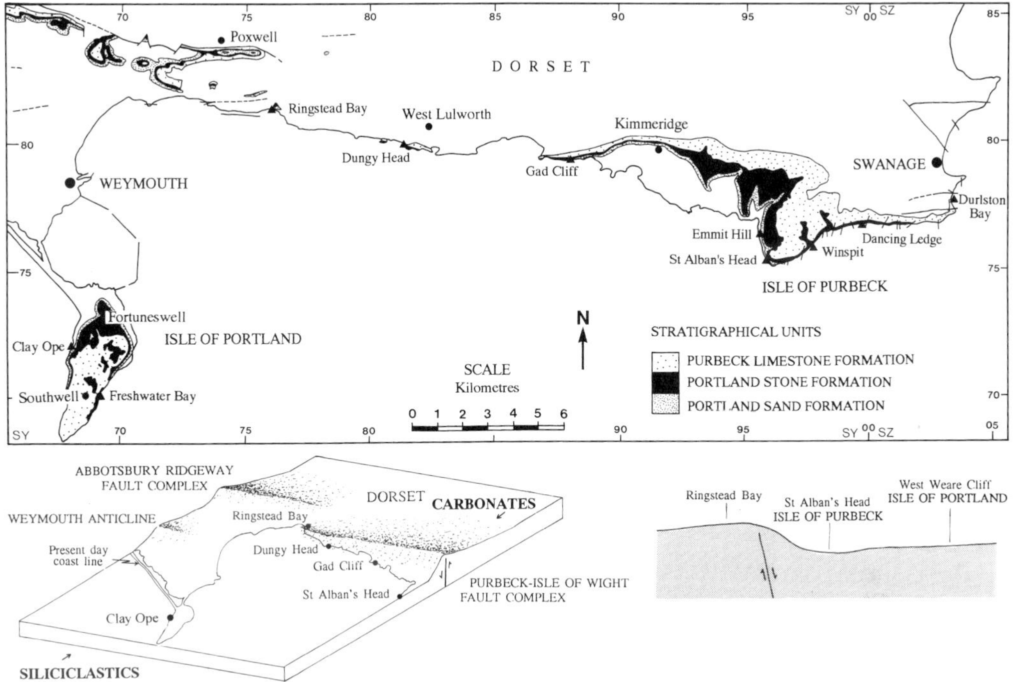

Fig. 5. Map of the outcrop of the Portland Beds and Purbeck Limestone in south Dorset, showing the position of the logs of Figs 7–9 and 16, constructed using data from British Geological Survey maps, sheet numbers 341, 342, 343, 327 and 328. Bottom left: sketch block diagram to show the relative position of the different sedimentary sections in relation to the known structural features of south Dorset. Bottom right: cross-section to show the positions of the Dorset sections relative to the faults.

downwards into the Kimmeridgian Stage, encompassing at least three ammonite zones (Cope 1978; Wimbledon & Cope 1978; Wimbledon 1980; Fig. 2). The more complete sections in the Vale of Wardour and in Dorset indicate that this gap is the result of two distinct periods of erosion. The basinal correlation of the first major erosion period occurs within the *fittoni* Zone of the Upper Kimmeridgian

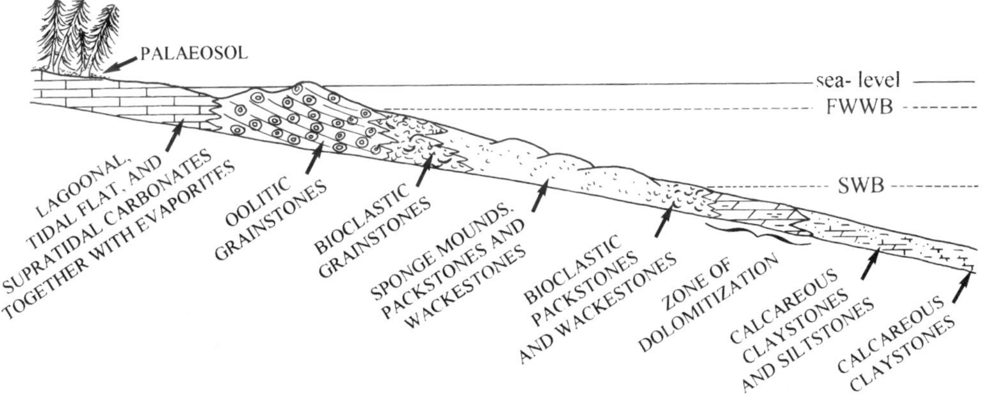

Fig. 6. Idealized carbonate ramp model and schematic cross-section to show the relative position of the different facies found in the Portland Beds in Dorset, modified after Townson (1971, 1975) and with additional information from Tucker & Wright (1990). See text for explanation of formation of dolomite. SWB, Storm wave base; and FWWB, fair weather wave base.

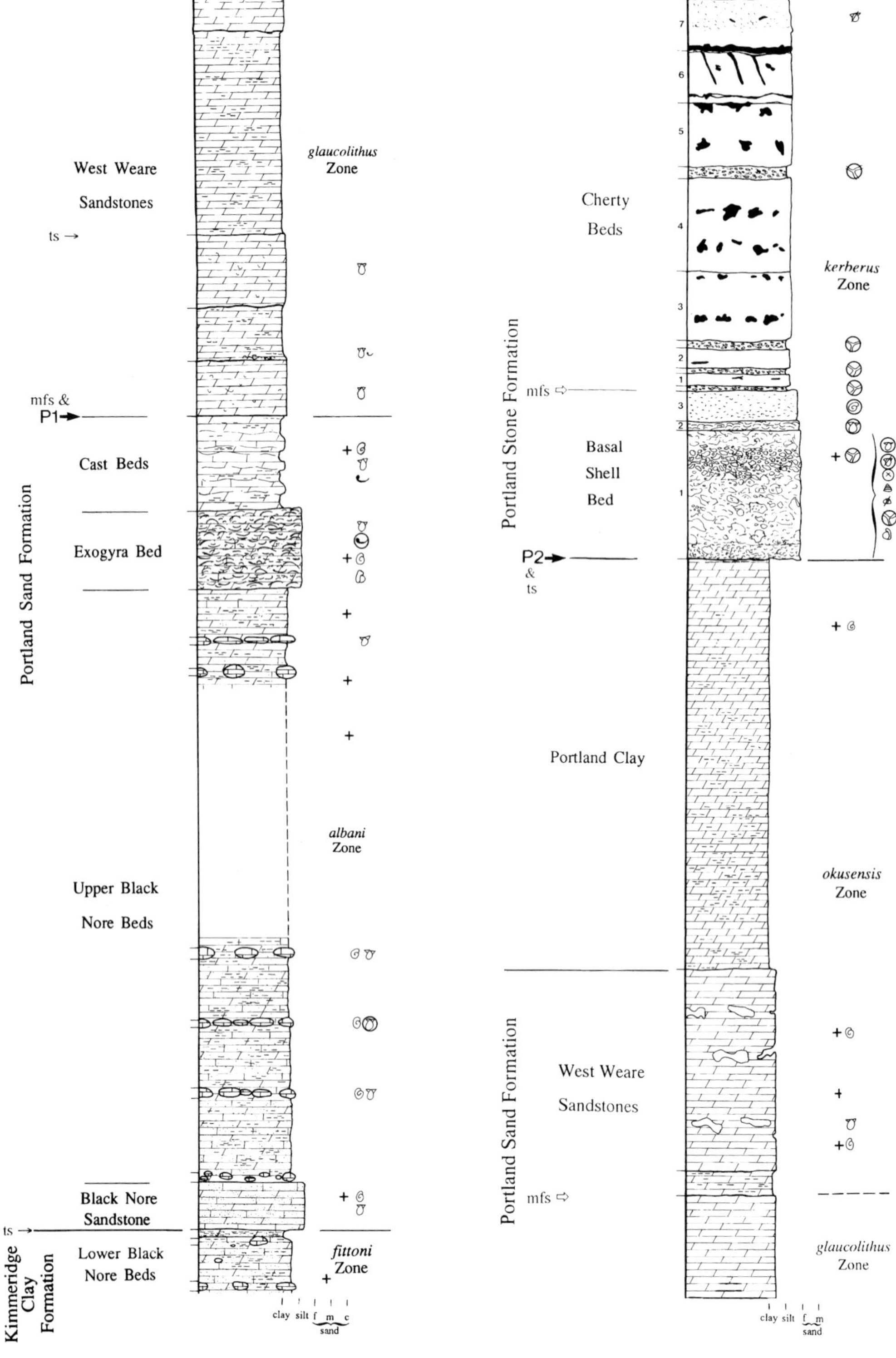
West Weare
Sandstones
glaucolithus Zone
ts →
mfs &
P1
Cast Beds
Exogyra Bed
Portland Sand Formation
albani Zone
Upper Black
Nore Beds
Black Nore
Sandstone
ts →
Lower Black
Nore Beds
fittoni Zone
Kimmeridge Clay Formation
clay silt f m c
sand
Cherty
Beds
kerberus Zone
Portland Stone Formation
mfs
Basal
Shell
Bed
P2
&
ts
Portland Clay
okusensis Zone
Portland Sand Formation
West Weare
Sandstones
mfs
glaucolithus Zone
clay silt f m
sand

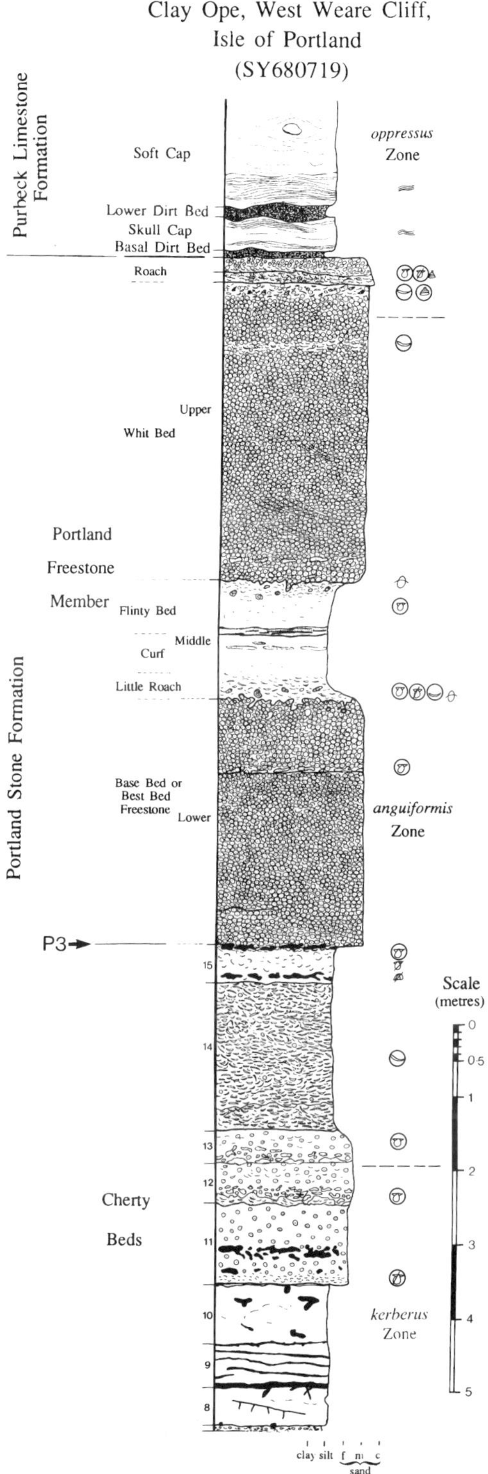

(Coe 1992) and the second erosion period is unconformity P1 belonging to the *glaucolithus* Zone.

South Dorset. Extensive sections through lower Portlandian strata occur at only a few places along the Dorset Coast; in the west, at Clay Ope on the Isle of Portland, Ringstead Bay and Dungy Head, and in the east, at Gad Cliff and between Emmit Hill and St Alban's Head on the Isle of Purbeck (Fig. 5). The facies types are very similar in all these sections, but the thicknesses of each facies type is variable. The lowest Portlandian strata (*albani* Zone) are bioturbated clay-rich siltstones and fine-grained sandstones with a variable amount of calcite and dolomite; the better-sorted units are cemented by calcite. These deposits comprise the Upper Black Nore Beds of the westerly sections and the Emmit Hill Marls of the easterly sections (Figs 7–9; see Fig. 10 for key to logs). A particularly thick (0.7–1.50 m) cemented unit marks the base of the Portlandian Stage in many of the sections and is known as the Black Nore Sandstone or the Massive Bed (Fig. 4). Passing up through the Upper Black Nore Beds and Emmit Hill Marls there is an increase in the abundance and diversity of fauna. This is particularly marked in the westerly sections, culminating with the Exogyra Bed and the overlying Cast Beds (Figs 7 and 9). The Exogyra Bed is about 1.15 m thick and packed full of disarticulated valves of the oyster *Nanogyra*, together with other bivalves, serpulids and sponge spicules. The overlying Cast Beds (1.4 m thick) are similar, but at most localities they are more clay-rich and originally aragonitic shells have been dissolved. The top of the Cast Beds coincides with the *albani/glaucolithus* zonal boundary (Wimbledon 1980, 1986; Wimbledon & Cope 1978). In the easterly sections, the top of the Emmit Hill Marls contains several indurated and fossiliferous horizons; the most prominent cemented horizon occurs at the top of the member and is termed the White Cementstone, the base of which marks the *albani/glaucolithus* zonal boundary (Wimbledon 1980, 1986; Wimbledon & Cope 1978; Fig. 8). It is overlain by the St Alban's Head Marls which comprise silty mudstones with variable amounts of calcite and dolomite (Townson

Fig. 7. Stratigraphical and sedimentological log for the Portland Sand and Portland Stone exposed at Clay Ope, Isle of Portland. Stratigraphical nomenclature and ammonite zonation from Wimbledon (1980, 1986) and Wimbledon & Cope (1978). Subdivisions of the Portland Freestone Member from Arkell (1947*a*, quarrymen's terms) and tripartite subdivision from Townson (1971, 1975). For key to symbols, see Fig. 10.

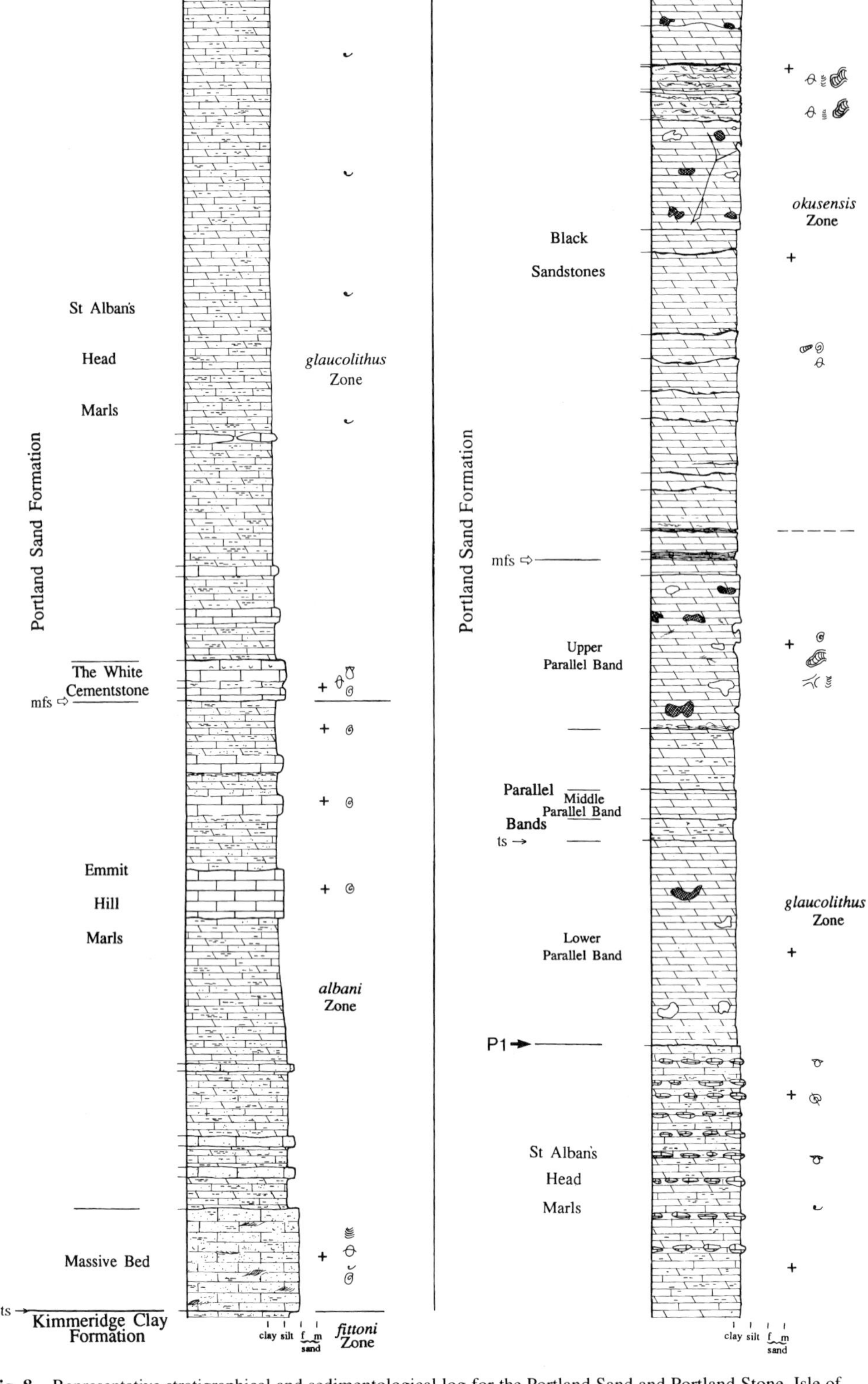

Fig. 8. Representative stratigraphical and sedimentological log for the Portland Sand and Portland Stone, Isle of Purbeck. Base of log to Bed F of the Lower Cherty Beds from St Alban's Head, Bed G of the Lower Cherty Beds to

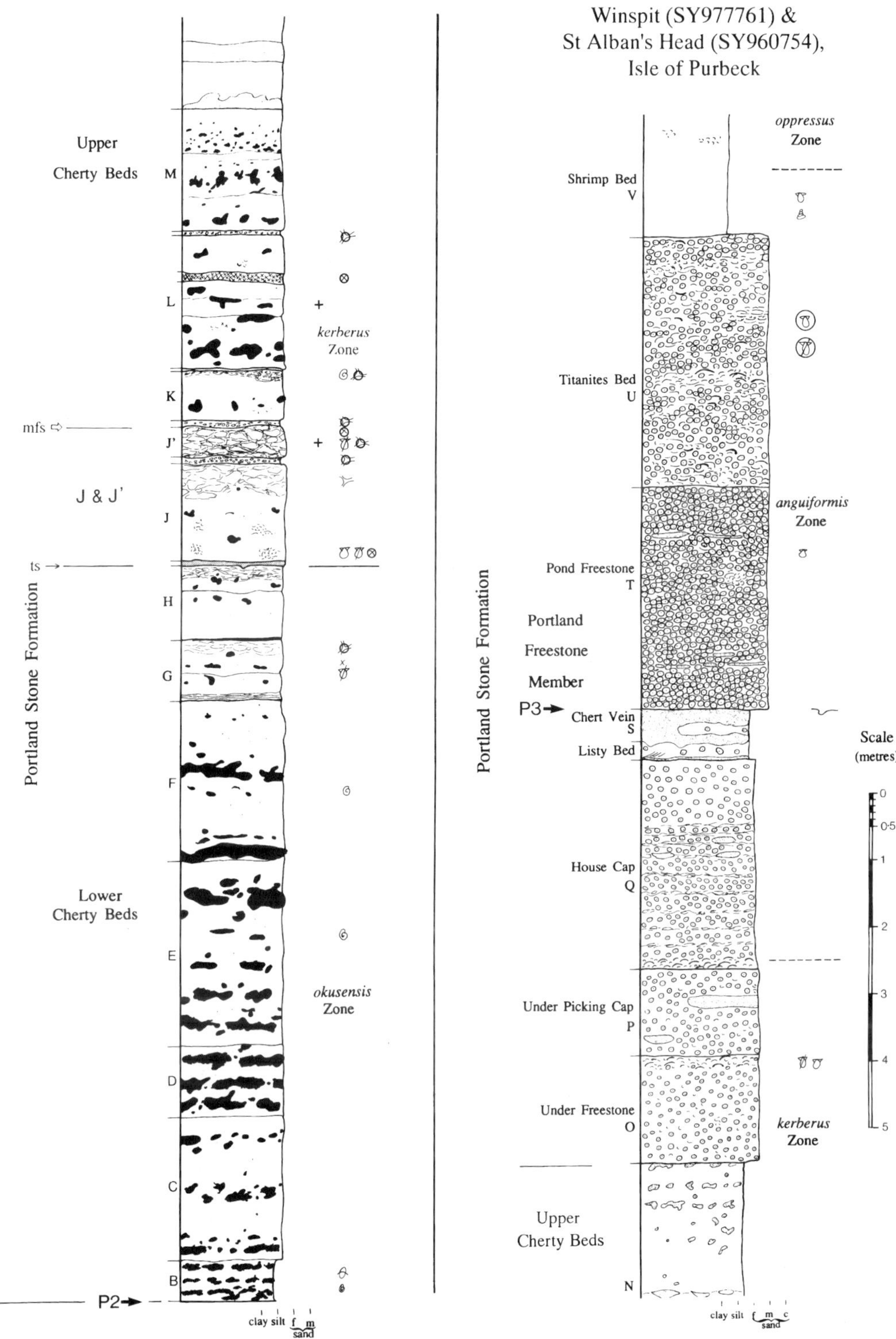

the Shrimp Bed from Winspit. Ammonite zonation from Wimbledon & Cope (1978) and Wimbledon (1980, 1986). Stratigraphical nomenclature from Arkell (1935) and Wimbledon (1980). For key to symbols, see Fig. 10.

1971). The marls also contain calcite-cemented bands with casts of bivalves and horizons rich in *Nanogyra*; these bands become more abundant towards the top of the member (Arkell 1935; Townson 1975).

Both the Cast Beds and the St Alban's Head Marls are overlain by finely crystalline dolomites containing horizons of *Thalassinoides* burrow systems, now infilled with black calcite (dedolomitized; Townson 1971; Figs 7–9). These carbonates comprise the West Weare Sandstones and Portland Clay in the westerly sections and the Parallel Bands and Black Sandstones in the easterly sections. They contain virtually no quartz sand, but small amounts of clay are present in some beds. The horizons of black calcite nodules are found near the base and top of the sedimentary package, whereas the middle part contains several thin organic-carbon-rich layers (Figs 7 and 8). Apart from ammonites (Wimbledon 1986), these members are almost barren of fossils (Townson 1971). The mode of formation of these unusual dolomites is described later.

A comparison of the succession from the top of the *albani* Zone to the middle of the *glaucolithus* Zone on the Isle of Purbeck (St Alban's Head; SY 959754) with the time-equivalent on the Isle of Portland (Clay Ope; SY 680719) reveals an important change from deposition of mixed siliciclastic and carbonate sediment to pure carbonate accumulation (Figs 7 and 8), which is interpreted as a shallowing upward trend (Fig. 6). This contrast in lithology led Townson (1971, 1975) to correlate lithostratigraphically the St Alban's Head Marls on the Isle of Purbeck with the Exogyra Bed and Cast Beds on the Isle of Portland. However, this lithological correlation is incompatible with the modern ammonite zonation (Wimbledon 1986; Wimbledon & Cope 1978), which demonstrates that the Cast Beds are coeval with the top of the Emmit Hill Marls (Figs 4, 7 and 8). However, both correlations can be reconciled if on the Isle of Portland there is no lateral equivalent of the 13 m of St Alban's Head Marls present on the Isle of Purbeck. This being the case, the lithological change from a mixture of clay, dolomite and sand to almost pure dolomite (base of the West Weare Sandstones and base of the Parallel Bands) is equivalent in both sections and clearly indicates a change in depositional conditions across the Dorset area. Townson also used the presence of *Nanogyra* at specific horizons for correlation, but these occur throughout all of the *albani* Zone and most of the *glaucolithus* Zone and thus are not stratigraphically diagnostic (Wimbledon 1986). Using the correlation proposed, comparison of the two sections shows that the successions are very similar in thickness and lithology below the White Cementstone/Cast Beds, and upwards from the base of the Parallel Bands/West Weare Sandstones. The new correlation presented with a stratal gap on the Isle of Portland (unconformity P1) is thus compatible with both the biostratigraphy and the lithostratigraphy (Fig. 4).

On the Isle of Portland, lack of sediment supply, together with increased erosion associated with the genesis of unconformity P1, were probably the main reasons for the formation of a stratal gap. On the Isle of Purbeck, erosion rates appear to have been lower and sediment was more readily deposited in the hanging-wall 'trough', resulting in accumulation of the St Alban's Head Marls. The top of the St Alban's Head Marls is interpreted as the correlative conformity of unconformity P1 and is dated as middle *glaucolithus* Zone (Fig. 8). The intervening sections at Gad Cliff (SY 885795), Dungy Head (SY 816799) and Ringstead Bay (Fig. 9; SY 762813) are also entirely consistent with this interpretation. The section at Gad Cliff is almost identical to that of the Isle of Purbeck, with a similar thickness of St Alban's Head Marls preserved. The sections at Dungy Head and Ringstead Bay (on and near the footwall of the fault system; Fig. 4) are the same as on the Isle of Portland as there is no facies representation of the St Alban's Head Marls, and the strata immediately below the P1 unconformity are condensed.

Vale of Wardour. In the Vale of Wardour the evidence for unconformity P1 is inconclusive due to poor exposure. The *albani* Zone and part of the *glaucolithus* Zone are represented by at least 10 m of alternating silty clays and cemented sandstones and siltstones of the Wardour Member (Wimbledon 1976, 1980; Fig. 4). At Chicksgrove Quarry (ST 962296; Fig. 11a) a borehole proved 7.5 m of siltstones and claystones with minor sandy units, but the base of the member was not reached. However, 3.5 m below the top of the member a horizon of black Carboniferous chert pebbles was found (Wimbledon 1976). These pebbles are interpreted as a lag deposit signifying a period of erosion and are probably partially equivalent to the Upper Lydite Bed found further to the north (Wimbledon 1976); this horizon is interpreted as unconformity P1. At Cuffs Lane (ST 949297; Fig. 11a) on the east side of Tisbury, M.J. Simms (pers. comm.) in 1982 recorded 12 m of Wardour Member comprising fine sand and silt with occasional ripples, overlain by fine-grained sand and silt, with three beds which he interpreted as palaeosols with calcified rootlets. However, their identity as palaeosols has not been confirmed and unfortunately this cryptic section is no longer exposed. Immediately above the first of these 'palaeosols' an indurated sand was exposed which contained abundant bivalves, fish debris and fragmented ammonites (Simms pers. comm.).

A further section of the Wardour Member at

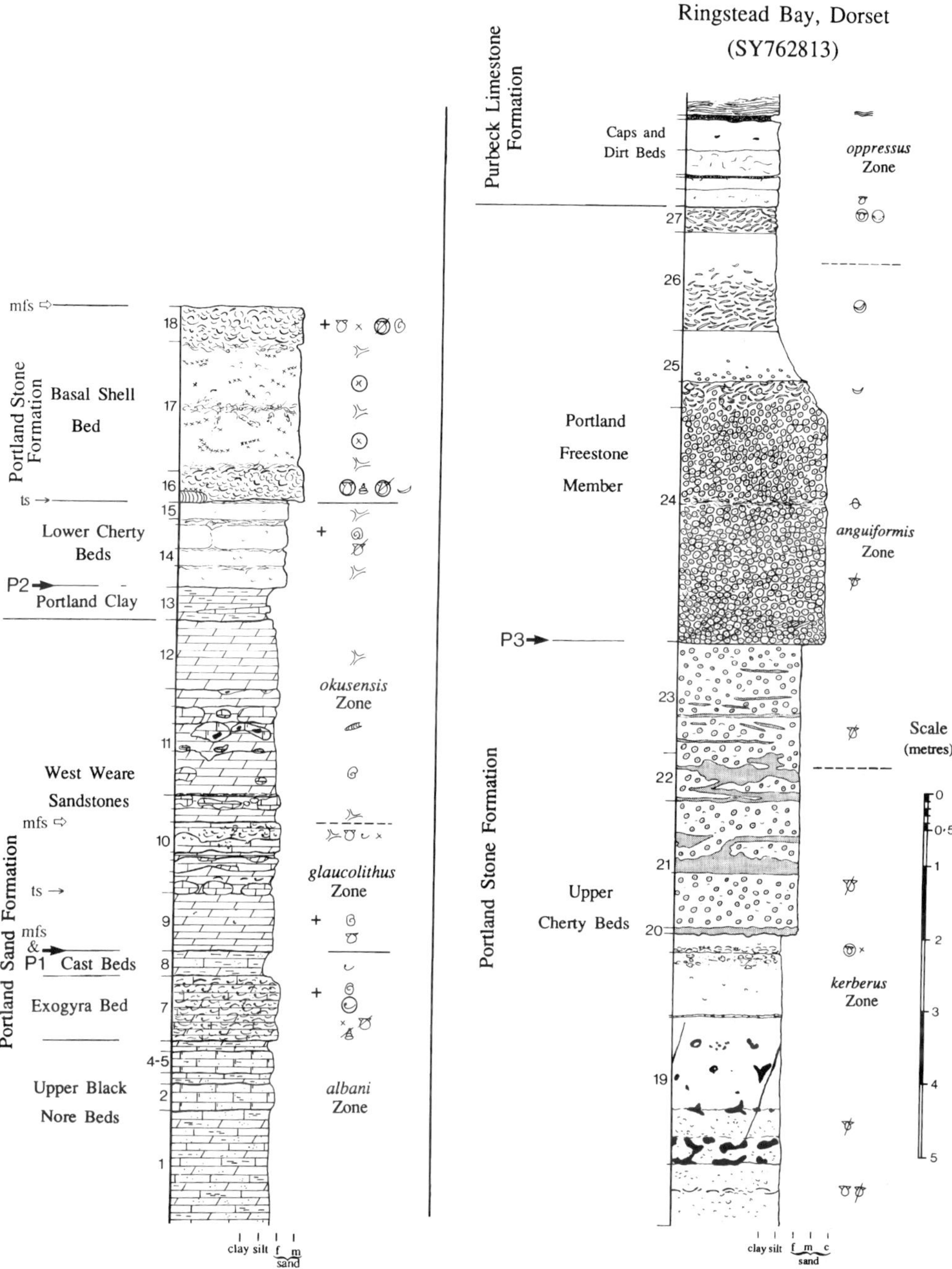

Fig. 9. Stratigraphical and sedimentological log for the Portland Sand and Portland Stone exposed at Ringstead Bay, Dorset. Most of the stratigraphical nomenclature and ammonite zonation from Wimbledon (1980, 1986) and Wimbledon & Cope (1978). The base of bed 11 forms the base of the Portland Stone as defined by Wimbledon (1980, 1986). For key to symbols, see Fig. 10.

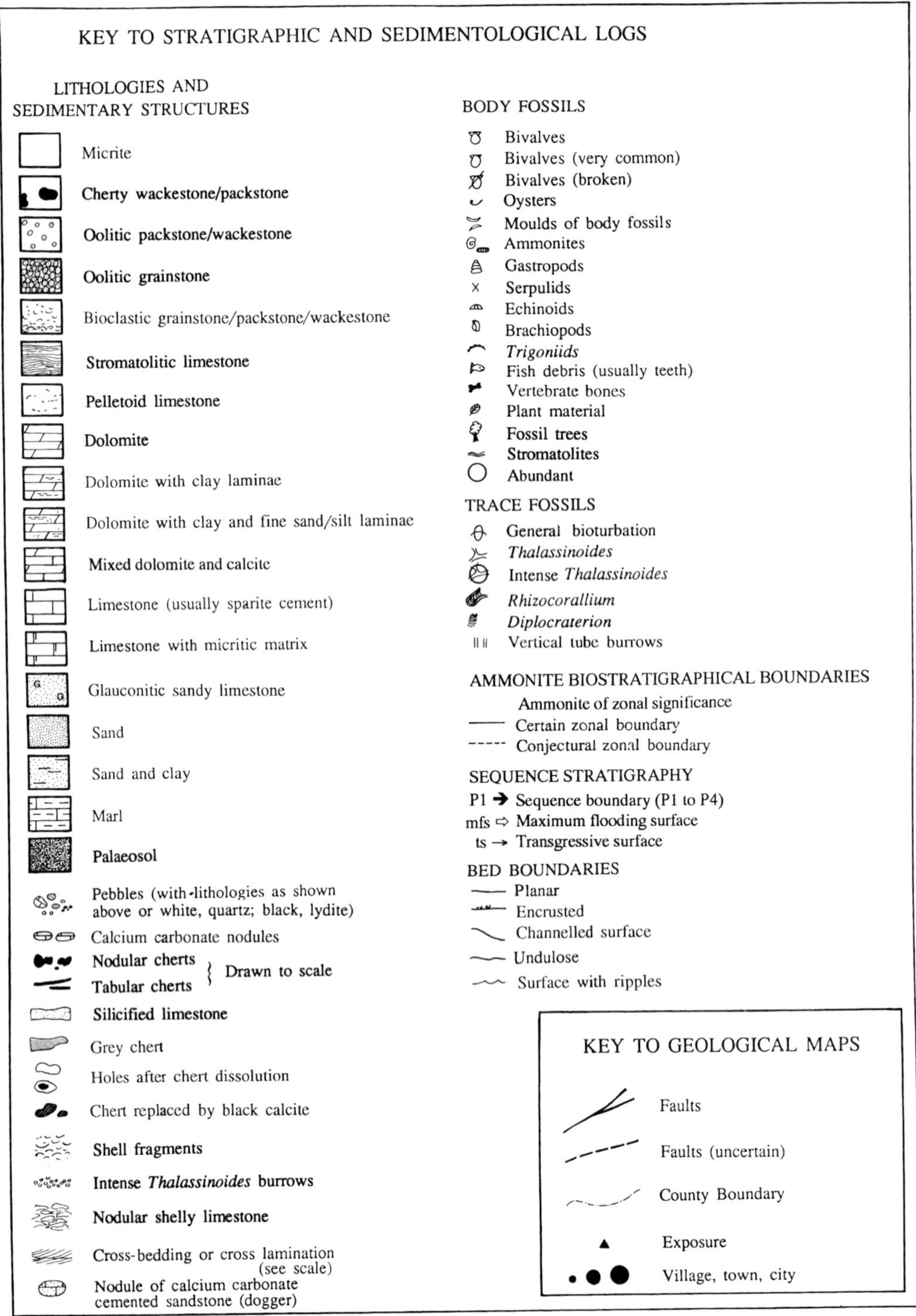

Fig. 10. Key to symbols used in the stratigraphical and sedimentological logs and locality maps of Figs 7–9, 11–16 and 18.

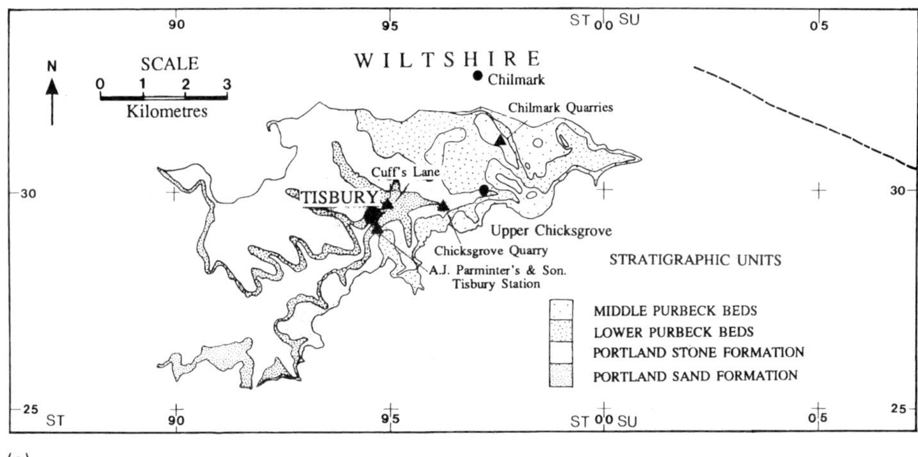

(a)

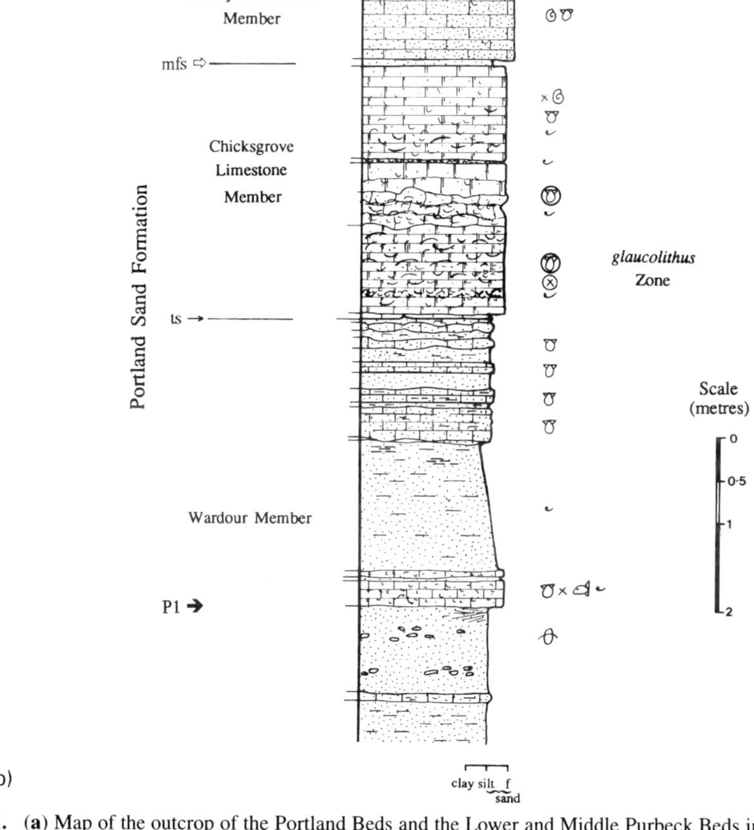

(b)

Fig. 11. (**a**) Map of the outcrop of the Portland Beds and the Lower and Middle Purbeck Beds in the Vale of Wardour and showing the position of the logs shown in (b) and Fig. 12. Compiled using data from British Geological Survey Maps, sheet numbers 297 and 298. (**b**) Stratigraphical and sedimentological log for the Portland Beds exposed behind the workshops of Parminters, near Tisbury Station, Vale of Wardour. Stratigraphical nomenclature from Wimbledon (1976) and M.J. Simms (pers. comm.). For key to symbols, see Fig. 10.

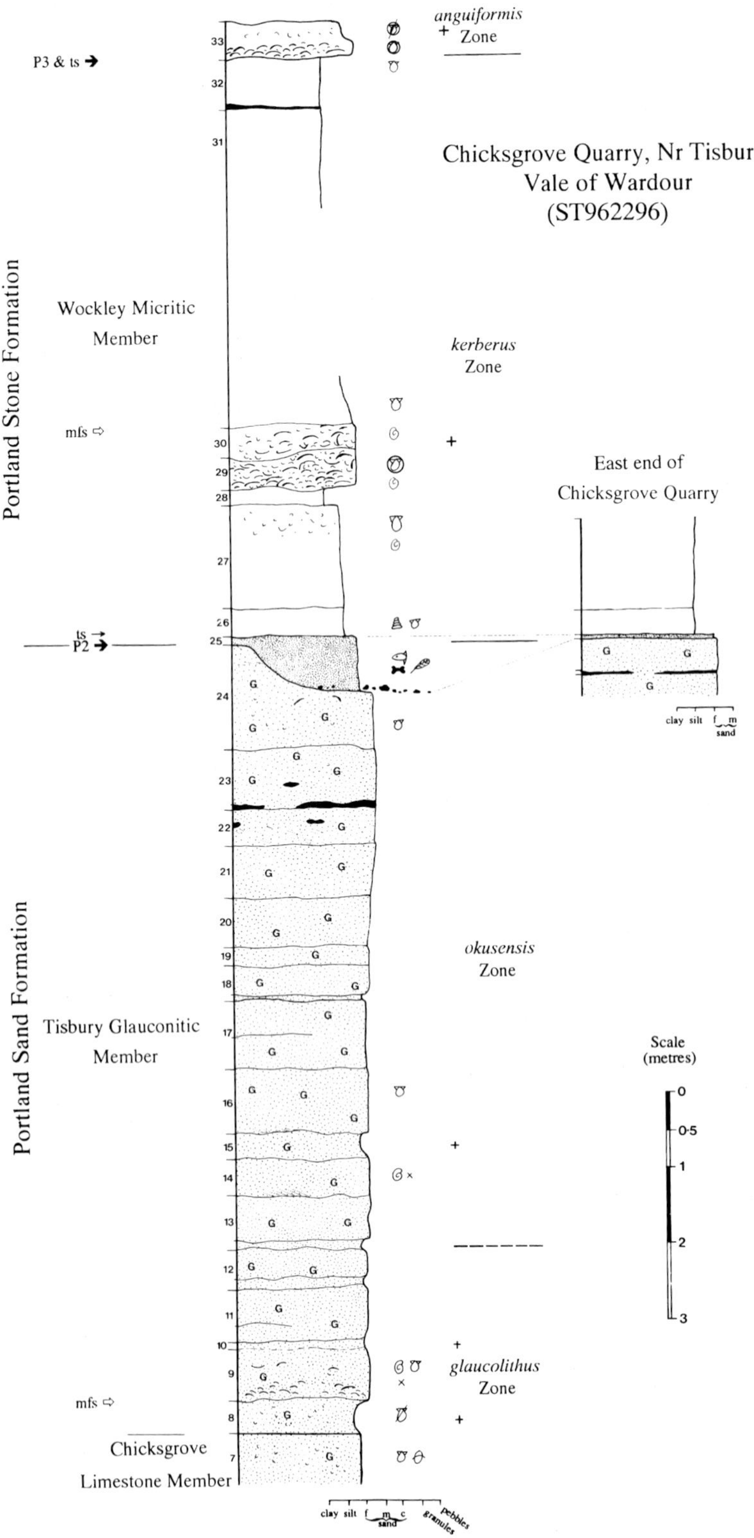
Chicksgrove Quarry, Nr Tisbury,
Vale of Wardour
(ST962296)
East end of
Chicksgrove Quarry
Portland Stone Formation
Wockley Micritic
Member
Portland Sand Formation
Tisbury Glauconitic
Member
Chicksgrove
Limestone Member
P3 & ts
mfs
ts
P2
mfs
anguiformis
Zone
kerberus
Zone
okusensis
Zone
glaucolithus
Zone
Scale
(metres)
0
0·5
1
2
3
clay silt f m c sand granules pebbles
clay silt f m sand

Tisbury Station (ST 945289) also contains a bed with fish debris and at this locality it is seen overlying a clay-rich fine-grained sand with small irregular calcareous nodules. These nodules have been interpreted as pseudomorphs after evaporites (Simms pers. comm.; Fig. 11b), but the evidence is unclear. At Tisbury Station these nodules are overlain by 4 m of fine silts and silty clays with flaser and lenticular bedding, structures which are normally associated with fairly shallow-water conditions of deposition.

At all localities, the Wardour Member is overlain by the Chicksgrove Limestone Member, which is well exposed at Tisbury Station where it consists of 2.9 m of medium-bedded biomicrites (Fig. 11). These contain a very rich fauna chiefly consisting of bivalves, predominantly *Nanogyra*. Ammonites are fairly common and indicate the *glaucolithus* Zone (Wimbledon 1980, 1986; Fig. 12). The member contains quartz sand and occasional lydite pebbles (Blake 1880; Hudleston 1883). The top of the member is exposed at Chicksgrove Quarry, where it is a bioturbated sandy biomicrite (Fig. 12).

The upper member of the sedimentary package overlying unconformity P1 is the Tisbury Glauconitic Member. This member is well exposed at Chicksgrove Quarry, where it comprises limestone beds containing quartz sand, glauconite, phosphate, peloids and bioclasts, the grain size increasing towards the top of the member (Fig. 12). Many beds are trough cross-stratified and occasional chert nodules are present. The top of the Tisbury Glauconitic Member has yielded a sparse assemblage of ammonites comparable in age with those in the top few metres of the West Weare Sandstones (Wimbledon 1976; Wimbledon & Cope 1978).

Vale of Pewsey and the London Platform. In the Vale of Pewsey and on the London Platform at Swindon and Oxford the *albani* Zone is absent. The first Portlandian deposit is a pebble bed, known as the Upper Lydite Bed (Figs 13–15). This bed consists mainly of 1–2 cm rounded to subrounded polished pebbles of vein quartz and Carboniferous chert (lydites), but also includes clasts of igneous rocks, Carboniferous corals, limestone and silicified oolite (Neaverson 1925). The composition of these pebbles and their absence in the more southerly outcrops of the Wessex Basin led Neaverson to conclude that they originally came from the north and, in part, from the Caledonides of Scotland. However, during Portlandian times the pebbles may simply have been reworked from older deposits on the nearby land masses (e.g. the London–Brabant massif). The matrix of the Upper Lydite Bed varies from yellow–orange mottled clay to micritic limestone (Bristow 1968). The Upper Lydite Bed contains an indigenous *glaucolithus* Zone ammonite fauna, together with derived faunas from the *albani* and other, Upper Kimmeridgian, zones (Cope 1978; Wimbledon & Cope 1978; Figs 2, 14 and 15).

The Upper Lydite Bed is overlain by the Glauconitic Beds, which are sandy micritic limestones rich in glauconite and phosphate, suggesting condensation. The fauna of the Glauconitic Beds includes many thick-shelled bivalves, such as *Trigonia*, and ammonites indicative of the *glaucolithus* Zone (Wimbledon & Cope 1978).

In the Vale of Pewsey the roadside section near Crookwood Farm described previously (Jukes-Brown 1905; Casey & Bristow 1964; Townson 1971) is now badly degraded. However, there is a good section in the stream just to the north of the Farm, where approximately 4 m of limestone are visible (SU 015582; Fig. 13). The lower beds are very shelly limestones rich in glauconite, phosphate and lydites, whereas the upper beds are rubbly and less glauconitic. The fauna includes the bivalves *Laevitrigonia gibbosa* (J. Sowerby), *Protocardia dissimilis* J. de C. Sowerby, *Camptonectes lamellosus* (J. Sowerby), *Modiolus, Eocallista, Pleuromya, Ostrea, Myophorella* and serpulids. From the upper rubbly beds, Wimbledon (1980) records a few poorly preserved ammonites which suggest the *okusensis* Zone. The remainder of this sedimentary package is missing in the Vale of Pewsey (Fig. 2).

At Swindon, the Glauconitic Beds are overlain by the Cockly Bed (Fig. 4). This is a condensed limestone divisible lithologically and biostratigraphically into two parts. The lower part is a buff-pink, grain-supported bioclastic sand with a micritic matrix. As in the underlying Glauconitic Beds, glauconite, scattered lydite pebbles and quartz sand are present, although the amount has decreased (Wimbledon 1976). Ammonites are very common and indicate the *okusensis* Zone. In contrast, the upper part of the Cockly Bed (*kerberus* Zone) contains only rare grains of glauconite and no quartz sand.

In the area between Oxford and Aylesbury the Glauconitic Beds are overlain by bioclastic limestones termed either the Rubbly Limestones or Aylesbury Limestone (Buckman 1909–1930 [1922]; Pringle 1926; Arkell 1944, 1947*b*; Ballance 1963; Barker 1966; Bristow 1968; Wimbledon

Fig. 12. Stratigraphical and sedimentological log for the Portland Beds exposed at Chicksgrove quarry, near Tisbury, Vale of Wardour. Stratigraphical nomenclature and ammonite zonation from Wimbledon (1976, 1980) and Wimbledon & Cope (1978). For key to symbols, see Fig. 10.

a)

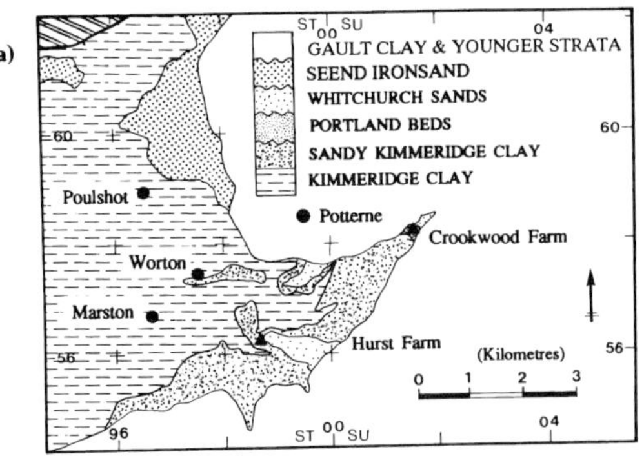

b)

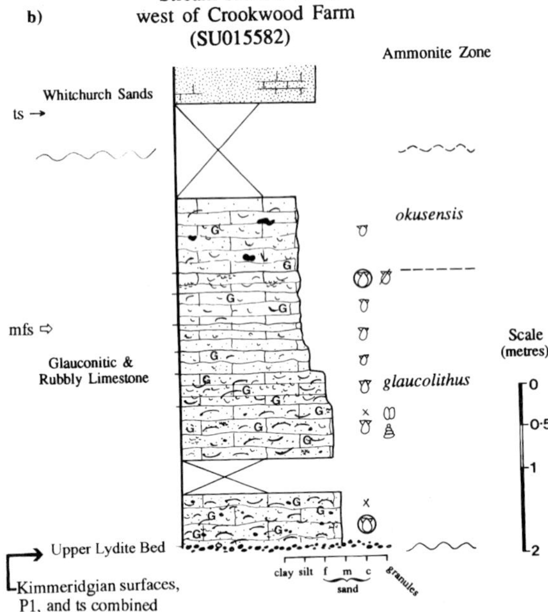

Fig. 13. (**a**) Map showing the outcrop of the Portland Beds and the Whitchurch Sands in the Vale of Pewsey. Compiled from Hesselbo *et al.* (1990) and British Geological Survey map 282. (**b**) Stratigraphical and sedimentological log for the Portland Beds exposed in the stream bank near to Crookwood Farm, Vale of Pewsey. For key to symbols, see Fig. 10.

1980; Figs 14 and 15). Horton *et al.* (in press) have recommended that all these limestone deposits are termed Aylesbury Limestone. However, for ease of comparison with published papers the old names are retained throughout this paper.

Formation of dolomite within the Portland Sand

Dolomite occurs within a well-defined interval of the Portland Sand. At its lowest occurrence it is associated with calcite and quartz sand in the Emmit Hill Marls or Black Nore Beds (Figs 4, 7 and 8), but it composes almost all of the overlying Black Sandstones or West Weare Sandstones. Dolomite disappears abruptly at the top of the Portland Clay in the west and at the top of the Black Sandstones in the east. No entirely satisfactory explanation for the occurrence of dolomite in the Portland Sand has previously been presented.

Petrographic studies reveal that the Black Nore Beds are composed of dolomite rhombohedra within clay laminae which alternate with laminae of quartz silt and sand. Selective replacement of former aragonite and high-Mg calcite mud can also

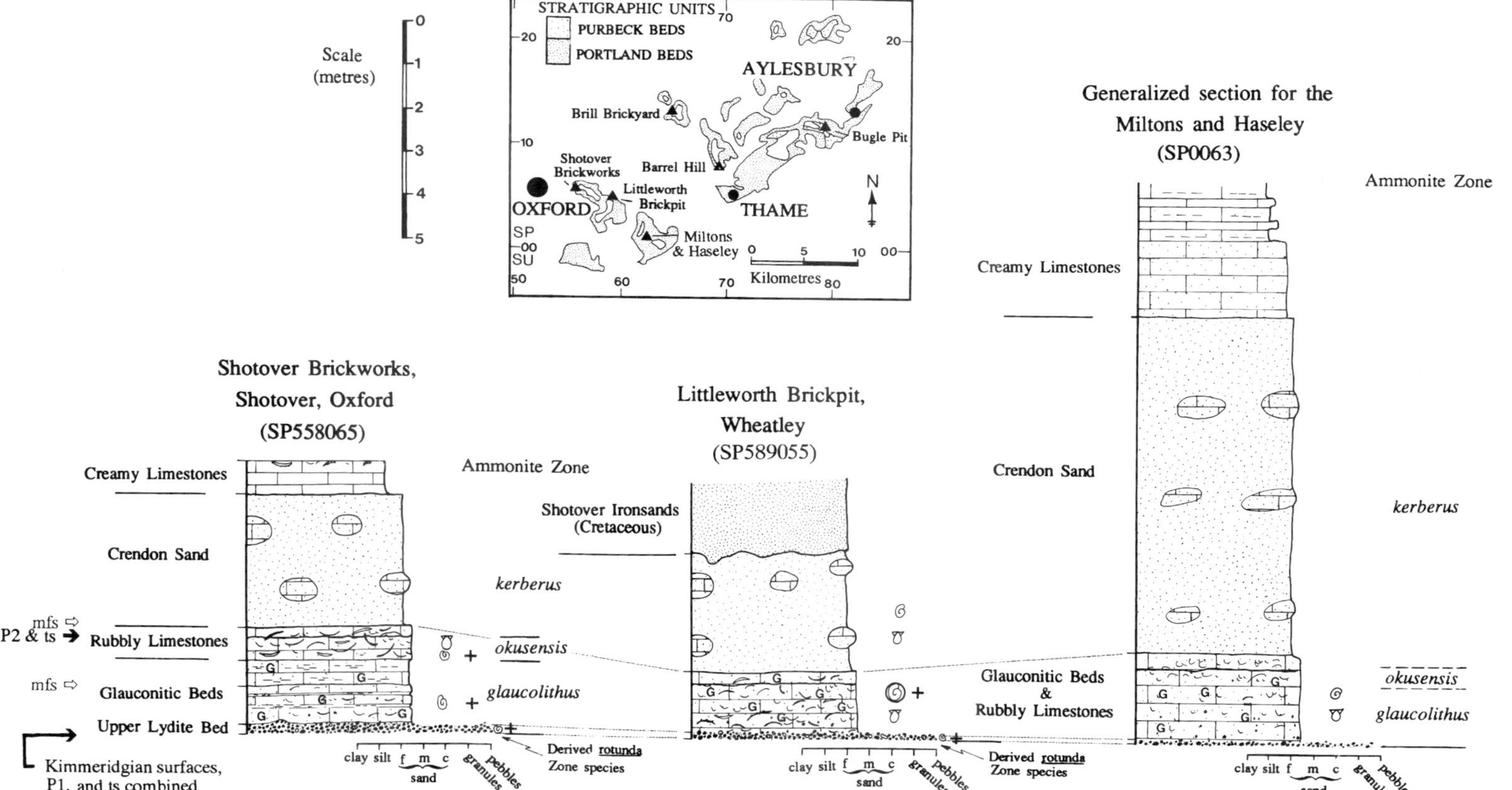

Fig. 14. Stratigraphical and sedimentological logs for the exposures of the Portland Beds between Shotover and Haseley. Shotover Brickworks, Oxford, compiled from Arkell (1947*b*) and Pringle (1926). Littleworth Brickpit, Wheatley, compiled from Pringle (1926), Arkell (1947*b*) and Coe (1992). Generalized log for the Miltons and Haseley compiled from Arkell (1944). Stratigraphical nomenclature from Wimbledon (1980). For key to symbols, see Fig. 10. Inset: map showing the outcrop of the Portland Beds and Purbeck Beds near Oxford and in Buckinghamshire. Compiled from British Geological Survey, Ten Mile Map, south sheet.

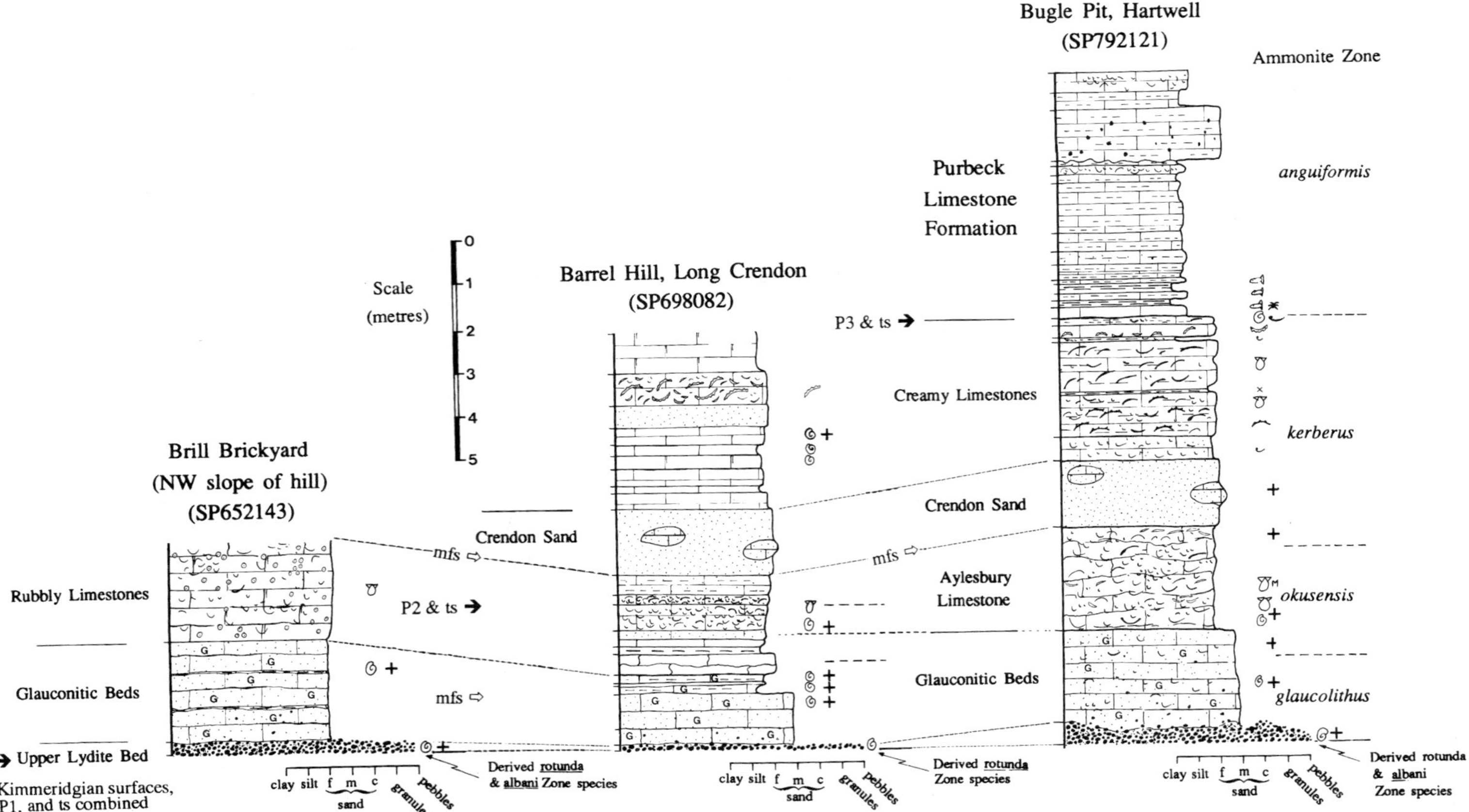

Fig. 15. Stratigraphical and sedimentological logs for exposures of the Portland Beds at selected exposures in Buckinghamshire. For location map, see Fig. 14. Brill Brickyard (northwest side of hill), compiled from Pringle (1926) and Arkell (1947*b*). Barrel Hill, Long Crendon, compiled from Buckman (1909–1930 [1922]) and Arkell (1947*b*), stratigraphical nomenclature from Wimbledon (1980) and modern ammonite zonation from Wimbledon & Cope (1978). Translation of Buckman's terms from Arkell & Tomkeieff (1953). Bugle Pit, Hartwell, section compiled from Arkell (1947*b*), Barker (1966) and Radley (1991), stratigraphical nomenclature from Wimbledon (1980) and modern ammonite zonation of Wimbledon & Cope (1978). For key to symbols, see Fig. 10.

be demonstrated (Townson 1971, 1975). All of the dolomitic rocks contain marine bivalves and ammonites and there are no sedimentary structures typical of sabkha deposits. Supratidal dolomitization can therefore be ruled out. Townson (1971, 1975) concluded that the dolomite was syn-sedimentary and proposed a model in which magnesium-enriched brine, concentrated by basin margin evaporation, sank downwards due to density contrasts, thereby dolomitizing the sedimented calcite mud.

Total organic carbon (TOC) values of these dolomites range between 0.7 and 1.4% TOC. Though not particularly high, these values are above average for marine limestones deposited in a moderate water depth. The presence of organic matter and anoxia has been shown to be linked with the formation of dolomite through syn-sedimentary replacement of calcite (Baker & Kastner 1981). Anoxic environments promote this process because they encourage microbial reduction of the SO_4^{2-} ion, which inhibits dolomitization. Thus the sediments affected have carbon isotope values which range from strongly negative to strongly positive (Baker & Kastner 1981). Selected carbon isotope data from the Portland Sand dolomites tend to support this model because there is a depletion of the ^{13}C isotope (average $\delta^{13}C = -3$‰PDB; Coe 1992). Kelts & McKenzie (1984) went one step further and showed that for anoxic dolomites in general the sedimentation rate directly affects $\delta^{13}C$. Under slow sedimentation rates, like those of the Portland Sand, the sediments remain in the sulphate reduction zone, resulting in depletion of the ^{13}C isotope. This occurs because the organic matter, relatively rich in the ^{12}C isotope, is oxidized by bacteria and combines with Ca or Mg to form carbonate. In contrast, under high sedimentation rates the sediments move into the methanogenesis zone where ^{12}C is expelled in the form of CH_4 resulting in enrichment of the ^{13}C isotope.

The Portland Sand samples are enriched in the ^{18}O isotope (average $\delta^{18}O = +2$‰PDB; Coe 1992). This can be explained in terms of the influence of evaporitic brines on dolomitization (high in ^{18}O isotope as the ^{16}O isotope is driven off by evaporation). The brines are envisaged to have formed in shallow marginal areas via gypsum precipitation and then to have sunk into the deeper basin, aiding dolomitization because of the high Mg:Ca ratio. Evaporites are preserved in the upper Portlandian at Durlston Bay, Dorset (Fig. 5) and there are other sedimentological and floral indications that an arid to semi-arid climate was well established by Portlandian times (Partington 1983; Quest 1985; Wignall & Ruffell 1990; Coe 1992; K. Myers & M. Partington pers. comm.). The well-defined occurrence of this fully marine dolomite is an important factor in the interpretation of unconformities P1 and P2.

Unconformity P2, middle okusensis *Zone*

Through all the outcrops in the Wessex Basin surface P2 is marked by a distinct upward facies change and in some sections it is demonstrably erosional (Figs 1–3). On the Isle of Portland, in the Vale of Wardour, Vale of Pewsey and on the London Platform, there are no strata representing the upper *okusensis* Zone because at all these sections, the unconformity is directly overlain by beds assigned to the *kerberus* Zone.

South Dorset. The most stratigraphically complete succession spanning this interval is exposed on the west side of St Alban's Head (Isle of Purbeck). Here, P2 is conformable and lies at the facies change from the dolomites of the Black Sandstones to the cherty wackestones of the Lower Cherty Beds (middle–upper *okusensis* Zone; Figs 4 and 8). This represents a small up-ramp facies shift with bioclastic packstones and wackestones absent (Fig. 6). In addition, the overlying strata shows a cyclic change from the chert-rich Lower Cherty Beds to the chert-poor beds J and J', which are condensed wackestones and packstones, and then back to a further unit of cherty wackestones forming the Upper Cherty Beds (Arkell 1935, 1947*a*; Cope & Wimbledon 1973; Wimbledon & Cope 1978). Similar successions may be observed at Gad Cliff, Dungy Head and Ringstead Bay, but the Lower Cherty Beds thin progressively towards the west. At Gad Cliff they are 9 m thick, thinning to 3.7 m at Dungy Head; further west at Ringstead Bay they are only 1 m thick. This thinning is probably the result of a lower carbonate supply (i.e. further away from the source) and lack of sediment accommodation space associated with the formation of unconformity P2. By comparison, the overlying beds J and J' (or equivalents) maintain a uniform thickness across the whole Dorset area (Coe 1992).

On the Isle of Portland there is no facies equivalent to the Lower Cherty Beds. The fine-grained dolomites of the Portland Clay are sharply overlain by condensed shelly packstones (Basal Shell Bed) similar to beds J and J' on the Isle of Purbeck and also belonging to the basal *kerberus* Zone (Figs 7, 8 and 16). Previous studies (Wimbledon 1980, 1986) have suggested that the Lower Cherty Beds are equivalent to the Portland Clay on the Isle of Portland. However, comparison of the Isle of Purbeck and Isle of Portland sections shows that the best biostratigraphical and lithostratigraphical correlation can be achieved if the Portland Clay is considered equivalent to the top of the Black Sandstones (Coe 1992). Consequently, on the Isle of Portland there is

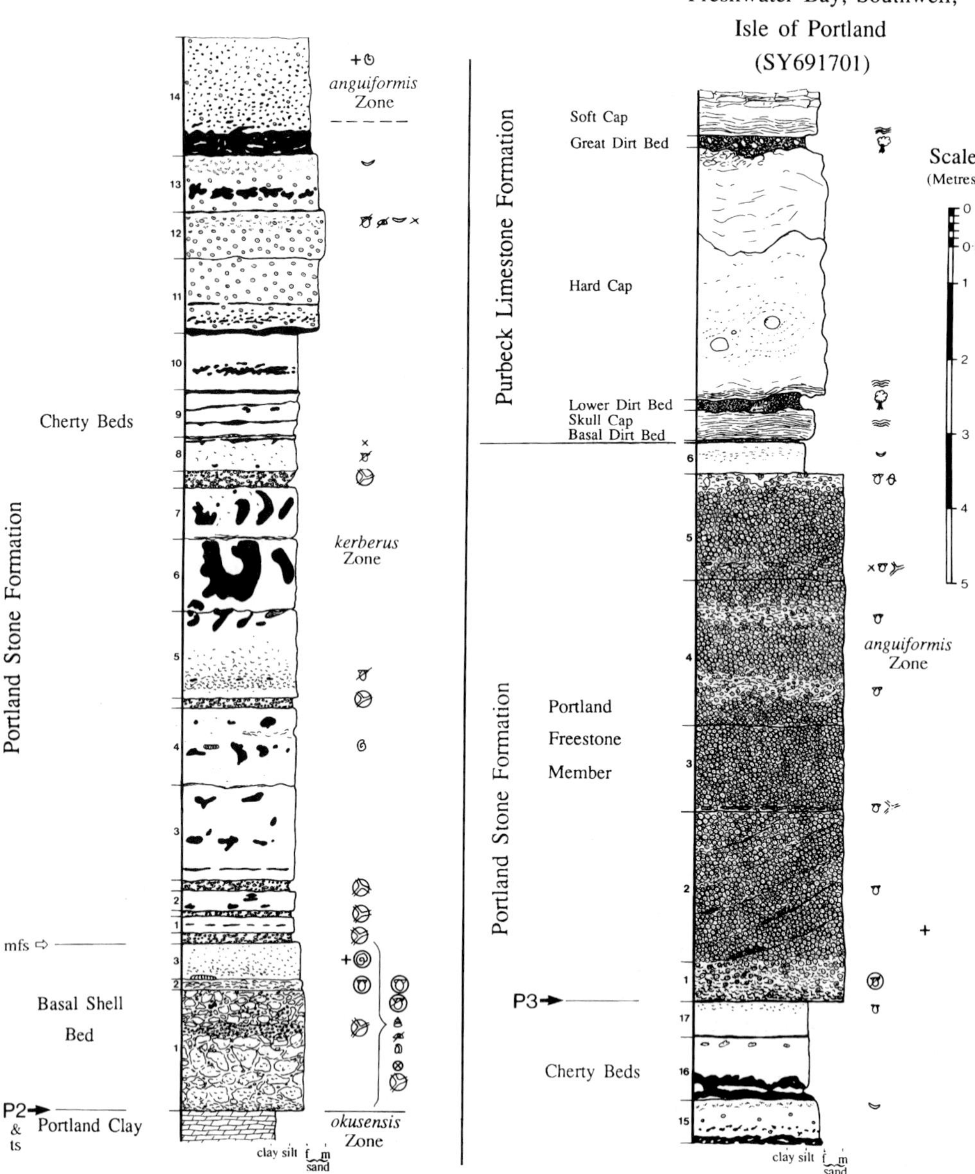

Fig. 16. Stratigraphical and sedimentological log for the Portland Stone exposed at Freshwater Bay, Isle of Portland. Portland Clay to base of bed 3 of the Portland Freestone Member from the sea cliff, upper part of the log from the small quarry at the top of the cliff. Biostratigraphy from Wimbledon & Cope (1978), stratigraphical nomenclature for the Portland Stone from Wimbledon (1986) and for the Purbeck Limestone Formation from Francis (1983). For key to symbols, see Figure 10.

no equivalent to the Lower Cherty Beds and P2 is unconformable (Fig. 4).

Interestingly, the first deposits overlying unconformity P2 (*kerberus* Zone) contain the greatest number of newly evolved ammonite species within the early Portlandian. If we assume gradual evolution, then the sudden increase in diversity would support the idea of a stratigraphical gap below this zonal boundary. Palaeomagnetic work by Ogg *et al.* (1995) on the Portland Sand and Portland Stone of

Dorset and Wiltshire is also suggestive of a gap at the base of the Basal Shell Bed on the Isle of Portland.

On the Isle of Purbeck and at Gad Cliff, Dungy Head and Ringstead Bay, the sedimentary package overlying P2 consists of buff-coloured cherty packstones and wackestones, the principal components of which are peloids and sponge spicules; these comprise the Lower Cherty Beds Member. The most prominent feature of the Lower Cherty Beds in the field is the black nodular chert formed from silica remobilized from sponge spicules during early diagenesis. The sponge spicules are mainly *Rhaxella* with some *Pachastrella* (Wilson 1966; Townson 1971, 1975). The nodular cherts appear to have formed in *Thalassinoides* systems; cherts exposed on bedding planes show the Y-shaped junctions typical of this burrow type. Lowered pH values caused by the concentration of organic matter within the burrow system probably account for preferential silica precipitation. Wilson (1966) demonstrated that the first phase of this silicification occurred before the first phase of calcite cementation. The sponge spicules must have been very abundant in these beds because most of the cherts have amalgamated to form long masses of black chert (Fig. 8).

At Ringstead Bay, Wimbledon (1980, 1986) documented two horizons of the Lower Cherty Beds separated by the Portland Clay; however, Townson (1971) interpreted the lower of these two horizons as part of the West Weare Sandstones. Townson's interpretation is accepted here because: (1) the cherts in the lower horizon (bed 11; Fig. 9) are pale grey silicified nodules very dissimilar to the black rounded nodules found in the Lower Cherty Beds elsewhere; (2) the overlying bed 12 (Fig. 9) is mainly composed of dolomite, a feature typical of the West Weare Sandstones rather than the Lower Cherty Beds; and (3) comparison between sections indicates that individual beds and lithologically distinct bed groups all correlate and also show proportionate thickness changes, whereas with Wimbledon's correlation they do not (Coe 1992).

On the Isle of Purbeck, the Lower Cherty Beds are overlain by Arkell's beds J and J' (1935, 1947*a*). These are buff-grey, chert-poor, shelly packstones (Fig. 8). Both beds show an intense network of *Thalassinoides* burrow systems, but the intensity and range in size of these systems is greatest in bed J'. The preserved body fossils consist mainly of ammonites together with varying amounts of broken bivalves and serpulids. At Dancing Ledge over 50 ammonites are concentrated in 1000 m^2 (Townson 1971); many of these are encrusted with serpulids and *Nanogyra*. Beds J and J' are overlain by the Upper Cherty Beds, which are very similar to the Lower Cherty Beds, though there are generally fewer chert nodules and subordinate amounts of shell debris (Fig. 8). The highest part of the sedimentary package consists of the lower 5–6 m of the Portland Freestone Member (beds O–S; Fig. 8). This is composed of oolitic wackestones with grey cherts, shell beds and many bioturbated horizons. Bed S (Arkell 1935; Fig. 8) is almost entirely composed of grey chert.

On the Isle of Portland, the lowest member of the sedimentary package overlying unconformity P2 is the Basal Shell Bed which, from lithostratigraphical and biostratigraphical similarities, has been shown to be equivalent to beds J and J' (Townson 1971; Cope & Wimbledon 1973; Wimbledon & Cope 1978; Fig. 2). The Basal Shell Bed itself is a complex condensed limestone, divisible into three units (Wimbledon 1974*a*; Figs 7, 9, 16 and 17). All three of these units contain a rich and diverse fauna comprising mainly bivalves and ammonites, but also including brachiopods, gastropods, serpulids and echinoids (Cox 1925). The lowest unit is packed throughout with randomly oriented shells, which are particularly concentrated at the base. Within this unit, clasts with several generations of encrustation are common and thin sections reveal extensive boring of the shells and other allochems. The whole unit is bioturbated and contains two or three bands of closely packed *Thalassinoides* burrow systems, together with incipient hardgrounds near the top (Fig. 17). Unit 2 of the Basal Shell Bed is a soft pinkish-buff limestone composed mainly of large, horizontally aligned, disarticulated and encrusted shells (Figs 7, 16, 17a and 17b). The uppermost unit (3) is a hard pale-grey, shelly-limestone; its most distinctive feature is the abundance of the large ammonite *Titanites*, which occur near the bottom and top of the unit, frequently encrusted with *Nanogyra* and serpulids (Fig. 17d).

A lateral transition between beds J and J' of the Isle of Purbeck sections and the Basal Shell Bed of the Isle of Portland is shown by the sections at Dungy Head and Ringstead Bay. At Dungy Head, the base of bed J is a shelly packstone similar to unit 1 of the Basal Shell Bed on the Isle of Portland; the top of beds J and J' are shelly wackestones with abundant *Thalassinoides* burrow systems. Bed K contains broken shell debris and many ammonites; it is comparable with unit 3 of the Basal Shell Bed. At Ringstead Bay the tripartite division of the Basal Shell Bed is recognized; the base of unit 1 is a shelly packstone, but the top two-thirds are almost entirely made up of the serpulid *Glomerula gordialis* (Schotheim); unit 2 is a very thin and intensely bioturbated wackestone and unit 3 is a shelly packstone (Fig. 9).

Above the Basal Shell Bed lies the Cherty Beds Member. The upper part of the Cherty Beds contains some tabular cherts whose origin is problematic: they are commonly found at the bases of beds

(a) (b) (c) (d)

Fig. 17. Features of the Basal Shell Bed at Freshwater Bay, Isle of Portland (Fig. 10). (a) Basal Shell Bed and part of the overlying Upper Cherty Beds; note how the bed thickness expands upwards in the Upper Cherty Beds. Height of cliff is 7 m. (b) The Basal Shell Bed, clearly showing the three part division and the intense ramifying network of *Thalassinoides* burrow systems in unit 1. At the base of unit 3 there is a large *Titanites* ammonite (length of hammer = 0.4 m). (c) Detail of the fabric of unit 1 of the Basal Shell Bed, showing the abundance of shell debris and the packing within *Thalassinoides* burrow systems. (d) Ammonite from the top of unit 3 intensely encrusted by *Nanogyra*. Diameter of ammonite is 55 cm.

(e.g. beds 11 and 17 at Freshwater Bay; Fig. 16), or else above large-scale, low-angle cross-stratification surfaces (e.g. within Bed 9). Both types of occurrence of the tabular chert can be explained in terms of replacement of winnowed concentrations of sponge spicules on a minor erosion surface. Some of the Cherty Beds are separated by thin, intensely bioturbated units (e.g. between beds 1 and

2, and 7 and 8; Figs 7 and 16). These bioturbated horizons can be correlated over the Isle of Portland (House 1970) and some can be correlated between the Isle of Portland and the Isle of Purbeck (Coe 1992). On the Isle of Portland, small (0.25–0.35 cm) ooids and superficial ooids first appear in bed 11 of the Cherty Beds Member, reaching their greatest abundance in bed 12 (*c.* 40%); they disappear above bed 15 (Fig. 16, beds 11–15). At Freshwater Bay, the top of beds 12, 13 and 15 all contain abundant moulds of bivalves and gastropods. The top of the Cherty Beds (beds 16 and 17 at Freshwater Bay) consist of cherty packstones and wackestones similar to beds 1–10, the allochems being tiny peloids (0.1 mm), sponge spicules and subordinate amounts of fine shell debris.

Comparison of the Cherty Beds of Freshwater Bay with those at Clay Ope shows that beds 1–13 of the Upper Cherty Beds are practically identical. However, beds 14 and 15 differ, as at Clay Ope they comprise shelly wackestones/packstones, rather than the cherty wackestones/packstones found at Freshwater Bay. The bioclasts in beds 14 and 15 are mainly of large bivalves, but gastropods are also common; all are preserved as casts. Beds 14 and 15 ('the Portland Roach') contain freshwater cements which, together with the shell dissolution described earlier, could be associated with emergence during the formation of the younger unconformity P3 (*anguiformis* Zone; Fig. 7).

Vale of Wardour. Near to Tisbury the only exposures of *okusensis* Zone and *kerberus* Zone deposits occur at Chicksgrove and Chilmark Quarries (Figs 2, 3 and 12). At Chicksgrove Quarry, unconformity P2 is well displayed as it comprises a channel (small incised valley) cut into the top surface of the marine quartzose–glauconitic sands of the Tisbury Glauconitic Member. The channel was infilled and then overstepped by the remainder of the Wockley Member (Ragstone Beds and Chalky Series; House 1958; Arkell 1933 and references cited therein; Fig. 12). The sediments of the Wockley Member which infill the channel are uncemented, fine-grained sands with occasional pebbles, containing a fairly rich non-marine fauna including plants, *Falcimytilus*, crocodile scutes, fish debris and reptile bones (Wimbledon 1976, 1980; Simms pers. comm.). Elsewhere in the quarry, beyond the limits of the channel, this sandy bed is only a few centimetres thick and tends to be more clay-rich with occasional serpulids. Two kilometres to the NNW at Chilmark Quarries, the part of the Tisbury Glauconitic Member assigned to the *okusensis* Zone is much thinner, being represented by only 2 m of section compared with the 7–8 m found at Chicksgrove (Wimbledon 1976; Figs 2 and 3). However, it is immediately overlain by the same type of plant- and vertebrate-bearing sand as seen at Chicksgrove (Wimbledon 1976). This suggests that either the Tisbury Glauconitic Member thinned depositionally from Chicksgrove to Chilmark or, more probably, because the overlying *kerberus* Zone strata hardly thin at all – that there is overstep and an angular unconformity.

At Chicksgrove Quarry the rest of the Wockley Member above the unconsolidated sand comprises a micritic limestone containing tiny brackish-water gastropods, overlain by very shelly micritic limestones with a marine bivalve fauna and *kerberus* Zone ammonites (House 1958; Wimbledon 1976). The upper part of the member is a chalky micrite containing fewer bivalves and with occasional bands of chert.

Vale of Pewsey. In the Vale of Pewsey there is an extensive stratal and faunal gap, within which unconformity P2 occurs (Figs 2 and 3). A few poorly preserved ammonites suggest the *okusensis* Zone (Wimbledon 1980) for the youngest deposits lying immediately below the hiatus (Wimbledon 1986; Figs 2, 3 and 13), whereas those lying above the hiatus are, at the oldest, of late Portlandian age (i.e. *lamplughi* Zone; Fig. 4; Casey & Bristow 1964; Wimbledon & Hunt 1983; Morter 1984; Allen & Wimbledon 1991; Coe 1992).

London Platform. At Swindon, unconformity P2 is at the boundary between two distinct condensed limestone facies which comprise the Cockly Bed. The upper part of the Cockly Bed is a micritic limestone containing matrix-supported bivalves; large trigoniids and *Protocardia* predominate, but there are common *Pleuromya* and *Camptonectes* (Wimbledon 1976). The smaller amount of shell debris and lack of quartz and glauconite distinguish it from the lower part of the Cockly Bed. Ammonites are fairly common throughout the Cockly Bed and indicate *okusensis* Zone for the lower part and *kerberus* Zone for the upper part. At Swindon, the Cockly Bed grades up into the Swindon Sand and Stone Member. This consists of fine-grained, sub-rounded to sub-angular, moderately to well-sorted sand, with calcareous nodules and bands. The beds show small-scale, low-angle cross-bedding and planar bedding (typical bed thickness is 3 cm); occasional vertical burrow tubes are also preserved (Fig. 18).

Near Oxford, and in Buckinghamshire, the unconformity is similar to that at Swindon, occurring within the condensed shelly limestones and marls of the Rubbly Limestones or Aylesbury Limestone (Figs 14 and 15). Ammonites indicative of the *kerberus* Zone have been collected from both members (Wimbledon & Cope 1978). At most localities these limestones and marls grade into the overlying

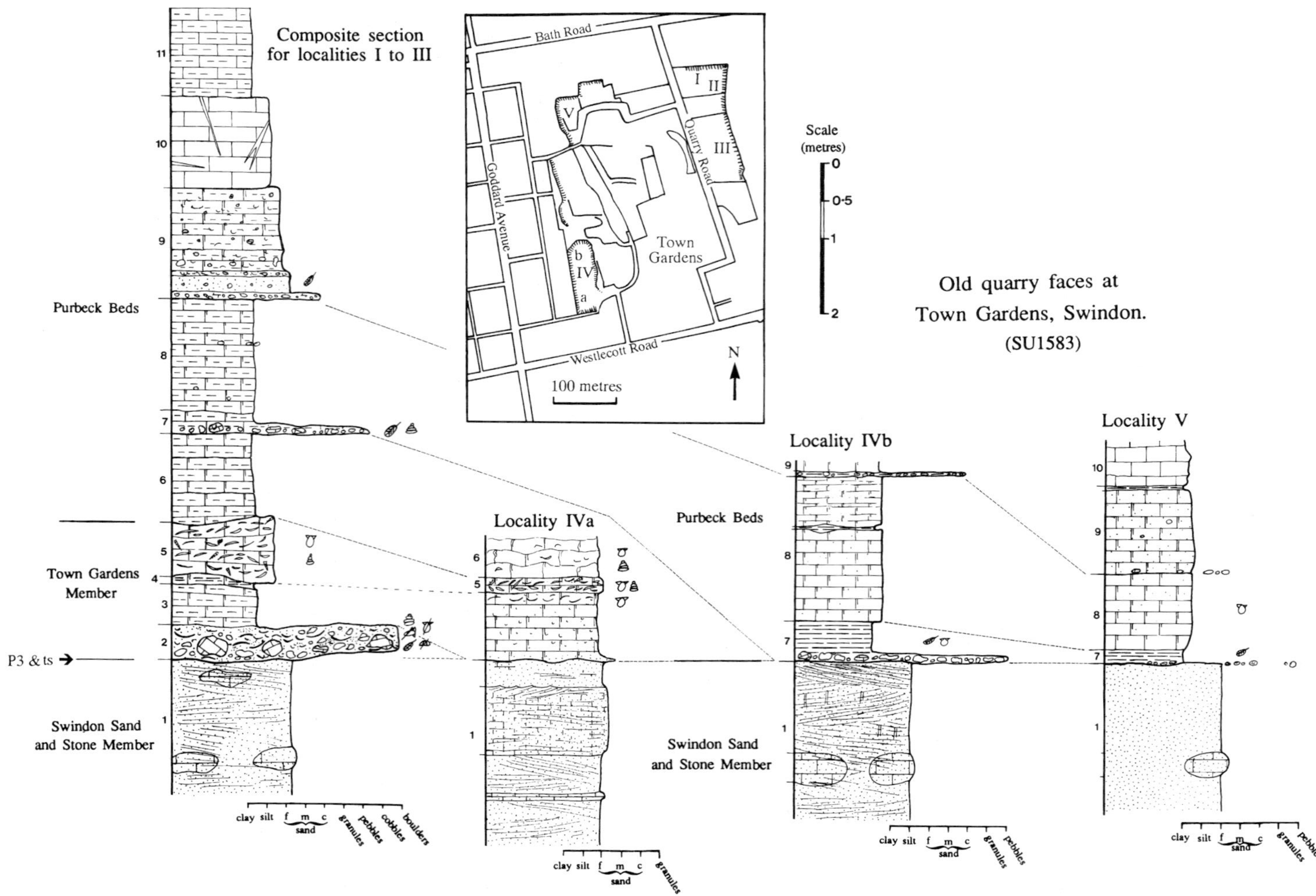
Composite section
for localities I to III
Old quarry faces at
Town Gardens, Swindon.
(SU1583)
Scale
(metres)
0
0·5
1
2
Bath Road
Quarry Road
Town
Gardens
Goddard Avenue
Westlecott Road
100 metres
N
I
II
III
IV
V
a
b
Locality IVa
Locality IVb
Locality V
Purbeck Beds
Town Gardens
Member
P3 & ts
Swindon Sand
and Stone Member
clay silt f m c sand granules pebbles cobbles boulders

fine- to medium-grained sands (Crendon Sand) which are, in turn, overlain by the Creamy Limestone, the latter comprising shelly, micritic limestones with some quartz sand and occasional clay horizons (Arkell 1947*b*; Ballance 1963; Townson 1971; Wimbledon 1974*b*; Figs 14 and 15). At some localities between Whitchurch and Haddenham, the gap associated with unconformity P2 becomes more marked, because the Crendon Sand (*kerberus* Zone) overlies the lower part of the Rubbly or Aylesbury Limestone (*okusensis* Zone; Fig. 4; Wimbledon 1980).

Unconformity P3, lower anguiformis *Zone*

The third unconformity, P3, occurs within the Portland Stone Formation close to the base of the *anguiformis* Zone. The unconformity is at the base of either oolitic grainstones or shelly packstones throughout Dorset and Wiltshire. On the London Platform unconformity P3 manifests itself as an erosion surface, which is demonstrably angular over short distances.

South Dorset. In south Dorset unconformity P3 separates massive oolitic grainstones and packstones from the underlying cherty packstones and wackestones. On the Isle of Portland it corresponds to the base of the Portland Freestone. However, to the east on the Isle of Purbeck, at Dungy Head, Ringstead Bay and other inland outcrops, it occurs within the Portland Freestone. Indeed, previous interpretations (Wimbledon 1980, 1986) suggest that the base of the oolites, and hence the base of the member, is diachronous: the base of the member belongs to the upper *kerberus* Zone on the Isle of Purbeck but lower *anguiformis* Zone on the Isle of Portland (Wimbledon & Cope 1978; Wimbledon 1980, 1986). However, re-logging and petrological studies of the sections (Coe 1992) revealed that there is no diachroneity of facies because there are in fact two sedimentary packages containing ooids throughout south Dorset. Unconformity P3 is positioned at the base of the upper, more major influx of ooids. The base of the lower, minor influx of ooids is within the *kerberus* Zone and the base of the upper influx is within the *anguiformis* Zone. The lower influx comprises laterally impersistent oolitic packstones, whereas the second is a thick, continuous package of oolitic grainstones with large well-developed ooids (0.35–0.55 cm in diameter).

At Freshwater Bay on the Isle of Portland, the surface of unconformity P3 is very sharp, modified only by occasional burrows that pipe ooids down a few centimetres into the top of the Cherty Beds (Fig. 16). The unconformity is overlain by a shell lag and then by massive cross-stratified oolitic grainstones. The beds are partially cemented by sparite and some shelly intervals occur, as well as horizons of *Thalassinoides* burrow systems. On the north side of Freshwater Bay, where the ammonite zonation was originally determined (Wimbledon & Cope 1978), the junction is sharp but very undulatory.

On the Isle of Purbeck, the base of the upper influx of ooids within the Portland Freestone is interpreted as the correlative surface of unconformity P3. Above this there are well-developed oolitic grainstones and packstones (Pond Freestone and Titanites Bed; Arkell 1935) similar to those in the Isle of Portland section. The Chert Vein occurs below unconformity P3 (Arkell 1935) and is a cherty packstone/wackestone, similar to beds 16 and 17 of the Cherty Beds at Freshwater Bay on the Isle of Portland. At Winspit (SY977761) large tool marks were found on the unconformity surface. The intervening sections at Gad Cliff, Dungy Head and Ringstead Bay are all similar to the Isle of Purbeck sections in that unconformity P3 lies within the Portland Freestone. The base of the Portland Freestone at each of these sections represents the first minor influx of ooids and the major influx is several metres higher within the *anguiformis* Zone. At Ringstead Bay the lower part of the Portland Freestone contains many grey chert nodules and horizons.

At Clay Ope and other sections on the west side of the Isle of Portland, the Portland Freestone can easily be divided into three parts (Townson 1971): the lower and upper parts are oolitic grainstones, whereas the middle part (1.5 m) consists of shelly micrites (Fig. 7). At several sections (e.g. Ringstead and Winspit; Figs 8 and 9), shelly beds occur near the top of the member. These contain a rich variety of bivalves, which are typically supported by a micritic matrix.

At all Dorset localities, the deposits at the top of the Portland Freestone and into the base of the Purbeck Limestone record a progressive shallowing trend, changing from subtidal oolitic grainstones, through intertidal and supratidal algal stromatolitic limestones to algal mats and palaeosols (West 1975)

Fig. 18. Stratigraphical and sedimentological logs for exposures of the Portland Beds and Purbeck Limestone exposed in the old quarries at Town Gardens, Swindon. Inset: map showing location of the sections, modified from Sylvester Bradley (1940). Composite section for localities I to III from Sylvester Bradley (1940), Arkell (1940, 1947*b*, 1948) and Chatwin & Pringle (1922). For key to symbols, see Fig. 10.

and eventually passing up into evaporites and freshwater limestones. The uppermost bed of the Portland Freestone is thin and variable; on the Isle of Portland it is a pelloidal limestone with a monospecific bivalve fauna; on the Isle of Purbeck it is a massive micrite, known as the Shrimp Bed (Figs 7, 8 and 16).

The base of the overlying Purbeck Limestone Formation consists of a series of palaeosols known as 'dirt beds', together with algal stromatolitic limestones and mats (Brown 1963; Pugh 1968; West 1975; Francis 1983). Three palaeosols can be recognized, the youngest of which is thick and laterally continuous and is known as the Great Dirt Bed. At some localities, a caliche breccia is developed immediately below this palaeosol (West 1975). Both the Great Dirt Bed and, to a lesser extent, the Lower Dirt Bed (Fig. 16) have yielded abundant remains of fossilized trees (Francis 1983). Palaeobotanical studies of the wood (Francis 1984) show that there were seasonal and yearly variations in climate. Above the palaeosols are fine-grained limestones with gypsum and halite pseudomorphs which, near the Purbeck–Isle of Wight disturbance, have been disrupted. The upper part of the sedimentary package is made up of biomicritic limestones, evaporites, marls, clays and freshwater cherty limestones which contain halite pseudomorphs, indicating an environment with fluctuating salinities (Clements 1969; Ensom 1985).

Vale of Wardour. In the Vale of Wardour there is a stratal gap and unconformity P3 coincides with the boundary between the *kerberus* and *anguiformis* Zones. Like unconformity P2, unconformity P3 displays an angular relationship between the sections at Chicksgrove and Chilmark Quarries (Figs 2 and 3). Unconformity P3 is exposed at the eastern end of Chicksgrove Quarry, where it occurs near the top of the Wockley Member. The surface is sharp; it is underlain by micritic limestones containing a thin tabular chert and overlain by shelly micritic packstone. The packstone contains some ooids and all the shells are aligned convex-up, indicating current activity (Fig. 12); this bed has also yielded a specimen of *Titanites* cf. *anguiformis* Wimbledon (Wimbledon 1976). At Chilmark Quarries, the deposits are very different, comprising an areally restricted oolitic sand body about 7 m thick (the Chilmark Oolite Member), the base of which has rarely been seen (Wimbledon 1986), but is assumed here to be correlative to unconformity P3. The ooid sands are cross-bedded and composed of 60% ooids, typically 0.8 mm in diameter, plus small gastropods and bivalves set in a micritic matrix. Unfortunately, the sections are rather poor (Wimbledon 1986) and difficult to access as they now lie within high-security RAF property. Thus the restricted nature of the oolitic sand body may be because it is infilling a palaeovalley cut during erosional processes associated with the unconformity P3, similar to the incised channel associated with unconformity P2 at Chicksgrove Quarry. Alternatively, the oolitic sand body may have been originally deposited as a lens.

Petrological work on the cherts found at the top of the Tisbury Member and the base of the Wockley Member at Chicksgrove Quarry (Fig. 12) revealed that the cherts were formed during a phase of meteoric-water circulation while the sediments were in the phreatic zone (Astin 1987). This may have been during the period of erosion and possible palaeosol formation associated with unconformity P3. Chert formation would have then ended as marine pore waters were introduced during deposition of the overlying shelly packstone (bed 33) at Chicksgrove (Fig. 12). However, the possibility that the silicification did not start until a slightly later phase associated with the formation of palaeosols at the base of the younger Lulworth Beds of the Purbeck Limestone cannot be excluded.

The remainder of the sedimentary package above unconformity P3 exposed in the Vale of Wardour is composed of stromatolitic limestones and palaeosols overlain by non-marine limestones, marls and clays; these are the Lulworth Beds and they are similar both palaeontologically and lithologically to those in Dorset (Fisher 1856; Andrews in Etheridge 1880; Andrews 1881; Hudleston 1883; Andrews & Jukes-Browne 1894; Arkell 1933; Simms pers. comm.).

Vale of Pewsey and London Platform. In the Vale of Pewsey unconformity P3 is amalgamated with unconformities P2 and P4 (Fig. 3). At Town Gardens Quarry, Swindon (SU 1583), on the London Platform, unconformity P3 is well displayed and is the boundary between the Swindon Sand and Stone and the overlying Town Gardens Member. There is good evidence for a significant hiatus associated with the unconformity because both members are thin, lithologically unrelated to each other and different beds of the Town Gardens Member onlap onto the erosion surface (Sylvester Bradley 1940; Blake 1880; Fig. 2). Towards the north end of the gardens, the Town Gardens Member is absent and the younger Purbeck Limestone Formation immediately overlies unconformity P3 (Fig. 18). Many parts of the section are now either covered or difficult to access, but the complete section has been described by Sylvester Bradley (1940; Fig. 18).

The unconformity can still be observed at several localities within the Town Gardens (locality numbers IVa, IVb and V; Fig. 18). At locality IVa the unconformity is gently undulating and overlain by micritic limestones with moulds of tiny bivalves.

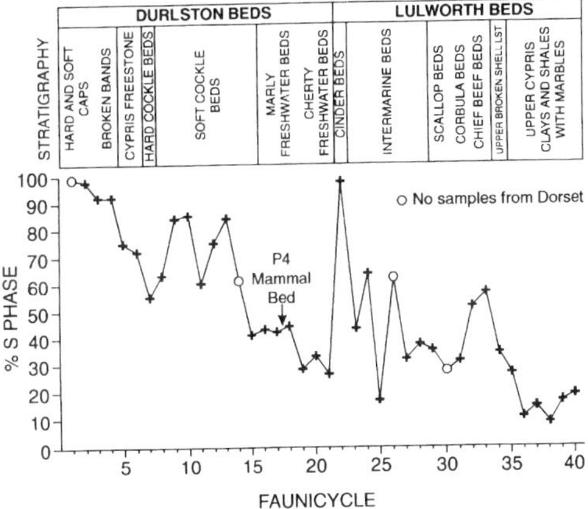

Fig. 19. Graph showing the variation in percentage of high-salinity tolerant ostracod faunas through the Purbeck Limestone as predicted from Anderson's (1985) ostracod faunicycles.

The basal few centimetres contain scattered quartz grains which were probably derived from the Swindon Sand and Stone immediately below. At locality IVb the unconformity is planar and the Middle Pebbly Bed of the Purbeck Limestone (bed 7, Sylvester Bradley 1940; Fig. 18) overlies the unconformity. The Middle Pebbly Bed is a grey and purple marl containing pebbles (2–10 cm) derived from the underlying limestone and sandstone, together with plant matter and degraded shell fragments. At locality V the unconformity is very sharp and distinctive: immediately beneath it is the Swindon Sand and Stone, the top of which is enriched with iron minerals and, similar to the situation at locality IVb, the unconformity is overlain by the Middle Pebbly Bed. At locality V the pebbles are very indurated and concentrated in depressions in the unconformity surface.

Unconformity P3 is highly undulatory and there are several erosion surfaces within the Purbeck Limestone. Thus the relationship between different beds over short distances is complex. These relationships were sketched out by Blake (1880) and re-illustrated by Sylvester Bradley (1940) and Arkell (1947*b*). From the evidence collected during the course of the present study, together with the previously published sketches and sections, the following can be deduced: (1) the top surface of the Swindon Sand and Stone (bed 1) was eroded, leaving an undulatory surface of long-wavelength peaks and troughs; (2) at the northeastern end of Swindon Town Gardens beds 2–5 (Town Gardens Member) and bed 6 (Purbeck Limestone) were deposited with bed 2 infilling an undulatory topography and beds 3–6 onlapping the unconformity (elsewhere in the Town Gardens beds 2–6 are mostly absent and were either deposited and then removed by subsequent erosion, or never deposited); (3) beds 7–9 were deposited with bed 7 infilling troughs in the topography of bed 1 and beds 8–9 onlapping onto the peaks in the topography of unconformity P3; (4) the surface beneath bed 10 cuts down through beds 8 and 9 in the northwest corner of the gardens.

In sections near Oxford and in Buckinghamshire the exact location of unconformity P3 is unclear because the sections are highly variable in lithology and generally poorly documented. However, it seems probable that in this region unconformity P3 is at the abrupt boundary between the Creamy Limestones (*kerberus* Zone; Wimbledon 1980; Wimbledon & Cope 1978) and the overlying non-marine Purbeck Limestone. This boundary was formerly well exposed at Bugle Pit, Hartwell, where sandy limestones with ammonites (Creamy Limestones) were sharply overlain by a marl containing fish teeth, insects and ostracods (Purbeck Limestone Formation; Fig. 15; Arkell 1947*b*; Ballance 1963; Barker 1966).

Unconformity P4, lamplughi *Zone*

Unconformity P4 is complex and less well defined than unconformities P1 to P3, as it occurs within non-marine carbonates. Consequently, correlation of these beds with the ammonite zonation relies on event stratigraphy (Casey 1963, 1967; 1973), ostracod 'faunicycles' (Anderson 1985), molluscan salinity cycles (Morter 1984), microfloral data (Norris 1969) and correlation of marker beds (Casey & Bristow 1964; Wimbledon & Hunt 1983; Morter 1984). There are two main lines of evidence for the existence of unconformity P4: (1) in the Vale of Pewsey and on the London Platform there are large biostratigraphical gaps inferred to span at least four ammonite zones (Fig. 2); and (2) the upper Portlandian rocks exposed in Dorset and Wiltshire show a marked upward decrease in marine influence.

South Dorset and the Vale of Wardour. The exact surface in Dorset which corresponds to the highest biostratigraphical gap on the London Platform is difficult to constrain. However, consideration of the non-marine lithological character and, particularly, the fauna of the Lulworth Beds and the basal part of the Durlston Beds, shows that there is a cyclic change. Anderson (1985 and references cited therein) noted that under the stress of a shallow-water environment varying in salinity, a large number of species and varieties of ostracod was produced, some of which had only limited salinity tolerance. Anderson recognized that there was a cyclicity between ostracod faunas with a high-salinity tolerance ('S' phase) and those with a preference for freshwater ('C' phase): these he called 'faunicycles'. A large number of specimens was collected and percentages of high-salinity tolerant species within each 'faunicycle' calculated. The high-salinity tolerant ostracods decrease to a minimum within the Marly Freshwater Beds and then increase sharply within the Cinder Bed before decreasing again through the Durlston Beds (Fig. 19). Assemblages of bivalves from the Weald Basin and adjacent areas also suggest several distinctive saline episodes, the first of which is the Cinder Bed (Morter 1984).

The Cinder Bed itself is packed with the oyster *Praexogyra distorta* (J. de C. Sowerby), but also includes other bivalves *Modiolus, Neomiodon* and pectinids, as well as *Hemicidaris* spines; these all indicate more saline conditions and are in marked contrast with the freshwater cherty limestones below (Clements 1969; Ensom 1985). Thus it is proposed that the lateral equivalent of the biostratigraphical gap seen on the London Platform is within the Marly Freshwater Beds in Dorset, corresponding to the maximum decrease in salinity determined from the ostracod 'faunicycles' (Fig. 19). A good candidate within the Marly Freshwater Beds is the Mammal Bed, this being a palaeosol with mammal bones exposed at Durlston Bay. A similar but thinner succession of non-marine limestone is exposed in the Vale of Wardour (Andrews 1881; Andrews & Jukes-Browne 1894; Woodward 1895).

Vale of Pewsey and London Platform. In the Vale of Pewsey, Swindon and Oxford areas, mid-Portlandian sedimentary rocks are unconformably overlain by the Whitchurch Sands (Casey & Bristow 1964). These sandstones have been ascribed to various stratigraphical intervals and not all the deposits may be contemporaneous; however, there is evidence to suggest that at least some of them are late Portlandian in age (Casey & Bristow 1964; Wimbledon & Hunt 1983; Morter 1984; Hesselbo *et al.* 1990; Allen & Wimbledon 1991). Thus if some of the Whitchurch Sands are of late Portlandian age there must be a biostratigraphical gap of at least three ammonite zones at the base of these sands, which encompasses unconformity P4. The Whitchurch Sands were interpreted by Casey & Bristow (1964) 'as the vestiges of a marine–brackish formation of Middle Purbeck age, indicating the transgressive front of an advancing sea thought to have come from the north and left its mark farther south in the Cinder Bed of Dorset'. The evidence for this interpretation is based on faunal similarities between the Whitchurch Sands and the Serpulite Horizon of Germany, the latter being regarded as equivalent to the Cinder Bed (earliest Ryazanian age). However, recent work on the palynology and bivalves by Wimbledon & Hunt (1983) and Morter (1984) suggests an earlier age for the Whitchurch Sands. This is partially confirmed by data from the Kingsclere borehole (SU 5475 5785), which provides the nearest available subsurface section to outcrops of the Whitchurch Sands; there is a typical development of the Cinder Bed and, 5 m below, a serpulite horizon with a faunal assemblage similar to that of the Whitchurch Sands. A study by Horton *et al.* (in press) suggests that the Whitchurch Sands are Valanginian in age. Their conclusion is based mainly on clay mineralogy from the deposits near Brill, but the data are not conclusive. It is possible that the Whitchurch Sands represent several different transgressive events, the oldest of which is possibly above unconformity P4, as described earlier.

The Weald. The succession in the Warlingham borehole on the northwestern edge of the Weald (TQ 3476 5719) appears to be complete and is correlatable with the boreholes and outcrop sections in Dorset and other parts of the Weald (Institute of Geological Sciences 1971; Worssam & Ivimey-Cook 1971; Lake & Holliday 1978; Morter 1984). However, the identification of the Cinder Bed in the Sussex outcrop has been questioned by Anderson & Bazley (1971) and Wimbledon & Hunt (1983). The oldest Purbeck rocks of the central Weald, which are evaporites, lie on Portlandian-age sandstone, which is no younger than the *kerberus* Zone (Wimbledon & Hunt 1983; Allen & Wimbledon 1991), hence if the evaporites found in Dorset and the Weald are age-equivalent (Worssam & Ivimey-Cook 1971; Lake & Holliday 1978; West 1975), then there is a hiatus of at least an ammonite zone at unconformity P3. However, palynological evidence from boreholes and outcrop material suggests that the hiatus is of much longer duration, with the age of the basal Purbeck evaporites as either late Portlandian or even early Ryazanian (Wimbledon & Hunt 1983; Allen & Wimbledon 1991), thus encompassing unconformities P3 and P4.

Sequence stratigraphical interpretation of the Portlandian succession

There have been two previous attempts to explain the deposition of the Portlandian strata of the Wessex Basin in terms of sea-level change, before the advent of sequence stratigraphy. Firstly, Townson (1971, 1975) described the Portland Group in terms of 'three cycles consisting of major regressive and minor transgressive phases superimposed on an overall regression'. Secondly, Wimbledon (1986) recognized many shallowing and deepening trends from the facies changes and concluded that although some events were isochronous others were more localized, but Wimbledon did not discriminate between different orders of magnitude of sea-level change.

In the present study, the four unconformities recognized within the Portlandian strata of the Wessex Basin are interpreted to be the result of erosional and non-depositional processes during relative sea-level falls. They therefore constitute sequence boundaries (Van Wagoner *et al.* 1988; Vail *et al.* 1991; Fig. 4). The overall shallowing trend from the Portland Sand to the Lulworth Beds is interpreted as part of a long-term regression, onto which four short-term cycles of relative sea-level change are superimposed.

Base of the Portlandian and the middle glaucolithus *Zone Sequence Boundary (P1)*

The base of the Massive Bed or Black Nore Sandstone, which corresponds to the base of the Portlandian Stage, is interpreted as a transgressive surface. Compared with the deposits above and below, the Massive Bed and Black Nore Sandstone are coarser grained and more poorly sorted; they also contain a diverse trace fossil assemblage. These are all features of condensation and reworking and are typical when sedimentation fails to keep up with an increase in sediment accommodation space during relative sea-level rise. In the early Portlandian (*albani* Zone), interbedded argillaceous silts and sands with carbonate bands were deposited. A faunal-abundance and diversity peak at the top of the *albani* Zone in all Dorset sections is interpreted as the maximum flooding surface. Sedimentation of interbedded silts, sands and carbonate bands continued into the *glaucolithus* Zone in the area between Gad Cliff and the Isle of Purbeck (i.e. on the main hanging wall of the Purbeck–Isle of Wight fault complex), whereas on the Isle of Portland and at Dungy Head and Ringstead Bay the sediment supply failed, causing simultaneous winnowing of the underlying sediments (Cast Beds). The siliciclastic sediment supply to these areas during early *glaucolithus* Zone times is believed to have failed because of a relative sea-level fall. This would have isolated Ringstead Bay and Dungy Head because they were located on the upthrown edge of the major fault block (northern block of the Purbeck–Isle of Wight fault). The fault structures affecting the Isle of Portland are not as clear, but the area could either have been topographically isolated or else the siliciclastic sediments could have been channelled away from the area, possibly along the base of the fault scarp, which, together with increased erosion as relative sea level was falling, resulted in the formation of a hiatus.

A fall in relative sea level in middle *glaucolithus* Zone times would also explain the establishment of a carbonate ramp depositional environment (Fig. 6) in the Dorset area and therefore the onset of carbonate production across much of the basin, a process which had been previously inhibited by the siliciclastic supply. Carbonate production then continued beyond the end of the Portlandian. The absence of major land-derived siliciclastic sediments at the base of this depositional sequence, and subsequent depositional sequences, is not surprising because there is strong evidence indicating that the climate was arid at this time; hence fluvial input would be at a minimum (see section on formation of dolomite within the Portland Sand).

In the Vale of Wardour the evidence for a sequence boundary of *glaucolithus* Zone age is inconclusive. However, whereas the lower part of the Wardour Member is open marine, the upper part exposed close to Tisbury Station contains sedimentary structures which are typical of subtidal–intertidal conditions, thus indicating a relative sea-level fall. These are then overlain by the open marine Chicksgrove Limestone. At Chicksgrove Quarry there is also evidence for a period of erosion in the form of a pebble bed. In the Vale of Pewsey and on the London Platform, the erosion was more intense and culminated in the deposition of the Upper Lydite Bed during the transgression above unconformity P2 (upper *glaucolithus* Zone).

Throughout south Dorset, the depositional sequence overlying sequence boundary P1 consists of bedded dolomites which show an increasing content of organic carbon towards the middle of the sequence and may be a reflection of high productivity during the high to maximum rate of relative sea- level rise. The beds are also thinner on average in the middle of the sequence, which is interpreted here as reflecting a low sedimentation rate in response to rapid sea-level rise. In the Vale of Wardour the base of the Chicksgrove Limestone is interpreted as a transgressive surface and the top, or just above the top, as the maximum flooding surface as these strata contain abundant ammonites and bivalves typical of sediments deposited during transgression (Coe 1992). In the Vale of Pewsey and on the London Platform the transgression culminated in the deposition of the Upper Lydite Bed followed by the overlying condensed, glauconitic and phosphatic sandy limestones of the Glauconitic Beds, the top of which forms the maximum flooding surface (Fig. 4).

Middle okusensis *Zone Sequence Boundary (P2)*

On the Isle of Purbeck, the change in lithology between the Black Sandstones and the Lower Cherty Beds suggests a rapid change in depositional conditions caused by a relative sea-level fall and the boundary is interpreted as a correlative conformity. The Lower Cherty Beds contain a very high proportion of cherts formed from sponge spicules, probably derived from the remains of sponge bioherms or a sponge forest (Boucot 1981). Spectacular sponge bioherms of a similar age are preserved in the Swabian Alps of southern Germany (Wendt 1980).

The absence of the Lower Cherty Beds on the Isle of Portland is thought to be mainly the result of limited sediment accommodation space during relative fall in sea level, coupled with distance from the site of carbonate production (Fig. 5). Thus the sharp, well-defined nature of unconformity P2 on the Isle of Portland, together with the obvious contrasting lithological characteristics above and below, suggests that the erosion surface here represents both the sequence boundary and the transgressive surface superimposed. This interpretation is further supported by the characteristics of the overlying Basal Shell Bed, as this shows faunal diversity and abundance together with intense bioturbation, all of which are features typical of a transgressive systems tract (Loutit *et al.* 1988). In addition, the constant thickness of the Basal Shell Bed and equivalent beds J and J' throughout Dorset is consistent with interpretation as a transgressive systems tract because, during transgression, sediments are generally deposited fairly uniformly over a large area when there is no limitation on sediment accommodation space. The fact that the stratal gap on the Isle of Portland is *above* the correlative conformity on the Isle of Purbeck tends to support the other evidence that the gap is mainly the result of a lack of sediment-accommodation space immediately after a relative fall in sea level, although the distance from the carbonate production area must also have had an effect because there is a thin succession of Lower Cherty Beds preserved at Dungy Head and Ringstead Bay. In contrast, the stratal gap associated with sequence boundary P1 on the Isle of Portland is *below* the correlative conformity on the Isle of Purbeck, but this gap is interpreted to be mainly the result of a failure of siliciclastic sediment supply, together with increased erosion in the Isle of Portland and Ringstead Bay areas during the start of a fall in relative sea level.

While the Lower Cherty Beds were being deposited on the Isle of Purbeck, sediments of the *okusensis* Zone were being eroded in both Wiltshire and Oxfordshire. This phase of erosion is exemplified by the incision of the channel exposed at Chicksgrove Quarry in the Vale of Wardour. The incised channel was then infilled during the subsequent transgression, initially with non-marine sand and, as the transgression continued, these sands were then overstepped by brackish and marine limestones of the Wockley Member (Fig. 4). The formation and infilling of this channel (small incised valley) gives clear and incontestable evidence of a fall and subsequent rise in relative sea level, in this area, at this time. On the London Platform, at Swindon and in Oxfordshire and Buckinghamshire, the sequence boundary lies between condensed, shallow-marine limestones of contrasting lithological character (the upper and lower part of the Cockly Bed).

The top of the Basal Shell Bed or bed J' is interpreted as the maximum flooding surface as the condensation features within unit 3 of the Basal Shell Bed of Portland, and its lateral equivalent bed J' on the Isle of Purbeck, can be explained in terms of minimum sediment input around the time of maximum transgression. This, in turn, explains the great abundance of ammonites in these particular horizons (Townson 1971). The overlying Cherty Beds of the Isle of Portland, and Upper Cherty Beds of the Isle of Purbeck, constitute the highstand systems tract because they contain a low diversity faunal assemblage dominated by the sponge *Rhaxella* and are fairly uniform in character throughout Dorset. Individual beds within the Cherty Beds can be correlated between the Isle of Portland and Isle of Purbeck. These features suggest that the Cherty Beds were deposited in some tens of metres of water depth under fairly quiet conditions (Boucot 1981; Rigby 1987; Coe 1992). This supposed background of quiet deposition was occasionally broken by currents, as demonstrated by the presence of shell beds. The Cherty Beds immediately overlying the Basal Shell Bed or bed J' contain only very small amounts of shell debris, which is consistent with a rapid increase in water depth, resulting in a slightly deeper water facies being deposited.

In the Vale of Wardour, the shelly beds of the Wockley Member ('Spangle') represent the most likely candidate for a maximum flooding surface. In the marginal areas of Wiltshire and Oxfordshire the maximum flooding surface is not well developed, but it most likely lies between the Cockly Bed/Rubbly Limestone and the overlying Swindon Sand/Crendon Sand, the latter showing the greatest areal extent, which is consistent with their being deposited during the early highstand systems tract.

Carbonate highstand systems tracts can be divided into two parts (Sarg 1988): the early and late highstand systems tract separated by a mid-highstand surface. The mid-highstand surface is thought to form in response to the onset of sea-level fall as sediment-accommodation space on the shelf decreases, resulting in better circulation, higher nutrient availability and greater agitation and hence increased carbonate productivity producing the late highstand systems tract. The base of the minor oolitic incursion is interpreted as the mid-highstand surface (i.e. that found in the uppermost part of the Cherty Beds on the Isle of Portland (base bed 11) and at the base of the Portland Freestone on the Isle of Purbeck, Ringstead Bay and Dungy Head; Fig. 4). The incursion probably reflects an increase in carbonate productivity associated with the onset of the sea-level fall which preceded the major change in relative sea-level that produced sequence boundary P3. In contrast with the second flood of ooids, the first incursion of ooids probably does not represent a major change in water depth because the ooids are typically small and show evidence of transportation. Furthermore, the ooids never reach a significant proportion of the sediment volume (only 25–35%) and the thickness of the minor oolite is laterally variable.

Lower anguiformis *Zone Sequence Boundary (P3)*

The general change from marine to non-marine conditions in the upper Portland Stone and lower Lulworth Beds is clearly shown by both the fauna and facies. The evidence that this trend is punctuated by a shorter term fall in relative sea level relies on the interpretation of rapid facies changes in Dorset and Wiltshire and on the well-defined angular unconformity in the Town Gardens, Swindon. Thus the surface in Dorset equivalent to the angular unconformity at Swindon is thought to be the erosion surface developed at the base of the upper (major) oolitic package – that is, at the base of the Portland Freestone on the Isle of Portland and within the Portland Freestone on the Isle of Purbeck. At Chilmark in the Vale of Wardour, the unconformity lies at the base of the areally restricted Chilmark Oolite; elsewhere in the Vale of Wardour, it is suggested that erosion occurred due to the lack of sediment-accommodation space and thus only thin, condensed, shelly limestones were deposited. This surface at the base of the oolites in Dorset and the Vale of Wardour has been selected as a sequence boundary rather than the palaeosols at the base of the Purbeck Limestone for the following reasons: (1) the change in estimated water depth between deposition of the cherty wackestones or packstones and the oolites is the most significant in the overall shallowing trend and is interpreted to be too rapid and widespread to be explained simply in terms of sediment progradation (in contrast, the facies at the top of the Portland Stone and base of the Purbeck Limestone show a very gradual change from subtidal to supratidal, a trend expected from sediment progradation, independent of water depth); (2) the facies model presented (cf. Townson 1971, 1975; Fig. 6) suggests that bioclastic grainstones are missing between the Cherty Beds and the Portland Freestone Member at most localities; (3) the timing of the unconformity event is more consistent across the basin; and (4) there is a greater probability that palaeosols will be preserved in the epicontinental basinal successions during a relative sea-level rise or stillstand, rather than during a fall when they would tend to be eroded.

At Swindon the erosion which occurred during the sea-level fall and early part of the rise is indicated by the occurrence of the pebble bed and onlap of the Town Gardens Member and Purbeck Limestone onto the Swindon Sand and Stone.

In Dorset, the overlying depositional sequence comprises oolitic grainstones and non-marine limestones and the succession is similar to, but thinner than, that in the Vale of Wardour. At Swindon, and in Oxfordshire and Buckinghamshire, the overlying depositional sequence is almost entirely composed of non-marine carbonates. The palaeosols developed at the base of the Purbeck Limestone are thought to represent the minimum relative sea level. The limestones and clays of the Cypris Freestone and Hard Cockle Beds (Clements 1969; Ensom 1985) are interpreted as the transgressive systems tract. Maximum flooding resulted in increased circulation of sea water into the brackish-water depositional environment typical of the Purbeck Limestone, thus providing an increase in the concentration of sulphates. This flooding, together with the high evaporation rate in the semi-arid climate that prevailed at the time, is interpreted to have resulted in the deposition of evaporites which form the basal beds of the Soft Cockle Member. An increase in salinity at the base of the Soft Cockle Member is also clearly indicated by the ostracod 'faunicycles' (Fig. 19).

lamplughi *Zone Sequence Boundary (P4)*

In Dorset and south Wiltshire the upper Portlandian and lower Ryazanian non-marine deposits and fauna clearly show a cycle of decreasing and then increasing marine influence. The ostracods (Fig. 19) show that the maximum rate of decrease in saline forms is within the Marly Freshwater Beds. The minimum is within the younger Cherty Freshwater Beds. In marine sediments it is at the maximum rate of relative sea-level fall that the sequence boundary is formed. Thus, by analogy, in these non-marine carbonates the sequence boundary is interpreted to fall within the Marly Freshwater Beds, probably at the level of the Mammal Bed.

In north Wiltshire and Oxfordshire there was a prolonged period of erosion and non-deposition until at least the late Portlandian; presumably because this area was land over most of this time period. Assuming that the late Portlandian age of at least some of the Whitchurch Sands deposits is correct, then they represent limited transgressive sediments deposited at the margins of the basin and derived from either the north (Boreal Realm; Casey 1963) or the south (Tethyan Realm; Wimbledon & Hunt 1983). These sands would then probably just post-date the sequence boundary, P4, and therefore fall within the trangressive systems tract. The Cinder Bed and its lateral equivalents are interpreted as the maximum flooding surface.

Summary and conclusions

The Portlandian rocks of the Wessex Basin can be divided into four depositional sequences. These represent smaller scale changes in regional relative sea level, which punctuate the overall marked sea-level fall between the mid-Kimmeridgian and the Early Cretaceous.

Within the northern sub-basins and at the northern basin margin the strata are thin and condensed and consequently each of the sequence boundaries is associated with a facies change and erosion surface. In the Vale of Wardour, sequence boundary P2 is associated with the cutting of a large channel which gives clear evidence of a fall in relative sea level. There is also similar, but less convincing, evidence (because of poor exposure) for erosion of a channel associated with sequence boundary P3 in the same area.

In south Dorset the structural positions of the different sections allows a detailed analysis of the sedimentary processes active in Portlandian times. The succession on the Isle of Purbeck is the most complete and all the sequence boundaries are interpreted as correlative conformities. Sequence boundary P1 is marked by a sharp surface and a stratal gap on the Isle of Portland and on the structural high at Ringstead Bay. The deposits above this sequence boundary are almost entirely carbonate across the whole Dorset area. There is also a stratal gap associated with sequence boundary P2 on the Isle of Portland, and the overlying depositional sequence contains a well-developed transgressive package capped by the maximum flooding surface. Sequence boundary P3 is marked by a sudden basinward shift in oolitic grainstones across the Dorset area. Sequence boundary P4 is difficult to constrain, but is thought to be near the top of the non-marine Lulworth Beds.

This analysis presents several new stratigraphical correlations across the Dorset area, emphasizing the importance of considering unconformities and possible stratal gaps in sedimentary successions. The new stratigraphical correlations indicate: (1) that on the Isle of Portland there are no equivalents of either the St Alban's Head Marls or the Lower Cherty Beds which are developed on the Isle of Purbeck and (2) that the base of the Portland Freestone is not diachronous between the Isle of Portland and the Isle of Purbeck, but that there are two sedimentary packages containing ooids, one in the upper part of the *kerberus* Zone and the other in the lower *anguiformis* Zone. The sequence stratigraphical interpretation provides a possible higher resolution correlation of the Portlandian strata across the Wessex Basin than that determined using the current ammonite biostratigraphy.

Funding from British Petroleum (EMRA award) is gratefully acknowledged. S. Hesselbo, H. Jenkyns, J. Kennedy and M. Widdowson are all thanked for useful comments and constructive criticisms during the preparation of this work. B. Cox and C. Wilson are thanked for reviewing this paper. Mike Simms is thanked for allowing me access to his unpublished BSc thesis (University of Bristol 1982).

References

ALLEN, P. & WIMBLEDON, W. A. 1991. Correlation of NW European Purbeck–Wealden (nonmarine Lower Cretaceous) as seen from the English type-areas. *Cretaceous Research*, **12**, 511–526.

ANDERSON, F. W. 1985. Ostracod faunas in the Purbeck and Wealden of England. *Journal of Micropalaeontology*, **4**, 1–68.

—— & BAZLEY, R. A. B. 1971. The Purbeck Beds of the Weald (England). *Bulletin of the Geological Survey of Great Britain*, **34**, 1–174.

ANDREWS, W. R. 1881. Note on the Purbeck Beds of Teffont. *Quarterly Journal of the Geological Society of London*, **37**, 248–253.

—— & JUKES-BROWNE, A. J. 1894. The Purbeck Beds in the Vale of Wardour. *Quarterly Journal of the Geological Society of London*, **50**, 55.

ARKELL, W. J. 1933. *The Jurassic System in Great Britain*. Clarendon Press, Oxford.

—— 1935. The Portland Beds of the Dorset Mainland. *Proceedings of the Geologists' Association, London*, **46**, 301–347.

—— 1944. Stratigraphy and structures east of Oxford, part 2: the Miltons and the Haseleys. *Quarterly Journal of the Geological Society of London*, **100**, 45–61.

—— 1947*a*. *The Geology of the Country Around Weymouth, Swanage, Corfe and Lulworth*. Memoir of the Geological Survey of Great Britain.

—— 1947*b*. *The Geology of Oxford*. Clarendon Press, Oxford.

—— 1948. A geological map of Swindon. *Wiltshire Archaeological and Natural History Magazine*, **52**, 195–212.

—— & TOMKEIEFF, S. I. 1953. *English Rock Terms as Used Chiefly by Miners and Quarrymen*. Oxford University Press, Oxford.

ASTIN, T. R. 1987. Petrology (including fluorescence microscopy) of cherts from the Portlandian of Wiltshire, UK – evidence of an episode of meteoric water circulation. *In*: MARSHALL, J. D. (ed.) *Diagenesis of Sedimentary Sequences*. Geological Society, London, Special Publications, **36**, 73–85.

BAKER, P. A. & KASTNER, M. 1981. Constraints on the formation of sedimentary dolomite. *Science*, **213**, 214–216.

BALLANCE, P. F. 1963. The Beds between the Kimmeridge and Gault Clays in the Thame-Aylesbury Neighbourhood. *Proceedings of the Geologists' Association, London*, **74**, 393–418.

BARKER, D. 1966. Ostracods from the Portland and Purbeck Beds of the Aylesbury district. *Bulletin of the British Museum (Natural History), Geology*, **11**, 458–487.

BLAKE, J. F. 1880. On the Portland rocks of England. *Quarterly Journal of the Geological Society of London*, **33**, 189–236.

BOUCOT, A. J. 1981. *Principles of Benthic Marine Paleoecology*. Academic Press, London.

BRISTOW, C. R. 1968. Portland and Purbeck Beds. *In*: SYLVESTER BRADLEY, P. C. & FORD, T. D. (eds) *Geology of the East Midlands*. Leicester University Press, 300–311.

BROWN, P. R. 1963. Algal limestones and associated sediments in the basal Purbeck of Dorset. *Geological Magazine*, **100**, 565–573.

BUCKMAN, S. S. 1909–1930. *Yorkshire Type Ammonites* (continued as) *Type Ammonites*. 7 vols. Published by the author, London and Thame.

CASEY, R. 1963. The dawn of the Cretaceous Period in Britain. *Bulletin of the S.E. Union of Scientific Societies*, **117**, 1–15.

—— 1967. The position of the Middle Volgian in the English Jurassic. *Proceedings of the Geological Society of London*, **1640**, 128–133.

—— 1973. The ammonite succession at the Jurassic–Cretaceous boundary in Eastern England. *In*: CASEY, R. & RAWSON, P. F. (eds) *The Boreal Lower Cretaceous*. Geological Journal Special Issue, **5**, 193–266.

—— & BRISTOW, C. R. 1964. Notes on some ferruginous strata in Buckinghamshire and Wiltshire. *Geological Magazine*, **101**, 116–128.

CHADWICK, R. A. 1986. Extensional tectonics in the Wessex Basin, southern England. *Journal of the Geological Society, London*, **143**, 465–488.

CHATWIN, C. P. & PRINGLE, J. 1922. The zones of the Kimmeridge and Portland rocks at Swindon. *Summary of the Progress of the Geological Survey for 1921*, 162–168.

CLEMENTS, R. G. 1969. The Purbeck Beds. *In*: TORRENS, H. G. (ed.). *International Field Symposium on the British Jurassic. Excursion No 1. Guide for Dorset and South Somerset*, University of Keele, A57–A64.

COE, A. L. 1992. *Unconformities within the Upper Jurassic of the Wessex Basin, Southern England*, DPhil Thesis, University of Oxford.

COPE, J. C. W. 1978. The ammonite faunas and stratigraphy of the upper part of the Upper Kimmeridge Clay of Dorset. *Palaeontology*, **21**, 469–533.

—— & WIMBLEDON, W. A. 1973. Ammonite faunas of the uppermost Kimmeridge Clay, the Portland Sand and the Portland Stone of Dorset. *Proceedings of the Ussher Society*, **2**, 593–598.

COX, L. R. 1925. The fauna of the Basal Shell Bed of the Portland Stone, Isle of Portland. *Proceedings of the Dorset Natural History and Antiquarian Field Club*, **46**, 113–172.

ENSOM, P. C. 1985. An annotated section of the Purbeck Limestone Formation at Worbarrow Tout, Dorset. *Proceedings of the Dorset Natural History and Archaeological Society*, **108**, 87–91.

ETHERIDGE, R. 1880. On a new species of Trigonia from the Purbeck Beds of the Vale of Wardour, with a note on the strata by Andrews, W.R. *Quarterly Journal of the Geological Society of London*, **37**, 246–248.

FISHER, O. 1856. On the Purbeck strata of Dorsetshire. *Transactions of the Cambridge Philosophical Society*, **9**, 555–581.

FRANCIS, J. E. 1983. The dominant conifer of the Jurassic Purbeck Formation, England. *Palaeontology*, **26**, 277–294.

—— 1984. The seasonal environment of the Purbeck (Upper Jurassic) fossil forests. *Palaeogeography, Palaeoclimatology, Palaeoecology*, **48**, 285–307.

HESSELBO, S. P., COE, A. L., BATTEN, D. J. & WACH, G. D. 1990. Stratigraphic relations of the Lower Greensand (Lower Cretaceous) of the Calne area, Wiltshire. *Proceedings of the Geologists' Association, London*, **101**, 265–278.

HORTON, A. H., SUMBLER, M. G., COX, B. M. & AMBROSE, K. A. *Geology of the Country Around Thame*. Memoir of the British Geological Survey, Sheet 237 (England and Wales), in press.

HOUSE, M. R. 1958. On the Portlandian zones of the Vale of Wardour and the use of *Titanites giganteus* as an Upper Jurassic zone fossil. *Proceedings of the Geologists' Association, London*, **69**, 17–19.

—— 1970. Portland Stone on Portland. *Proceedings of the Dorset Natural History and Archaeological Society*, **91**, 38–39.

—— 1989. *The Geology of Dorset*. Geologists' Association Field Guide, Geologists' Association, London.

HUDLESTON, W. H. 1883. On the geology of the Vale of Wardour (Portland and Purbeck Beds). *Proceedings of the Geologists' Association, London*, **7**, 161–185.

INSTITUTE OF GEOLOGICAL SCIENCES 1971. *Annual Report for 1970*. Institute of Geological Sciences, London, 22.

JUKES-BROWNE, A. J. 1905. *Geology of the Country South and East of Devizes*. Memoir of the Geological Survey, Sheet 282.

KARNER, G. D., LAKE, S. D. & DEWEY, J. F. 1987. The thermal and mechanical development of the Wessex Basin, southern England. *In*: COWARD, M. P., DEWEY, J. F. & HANCOCK, P. L. (eds) *Continental Extensional Tectonics*. Geological Society, London, Special Publications, **28**, 517–536.

KELTS, K. & MCKENZIE, J. 1984. A comparison of anoxic dolomite from deep-sea sediments: Quarternary Gulf of California and the Messinian Tripoli Formation of Sicily. *In*: GARRISON, R. E., KASTNER, M. & ZENGER, D. H. (eds) *Dolomites of the Monterey Formation and Other Organic-rich Units*. Society of Economic Paleontologists and Mineralogists, 20–41.

LAKE, R. D. & HOLLIDAY, D. W. 1978. Purbeck Beds of the Broadoak Borehole, Sussex. *Report of the Institute of Geological Sciences*, **78/3**, 1–12.

LAKE, S. D. & KARNER, G. D. 1987. The structure and evolution of the Wessex Basin, southern England: an example of inversion tectonics. *Tectonophysics*, **137**, 347–378.

LOUTIT, T. S., HARDENBOL, J. & VAIL, P. R. 1988. Condensed sections: the key to age dating and correlation of continental margin sequences. *In*: WILGUS, C. K., HASTINGS, B. S., KENDALL, C. G. ST. C., POSAMENTIER, H. W., ROSS, C. A. & VAN WAGONER, J. C. (eds) *Sea-level Changes: an Integrated Approach.* Society of Economic Paleontologists and Mineralogists, Special Publications, **42**, 183–213.

MORTER, A. A. 1984. Purbeck–Wealden Beds Mollusca and their relationship to ostracod biostratigraphy, stratigraphical correlation and palaeoecology in the Weald and adjacent areas. *Proceedings of the Geologists' Association, London*, **95**, 217–234.

NEAVERSON, E. 1925. The petrology of the Upper Kimmeridge Clay and Portland Sand in Dorset, Wiltshire, Oxfordshire and Buckinghamshire. *Proceedings of the Geologists' Association, London*, **36**, 240–256.

NORRIS, G. 1969. Miospores from the Purbeck Beds and marine Upper Jurassic of Southern England. *Palaeontology*, **12**, 574–620.

OGG, J. G., HASENYAGER II, R.W. & WIMBLEDON, W. A. 1995. Jurassic–Cretaceous boundary: Portland–Purbeck magnetostratigraphy and correlation to the Tethyan faunal realm. *Géobios Memoir Speciale*, **17**, 519–527.

PARTINGTON, M. A. 1983. *The stratigraphy, distribution and phylogeny of some Lower Cretaceous* Circumpolles *from southern England.* PhD Thesis, University of Aberdeen.

PRINGLE, J. 1926. *The Geology of the Country Around Oxford.* Memoir of the Geological Survey, England.

PUGH, M. E. 1968. Algae from the Lower Purbeck Limestones of Dorset. *Proceedings of the Geologists' Association, London*, **79**, 513–523.

QUEST, M. 1985. *Petrographical and geochemical studies of the Portland and Purbeck strata of Dorset.* PhD Thesis, University of Birmingham.

RADLEY, J. D. 1991. Palaeoecology and deposition of Portlandian (Upper Jurassic) strata at the Bugle Pit, Hartwell, Buckinghamshire. *Proceedings of the Geologists' Association, London*, **102**, 241–250.

RIGBY, J. K. 1987. Phylum Porifera *In*: BOARDMAN, R. S., CHEETHAM, A. H. & ROWELL, A. J. (eds) *Fossil Invertebrates.* Blackwell Scientific Publications, Oxford, 116–139.

SARG, J. F. 1988. Carbonate sequence stratigraphy. *In*: WILGUS, C. K., HASTINGS, B. S., KENDALL, G. ST. C., POSAMENTIER, H. W., ROSS, C. A. & VAN WAGONER, J. C. (eds) *Sea-level Changes: an Integrated Approach.*. Society of Economic Palaeontologists and Mineralogists, Special Publications, **42**, 155–183.

SYLVESTER BRADLEY, P. C. 1940. The Purbeck Beds of Swindon. *Proceedings of the Geologists' Association*, **51**, 349–372.

TOWNSON, W. G. 1971. *Facies analysis of the Portland Beds.* DPhil Thesis, University of Oxford.

—— 1975. Lithostratigraphy and deposition of the type Portlandian. *Journal of the Geological Society, London*, **131**, 619–638.

TUCKER, M. E. & WRIGHT, V. P. 1990. *Carbonate Sedimentology.* Blackwell Scientific Publications, Oxford.

VAIL, P. R., AUDEMARD, F., BOWMAN, S. A., EISNER, P. N. & PEREZ-CRUZ, C. 1991. The stratigraphic signatures of tectonics, eustacy and sedimentology – an overview. *In*: EINSELE, G., RICKEN, W. & SEILACHER, A. (eds) *Cycles and Events in Stratigraphy.* Springer-Verlag, Berlin, 617–659.

VAN WAGONER, J. C., POSAMENTIER, H. W., MITCHUM, R. M. JR, SARG, J. F., LOUTIT, T. S. & HARDENBOL, J. 1988. The key definitions of sequence stratigraphy. *In*: WILGUS, C. K., HASTINGS, B. S., KENDALL, C. G. ST. C., POSAMENTIER, H. W., ROSS, C. A. & VAN WAGONER, J. C. (eds). *Sea-level Changes: an Integrated Approach.* Society of Economic Paleontologists and Mineralogists, Special Publications, **42**, 39–45.

WEST, I. M. 1975. Evaporites and associated sediments of the basal Purbeck Formation (Upper Jurassic) of Dorset. *Proceedings of the Geologists' Association, London*, **86**, 205–225.

WENDT, J. 1980. Sponge reefs of the German Upper Jurassic. *In*: HARTMANN, W. D., WENDT, J. & WEIDENMAYER, F. (eds) *Living and Fossil Sponges.* Notes for a Short Course. Sedimenta, **8**, 122–130.

WHITTAKER, A. (ed.) 1985. *Atlas of Onshore Sedimentary Basins in England and Wales: Post-Carboniferous Tectonics and Stratigraphy.* Blackie, Glasgow.

WIGNALL, P. B. & RUFFELL, A. H. 1990. The influence of a sudden climatic change on marine deposition in the Kimmeridgian of northwest Europe. *Journal of the Geological Society, London*, **147**, 365–372.

WILSON, R. C. L. 1966. Silica diagenesis in the Upper Jurassic limestones of southern England. *Journal of Sedimentary Petrology*, **36**, 1036–1049.

WIMBLEDON, W. A. 1974*a*. The Basal Shell Bed of the Portland Stone *In*: COPE, J. C. W. (ed.) *Geology in the Natural Sciences 1973.* Proceedings of the Dorset Natural History and Archaeological Society, **95**, 105–106.

—— 1974*b*. *The Stratigraphy and Ammonite Faunas of the Portland Stone of England and Northern France.* PhD Thesis, University of Wales.

—— 1976. The Portland Beds (Upper Jurassic) of Wiltshire. *Wiltshire Archaeological and Natural History Magazine*, **71**, 3–11.

—— 1980. The Portlandian. *In*: COPE, J. C. W., DUFF, K. L., PARSONS, C. F., TORRENS, H. S., WIMBLEDON, W. A. & WRIGHT, J. K. 1980. *A Correlation of Jurassic Rocks of the British Isles. Part Two: Middle and Upper Jurassic.* Geological Society, London, Special Reports, **15**, 85–93.

—— 1986. Rhythmic sedimentation in the Late Jurassic–Early Cretaceous. *Proceedings of the Dorset Natural History and Archaeological Society*, **108**, 127–133.

—— & COPE, J. C. W. 1978. The ammonite faunas of the English Portland Beds and the zones of the Portlandian Stage. *Journal of the Geological Society, London*, **135**, 183–190.

—— & HUNT, C. O. 1983. The Portland–Purbeck junction (Portlandian–Berriasian) in the Weald, and correlation of latest Jurassic–early Cretaceous rocks in southern England. *Geological Magazine*, **120**, 267–280.

WOODWARD, H. B. 1895. *The Middle and Upper Oolitic Rocks of England (Yorkshire excepted)*. Memoir of the Geological Survey, Great Britain, Vol. 5 of the Jurassic Rocks of Britain.

WORSSAM, B. C. & IVIMEY-COOK, H. C. 1971. The stratigraphy of the Geological Survey Borehole at Warlingham, Surrey. *Bulletin of the Geological Survey, Great Britain*, **36**, 1–111.

Slump and debris-flow dominated basin-floor fans in the North Sea: an evaluation of conceptual sequence-stratigraphical models based on conventional core data

G. SHANMUGAM[1], R. B. BLOCH[1], S. M. MITCHELL[2], J. E. DAMUTH[3], G. W. J. BEAMISH[4], R. J. HODGKINSON[5], T. STRAUME[6], S. E. SYVERTSEN[6], & K. E. SHIELDS[1]

[1]*Mobil Exploration and Producing Technical Center PO Box 650232, Dallas, Texas 75265-0232, USA*

[2]*Mobil Exploration and Producing U.S. Inc., PO Box 650232, Dallas, Texas 75265-0232, USA*

[3]*Earth Resource and Environment Center The University of Texas at Arlington PO Box 19049, Arlington, Texas 76019, USA*

[4]*Mobil North Sea Ltd, 3 Clements Inn, London WC2A 2EB, UK*

[5]*Mobil New Business Development, PO Box 650232, Dallas, Texas 75265-0232, USA*

[6]*Mobil Exploration Norway Inc., PO Box 510, 4001 Stavanger, Norway*

Abstract: The sequence-stratigraphical concept of the basin-floor fan within the Vail/Exxon conceptual sea-level model is popular in hydrocarbon exploration because this model predicts that basin-floor fans are composed of sand-rich turbidites with laterally extensive reservoir geometries. Many producing deep-marine facies in the Palaeogene and Cretaceous of the North Sea have been interpreted as basin-floor fans based on their seismic (e.g. mounded forms) and wire-line log signatures (e.g. blocky log motifs). We have examined nearly 3700 m (12000 feet) of conventional core from features interpreted as basin-floor fans in the North Sea and adjacent regions to determine the sedimentary facies, which actually comprise these features. Our study reveals that basin-floor fans are predominantly composed of mass-transport deposits (mainly slumps and debris flows); turbidites are extremely rare. Sedimentary features indicative of slump and debris-flow origin include sand units with sharp upper contacts, slump folds, discordant, steeply dipping layers (up to 60°), glide planes, shear zones, brecciated clasts, clastic injections, floating mudstone clasts, planar clast fabric, inverse size grading, and moderate to high matrix content (5–30%). Calibration of these facies in long cores (120–210 m or 400–700 ft) with seismic and wire-line log signatures suggests that seismic mounds and blocky log motifs are actually a manifestation of amalgamated sand units emplaced by these mass-transport processes. The cores show that in some cases, sequence boundaries on seismic profiles actually correspond to primary glide planes (decollement) between slump sheets in cores examined. Unlike laterally continuous turbidites, slump- and debris-flow-emplaced sands are often laterally discontinuous, thus making reservoir geometry and continuity much harder to predict. Nevertheless, these mass-transport deposits still form important hydrocarbon-producing reservoirs (e.g. the Frigg Field). Because both sands and muds deposited by slumps and debris flows form features that are routinely interpreted as basin-floor fans on seismic and log data, process-sedimentological interpretation of conventional core is critical to the establishment of the true origin and distribution of sands contained in such features and thereby evaluate their hydrocarbon potential.

In a sequence-stratigraphical framework within the Vail/Exxon conceptual sea-level model, the *lowstand systems tract* is composed of the *basin-floor fan*, the *slope fan* and the *prograding wedge complex* (Vail 1987; Posamentier & Vail 1988; Vail *et al.* 1991). Basin-floor fans and slope fans have also been termed 'lowstand fans' and 'early lowstand wedges,' respectively (Posamentier & Erskine 1991). Basin-floor fans are considered attractive hydrocarbon prospects because this model predicts that they are very sand-rich features (Posamentier & Vail 1988, p. 140) dominated by turbidites, which occur as 'sheet mounds made up of massive sands deposited as lobes or sheets in a deep marine setting' (Vail *et al.* 1991, p. 646; Vail & Wornardt 1990, fig. 3). According to Mitchum (1993, p. 167), 'basin-floor fans consist lithologically of relatively clean, well-sorted sandstone with good reservoir quality'. Thus, this conceptual model is being increasingly applied in the petroleum industry to predict reservoir facies in frontier and mature basins.

From HESSELBO, S. P. & PARKINSON, D. N. (eds), 1996, *Sequence Stratigraphy in British Geology*, Geological Society Special Publication No. 103, pp. 145–175.

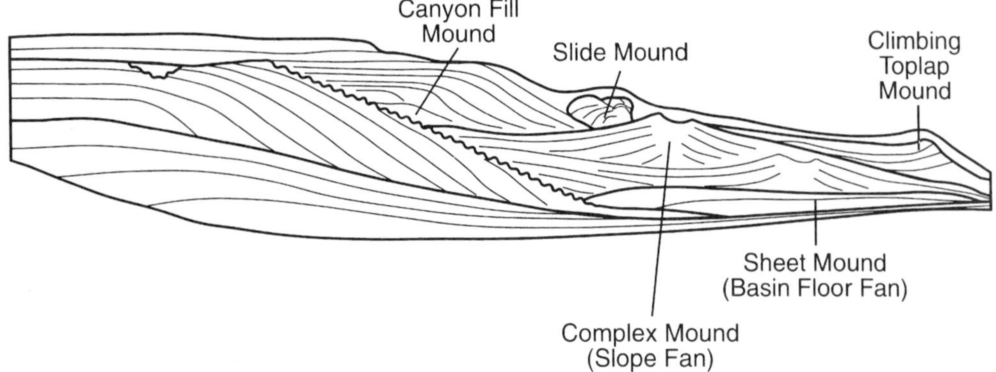

Fig. 1. Schematic diagram of a depositional sequence within the Vail/Exxon conceptual sea-level model showing stratal patterns of mounded deposits such as basin-floor fans (sheet mound) and slope fans (complex or gull-wing mound). Simplified from Vail *et al.* (1991).

Basin-floor fans have routinely been recognized based on seismic profiles (Jackson *et al.* 1992) and wire-line log motifs (Mitchum *et al.* 1990, 1993; Vail & Wornardt 1990). To date, little attempt has been made to document systematically the actual sedimentary facies of these features and the processes that form them using conventional core data (Posamentier & Weimer 1993). In fact, we are not aware of a single published example of a basin-floor fan that has had its predicted turbidite facies documented using conventional core data, or its laterally continuous, sheet-like, sand-body geometry documented by correlation of sand units between closely spaced wells, or by drill-stem tests. As a result, the validity of this conceptual model and it predictive capabilities remain questionable (Shanmugam *et al.* 1995).

The primary objective of our study is to evaluate the merits of the basin-floor fan conceptual model utilizing conventional core data from seismic features that have previously been interpreted as basin-floor fans by other investigators, by determination of the sedimentary facies that comprise these features. A second important objective is to calibrate long cores (up to 210 m or 700 ft) with log and seismic data to begin to understand the relationship between process sedimentology (core-scale features) and seismic stratigraphy (seismic-scale features).

Characteristics of basin-floor fans and slope fans

Basin-floor fans theoretically represent deposits of unconfined sandy turbidity flows, which flow outward across the basin floor; whereas, slope fans comprise deposits created mainly by muddy channelized turbidity currents which overflow channel-levee systems and are mainly deposited as overbank flows on slopes. Therefore, these two systems are predicted to develop distinctly different sand-body geometries and reservoir quality. Standard criteria have been proposed to recognize basin-floor fans, as well as slope fans in seismic profiles and wire-line logs (Vail & Wornardt 1990; Posamentier & Erskine 1991; Vail *et al.* 1991; Mitchum *et al.* 1993).

The characteristics of basin-floor fans include: (1) mound-shaped forms on seismic profiles that downlap or onlap onto Type 1 sequence boundaries (unconformities) on the basin floor (Fig. 1); (2) hummocky to chaotic internal reflection character or seismic facies; (3) high-amplitude reflections that show bi-directional downlap at the base of the mound (Fig. 1); (4) lateral pinch-out geometry; and (5) blocky wire-line-log motifs. According to the conceptual model, basin-floor fans are predicted to be sand-rich turbidites with sheet-like geometries (Vail & Wornardt 1990; Mitchum *et al.* 1990, 1993; Vail *et al.* 1991).

Slope fans are deposited after basin-floor fans and, thus, normally downlap onto them (Fig. 1). Slope fans: (1) exhibit 'gull-wing' mounds or 'bow-tie' reflections on seismic profiles (complex mound in Fig. 1); (2) display zones of chaotic internal reflections, and (3) show crescent-shaped, wire-line log motifs. In contrast to sand-rich basin-floor fans, slope fans are predicted to be predominantly mud-rich systems characterized by muddy channel-levee complexes and mass-transport deposits (Vail & Wornardt 1990; Mitchum *et al.* 1990; Vail *et al.* 1991). The modern Amazon and Mississippi deep-sea fans are often cited as examples of large slope fans (e.g. Mitchum *et al.* 1990; Weimer 1990; P.R. Vail, pers. comm. 1993).

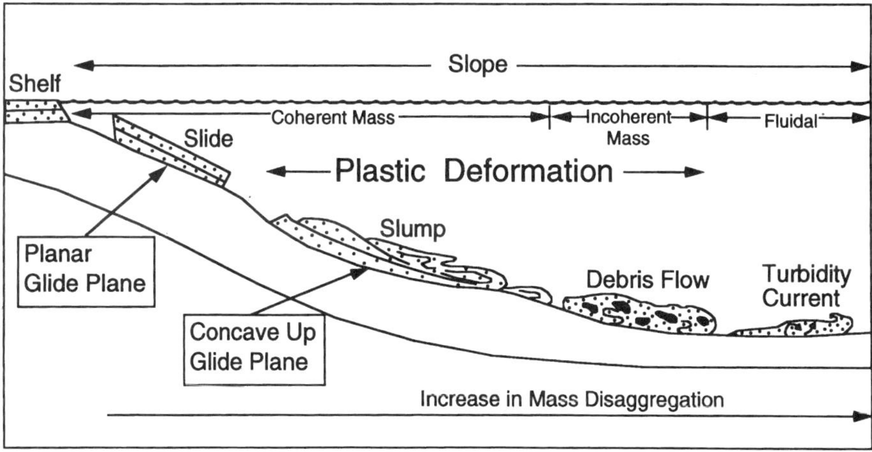

Fig. 2. Schematic diagram showing end member types of gravity-driven processes that transport sediment into deep marine basins. A slide represents a coherent translational mass movement of a block or strata on a planar glide plane (shear surface) without internal deformation and may be transformed into a slump, which represents a coherent rotational mass movement of a block of strata on a concave-up glide plane (shear surface) with internal deformation. Upon addition of fluid during downslope movement, slumped material may evolve into a debris flow, which transports sediment as an incoherent viscous mass in which inter-granular movements predominate over shear-surface movements (i.e. plastic flow). As fluid content increases in a debris flow, the flow may evolve into a fluidal turbidity current (i.e. turbulent flow). The diagram does not imply that all turbidity currents evolved from debris flow. Terminological schemes of Varnes (1958) and Dingle (1977) were followed in constructing this conceptual model. From Shanmugam *et al.* (1994).

Deep-marine processes and facies

A variety of gravity-driven processes, such as slides, slumps, debris flows and turbidity currents, transport sediments downslope from the shelf and deposit them in deep-water slope and basin environments (Fig. 2). In this paper, we classify these processes following terminological schemes originally proposed by Varnes (1958) and Dingle (1977) with some modification (see Shanmugam *et al.* 1994).

Slumps and slides

Slumps and slides sometimes are not distinguished from each other in the literature (Woodcock 1979) because both processes involve downslope movement of sediment as a coherent mass (Fig. 2). We term a mass or block that moves downslope on a planar glide plane and shows no internal deformation a *slide*; whereas, a *slump* is a block that moves downslope on a concave-up glide plane (Dingle 1977) and undergoes rotational movements causing internal deformation (Fig. 2). Thus, slides represent translational movements and slumps represent rotational movements along shear surfaces. The type of glide plane present usually can not be determined from cores alone; hence, slump deposits often cannot be distinguished from slide deposits in cores, unless deformation features (e.g. slump folds) are present. Slump and slide facies can be recognized in cores using the following criteria: (1) soft-sediment folds and contorted bedding; (2) steeply dipping layers (up to 60° in cores examined for this study) with various discordant orientations; (3) basal (primary) glide planes and shear zones; (4) internal (secondary) glide planes; (5) abrupt changes in fabric; (6) inclined dish structures; (7) brecciated zones, and (8) clastic injections (Helwig 1970; Woodcock 1976; Jacobi 1976; Gawthorpe & Clemmey 1985; Martinsen 1989; Shanmugam *et al.* 1994).

Debris flows

An increase in mass disaggregation and mixing of sediment with water as a slump moves downslope can transform a slump into a *debris flow* (Fig. 2) in which sediment is transported as an incoherent, viscous mass via plastic (laminar) flow (Fig. 2). In debris flows, inter-granular movements predominate over shear-surface movements. Debris-flow facies are characterized in cores by: (1) units with sharp and irregular upper contacts; (2) floating mudstone or rock clasts; (3) planar clast fabric; (4) inverse grading; and (5) high matrix content (Johnson 1970; Fisher 1971; Hampton 1972; Middleton & Hampton 1973; Shanmugam & Benedict 1978; Embley 1980).

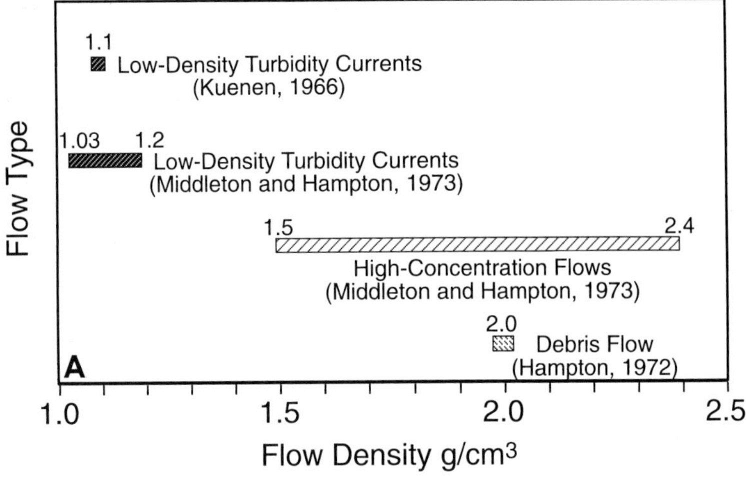

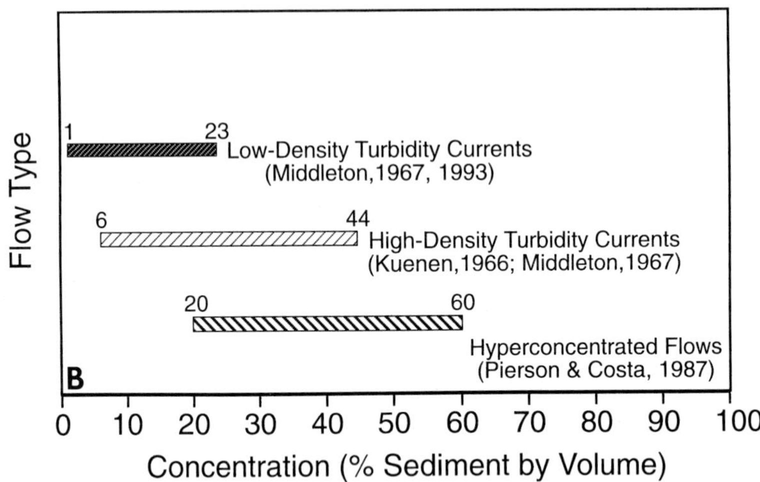

Fig. 3. (**a**) Plot of flow density for different flow types. Note overlapping density values between high-concentration flows that include high-density turbidity currents and debris flows. (**b**) Plot of flow concentration for different flow types. A flow with 20% sediment concentration can be classified as any one of the three types.

Low- and high-density turbidity currents

As fluid content increases within a debris flow as it moves downslope, the plastic debris flow may evolve into a turbulent fluidal flow termed a *turbidity current* or *low-density turbidity current*. Low-density turbidity-current deposits can be recognized in cores by: (1) normal size grading; (2) sharp basal contacts; (3) gradational upper contacts; and (4) Bouma divisions (Bouma 1962).

The concept of *high-density turbidity current* was envisaged by Kuenen (1950). The distinction between 'low' and 'high' density currents was set at a density value of 1.1 g cm^{-3} (Kuenen 1966) (Fig. 3a). High-concentration flows that include high-density turbidity currents have a density range of 1.5–2.4 g cm^{-3} (Middleton & Hampton 1973). Debris flows used in Hampton's (1972) experiments had a density of 2.0 g cm^{-3} (Fig. 3a). Because debris flows and high-density turbidity currents have similar density values, it is not practical to distinguish these processes in the rock record (Shanmugan 1996).

Other problems arise because disagreements exist on the sediment concentration value that separates low and high density flows (Fig. 3b). According to Kuenen (1966), for example, high-density behaviour begins at a concentration value of 6% (see Pickering

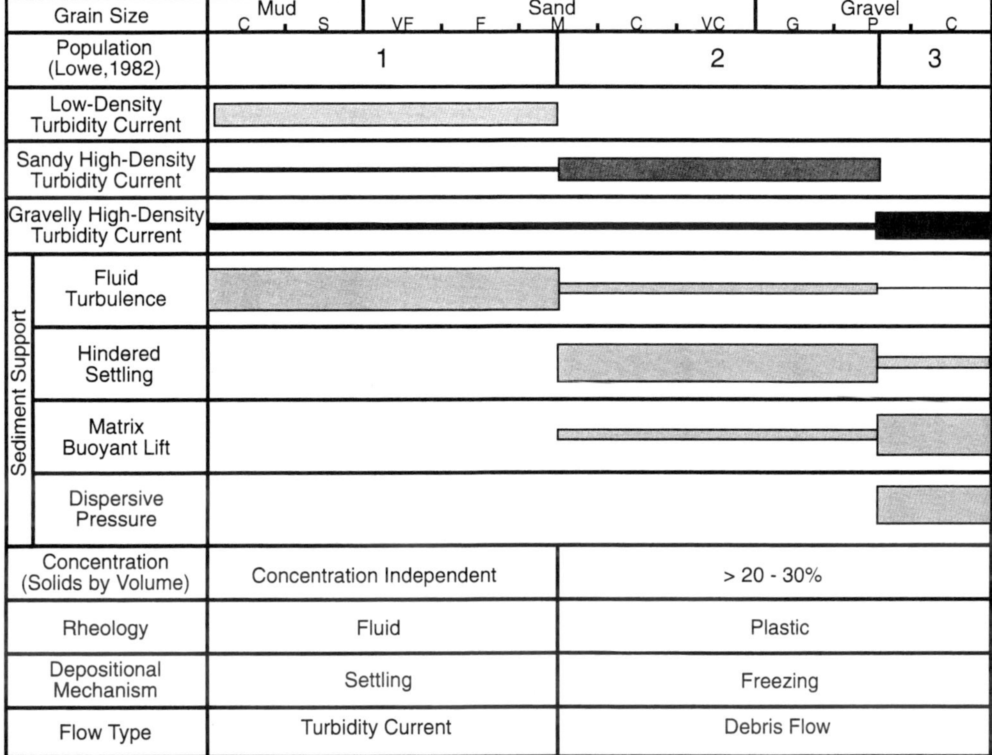

Fig. 4. Lowe's (1982) classification of turbidity currents and their grain-size populations, sediment support mechanisms, and particle concentrations. Population 1 – low-density turbidity currents; population 2 – sandy high-density turbidity currents; population 3 – gravely high-density turbidity currents. We have added our inferences on rheology, depositional mechanism, and flow type for clarification. Because hindered settling, matrix buoyant lift and dispersive pressure are important sediment-support mechanisms in Lowe' s (1982) high-density turbidity currents, we consider these high-density flows to be more like plastic debris flows than fluidal turbidity currents, and have classified them as such.

et al. 1989, p. 17). In Middleton's (1967) experiments, high-density flows require a concentration value of 44% solids by volume (Fig. 3b). Lowe (1982) considers a concentration value of above 20 or 30% for the onset of high-density behaviour. According to Pierson & Costa (1987), hyper-concentrated flow, which is a type of debris flow in some classifications (Qian *et al.* 1980), has a concentration value of 20–60%. In other words, flows with 20% sediment concentration would be considered as low-density flows by Middleton (1967), high-density flows by Kuenen (1966), and hyper-concentrated flows (i.e. debris flows) by Pierson & Costa (1987). Because of this conceptual problem, many deep-water sands with features that indicate a plastic rheology (e.g. rafted clasts) have been classified as high-density turbidites (e.g. North Sea Gryphon Field; Newman *et al.* 1993). The Gryphon example is discussed in detail below.

Lowe (1982) classified turbidity currents into two principal types, namely low-density flows and high-density flows based on grain size populations, particle concentrations, and sediment support mechanisms (Fig. 4). Low-density turbidity currents are composed of population 1 grains (clay to medium-grained sand) in which the sediment-support mechanism (i.e. turbulence) is independent of particle concentration. These flows are true turbidity currents in terms of flow rheology (i.e. fluid) and sediment-support mechanism (i.e. flow turbulence) as defined by Lowe (1982) and Middleton & Hampton (1973). It is important to note, however, that turbidity currents can and do form in all grain sizes (e.g. fine-grained turbidites and coarse-grained turbidites) and that turbulence is always the principal sediment support mechanism.

In high-density turbidity currents, sediment support is concentration dependent (above 20–30%). In sandy high-density flows composed of population 2 (coarse sand to small pebble), grains are supported by hindered settling and turbulence (Fig. 4). In gravely high-density flows containing population 3 (pebble and cobble), grains are supported mainly by matrix buoyant lift and dispersive pressure (Lowe 1982, p. 283). Because hindered settling, matrix buoyant lift and dispersive pressure are the predominant sediment-support mechanisms in high-density turbidity currents (Lowe 1982, p. 282), we would argue that such high-density flows are more closely related to plastic debris flows than to fluidal turbidity currents.

Sandy debris flows

We apply the concept of sandy debris flows, which represents a continuous spectrum of processes between cohesive and cohesionless debris flows (Shanmugam *et al.* 1995; Shanmugam 1996; see Shulz 1984), to the North Sea reservoirs in this study. Sandy debris flows are rheologically pseudo plastic and their sediment support mechanisms include matrix strength, dispersive pressure, and buoyant lift. These flows are characterized by a moderate to high grain/clast concentration, a low to moderate content of mud matrix, and a lack of turbulence. Such flows can explain a wide range of problematical submarine 'massive' sands, which contain features indicative of plastic rheology and flow strength, observed by us in the North Sea examples. The concept of sandy debris flow also alleviates the problem of misusing the term 'high-density turbidity current' for non-turbulent plastic flows. For these reasons, we do not use the term high-density turbidity current, which creates confusion about the nature of the flow. We recognize that other investigators may choose to employ the term high-density turbidity current as opposed to sandy debris flow; however, the sedimentary features that we have observed in the North Sea examples convince us that these deposits should more properly be classified as sandy debris flows.

In general, turbidites tend to be dominant in more distal (basinal) settings; whereas, mass-transport deposits (slumps, slides and debris flows) tend to dominate more proximal (slope) settings. However, either type of deposit can occur at any location; for example, debris flows have been documented to travel hundreds of kilometres down slopes of less than 1° (Embley & Jacobi 1977; Damuth & Embley 1981). Distinguishing deposits of debris flows from those of true turbidity currents is important to petroleum exploration and production because plastic flows (e.g. debris flows) emplace sediment by freezing, which can result in laterally discontinuous, disconnected sand bodies. In contrast, turbidity currents can deposit more laterally continuous, interconnected sand bodies.

Bottom currents

In addition to these gravity-driven downslope processes, sediments can also be eroded, transported and deposited by deep thermohaline- or wind-driven currents which flow roughly parallel to the basin slope and impinge on the sea floor. Bottom-current reworked sands are commonly composed of thin lenticular layers and can be recognized by their traction structures, such as horizontal lamination, ripple-cross lamination, and cross lamination (Hollister 1967; Hollister & Heezen 1972; Shanmugam *et al.* 1993).

Sedimentary facies of basin-floor fans

Study areas and methods

This study utilized conventional cores from fields in the North Sea and the continental margins of the United Kingdom and Norway, because sequence-stratigraphical models and concepts were developed from observations of such divergent continental margins (Vail *et al.* 1991). We examined more than 3700 m (12000 feet) of conventional core from 50 different wells in 10 different areas or fields. From this large data base, we have selected the following four examples for discussion here because we were granted permission to publish data from the released wells in these areas: (1) Faeroe Basin area, northern UK continental margin northwest of the Shetland Islands (Esso 214/28-1 and Shell/Britoil 206/1-2); (2) Agat area in the Norwegian North Sea (Saga 35/3-2, Saga 35/3-4, and Saga 35/3-5); (3) Frigg area in the Norwegian North Sea (Elf 25/2-8); and (4) Gryphon–Forth area in the UK North Sea (Kerr–McGee 9/18b-7, Britoil 9/23b-7, and Conoco 9/18a-15). The cored intervals studied are Palaeogene and Cretaceous clastic sections, which are, for the most part, hydrocarbon bearing.

In addition, detailed regional sequence-stratigraphical analyses of the Faeroe (Mitchell *et al.* 1993), Agat (Gulbrandsen 1987; Skibeli *et al.* 1992), Frigg (McGovney & Radovich 1985), and Gryphon–Forth (Newman *et al.* 1993; Timbrell 1993) areas have previously been published. These studies indicate that the cored intervals in all these areas are deep-marine bathyal deposits within low-stand systems tracts. Further, the published descriptions based on seismic and wire-line log data indicate that the cored intervals in these four examples fit the criteria for basin-floor fans.

Conventional cores (13 cm wide in most cases) with fibreglass liners were used in this study. All cored wells are straight holes. We conducted detailed megascopic and microscopic examinations of both the resinated slabs and unresinated 2/3 cuts

Table 1. *Types of lithofacies, their characteristics, and interpreted depositional processes*

Facies	Sedimentary features*	Depositional process
1	Gravel and pebbly sand (coarse to medium grained), amalgamated units, contorted bedding, floating clasts	Gravely slump, slide, debris flow (mass transport)†
2	Fine- to medium-grained sand, amalgamated units, 'massive' appearing, sharp and irregular top, basal (primary) glide planes, internal (secondary) glide planes, slump folds, contorted bedding, steep layers, shear zones, floating clasts, planar clast fabric, inclined dish structures, brecciated zones, clastic injections, synsedimentary faulting	Sandy slump, slide, debris flow (mass transport)†
3	Mudstone, sharp and irregular top, contorted bedding, steep layers, slump folds, shear zones, floating clasts, brecciated zones, planar clast fabric, clastic injections	Muddy slump, slide, debris flow (mass transport)†
4	Fine-grained sand and silt, current ripples, horizontal lamination, mud offshoots	Bottom-current reworking (traction transport)
5	Fine-grained sand and silt, sharp base, gradational top, normal size grading, Bouma divisions	Turbidity currents (suspension and traction transport)
6	Mudstone, horizontal lamination, shell fragments, bioturbation, glauconite	Pelagic and hemipelagic (suspension settling)

* Sedimentary features may be used to infer processes that prevailed during the final stages of deposition, but these features may not necessarily relate to the processes of transport (Middleton & Hampton 1973).
† Cored intervals contain discrete units recognized as slides, slumps or debris flows, as well as transitional or mixed facies.

Table 2. *Distribution of lithofacies computed as percentage of cored interval in the Palaeocene, Faeroe Basin*

Well	Esso 214/28–1 Sequence 70	Esso 214/28–1 Sequence 20	Shell/Britoil 206/1–2 Sequence 30
Core thickness (ft)	115.5	60.5	194
Mud	15%	31%	41%
Sand	85%	69%	59%
Fine sand	100%	100%	100%
Facies			
1	–	–	–
2	69%	52%	58%
3	15%	3%	6%
4	4%	12%	1%
5	–	–	1%
6	–	–	1%
2&3	11%	–	–
2&4	–	5%	–
3&4	–	28%	–
3&6	–	–	35%
4&5	1%	–	–

Total core: 370 ft (113 m)

of these cores to determine process sedimentology. Most cores were logged at a scale of 1:20 or 1:50 to resolve decimetre-thick variations in facies. Sedimentary structures and features were documented in detail by close-up colour photography (see accompanying figures). Depths of cores are given as driller's depths.

Six lithofacies have been recognized on the basis of lithological variations and physical sedimentary features in the 3700 m of core from the four study areas (Table 1). Distribution of depositional facies as a percentage of cored interval in each well is given in Tables 2, 3 and 4. The following detailed descriptions of our four study areas (Faeroe, Agat, Frigg, and Gryphon–Forth) provide more detailed examples of these sedimentary facies.

Table 3. *Distribution of lithofacies computed as percentage of cored interval in the Lower Cretaceous Agat Formation, Agat area*

Well	Saga 35/3–2	Saga 35/3–4	Saga 35/3–5
Core thickness (ft)	192	302	213
Mud	51%	43%	15%
Sand	49%	57%	85%*
Fine sand	100%	100%	91%
Medium sand	–	–	6%
Course sand and gravel	–	–	3%
Facies			
1	–	–	Trace
2	46%	54%	74%
3		–	3%
4	3%	1%	6%
5	–	–	1%
6	Trace	–	1%
1&2	–	–	3%
3&6	51%	43%	10%
4&2	Trace	2%	2%
5&6	–	–	–

Total core: 707 ft (215 m)
* Includes gravel

Table 4. *Distribution of lithofacies computed as percentage of cored interval in the Palaeogene, Frigg and Gryphon–Forth areas*

Well	Elf 25/2–8 (Frigg East)	Kerr-McGee 9/18b–7 (Gryphon)	Britoil 9/23b–7 (Forth)	Conoco 9/18a–15
Core thickness (ft)	218	398	714	296
Mud	19%	10%	28%	65%
Sand	81%*	90%	72%	35%
Fine sand	65%	–	40%	100%
Medium sand	30%	100%	60%	tr
Gravel	5%	–	–	–
Facies				
1	4%	–	–	–
2	76%	90%	72%	35%
3	6%	10%	16%	51%
4	2%	–	–	–
5	–	–	–	–
6	–	–	4%	–
3/2	–	–	–	–
3&6	–	–	8%	14%
3&4	3%	–	–	–
5&6	9%	–	–	–

Total core: 1626 ft (496 m)
* Includes gravel

Faeroe basin area, UK continental margin, west of the Shetland Islands

Mitchell *et al.* (1993) recognized 11 Palaeocene and four Eocene sequences, each separated by Type 1 sequence boundaries, in the Palaeogene section of the Faeroe Basin (Fig. 5). Their Sequences 20, 30, and 70 are discussed here. Seismic-facies units within Palaeocene sequences exhibit well developed external mounded forms with internal bidirectional downlap onto Type 1 sequence boundaries (Figs 7a & b). Some of these mounded forms, which have been interpreted to occur within lowstand systems tracts (Fig. 7a), were cored in the

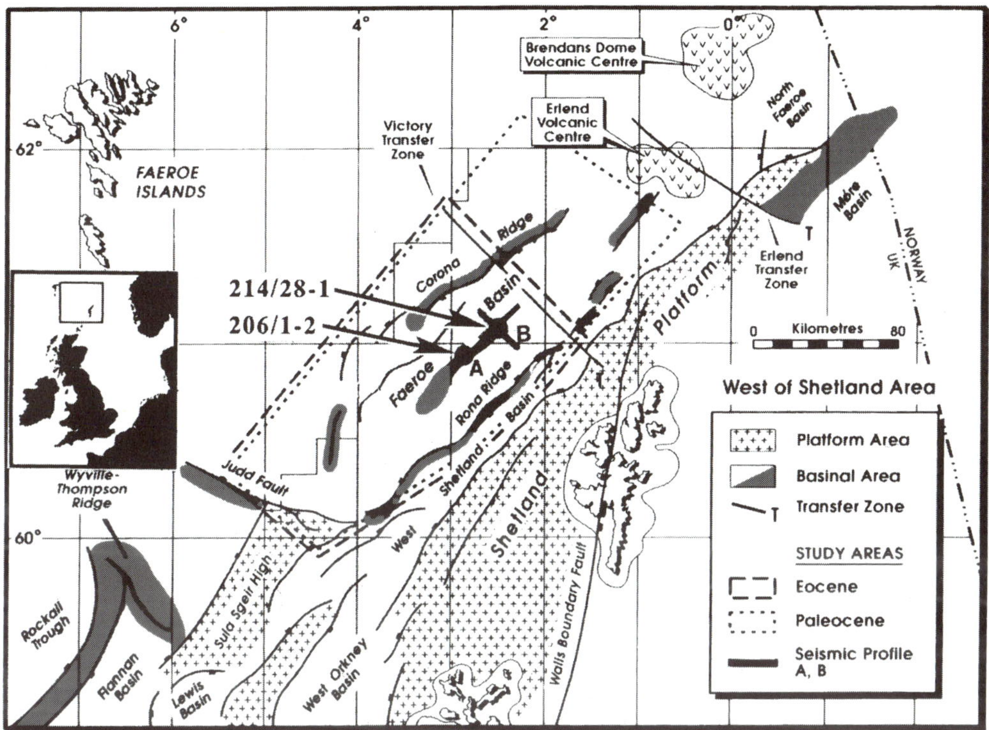

Fig. 5. Faeroe study area showing eastern platform and western basinal areas. Arrows indicate position of Esso 214/28-1 and Shell/Britoil 206/1-2 wells. Seismic profiles A and B are shown in Fig. 7.

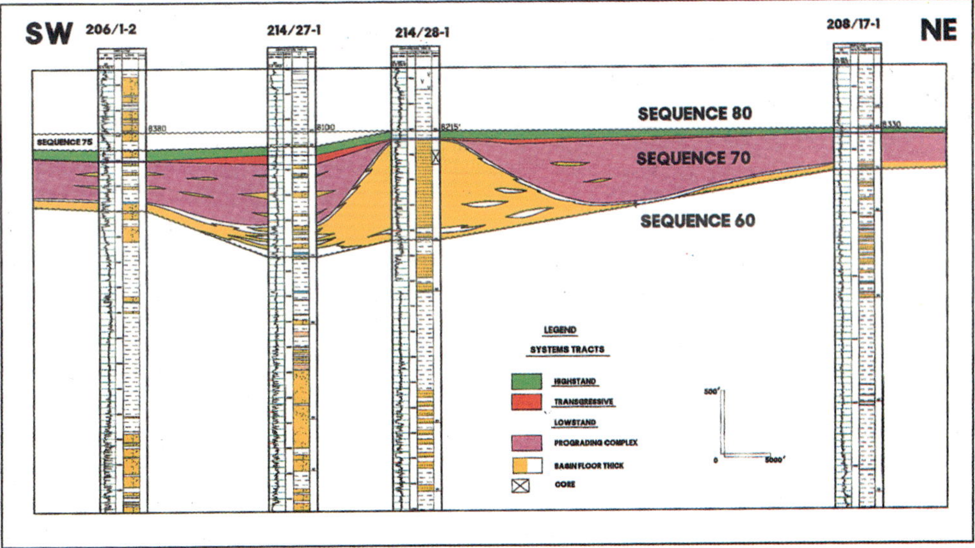

Fig. 6. A NE–SW cross-section through a seismic mound in sequence 70. Note rapid lateral change in thickness of seismic mound. The described core was taken from the upper part of the mound in the 214/28-1 well.

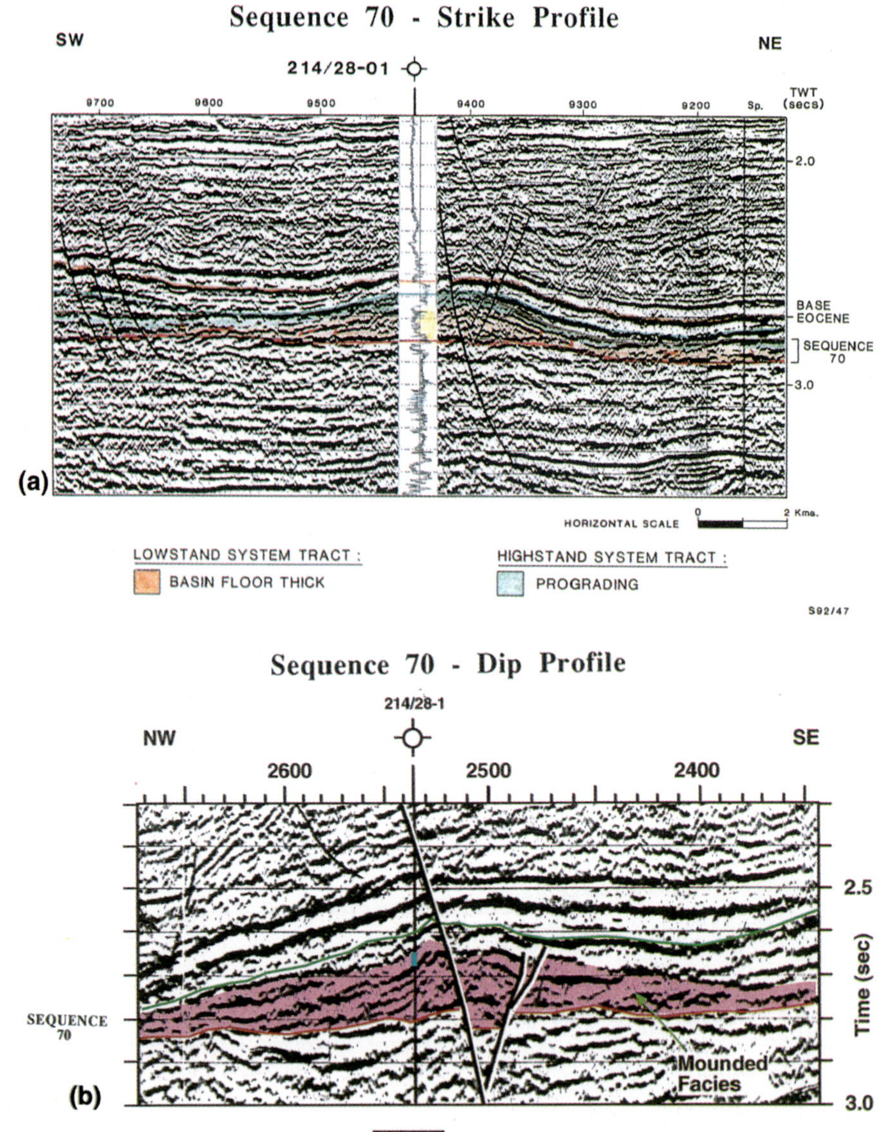

Fig. 7. Strike (**a**) and Dip (**b**) oriented seismic profiles from the Faeroe Basin showing an external mounded form (i.e. basin-floor 'thick') with internal bidirectional downlap in sequence 70 (termed Sequence 50 in Mitchell *et al.* 1993). Note blocky log motif within sequence 70 at the Esso 214/28-1 well location. Note lateral pinch-out geometries of seismic mounds. Note cored interval (Cores 1 and 2, Fig. 8) near the top of mound. Location of profiles shown in Fig. 5.

Esso 214/28-1 (Figs 7, 8 & 9) and Shell/Britoil 206/1-2 (Fig. 10) wells. The gamma-ray logs of the sandstone intervals show well developed blocky motifs (Figs 8a & 10a). These mounded forms were interpreted as basin-floor fans based on their seismic and wire-line log characteristics, plus their occurrence in basinal positions on Type 1 sequence boundaries (Mitchell *et al.* 1993). Mitchell *et al.* (1993) termed these mounded forms as basin-floor thicks (see Fig. 7a), but recognized that these features were equivalent to basin-floor fans in a sequence stratigraphical framework.

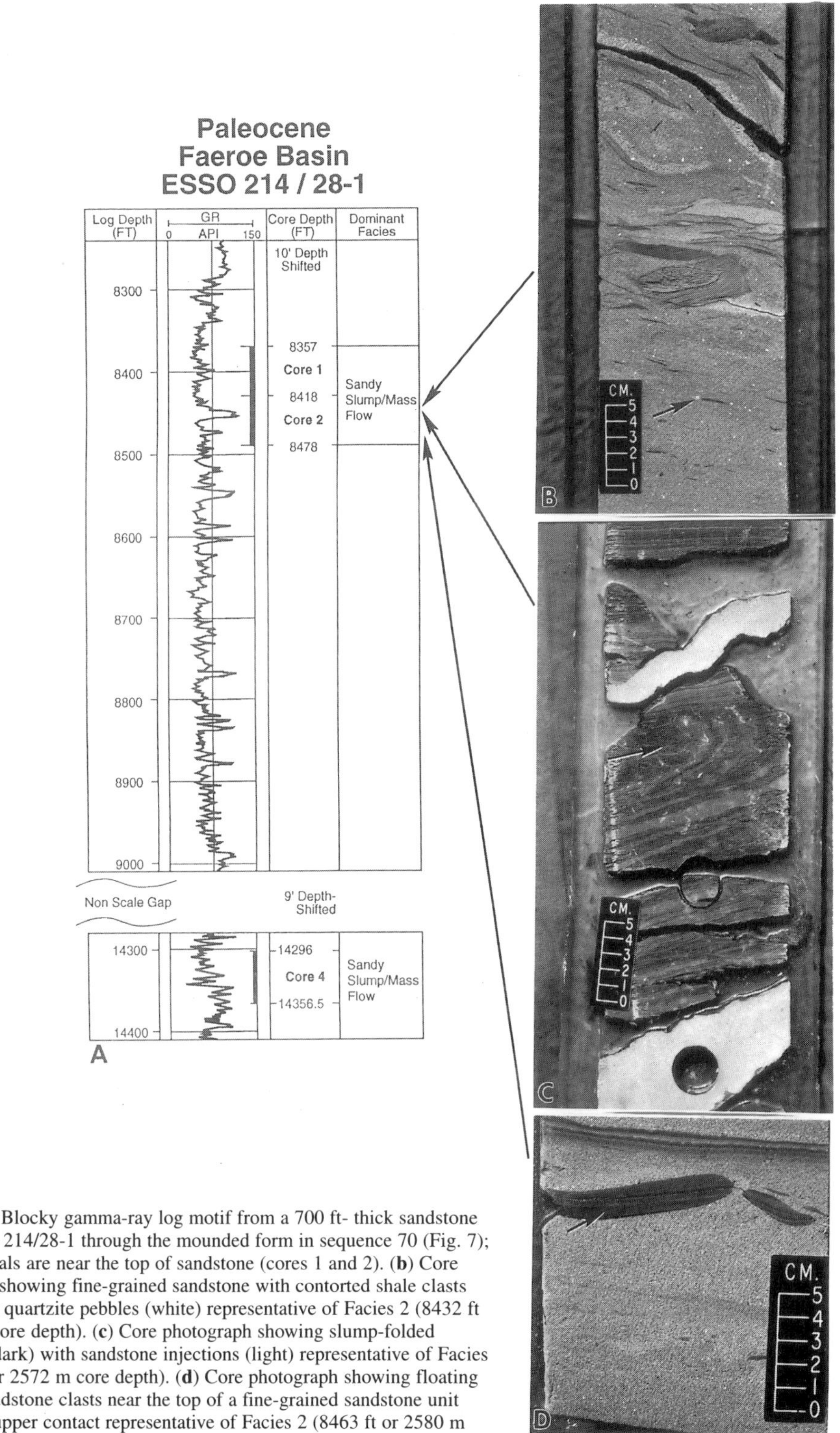

Fig. 8. (**a**) Blocky gamma-ray log motif from a 700 ft- thick sandstone unit in Esso 214/28-1 through the mounded form in sequence 70 (Fig. 7); cored intervals are near the top of sandstone (cores 1 and 2). (**b**) Core photograph showing fine-grained sandstone with contorted shale clasts and floating quartzite pebbles (white) representative of Facies 2 (8432 ft or 2571 m core depth). (**c**) Core photograph showing slump-folded mudstone (dark) with sandstone injections (light) representative of Facies 3 (8437 ft or 2572 m core depth). (**d**) Core photograph showing floating elongate mudstone clasts near the top of a fine-grained sandstone unit with sharp upper contact representative of Facies 2 (8463 ft or 2580 m core depth). Note planar clast fabric.

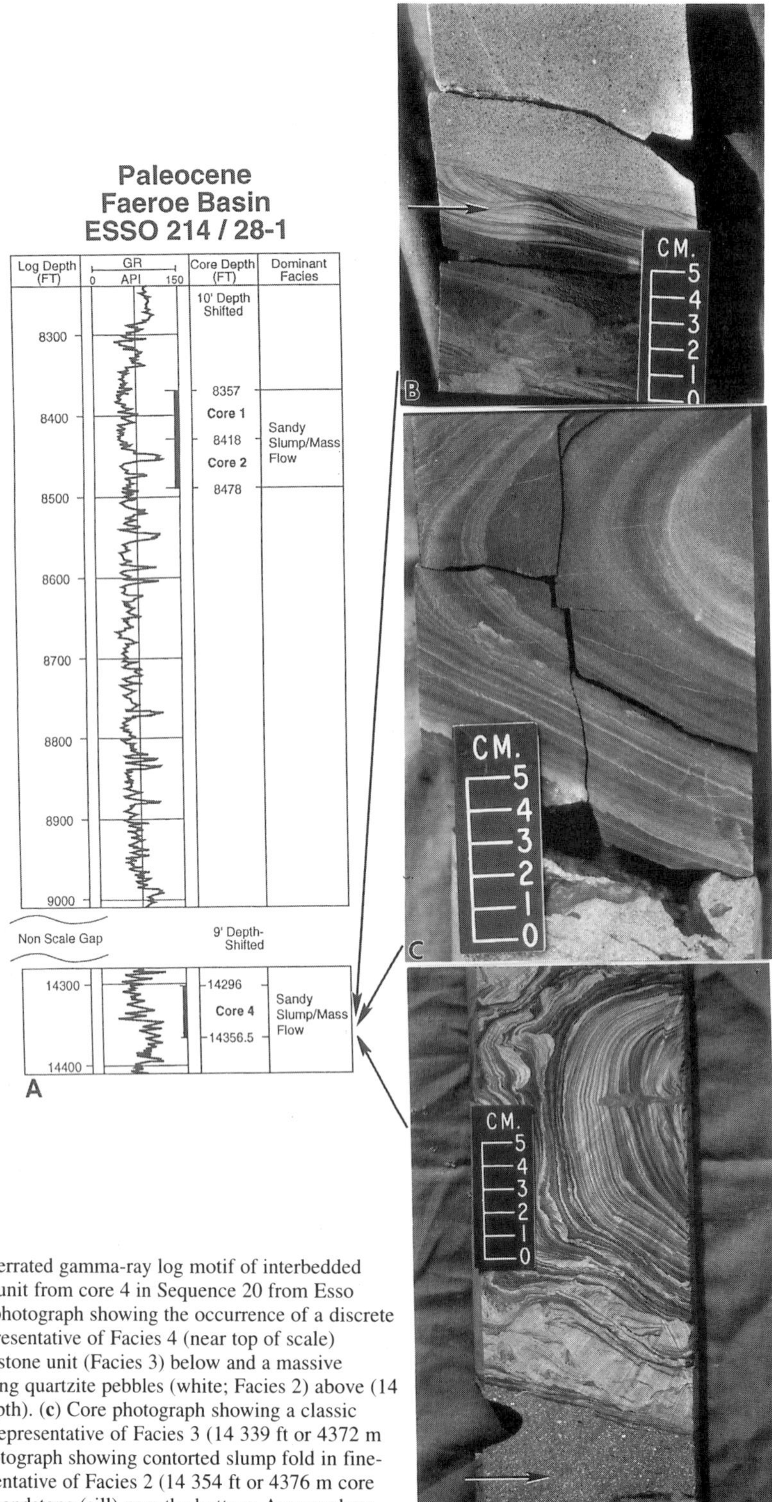

Fig. 9. (**a**) Crescent to serrated gamma-ray log motif of interbedded sandstone and mudstone unit from core 4 in Sequence 20 from Esso 214/28-1 well. (**b**) Core photograph showing the occurrence of a discrete ripple-laminated unit representative of Facies 4 (near top of scale) between a contorted mudstone unit (Facies 3) below and a massive sandstone unit with floating quartzite pebbles (white; Facies 2) above (14 300 ft or 4360 m core depth). (**c**) Core photograph showing a classic slump fold in mudstone representative of Facies 3 (14 339 ft or 4372 m core depth). (**d**) Core photograph showing contorted slump fold in fine-grained sandstone representative of Facies 2 (14 354 ft or 4376 m core depth). Note an injected sandstone (sill) near the bottom. Arrows show stratigraphical position of core photographs.

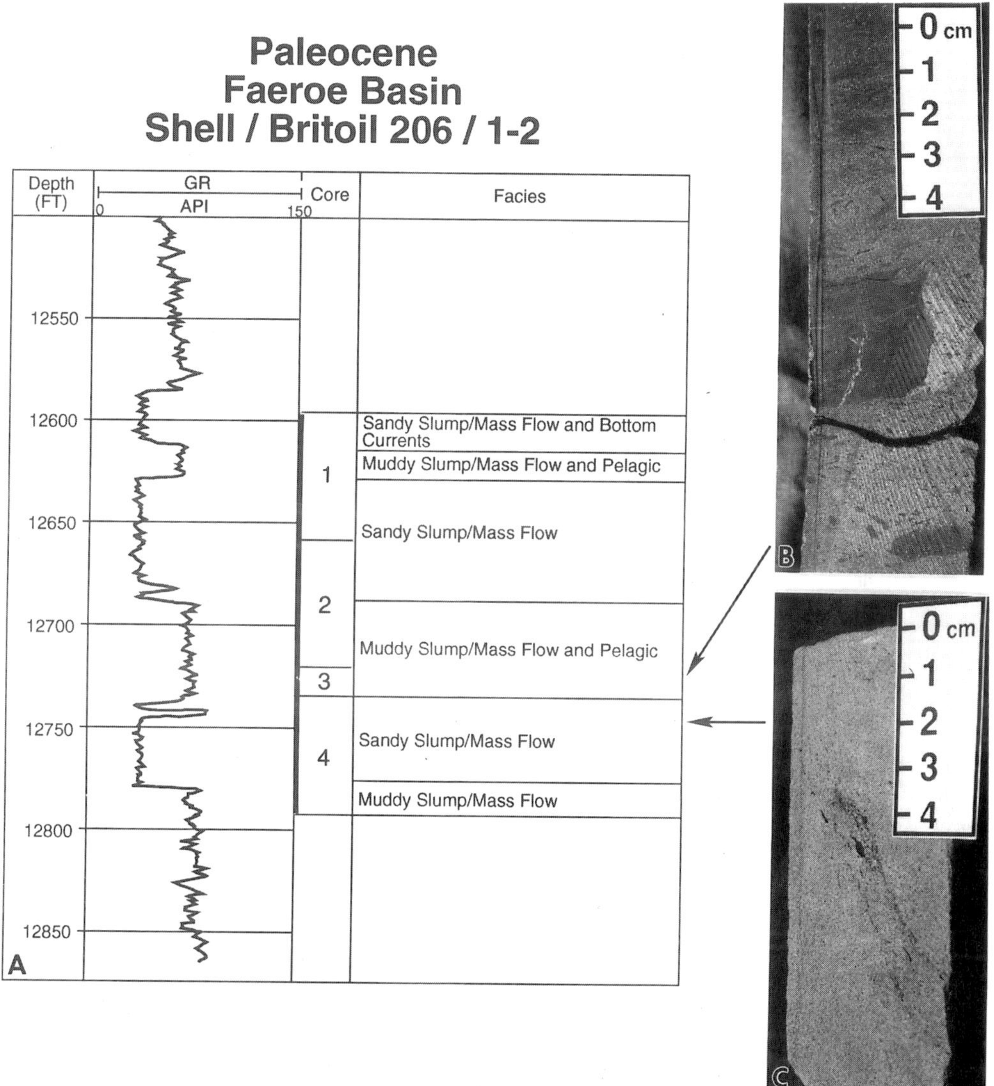

Fig. 10. (**a**) Blocky gamma-ray log motif composed of slump/mass-flow deposits in sequence 30 from Shell/Britoil 206/1-2 well in the Faeroe Basin. (**b**) Core photograph showing large mudstone clasts floating in a muddy matrix (Facies 3) (12 722.6 ft or 3879 m core depth). (**c**) Core photograph showing steeply dipping (left to right), contorted layers with mudstone clasts in fine-grained sandstone (Facies 2), (12 742.2 ft or 3885 m core depth). Arrows show stratigraphical position of core photographs.

The seismic mound in Sequence 70 shows high relief (Figs 6 & 7) with slopes of approximately 9° on its flanks. There is no evidence of erosion at its base. The mound is approximately 10 km by 7 km in size, and is about 247 m (810 feet) thick in the well. Rapid lateral pinch-out geometry is well developed (Figs 6 & 7). The thickest part of this mound is penetrated by the 214/28-1 well, and the upper part of the mound was cored (Figs 6 & 7b). This cored interval is composed of consolidated sandstone and mudstone. The sandstones (Facies 2) show: (1) sharp upper contacts; (2) floating mudstone clasts (Fig. 8d); (3) planar clast fabric (Fig. 8d); (4) floating quartzite pebbles in fine-grained

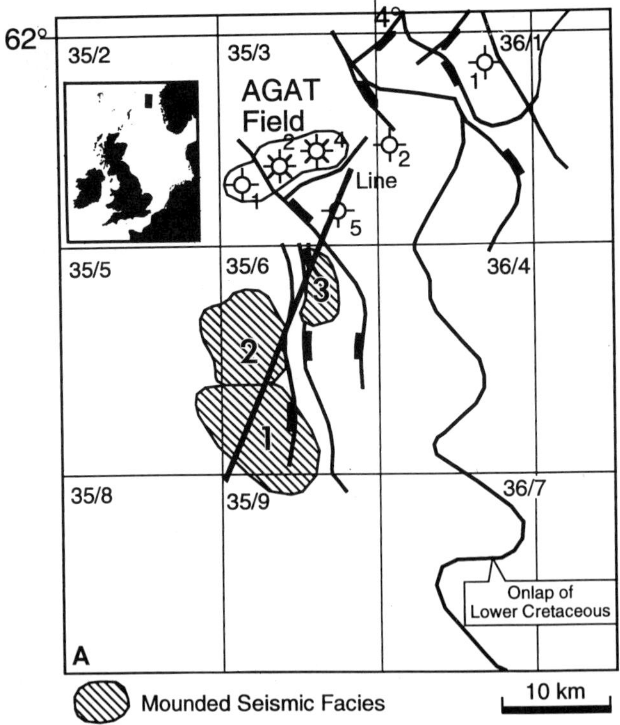

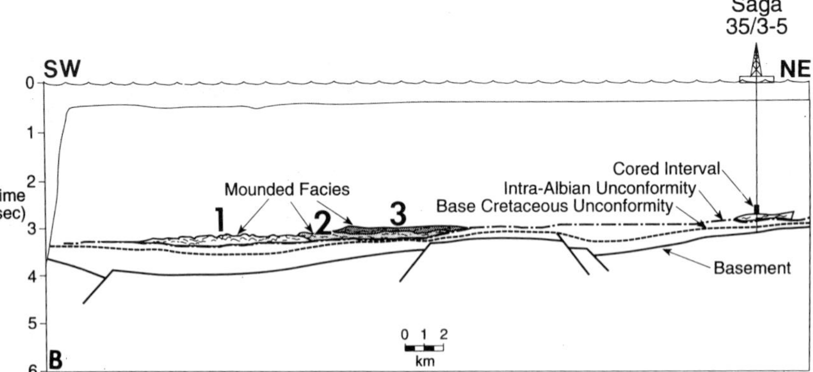

Fig. 11. (**a**) Distribution of three (1, 2 and 3) Lower Cretaceous seismic mounds in Block 35/6, south of Agat Field. The NE–SW-trending line through the Saga 35/3-5 well represents the location of seismic profile shown in (**b**). Conventional cores from Saga 35/3-2, Saga 35/3-4 and Saga 35/3-5 wells are used in this study. (**b**) Line tracing from a seismic profile showing three distinct seismic mounds downlapping onto Intra-Albian Unconformity (see Shanmugam *et al.* 1994 for original seismic data). Note mounded (convex-upward) external forms and hummocky/chaotic internal reflection patterns.

sandstone (Fig. 8b); (5) sandstone units with inverse size grading, and (6) moderate (5–10%) to high (20–30%) matrix content. These features indicate deposition by debris-flow mechanisms (Table 1).

Slump deposits (Facies 2 and 3) are also prevalent in this mound. Evidence for sandy slumps includes contorted bedding (Fig. 8b), slump folding (Figs 8c, 9c, 9d) and associated clastic injection (Figs 8c & 9d). Calibration of the blocky log motif with the sedimentological log suggests that the sharp bases of blocky log intervals correspond to primary glide planes (decollements). For example, a primary glide plane is overlain by a shear zone, which is 3 cm thick at 4371.28 m (14 337.8 ft) in the Esso 214/28-1 well. In the shear zone, dark coloured layers tend to dip in one direction and form an anastomosing or contorted network. Such shear zones are common along the bottom of slump sheets (Maltman 1987). Steeply dipping layers that occur within the sand provide additional evidence of slumping in these mounds (e.g. 3892 m or 12 766 ft core depth, Shell/Britoil 206/1–2). The intervening mudstone intervals were also primarily deposited by slumps and debris flows (Fig. 10b).

Overall, approximately 95% of the cored interval (175 ft or 53 m) in the 214/28-1 well is composed of slumps and debris flows (Table 2). Thin, graded beds (less than 5 cm thick) deposited from turbidity currents are extremely rare. Discrete ripple-laminated sandstone units often occur above contorted mudstone units and apparently represent reworking of slump sheets by bottom currents (Fig. 9b). These bottom-current reworked facies comprise 4–12% of the cored sediments in the 214/28–1 well.

Some investigators may interpret the rippled intervals as levee deposits. However, we do not ascribe them to channel-levee systems for the following reasons: (1) modern channel-levee complexes, such as the Amazon Fan (Damuth *et al.* 1988), are mud-rich, whereas cored intervals in our study area are sand-rich; (2) channel-levee complexes are characterized by channel-fill turbidite sands, whereas cored sand intervals are products of sandy slumps and debris flows; and (3) rippled units of levee deposits are commonly associated with normally graded turbidite units, whereas rippled units in our study areas are discrete and unassociated with turbidites. Because of their plastic rheology, sandy debris flows do not overbank and form levee deposits as do fluidal turbidity currents. Therefore, we prefer bottom current interpretation for the rippled zones. Damuth & Olson (1993) reported evidence for strong contour-current (i.e. bottom current) activity in the Faeroe Basin back to the Palaeogene.

In summary, although the Palaeocene mounds in the Faeroe Basin exhibit properties of basin-floor fans on seismic and log data, conventional core data show that these sequences are predominantly composed of slump and debris-flow deposits. Turbidites are extremely rare, even though the conceptual model predicts that basin-floor fans such as these should be predominantly composed of turbidites. In short, the cored intervals do not support a conventional submarine fan setting with an organized distribution of sandstones, and therefore the application of fan concepts with proximal and distal 'fan' areas in the Faeroe Basin is not meaningful in terms of sand distribution.

Agat area, Norwegian North Sea

The Agat Field is located within the Maloy fault blocks in the northern North Sea (Fig. 11a). The Cretaceous interval is relatively undeformed, and onlaps westerly dipping Jurassic strata. The hydrocarbon trap is primarily stratigraphical. The Lower Cretaceous (Albian) Agat Formation has been interpreted as a lowstand systems tract within a sequence stratigraphical framework (Skibeli *et al.* 1992). Well-developed mounded forms within the Albian sequence in Block 35/6 (Fig. 11a) are characterized by convex-upward external forms and hummocky to chaotic internal reflection patterns on seismic profiles (Fig. 11b). These mounds downlap onto the Intra-Albian Unconformity, a Type 1 sequence boundary (Fig. 11b), and thus fit the criteria for basin-floor fans based on their shape and stratigraphical position. They also exhibit blocky log motifs on wire-line logs (Fig. 12a). The tie between these seismic mounds in Block 35/6 and the best sand interval in the 35/3-5 well just to the northeast suggests that this sand interval occurs in the same lowstand systems tract as these mounds in Block 35/6, and thus, may be an equivalent basin-floor fan feature. This sand interval has been cored in the 35/3-5 (Fig. 11b), 35/3-2 and 35/3-4 wells.

These cores show that the Agat Formation is composed of consolidated sandstone and mudstone. Facies 2 (contorted sand) is the dominant facies in the cores from all three wells (Table 3), and comprises of light grey, micaceous, poorly sorted, fine-grained sandstone. Laminae often dip at variable, but high angles (up to 60°). The interval exhibiting a blocky log motif in the 35/3-2 well (Fig. 12a) is composed of sandstone with steeply dipping discordant layers (Fig. 12c). Contorted sandstone units with steeply dipping layers (Fig. 12c) commonly are interbedded with undisturbed mudstone units. Mud clasts (0.5–2 cm) and carbonaceous fragments are present throughout, but are most common near the tops of the units. Larger subangular clasts (up to 10 cm in diameter) occur rarely. A planar fabric of mud clasts and water-escape structures is present. Intervals can be argillaceous in nature, having a moderate to high mud matrix (5–10%). The basal contacts of some of these units show evidence of shearing.

Vertical injection of sand discordant to normal bedding (i.e. sand dykes) and horizontal injection of sand concordant to normal bedding (i.e. sand sills) are common in the Agat Formation (Shanmugam *et al.* 1994). This interval of sand dykes and sills shows serrated motifs in gamma-ray logs. Thicker intervals of sand injections tend to generate blocky log motifs.

Based on these observations, Shanmugam *et al.* (1994) interpreted the Agat Formation as predominantly slump and debris-flow facies deposited in an

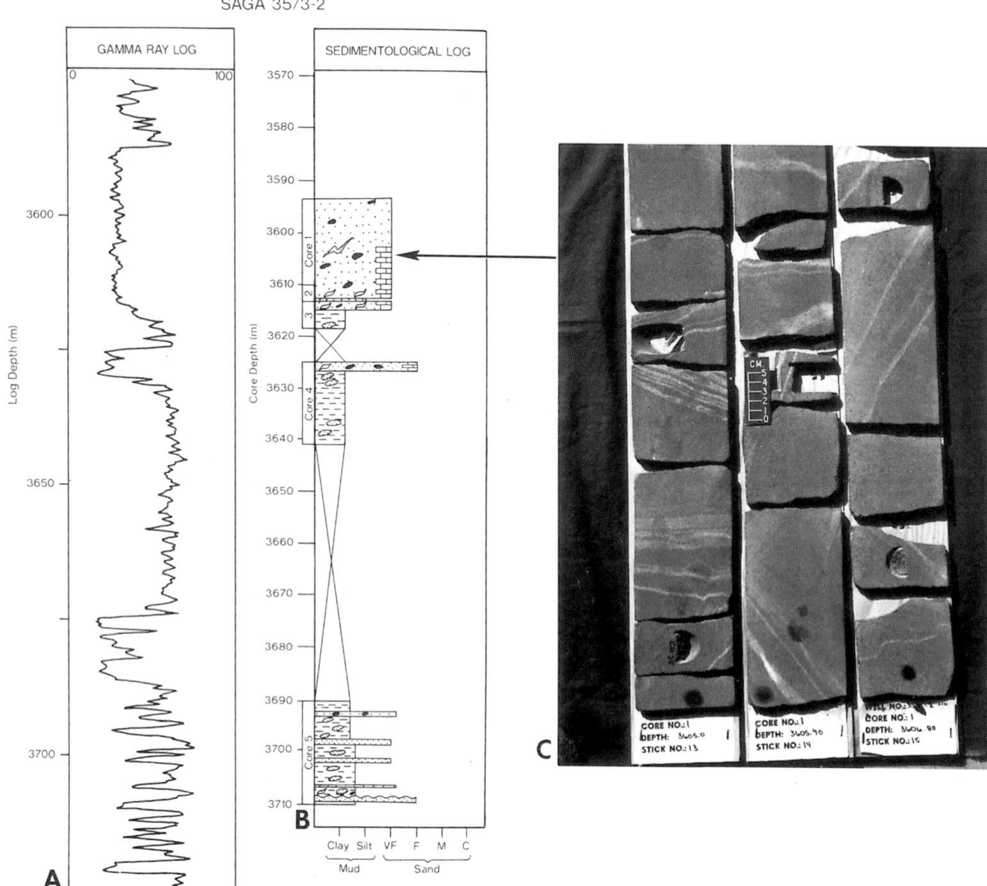

Fig. 12. (**a**) Blocky log motif (near 3600 m) in the Agat Formation, Albian, Saga 35/3-2. (**b**) Depth-tied sedimentological log. See Fig. 16 for explanation of symbols. (**c**) Core photograph showing steep layers and changes in fabric within massive sandstone unit (Facies 2), Agat Formation, Albian, Saga 35/3-5, 3604 m (upper left)–3606.8 m (lower right). Arrow shows stratigraphical position of core photograph.

upper slope setting. Regional seismic data further indicate a slope setting. Their proposed mass-transport model is quite different from the conventional submarine-fan model previously proposed for the Agat Formation (Gulbrandsen 1987). Conventional submarine fan models imply base-of-slope to basin deposition dominated by turbidites that form predictable, laterally continuous, sheet-like sand bodies in outer fan environments (Mutti 1977). In contrast, the mass-transport model implies that sands were deposited by more random slumps and debris flows on the upper slope. The sands in these deposits are apt to be more discontinuous and, thus, less predictable than those in a submarine fan. This mass-transport model thus has important implications for reservoir geometry and correlation. For example, cored intervals show major differences in stratigraphical position, thickness and frequency of occurrence of the Agat sandstone units among the three wells. Therefore, the sandstone units are not correlatable between wells and may not be interconnected (Shanmugam *et al.* 1994); this is consistent with the mass-transport model. In summary, although on seismic data the Agat Formation exhibits the seismic facies and sequence-stratigraphical position consistent with basin-floor fans, turbidites and continuous sheet-like sand bodies predicted for basin-floor fans are not observed in cores and are not substantiated by drill-stem tests (see Gulbrandsen 1987 and Shanmugam *et al.* 1994 for details of differences in gas–water contacts and pore pressures between wells).

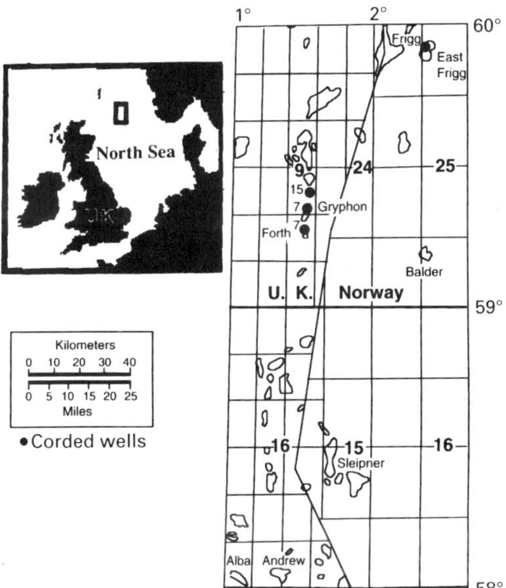

Fig. 13. Location map showing cored wells (solid dots) described in this study from Frigg (Elf 25/2-8) and Gryphon–Forth (Kerr–McGee 9/18b-7, Britoil 9/23b-7, Conoco 9/18a-15) areas.

On the modern Norwegian continental margin, large mass-transport deposits occur on the northern and southern flanks of the Voring Plateau (Damuth 1978; Bugge 1983; Jansen *et al.* 1987) and provide possible modern analogs for the Cretaceous Agat Formation. The large Storegga Slide on the southern flank of the plateau exhibits mounded seismic patterns in sparker profiles and the core is composed primarily of debris-flow deposits (Bugge 1983; Jansen *et al.* 1987). Although the dimensions of the Storegga Slide are large (>800 km in length, 290 km in width, and 5580 km^3 in volume), its depositional processes and setting are probably comparable to those associated with the Cretaceous Agat Formation. The Storegga Slide lies along the trend of recent earthquake epicentres, which coincide with the Jan Mayan Fracture Zone and its landward extension. The slide occurred in three phases (30 000, 8000 and 5000 years BP) and was apparently triggered by earthquakes (Bugge 1983; Jansen *et al.* 1987). This observation is important because mass-transport deposits initiated by earthquakes are not related to sea-level changes (eustatic or relative). In fact, when compared to the SPECMAP $\delta^{18}O$ sea-level curve and time scale (Imbrie *et al.* 1984), depositional events that emplaced the Storegga slide apparently occurred during all phases of sea-level, including fall (lowstand systems tract, 30 ka), rise (transgressive systems tract, 8 ka) and maximum flooding (transgressive/highstand systems tracts, 5 ka), during the last fourth order sea-level cycle of the latest Pleistocene (late Wisconsin) to Holocene. The Storegga Slide demonstrates that mass-transport deposits can be emplaced at anytime during a sea-level cycle. Thus, mass-transport deposits in the rock record may be misinterpreted as basin-floor fans deposited in lowstand systems tracts, if they are interpreted strictly in the context of the conceptual sea-level model.

Frigg area, Norwegian North Sea

The Frigg Field and its associated satellites form a giant complex of gas fields (8 tcf) in the North Sea (Fig. 13). Based on regional sequence-stratigraphical analyses, the Lower Eocene Frigg Formation has been traditionally considered an example of a submarine-fan system composed of turbidites (Heritier *et al.* 1979; McGovney & Radovich 1985). In the Vail *et al.* (1991) conceptual sea-level model, the 'sheet mounds' that characterize the Frigg Formation (Fig. 14a) fit the seismic criteria for basin-floor fans. The upper part (1920–1980 m) of the cored interval in the 25/2-8 well shows a blocky log motif (Fig. 15a) that is consistent with a basin-floor fan interpretation; whereas, the lower part (2070–2210 m) shows a serrated motif. McGovney & Radovich (1985) interpreted the upper member with blocky log motif as channel sands, and the lower member with serrated log motif as sheet-like turbidites. According to the conceptual sea-level model, however, blocky log motifs would be interpreted as basin-floor fans with sheet-like turbidites (e.g. Vail & Wornardt 1990; Vail *et al.* 1991).

We have examined conventional cores from six wells in the main Frigg and Frigg East Alpha Fields. The Elf 25/2-8 well in the Norwegian Sea was studied in detail owing to the pristine preservation of sedimentary structures in the core. The Frigg Formation is primarily comprised of unconsolidated sand. Although the sand units of the Upper Frigg Formation were described as 'massive' (Brewster 1991), our examination disclosed steeply dipping (up to 30°) discordant layers, slump folds, abrupt changes in fabric, inclined dish structure, and internal glide planes. These features indicate emplacement by slumps (Facies 2). Depositional fabric (i.e. orientation of grains) changes drastically in the adjacent mudstone intervals (Facies 3); opposing fabrics show variation in dip of up to 60°. In many cases, contorted sandstone units (Facies 2) with steep layers are sandwiched between undisturbed mud units. This is perhaps the most convincing evidence for synsedimentary deformation by slumping (Woodcock 1976). Massive sand intervals exhibit planar clast fabric (Fig. 15c), indicating deposition

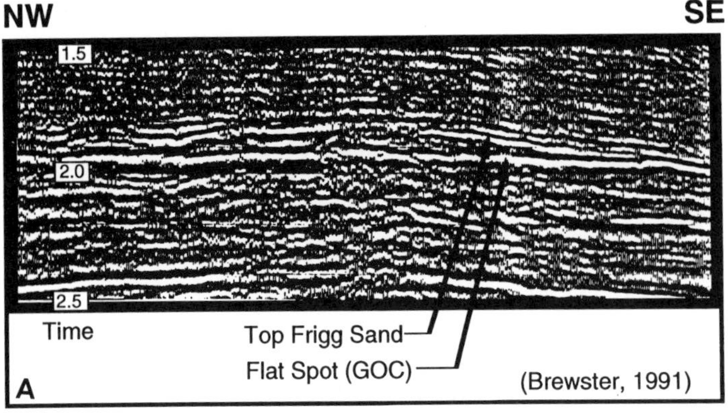

Fig. 14. (**a**) Seismic profile across the Frigg Field showing sheet mound geometry (see the top Frigg sand) (Line 73-F-9, SSL Survey, 1973). (**b**) General stratigraphical model for the Frigg Field (simplified from Brewster 1991). Core from the Elf 25/2-8 well is described in this study. Note discontinuous nature of debris-flow emplaced sands.

from laminar flows, such as debris flows. Individual units of sandy slumps and debris flows range in thickness from 1 to 4 m.

The sand units of the Lower Frigg Formation are poorly sorted with 5–10% matrix, show sharp upper contacts, water-escape structures and floating mudstone clasts (Fig. 15d). These sands apparently represent debris-flow deposits (Facies 2). Slump and debris-flow deposits account for nearly 86% of the cored interval in the 25/2-8 well (Table 4). Turbidites with normal size grading are extremely rare and comprise less than 1% of these cores. Turbidite units range in thickness from 2 to 3 cm and occur mainly in mudstone intervals.

If published submarine-fan models for Frigg Field (Heritier *et al.* 1979; McGovney & Radovich 1985) were correct, these reservoir sands should be largely composed of turbidites. However, our core data clearly indicate deposition was predominantly by mass-transport processes. A more recent sequence stratigraphical study also concluded that the Frigg reservoir represents a massive slump, not a turbidite fan (Battie *et al.* 1994). According to Battie *et al.* (1994), the Frigg slump was deposited approximately 50 km from the palaeo-shelf edge to the west, and the areal closure of the Frigg Field (115 km^2) is quite close in dimensions to the area of the slump scar (80 km^2).

In spite of their slump and debris flow origin, sands of the Frigg Formation do contain, thick, amalgamated units. They also exhibit high porosity (27–32%) and permeability (900–4000 mD). This case demonstrates that slump/debris-flow emplaced sands can be of excellent reservoir quality, although their lateral continuity may be limited and, therefore harder to predict (Fig. 14b).

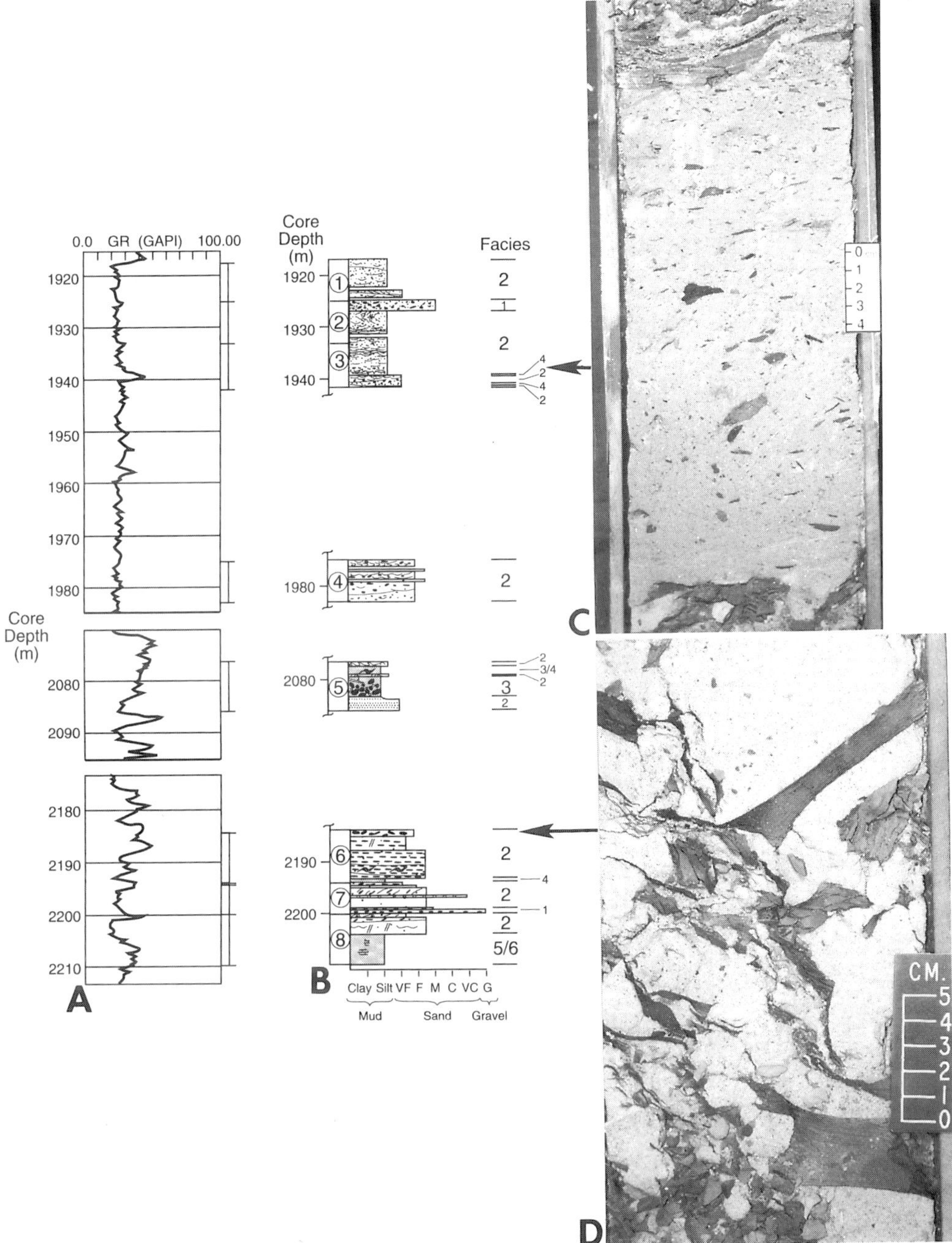

Fig. 15. (**a**) Upper 'blocky' log motif (1920–1980 m) and lower 'serrated' log motif (2070–2210 m) in the lower Eocene Frigg Formation, Elf 25/2-8 well. (**b**) Depth-tied sedimentological log showing facies distribution. See Table 1 for explanation of facies. See Fig. 16 for symbols. (**c**) Core photograph showing planar clast fabric (upper part of photo) and random orientation of floating mudstone clasts (lower part of photo) in fine-grained sand (Facies 2) from Lower Eocene Frigg Formation, Elf 25/2-8, 1939 m. (**d**) Core photograph showing random orientation of floating mudstone clasts in fine-grained sand (Facies 2) from Lower Eocene Frigg Formation, Elf 25/2-8, 2185 m. Note that Facies 2 causes both blocky and serrated log motifs. Arrows show stratigraphical position of core photographs.

Kerr-McGee 9/18b-7
Lower Eocene, Gryphon Field

Legend

Lithology			Depositional Facies	
Gravel	Injected Sand	Nodules	Facies	Interpreted Environment
Sand	Sand/Silt Lenses	Fractures	1	Gravelly mass flow slump
Silt	Mud Layers	Normal Fault	2	Sandy slump/mass flow
Clay	Steep Layers	Erosion/Irregular Contact	3	Muddy slump/mass flow
Normal Size Grading	Flame Structures	Sharp Contact	4	Bottom current reworking
Inverse Size Grading	Water-Escape Structures	Gradational Contact	5	Turbidity currents
Horizontal Lamination	Bioturbation	30° Dip - From Core Horizontal	6	Pelagic/hemipelagic
Current Ripples	Calcite Cement	RF Rock Fragments		
Contorted Bedding	Mud Clasts/Chips	C Carbonaceous Fragments		
		G Glauconite		
		M Mica		
		S Siderite		
		CA Calcite		
		P Pyrite		
		F Fault		
		E Erosion		

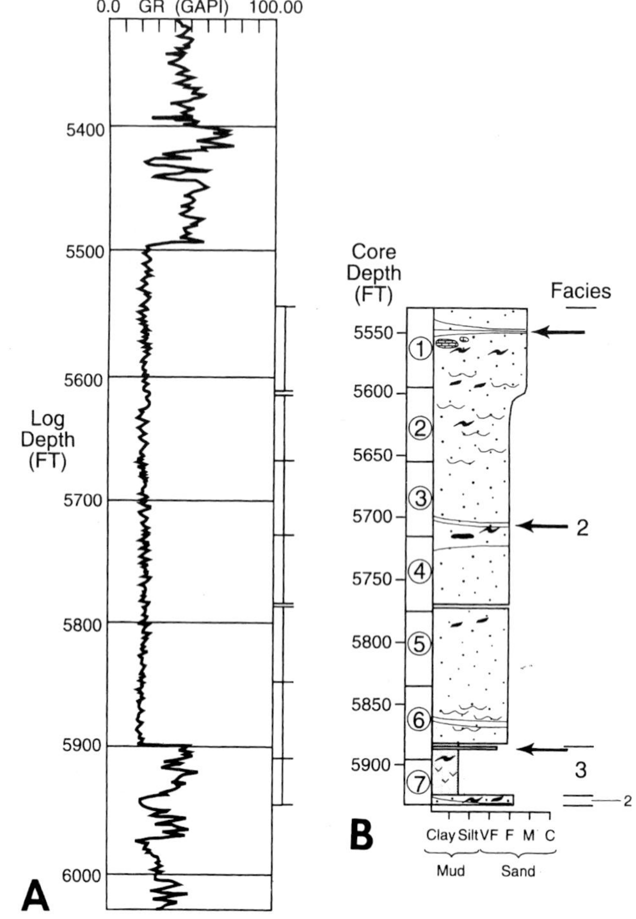

Fig. 16. (**a**) Well-developed blocky wireline-log motif of a 122 m (400 ft) thick sandstone unit, lower Eocene, Gryphon Field, Kerr–McGee 9/18b-7. (**b**) Depth-tied sedimentological log from this sandstone unit showing distribution of facies 2 and 3. Arrows denote positions of 3 shear planes (glide planes) observed in the core (note locations of these shear planes in Fig. 17). The basal shear plane is a decollement that coincides with a major sequence boundary. Core numbers are shown within circles. See Table 1 for explanation of facies.

Gryphon-Forth area, UK North Sea

The Gryphon and Forth (recently renamed Harding) fields lie southwest of the Frigg Field in the Viking Graben, UK Quadrant 9 (Fig. 13). We described conventional core totaling 823 m (2700 feet) from 20 wells in these fields and the surrounding area. We describe three wells with long cores here: (1) Kerr-McGee 9/18b-7 (Gryphon Field); (2) Britoil 9/23b-7 (Forth Field), and (3) Conoco 9/18a-15. The Upper Palaeocene/Lower Eocene units (Balder/Frigg) contain the principal reservoirs. A major source of sediment supply during the Palaeogene was from the East Shetland Platform and the Scottish mainland (Stewart 1987).

In the Gryphon Field, the Lower Eocene reservoir exhibits well-developed blocky log motifs (Fig. 16). On seismic data, the reservoir also shows external mounded forms with lateral pinch-out geometries resting upon on an unconformity surface (blue sequence boundary, Fig. 17a & b). Based on these seismic and log signatures, this reservoir can be interpreted as a basin-floor fan. In fact, Newman *et al.* (1993) previously interpreted the reservoir in the Gryphon Field as lowstand fans (i.e. basin-floor fans); however, Timbrell (1993) interpreted the same reservoir as slope fans. Reynolds & Mackay (1992) considered the reservoir in the Gryphon Field as a product of post-depositional sand injection.

Our examination of conventional core shows that the Lower Eocene reservoir comprises unconsolidated sand. Facies 2 (contorted sand) and 3 (contorted mud) account for 60–100% of the cored interval (Table 4). Massive sand beds are common. Sands are poorly sorted with 5–15% matrix, and show sharp upper contacts, floating mudstone clasts (up to 15 cm diameter, Fig. 18c), primary (basal) glide planes, steep internal shear planes (Fig. 18d), and water-escape structures. These features suggest deposition from slumps and debris flows.

Newman *et al* (1993) interpreted the massive sand intervals in the Gryphon area as deposits of high-density turbidity currents. However, we did not observe a sequence of structures, which Lowe (1982) considers characteristic of sandy high-density turbidites; i.e., S_1 (traction), S_2 (traction carpet), and S_3 (suspension) structures in ascending order in the massive fine-grained sands in the Gryphon area. This is not surprising because traction carpets do not develop in fine-grained sands. Lowe (1982, p. 289), for example, states that: 'Flows composed largely of fine- and very fine-grained sand will not deposit traction carpet layers because of the negligible dispersive pressure between such fine grains'. Furthermore, the validity of the traction carpet concept has recently been questioned by Hiscott (1994). In addition, high-density turbidity currents are characteristic of grain-size populations 2 (coarse sand to small pebble) and 3 (pebble and cobble) of Lowe (1982, p. 283), whereas massive sands in the Gryphon area predominantly are fine-grained sand (i.e. population 1 of Lowe 1982). Although population 1 is expected to behave like low-density turbidity currents and develop normal size grading and Bouma divisions (Lowe 1982), most massive sand units in the Gryphon area are ungraded. This is in spite of their high matrix content (up to 15%) and poor sorting. More importantly, these massive sands reveal diagnostic features of slumping and debris flows (Fig. 18c & d). As mentioned earlier, high-density turbidity currents, as defined by Lowe (1982), reflect properties of plastic debris flows, rather than fluidal turbidity currents (Fig. 4).

The nearly 122 m (400 ft) of continuous core in well 9/18b-7 makes it possible to calibrate core-scale features with seismic-scale features using synthetic seismograms (Fig 17a and b). The cored interval penetrates most of the thickness of the pink and orange seismic mounds (Fig. 17a). We have recognized three distinct shear planes in the core (see three arrows in Fig. 16). A basal shear plane recognized in the core corresponds to the base of blocky log motif, as well as to the basal contact of the pink seismic mound (Fig. 17). The blue sequence boundary in seismic profiles (Fig. 17a & b) is coincident with the primary glide plane or the decollement of a slump sheet recognized in the core. The decollement is the basal contact of the 120 m (400 ft) thick sand with underlying Balder Tuff. The chalk-like texture of the Balder Tuff apparently provided a slippery shear surface for mass movements of slump sheets composed of sands.

The middle shear plane (or secondary glide plane) and associated mudstone clasts recognized in the core (Fig. 18c and d) coincides with the high-amplitude reflection that separates the pink and orange mounds (Fig. 18e). We infer that the cored interval is composed of at least two slump sheets separated by an internal (middle) shear plane. Amalgamation of these two slump sheets (pink and orange) has created a 120 m thick sand unit with a well-developed blocky log motif (Fig. 18a). Thus mounded forms observed on seismic data from the Gryphon area apparently result from accumulation of multiple packages of slumps and debris flows (Fig. 18e). Slope angles of up to 15° occur on the flanks of seismic mounds and further support a slump interpretation. Seismic data show erosional features upslope from the mounds that may represent slump scars (Fig. 18e).

In the mass-transport model we propose, slumped sands are areally restricted and extremely complex in their geometry (Fig. 18e). Rapid lateral pinch out of the 122-m (400-ft) thick reservoir sand

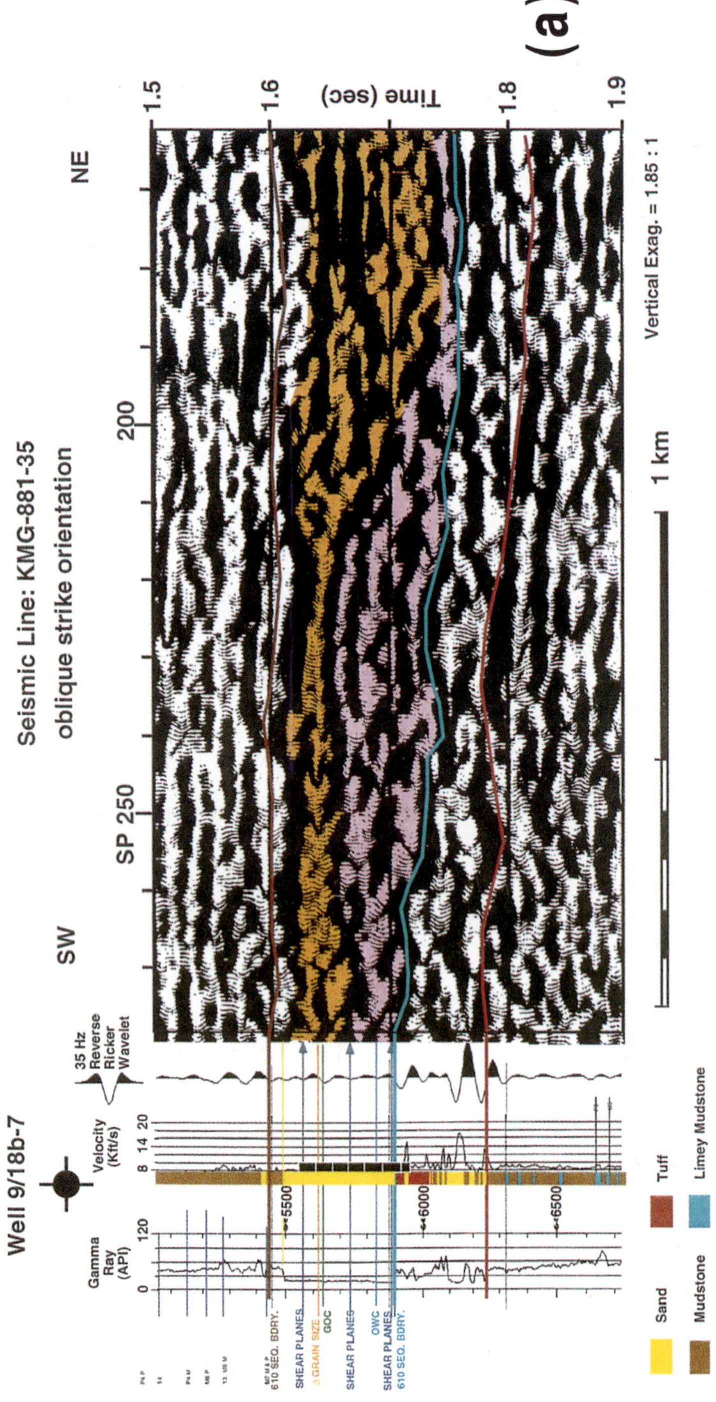
(a)
Well 9/18b-7
Seismic Line: KMG-881-35
oblique strike orientation
SW
SP 250
200
NE
Time (sec)
1.5
1.6
1.8
1.9
Gamma Ray (API)
0
120
Velocity (Kft/s)
8
14
20
35 Hz Reverse Ricker Wavelet
5500
6000
6500
610 SEQ. BDRY.
SHEAR PLANES
GRAIN SIZE
GOC
SHEAR PLANES
OWC
SHEAR PLANES
610 SEQ. BDRY.
1 km
Vertical Exag. = 1.85 : 1
Sand
Mudstone
Tuff
Limey Mudstone

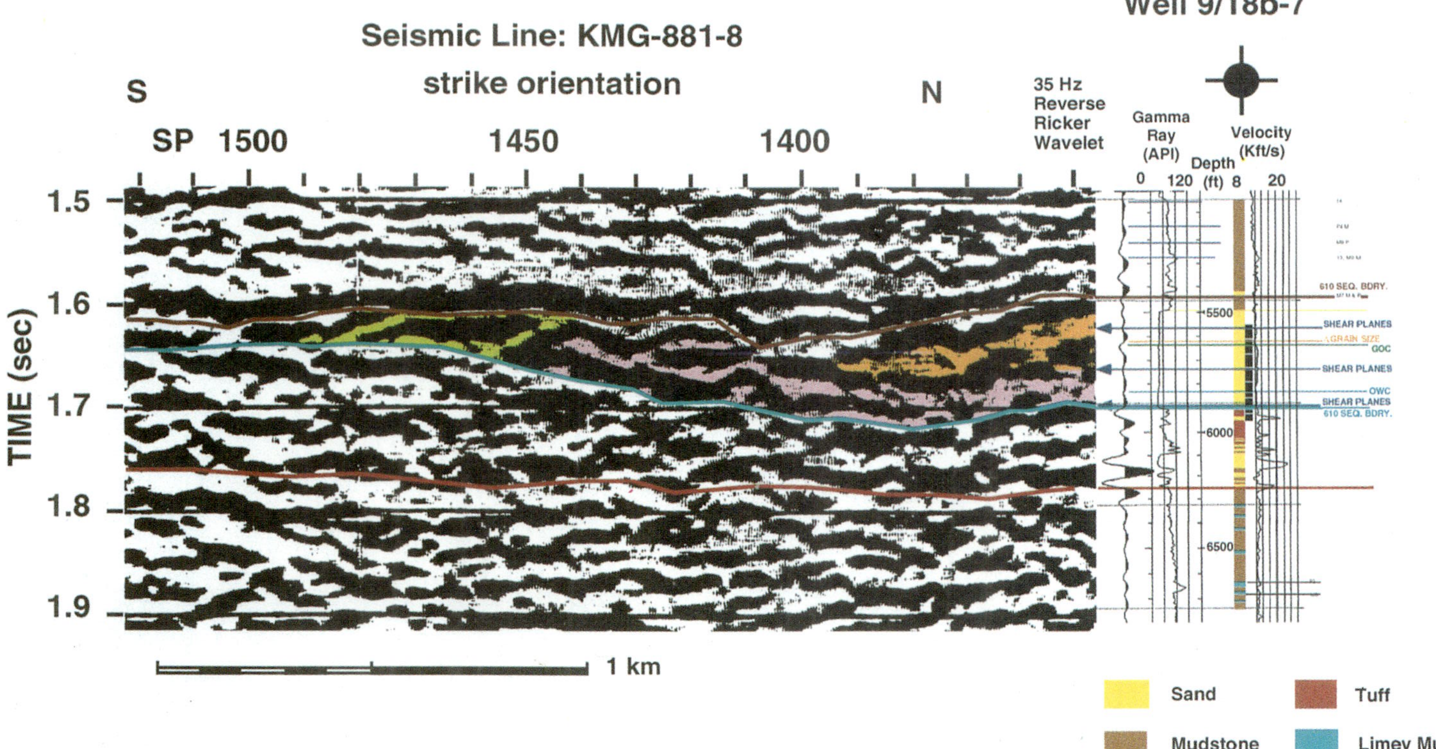

Fig. 17. (**a**) Seismic line (KMG-881-35) showing mounded geometries of pink and orange seismic packages with irregular upper surfaces and hummocky to chaotic internal reflections. Note lateral pinch out of the pink mound and onlapping of the overlying orange package on the right-hand side of the pink mound. Calibration of well data with seismic data using a synthetic seismogram shows that cored intervals (solid black bars) span most of the thickness of the mounded facies. Note that the basal shear plane coincides with the base of the blocky log interval, as well as the blue sequence boundary at the base of pink package. The middle shear plane corresponds to a high-amplitude reflection that separates the pink and orange packages. (**b**) Seismic line (KMG-881-8) showing stratigraphical relationships between three seismic mounds (pink, orange and green) which downlap onto the blue sequence boundary. Note lateral pinch out geometries of seismic mounds. See Fig. 18e for schematic diagram based on seismic lines KMG-881-35 and KMG-881-8.

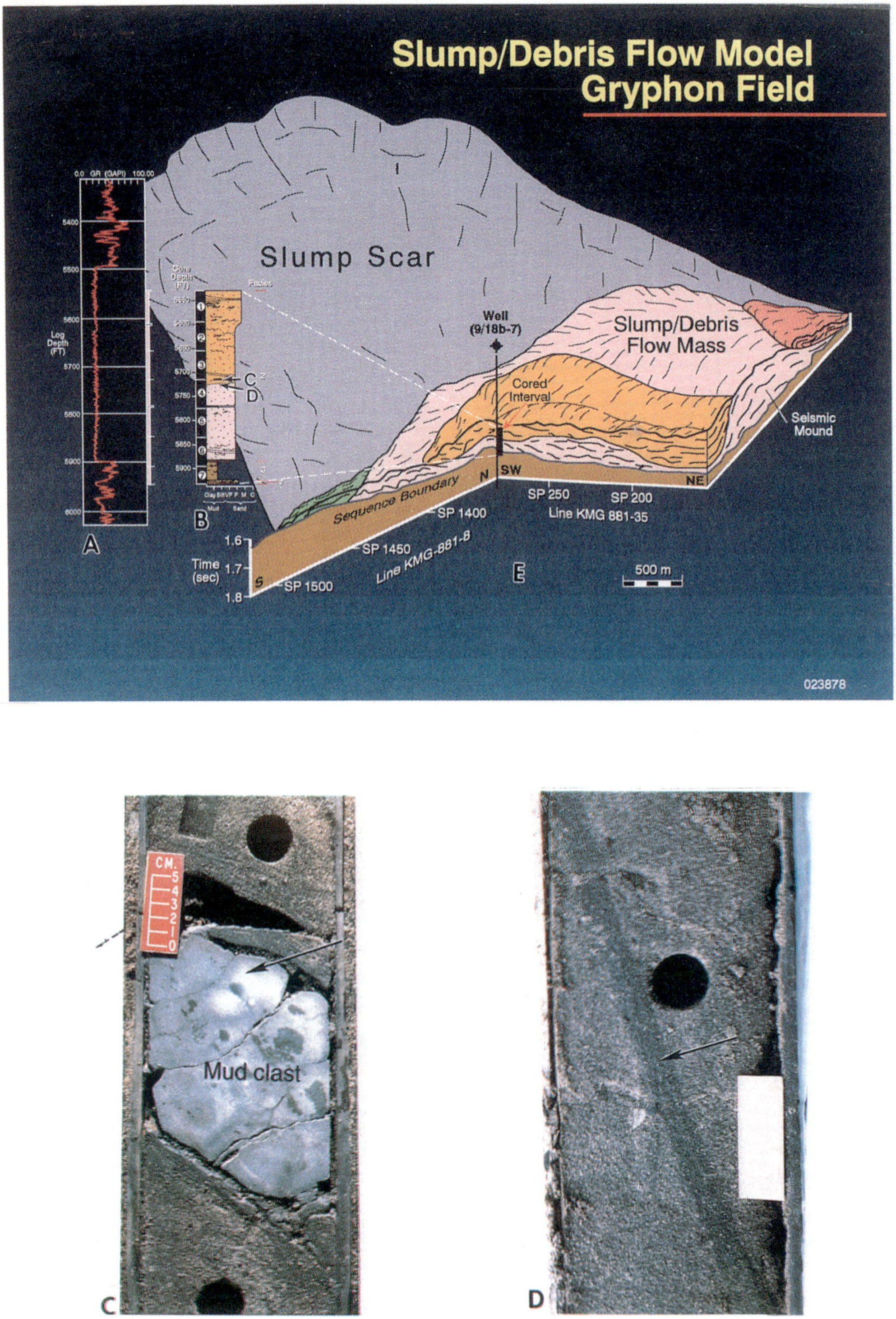

Slump/Debris Flow Model
Gryphon Field
0.0 GR (GAPI) 100.00
Log Depth (FT)
Slump Scar
Core Depth (FT)
Facies
C
D
Well (9/18b-7)
Slump/Debris Flow Mass
Cored Interval
Seismic Mound
N
SW
NE
Sequence Boundary
SP 1400
SP 1450
SP 1500
SP 250
SP 200
Line KMG 881-35
Line KMG-881-8
Time (sec)
1.6
1.7
1.8
S
A
B
E
500 m
023878

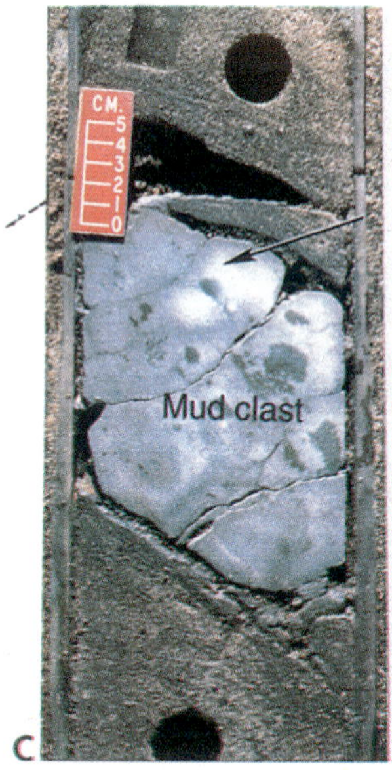
CM.
5
4
3
2
1
0
Mud clast
C

D

in the Gryphon area has been previously proposed by Newman *et al.* (1993). Outcrop studies in other areas also indicate that slump and debris-flow facies are laterally restricted with a thickness-to-width ratio of 1:30–50 (Cook 1979). Macdonald *et al.* (1993) described a seismic-scale slump sheet (350 m thick and 3.5 km wide) with a thickness-to-width ratio of 1:10 from the Pluto Glacier Formation (Aptian–Albian) exposed in Alexander Island of Antarctica. In such cases, long-distance correlation of sands showing blocky log motifs (Fig. 16a) under the premise that they represent sheet-like turbidites of basin-floor fans, would result in an erroneous estimate of distribution and continuity of sand in a reservoir.

Seismic mounds with blocky wire-line log motifs similar to those observed in the Gryphon area are also observed in the adjacent Forth Field (Shanmugam *et al.* 1995). Long continuous cores from the Forth Field indicate that the lower blocky log interval is dominated by slump and debris-flow facies (Table 4); whereas, the upper serrated log interval is dominated by mudstone with sandstone dykes and sills. Although the reservoir sand in the Forth Field has been interpreted to represent a submarine channel system filled with stacked turbidites (Jager *et al.* 1993), our core data suggest that slump and debris-flow facies are predominant. Based on our observations, we infer a non-channelized slope environment for the Gryphon–Forth area. Regional mapping shows that the Gryphon–Forth area sands occur about 15–20 km seaward of the age-equivalent palaeo-shelf-edge. This slope setting would be expected to be the site of major sediment failures that generated slides, slumps and debris flows (see Fig. 2).

An evaluation of the basin-floor fan concept based on conventional core data

Sedimentary facies of basin-floor fans

In the context of the Vail/Exxon conceptual sea-level model basin-floor fans are predicted to be very sand-rich features composed mainly of turbidites, which occur as laterally consistent lobes or sheets in a basinal setting (Posamentier & Vail 1988; Vail *et al.* 1991; Vail & Wornardt 1990; Mitchum *et al.* 1993). In contrast, our core studies reveal that turbidites are extremely rare (<1%) in the basin-floor fan features we have studied. Mass-transport deposits, especially slumps, slides and debris flows, are clearly the predominant deposits in these features (Fig. 19). Sands with contorted bedding (Facies 2) comprise nearly 90% of all the cored intervals we have examined (see Tables 2, 3, and 4). Bottom current deposits (Facies 4) are of minor importance and turbidites (Facies 5) are essentially absent (Fig. 19). Most of the sediment deposition took place in slope environments; basinal environments dominated by submarine-fan deposition (i.e. turbidites) were either absent or greatly restricted.

It could be argued that some of the sands in the mass-transport deposits of our study areas were originally emplaced by turbidity currents, then later remobilized by sliding and slumping. If this were the case, we would expect some of the sands in the slump and slide units to retain original turbidite features (e.g. normal grading) because many portions of slides and slumps moved as coherent, relatively undeformed masses (see Fig. 2), and pre-slump depositional features, such as turbidites, would be preserved in these deformed units. However, we have observed no sedimentary evidence of turbidites in these sections. In any case, the question of whether these sands were originally emplaced as turbidites prior to slumping is irrelevant to hydrocarbon exploration and production because the present sand-body geometry is strictly controlled by the most recent depositional event, which is slumping.

Interpretation of basin-floor and slope fans from mounded forms on seismic profiles

Basin-floor fans. Basin-floor fans generally appear on seismic profiles as mound-shaped forms in lowstand systems tracts that downlap onto sequence boundaries. They are generally portrayed as mound-like features in conceptual diagrams (Fig. 1). For

Fig. 18. (**a**) Well-developed blocky log motif from a lower Eocene sand unit in the Gryphon Field, Kerr–McGee 9/18b-7 (see Fig. 16a). (**b**) Depth-tied sedimentological core log showing facies distribution (see Fig. 16b). See Tables 1 and 3 for explanation and distribution of facies. Stratigraphical positions of core intervals shown in (c) and (d) are shown by arrows. (**c**). Core photograph showing a large mudstone clast (arrow) in fine-grained sand (Facies 2). Note irregular upper surface of mudstone clast (5725 feet or 1746 m). (**d**) Core photograph showing steeply dipping layer (arrow) in Facies 2, interpreted as an internal shear plane at 5726.7 feet (1746.6 m). The interval of shear plane and associated large mudstone clasts correlates with the high-amplitude reflection that separates the pink and orange mounded seismic packages (see Fig. 17). (**e**) Schematic diagram showing the depositional model proposed for the Gryphon Field based on integration of core, log and seismic data. The mounded seismic facies represent slumps and debris flows. Note that the 9/18b-7 well is located at the intersection of seismic lines KMG-881-35 and KMG-881-8 (shown by sketches) and that the cored interval is shown by a solid black bar. From Shanmugam *et al.* (1995, reprinted by permission).

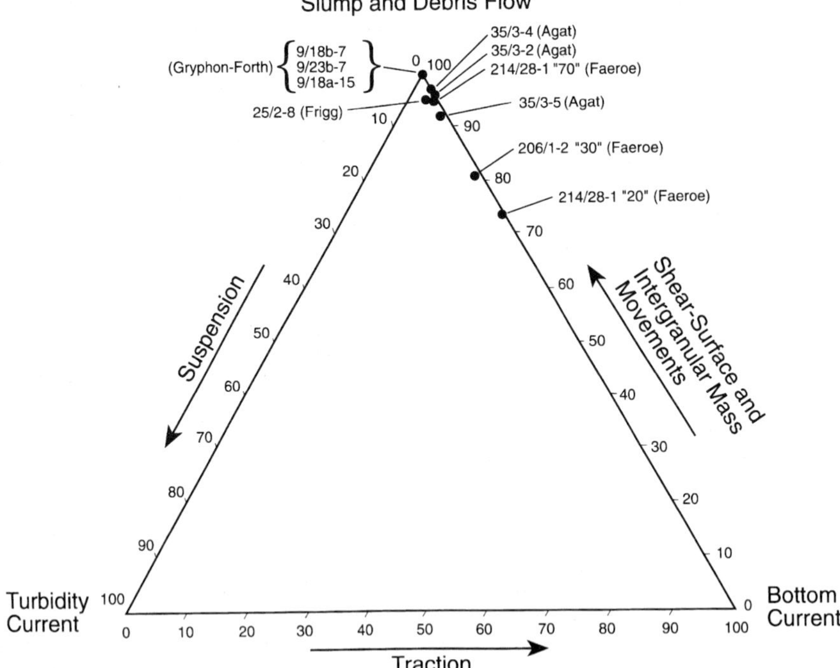

Fig. 19. Ternary diagram showing the volumetric abundance of mass-transport facies in our study areas. Plots are based on normalized percentages of resedimented facies from Tables 2, 3, and 4. Note the influence of bottom currents (contourite) in Sequences '20' and '30' in the Faeroe Basin.

example, mounded forms that downlap onto sequence boundaries in the Faeroe, Agat, Frigg, and Gryphon–Forth areas exhibit published criteria for basin-floor fans in seismic and log data; consequently, some of these features have been interpreted as basin-floor fans by previous investigators. In published literature such seismic mounds identified as basin-floor fans have usually been interpreted as submarine fans, or subenvironments of submarine fans. For example, Sarg & Skjold (1982) interpreted mounded forms in the Balder field as a submarine-fan complex. An individual mound was interpreted as a 'suprafan lobe' containing an 'inner-fan' with fining upward channelized deposits and an outer 'fan-fringe' with coarsening upward sheet sands. McGovney & Radovich (1985) proposed essentially the same submarine-fan model for mounded forms in the Frigg Field. Mitchum (1985) proposed an integrated fan model in which sand-rich mounded forms, such as those of the Balder and Frigg fields, are actually sandy depositional lobes of the lower fan subenvironment. Such lobes occur at the downslope ends of more mud-rich leveed distributary channels, which extend downslope across the upper and middle fan. In the sequence-stratigraphical conceptual model (Fig. 1), Mitchum's lower-fan lobes would correspond to the basin-floor fan; whereas, his upper-fan leveed channels would correspond to the slope fan.

It is not clear why sheet-like sand bodies or lobes on the lower portion of a submarine fan (such as those in Mitchum's models) would appear as mounded features with relatively steep (up to 15°) convex-upward upper surfaces in seismic sections (Shanmugam 1990; Shanmugam & Moiola 1991). It has been postulated that differential compaction between sand-rich and adjacent mud-rich portions of fans might increase the steepness of these features to produce the observed mounded geometry (e.g. Heritier *et al.* 1979; Mitchum 1985; Swarbrick 1991). In the models of Sarg & Skjold (1982), McGovney & Radovitch (1985), and Mitchum (1985), an entire mound consists of sand. Therefore, a large amount of differential compaction throughout an individual mound should not be expected. This suggests that the relatively steep flanks of a mound were formed during deposition, although they could have subsequently been steepened through erosion by contour currents (e.g. Vail *et al.* 1991; Normark *et al.* 1993).

Unconfined turbidity currents, whether flowing unconfined down a slope to the basin floor or spreading laterally outward from the mouth of a fan channel across the lower fan, will, because of their fluidal nature, seek the areas of lowest topography and pond in them (e.g. compensation cycles of Mutti & Sonnino 1981). Therefore, it is difficult to explain how unconfined turbidity currents could deposit steep mound-shaped forms with several to tens of metres of relief above the existing sea floor. This is further evidenced by the sand-rich turbidite deposits or lobes on the lower portions of modern deep-sea fans, which would be equivalent to the lower-fan mounded deposits in Mitchum's (1985) model. These lower-fan deposits show no evidence of mounded geometry or sea-floor relief on seismic data of any scale or resolution (e.g. Damuth *et al.* 1988; O'Connell *et al.* 1991; Shanmugam & Moiola 1991). These observations suggest that basinal or lower-fan deposits dominated by sand-rich turbidites are unlikely to form mounded forms with significant relief and relatively steep slopes (up to 15°) similar to those observed on standard industry seismic, which have been interpreted as basin-floor fans.

A more plausible explanation for the origin of many mounded forms observed on seismic data is that they represent mass-transport deposits. Our core studies, including the examples presented above, suggest that seismic mounds in our study areas are composed primarily of slumps (Figs 8–10 & 18) and debris-flow deposits (Figs 8, 10 & 16). The predominance of slumps and debris flows in many areas we studied suggests that deep-water mounded seismic facies in other areas of the North Sea may also be mass-transport deposits. Unlike fluidal turbidity currents, the plastic rheology and flow behaviour (like wet concrete) of slumps and debris flows allows them to form deposits with topographic relief above the existing sea floor (e.g. Jacobi 1976; Embley & Jacobi 1977; Embley 1980). Quaternary debris-flow slices have been reported to show mounded forms in seismic profiles (Hiscott & Aksu 1994). This might explain the unusually high angles of slope (up to 15°) observed on the flanks of seismic mounds (Fig. 17) as well as the rapid lateral changes in sediment thickness.

Because of these complexities, interpretation of seismic mounds as sand-rich basin-floor fans composed of laterally consistent turbidites needs to be re-evaluated. At present, our understanding of the origin and sedimentary composition of seismic mounds is poor because of insufficient core studies of these features. As our core studies reported here show, seismic mounds appear to be composed largely of mass-transport deposits. Only through additional studies involving systematic integration of core, wire-line log and seismic data, can we begin to understand seismic mounds in terms of their sand-prone nature, geometry, and depositional processes.

Slope fans. Vail *et al.* (1991) consider 'complex seismic mounds' with steep flanks (Fig. 1), also termed 'gull-wing' mounds and 'bow-tie' reflections, indicative of slope fans. These mounds within the slope fan have generally been interpreted as mud-rich distributary channel-levee systems on the upper and middle portions of submarine fans (e.g. Mitchum 1985; Mitchum *et al.* 1990, 1993; Vail & Wornardt 1990; Vail *et al.* 1991). The concept of associating 'gull-wing' mounds with channel-levee systems was derived from observations of channel-levee systems on modern fans, such as, the Amazon (e.g. Fig. 20; Damuth *et al.* 1988; Flood *et al.* 1991). The 'gull-wing' mounded shape associated with such features is readily recognizable in seismic profiles from modern fans, which are normally displayed at high vertical exaggerations (e.g. VE = 20 :1, top profile, Fig. 20). However, vertical exaggerations are much lower (VE < 5:1; Fig. 20) on standard industry seismic data causing the 'gull-wing' shaped mound to be either very subtle or no longer resolvable. This is because the channel-levee systems which give rise to these mounded forms are actually features of very subtle relief (Fig. 20). Although locally slopes on the backsides of levees can be up to several degrees, the overall slopes of levees generally average about 1° (Damuth *et al.* 1988). Furthermore, the 'gull-wing' shape of a channel-levee system is the manifestation of overbank mud build-up to form the levees. After burial and compaction, deep-marine channel-levee systems may be reduced to interbedded sandstone and mudstone in the rock record (e.g. Eocene Peira-Cava Sandstone, France; see Shanmugam & Moiola 1991, fig. 9), which may not show the presence of 'gull-wing' shaped mounds when imaged in seismic profiles.

Interpretation of relatively steep-sided mounded forms ('gull-wing' and otherwise) on industry seismic data as channel-levee systems within slope fans has to be made with caution. When making such interpretations, the true relief, size and scale of real fan channel-levee systems, such as the one illustrated in Fig. 20, must be considered. Examples of mounds interpreted as channel-levee systems are beginning to appear in the published literature (e.g. Pacht *et al.* 1990, fig. 7a) which have flanks that appear to be far too steep to be actual channel-levee systems, even if differential compaction has occurred. Such steep-sided, high relief mounds are unlikely to be channel-levee systems. We suggest that such seismic mounds with steep flanks are actually deposits of slumps and debris flows in most cases. Slumps are more likely to create mounded forms with steep flanks than any other deep-marine deposits. Perhaps misinterpretation of channel-levee systems is, in part, due to the fact that on conceptual diagrams such as Fig. 1, channel-levee systems are always represented as very large,

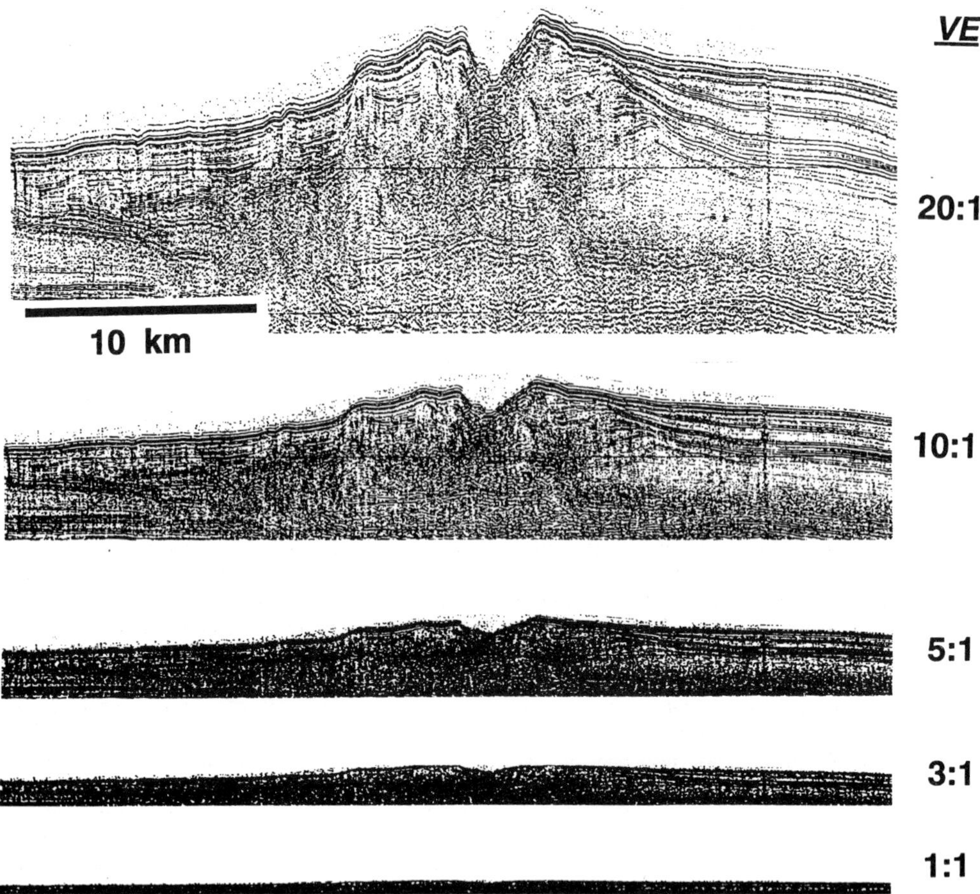

Fig. 20. Watergun seismic reflection profile of a large channel-levee system on the Amazon deep-sea fan displayed at various vertical exaggerations including 1:1, which demonstrates the very subtle sea-floor relief of fan channel-levee systems (from Damuth *et al.* 1995). Top profile (VE = 20:1) shows well developed 'gull-wing' mounded form. Note that at VE =3:1, which is the vertical exaggeration commonly used to display multifold seismic data utilized by industry, the 'gull-wing' mound shape is nearly unresolvable.

steep-sided features with high relief. This misleads interpreters into looking for very large, steep-sided mounds on seismic. Interpreters should take into account the true scale and geometry of channel-levee systems when making interpretations (e.g. Fig. 20; see also Damuth *et al.* 1995). In addition, the feature being interpreted should be mapped in plan view to determine if it has a sinuous planform consistent with that expected for a channel-levee system.

Interpretation of basin-floor fans from wireline-log motifs

In the conceptual sea-level model, the sand-rich turbidites of the basin-floor fan generate a blocky log motif on wire-line logs (Mitchum *et al.* 1990, 1993; Vail & Wornardt 1990; Vail *et al.* 1991). However, our core study indicates that a variety of depositional facies and processes, not just sandy turbidites, can generate blocky log motifs (Figs 8, 10,

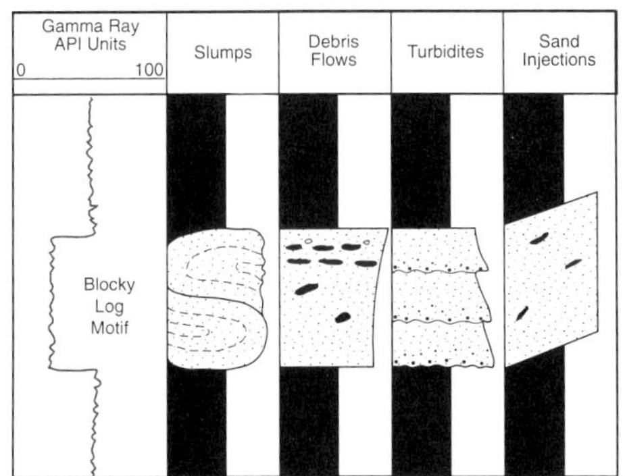

Fig. 21. Conceptual diagram showing that a variety of deep-water deposits or facies can return blocky log motifs.

16). Conversely, a single depositional facies (e.g. sandy slump) can generate a variety of log motifs including serrated and blocky motifs (Figs 8, 15 and 16). Furthermore, facies formed by post-depositional processes such as sand injections can generate blocky log motifs (Fig. 21). In summary, wire-line log motifs in deep-water facies simply reflect changes in lithology and fluid content, and a single motif may actually represent more than one depositional process or facies (Fig. 21). To date, there has been no systematic calibration of depositional facies observed in conventional cores with wire-line log data and motifs. Until such calibration is completed, the validity of using log motifs to predict depositional facies in basin-floor fans or 'slope fans' will remain uncertain.

Conclusions

(1) Process sedimentological analysis of siliciclastic hydrocarbon reservoirs of deep-water origin in the North Sea indicates that they are predominantly composed of mass-transport deposits emplaced by sandy slumps, slides, and debris flows. Classic or low-density turbidites are extremely rare. Other investigators might classify our sandy debris-flow deposits as high-density turbidites, however, we contend that these deposits are more appropriately termed sandy debris flows because of the following properties: (a) concentration of rafted mudstone clasts near the tops of sandstone beds (flow strength, rigid plug); (b) inverse grading of clasts (flow strength and buoyant lift); (c) floating quartz granules in fine-grained sandstone (pseudo plastic flow); (d) planar clast fabric (laminar flow); (e) sharp and irregular upper contacts observed in core and lateral pinch-out geometries observed in seismic profiles (freezing of primary relief), and (f) primary detrital matrix (plastic flow).

(2) Distinguishing deposits of plastic flows (i.e. debris flows) from deposits of fluidal flows (i.e. turbidity currents) is extremely important to the petroleum industry for prediction of reservoir geometry. Plastic flows emplace sediment by freezing, which can result in the formation of discontinuous sand bodies that are harder to predict. In contrast, fluidal turbidity currents can deposit laterally continuous sheet-like sand bodies that are predictable.

(3) The conceptual sequence stratigraphical model predicts that basin-floor fans and slope fans form during specific phases of a sea-level cycle. This model further predicts that basin-floor fans and slope fans contain very distinctive and, thus, predictable, sediment facies associations and elements, which are controlled by the position and rate of relative sea-level fall. Our studies of actual sedimentary facies cored from features interpreted as basin-floor fans indicate that these features are composed of facies that are substantially different and more variable than those predicted by the conceptual model. Sand content varies from 35% to 90% of the cored intervals. Instead of laterally consistent sheets of sand-rich turbidites predicted by this model, our studies show that the basin-floor fans we have studied to date consist predominantly of a variety of mass-transport deposits, especially slumps and debris flows. The sandy deposits of basin-floor fans are much more complex, less laterally extensive, and therefore are much harder to predict.

(4) Our core study underscores the complexities of deep-water depositional systems and indicate that model-driven interpretation of remotely sensed data, such as mounded forms on seismic and blocky motifs on wireline logs (e.g. basin-floor fans), in terms of specific sedimentary facies (e.g. turbidites) and depositional features (e.g. submarine fans) should proceed with caution. While features classified as basin-floor fans may occur at specific and predictable stratigraphical positions and exhibit characteristic seismic facies and wireline-log motifs, our study indicates that these features do not necessarily represent specific and therefore predictable lithologies (i.e., sand-prone vs. mud-prone) or depositional facies (i.e., turbidite vs. debris flow).

We thank S.P. Hesselbo and D.N. Parkinson for inviting us to participate in the conference 'Progress in Sequence Stratigraphy' held at the Geological Society in London, March 1994. We also thank K. T. Pickering and T. Marjanac for critical review of the manuscript; R.E. Dunay, K.P. Dean, S. H. Gabay, H. C. Olson, C. E. Sheppard, J. F. Sarg and R. J. Moiola for comments on an earlier version of this manuscript; L.M. Murdoch, G.A. Hird, E. M. Leavitt, M. A. Northam, K. C. King, O. K. Johansen, and G. K. Baker for their managerial support; and M. K. Lindsey and A. F. Long for drafting. We thank I.R. Hatton, Exploration Manager, Kerr-McGee (UK) and their partners for permission to publish their seismic profiles from the Gryphon Field area and J. Brewster for granting permission to publish seismic profiles from the Frigg Field area. We thank Mobil for granting permission to publish this paper and Mobil Exploration Norway Inc for paying the printing cost of colour figures. The opinions expressed herein are solely those of the authors, and do not necessarily reflect the views of Mobil Oil Corporation or its affiliates.

References

BATTIE, J. E., SATERSMOEN, B. H. & AUBERT, K. 1994. Frigg Field – A turbidite mass flow or a massive slump? *European Association of Exploration Geophysicists, 56th Meeting and Technical Exhibition*, Extended Abstract, P077.

BOUMA, A. H. 1962. *Sedimentology of some flysch deposits: A graphic approach to facies interpretation*. Elsevier, Amsterdam.

BREWSTER, J. 1991. The Frigg field, Block 10/1 UK North Sea and 25/1, Norwegian North Sea. *In*: ABBOTTS, I. L. (ed.) *United Kingdom Oil and Gas Fields, 25 Years Commemorative Volume*. Geological Society, London, Memoirs, **14**, 117–126.

BUGGE, T. 1983. *Submarine slides on the Norwegian continental margin, with special emphasis on the Storegga area*. IKU Continental Shelf Institute Publication, **110**, Norway.

COOK, H. E. 1979. Ancient continental slope sequences and their value in understanding modern slope development. *In*: DOYLE, L. J. & PILKEY, O. H. (eds) *Geology of Continental Slopes*. Society of Economic Paleontologists and Mineralogists, Special Publications, **27**, 287–305.

DAMUTH, J. E. 1978. Echo character of the Norwegian-Greenland Sea: Relationship to Quaternary sedimentation. *Marine Geology*, **28**, 1–36.

—— & EMBLEY, R. W. 1981. Mass-transport processes on Amazon Cone: Western Equatorial Atlantic. *American Association of Petroleum Geologists Bulletin*, **65**, 629–643.

—— & OLSON, H. C. 1993. Preliminary observations of Neogene-Quaternary depositional processes in the Faeroe–Shetland Channel revealed by high-resolution seismic facies analysis. *In*: PARKER, J. R. (ed.) *Petroleum Geology of Northwest Europe: Proceedings of the 4th Conference*. Geological Society, London, 1035–1045.

——, FLOOD, R. D. KOWSMANN, R. O. BELDERSON, R. H. & GORINI, M. A. 1988. Anatomy and growth-pattern of Amazon deep-sea fan revealed by long-range side-scan sonar (GLORIA) and high-resolution seismic studies. *American Association of Petroleum Geologists Bulletin*, **72**, 885–911.

——, ——, PIRMEZ, C. & MANLEY, P. L. 1995. Architectural elements and depositional processes of Amazon Deep-Sea Fan imaged with long-range side-scan sonar (GLORIA), bathymetric swath-mapping (SeaBEAM) and high-resolution seismics. *In*: PICKERING, K. T., HISCOTT, R. N., KENYON, N. H., RICCI LUCCHI, F. & SMITH, R. D. A. (eds) *Atlas of Deep-Water Environments: Architectural Style in Turbidite Systems*. Chapman and Hall, London, 105–121.

DINGLE, R. V. 1977. The anatomy of a large submarine slump on a sheared continental margin (SE Africa). *Journal of Geological Society, London*, **134**, 293–310.

EMBLEY, R. W. 1980. The role of mass transport in the distribution and character of deep-ocean sediments with special reference to the North Atlantic. *Marine Geology*, **38**, 23–50.

—— & JACOBI, R. D. 1977. Distributions and morphology of large sediment slides and slumps on Atlantic continental margins. *Marine Geotechnology*, **2**, 205–228.

FISHER, R. V. 1971. Features of coarse-grained, high-concentration fluids and their deposits. *Journal of Sedimentary Petrology*, **41**, 916–927.

FLOOD, R. D., MANLEY, P. L., KOWSMANN, R. O., APPI, C. J. & PIRMEZ, C. 1991. Seismic facies and Late Quaternary growth of Amazon submarine fan. *In*: WEIMER, P. & LINK, M. H. (eds) *Seismic Facies and Sedimentary Processes of Submarine Fans and Turbidite Systems*. Springer-Verlag, New York, 415-433.

GAWTHORPE, R. L. & CLEMMEY, H. 1985. Geometry of submarine slides in the Bowland Basin (Dinantian) and their relation to debris flows. *Journal of the Geological Society, London*, **142**, 555–565.

GULBRANDSEN, A. 1987. Agat. *In*: SPENCER, A. M., HOLTER, E., CAMPBELL, C. J., HANSLIEN, S. H., NELSON, P. H. H., NYSAETHER, E. & ORMAASEN, E. G. (eds) *Geology of the Norwegian Oil and Gas Fields*. Graham & Trotman, London, 363–370.

HAMPTON, M. A. 1972. The role of subaqueous debris flows in generating turbidity currents. *Journal of Sedimentary Petrology*, **42**, 775–793.

HELWIG, J. 1970. Slump folds and early structures, northeastern Newfoundland Appalachians. *Journal of Geology*, **78**, 172–187.

HERITIER, F. E., LOSSEL, P. & WATHNE, E. 1979. Frigg field – large submarine-fan trap in lower Eocene rocks of North Sea. *American Association of Petroleum Geologists Bulletin*, **63**, 1999–2020.

HISCOTT, R. N. 1994. Traction-carpet stratification in turbidites – fact or fiction?. *Journal of Sedimentary Research*, **A64**, 204–208.

—— & AKSU, A. E. 1994. Submarine debris flows and continental slope evolution in front of quaternary ice sheets, Baffin Bay, Canadian Arctic. *American Association of Petroleum Geologists Bulletin*, **78**, 445–460.

HOLLISTER, C. D. 1967. *Sediment distribution and deep circulation in the western North Atlantic*. PhD thesis, Columbia University, New York.

—— & HEEZEN, B. C. 1972. Geologic effects of ocean bottom currents: western north Atlantic. *In*: GORDON, A. L. (ed.) *Studies in Physical Oceanography, Volume 2*. Gordon and Breach Science Publishers, New York, 37–66.

IMBRIE, J., HAYS, J. D., MARTINSON, D. G., MCINTYRE, A., MIX, A. C., MORLEY, J. J., PISIAS, N. G., PRELL, W. L. & SHACKLETON, N. J. 1984. The orbital theory of Pleistocene climate: Support for a revised chronology of the marine $\delta^{18}O$ record. *In*: BERGER, A. L., IMBRIE, J., HAYS, J. D., KUKLA, G. & SALTZMAN, B. (eds). *Milankovitch and Climate, Part I*. NATO ASI

Series C: Mathematical and Physical Sciences, **126**, 269–305.

JACKSON, M. J., DIEKMAN, L. J., KENNARD, J. M., SOUTHGATE, P. N., O'BRIEN, P. E. & SEXTON, M. J. 1992. Sequence stratigraphy, basin-floor fans and petroleum plays in the Devonian–Carboniferous of the northern Canning basin. *Journal of the Australian Petroleum Exploration Society*, **32**, 214–230.

JACOBI, R .D. 1976. Sediment slides on the northwestern continental margin of Africa. *Marine Geology*, **22**, 157–173.

JAGER, D. D., GILES, M. R. & GRIFFITHS, G. R. 1993. Evolution of Paleogene fans of the North Sea in space and time. *In*: PARKER, J. R. (ed.) *Petroleum Geology of Northwest Europe: Proceedings of the 4th Conference*. Geological Society, London, 59–71.

JANSEN, E., BEFRING, S., BUGGE, T., EIDVIN, T., HOLTEDAHL, H. & PETTER SEJRUP, H. 1987. Large submarine slides on the Norwegian continental margin: Sediments, transport and timing. *Marine Geology*, **78**, 77–107.

JOHNSON, A. M. 1970. *Physical Processes in Geology* Freeman, Cooper and Co., San Francisco, 577.

KUENEN, P. H .H. 1950. Turbidity currents of high density. *18th International Geological Congress, London, Reports Part 8*, 44–52.

—— 1966. Matrix of turbidites: experimental approach. *Sedimentology*, **7**, 267–297.

LOWE, D. R. 1982. Sediment gravity flows: II. Depositional models with special reference to the deposits of high-density turbidity currents. *Journal of Sedimentary Petrology*, **52**, 279–297.

MACDONALD, D. I. M., MONCRIEFF, A. C. M. & BUTTERWORTH, P. J. 1993. Giant slide deposits from a Mesozoic fore-arc basin, Alexander Island, Antarctica. *Geology*, **21**, 1047–1050.

MALTMAN, A. J. 1987. Microstructures in deformed sediments, Denbigh Moors, North Wales. *Geological Journal*, **22**, 87–94.

MARTINSEN, O. J. 1989. Styles of soft-sediment deformation on a Namurian (Carboniferous) delta slope, Western Irish Namurian Basin, Ireland. *In*: WHATELEY, K. G. & PICKERING, K. T. (eds) *Deltas: Sites and Traps for Fossil Fuels*. Geological Society, London, Special Publications, **41**, 167–177.

MCGOVNEY, J. E. & RADOVICH, B. J. 1985. Seismic stratigraphy and facies of the Frigg fan complex. *In*: BERG, O. R. & WOOLVERTON, D. G. (eds) *Seismic stratigraphy II: An Integrated Approach to Hydrocarbon Exploration*. American Association of Petroleum Geologists, Memoirs, **39**, 139–154.

MIDDLETON, G. V. 1967. Experiments on density and turbidity currents: III. Deposition of sediment. *Canadian Journal of Earth Sciences*, **4**, 475–505.

—— & HAMPTON, M. A. 1973. Sediment gravity flows: Mechanics of flow and deposition. *In*: MIDDLETON, G. V. & BOUMA, A. H. (eds) *Turbidites and deep-water sedimentation*. Pacific Section, Society of Economic Paleontologists and Mineralogists, Los Angeles, 1–38.

MITCHELL, S. M., BEAMISH, G. W. J., WOOD, M. V., MALECEK, S. J. ARMENTROUT, J., DAMUTH, J. E. & OLSON, H. C. 1993. Paleogene sequence stratigraphic framework of the Faeroe Basin. *In*: PARKER, J. R. (ed.) *Petroleum Geology of Northwest Europe: Proceedings of the 4th Conference*. Geological Society, London, 1011–1023.

MITCHUM, R. M., JR. 1985. Seismic stratigraphic expression of submarine fans. *In*: BERG, O. R. & WOOLVERTON, D. G. (eds) *Seismic stratigraphy II: An Integrated Approach to Hydrocarbon Exploration*. American Association of Petroleum Geologists, Memoirs, **39**, 117–136.

——, SANGREE, J. B., VAIL, P. R. & WORNARDT, W. W. 1990. Sequence stratigraphy in Late Cenozoic expanded sections, Gulf of Mexico. *11th Annual Research Conference, Gulf Coast Society of Economic Paleontologists and Mineralogists Foundation*, 237–256.

——, ——, —— & —— 1993. Recognizing sequences and systems tracts from well logs, seismic data, and biostratigraphy: Examples from Late Cenozoic of the Gulf of Mexico. *In*: WEIMER, P. & POSAMENTIER, H. (eds) *Siliciclastic Sequence Stratigraphy: Recent Developments and Applications*. American Association of Petroleum Geologists, Memoirs, **58**, 163–197.

MUTTI, E. 1977. Distinctive thin-bedded turbidite facies and related depositional environments in the Eocene Hecho Group (south-central Pyrenees, Spain). *Sedimentology*, **24**, 107–131.

—— & SONNINO, M. 1981. Compensation cycles: a diagnostic feature of sandstone lobes. *International Association of Sedimentologists, 2nd European Meeting, Abstracts, Bologna, Italy*, 120–123.

NEWMAN, M. ST., LEEDER, M. L., WOODRUFF, A. H. W. & HATTON, I. R. 1993. The geology of the Gryphon Field. *In*: PARKER, J. R. (ed.) *Petroleum Geology of Northwest Europe: Proceedings of the 4th Conference*. Geological Society, London, 123–133.

NORMARK, W. R., POSAMENTIER, H. W. & MUTTI, E. 1993. Turbidite systems: state of the art and future directions. *Review of Geophysics*, **31**, 91–116.

O'CONNELL, S., RYAN, W. B. F. & NORMARK, W. R. 1991. Evolution of a fan channel on the surface of the outer Mississippi Fan: Evidence from side-looking sonar. *In*: WEIMER, P. & LINK, M. H. (eds) *Seismic Facies and Sedimentary Processes of Submarine Fans and Turbidite Systems*. Springer-Verlag, New York, 365–381.

PACHT, J. A., BOWEN, B. E., SHAFFER, B. L. & POTTORF, B. R. 1990. Sequence stratigraphy of Plio-Pleistocene strata in the offshore Louisiana Gulf Coast: Applications to hydrocarbon exploration. *In*: *Sequence Stratigraphy as an Exploration Tool, Concepts and Practices in the Gulf Coast*. 11th Annual Research Conference, Gulf Coast Society of Economic Paleontologists and Mineralogists Foundation, 269–285.

PICKERING, K. T. HISCOTT, R. N. & HEIN, F. J. 1989. *Deep-Marine Environments*. Unwin Hyman, London, 416.

PIERSON, T. C. & COSTA, J. E. 1987. A rheologic classification of subaerial sediment-water flows, *In*: COSTA, J. E. & WIECZOREK, G. F. (eds) Debris flows/ Avalanches: Process, Recognition, and Mitigation: *Geological Society of America Reviews in Engineering Geology*, **VII**, 1–12.

POSAMENTIER, H. W. & VAIL, P. R. 1988. Eustatic controls

on clastic deposition II – Sequence and systems tract models. *In*: WILGUS, C. K., HASTINGS, B. S., KENDALL, C. G. ST. C., POSAMENTIER, H. W., ROSS, C. & VAN WAGONER, J. C. (eds) *Sea-Level Changes: An Integrated Approach*. Society of Economic Paleontologists and Mineralogists, Special Publications, **42**, 125–154.

—— & ERSKINE, R. D. 1991. Seismic expression and recognition criteria of ancient submarine fans. *In*: WEIMER, P. & LINK, M. H. (eds) *Seismic Facies and Sedimentary Processes of Submarine Fans and Turbidite Systems*. Springer-Verlag, New York, 197–222.

—— & WEIMER, P. 1993. Siliciclastic sequence stratigraphy and petroleum geology – where to from here? *American Association of Petroleum Geologists Bulletin*, **77**, 731–742.

QIAN, Y. YANG, W. ZHAO, W. CHENG, X. ZHANG, L. & XU, W. 1980. Basic characteristics of flow with hyperconcentrations of sediment, *In*: *Proceedings of the International Symposium on River Sedimentation*. Beijing, Chinese Society of Hydraulic Engineering, 175–184.

REYNOLDS, J. & MACKAY, T. 1992. Post depositional modification of deep water sandstones – reservoir geometry of the Gryphon sand, Gryphon Field, U.K. Sector 9/18b. *In*: *Deep Water Massive Sands*. Arthur Holmes European Research Conferences Programme, 19.

SARG, J. F. & SKJOLD, L. J. 1982. Stratigraphic traps in Paleocene sands in the Balder area, North Sea. *In*: HALBOUTY, M. T. (ed.) *The Deliberate Search for the Subtle Trap*. American Association of Petroleum Geologists, Memoirs, **32**, 197–206.

SHANMUGAM, G. 1990. Deep-marine facies models and the interrelationship of depositional components in time and space. *In*: BROWN, G. C., GORSLINE, D. S. & SCHWELLER, W. J. (eds) *Deep-Marine Sedimentation: depositional models and case histories in hydrocarbon exploration and development*. Society of Economic Paleontologists and Mineralogists, Pacific Section, San Francisco, 199–246.

—— 1996, High-density turbidity currents: are they sandy debris flows? *Journl of Sedimentary Research*, **A66**, in press.

—— & BENEDICT, G. L. 1978. Fine-grained carbonate debris flow, Ordovician basin margin, Southern Appalachians. *Journal of Sedimentary Petrology*, **48**, 1233–1240.

—— & MOIOLA, R. J. 1991. Types of submarine fan lobes: models and implications. *American Association of Petroleum Geologists Bulletin*, **75**, 156–179.

——, BLOCH, R. B., MITCHELL, S. M., BEAMISH, G. W. J. HODGKINSON, R. J., DAMUTH, J. E., STRAUME, T., SYVERTSEN, S. E. & SHIELDS, K. E. 1995. Basin-floor fans: Sequence stratigraphic models vs. sedimentary facies. *American Association of Petroleum Geologists Bulletin*, **79**, 477–512.

——, LEHTONEN, L. R., STRAUME, T., SYVERSTEN, S. E., HODGKINSON, R. J. & SKIBELI, M. 1994. Slump and debris flow dominated upper slope facies in the Cretaceous of the Norwegian and Northern North Seas (61°–67°): Implications for Sand distribution. *American Association of Petroleum Geologists Bulletin*, **78**, 910–937.

——, SPALDING, T. D. & ROFHEART, D. H. 1993. Process sedimentology and reservoir quality of deep-marine bottom-current reworked sands (sandy contourites): an example from the Gulf of Mexico. *American Association of Petroleum Geologists Bulletin*, **77**, 1241–1259.

SHULZ, A. W. 1984. Subaerial debris-flow deposition in the Upper Paleozoic Cutler Formation, Western Colorado. *Journal of Sedimentary Petrology*, **54**, 759–772.

SKIBELI, M., BARNES, K., STRAUME, T., SYVERTSEN, S. E. & SHANMUGAM, G. 1992. A sequence stratigraphy study of the Lower Cretaceous deposits of the Norwegian North Sea. *Sequence Stratigraphy Forum*, Stavanger.

STEWART, I. J. 1987. A revised stratigraphic interpretation of the Early Palaeogene of the Central North Sea. *In*: BROOKS, J. & GLENNIE, K. W. (eds) *Petroleum Geology of North West Europe, Volume 1*. Graham & Trotman, London, 555–576.

SWARBRICK, R. E, 1991, Geological mounds and their seismic expression. *American Association of Petroleum Geologists Annual Convention Official Program*, 214.

TIMBRELL, G. 1993. Sandstone architecture of Balder Formation depositional system U.K. Quad 9 and adjacent areas. *In*: PARKER, J. R. (ed.) *Petroleum Geology of Northwest Europe: Proceedings of the 4th Conference*. Geological Society, London, 107–121.

VAIL, P. R. 1987. Seismic-stratigraphy interpretation using sequence stratigraphy. Pt. 1: Seismic stratigraphy interpretation procedures. *In*: BALLY, A. W. (ed.) *Atlas of Seismic Stratigraphy*. American Association of Petroleum Geologists Studies in Geology, **27**, 1, 1–10.

—— & WORNARDT, W. W. 1990. Well log–seismic sequence stratigraphy: an integrated tool for the 90's. *In*: *Sequence Stratigraphy as an Exploration Tool, Concepts and Practices in the Gulf Coast*. 11th Annual Research Conference, Gulf Coast Society of Economic Paleontologists and Mineralogists Foundation, 379–388.

——, AUDEMARD, F., BOWMAN, S. A., EISNER, P. N. & PEREZ-CRUZ, C. 1991. The stratigraphic signatures of tectonics, eustacy and sedimentology – an overview. *In*: EINSELE, G., RICKEN, W. & SEILACHER, A. (eds) *Cycles and Events in Stratigraphy*. Springer-Verlag, Berlin, 618–659.

VARNES, D. J. 1958. Landslide types and processes. *In*: ECKEL, E. D. (ed.) *Landslide and Engineering Practice*. Special Reports of the Highway Research Board, **29**, 20–47.

WEIMER, P. 1990. Sequence stratigraphy, facies geometries, and depositional history of the Mississippi Fan, Gulf of Mexico. *American Association of Petroleum Geologists Bulletin*, **74**, 425–453.

WOODCOCK, N. H. 1976. Structural style in slump sheets: Ludlow Series, Powys, Wales. *Journal of the Geological Society, London*, **132**, 399–415.

—— 1979. Sizes of submarine slides and their significance. *Journal of Structural Geology*, **1**, 137–142.

Turonian correlation and sequence stratigraphy of the Chalk in southern England

ANDREW S. GALE

Department of Palaeontology, Natural History Museum, Cromwell Road, London SW7 5BD and Department of Geology, Imperial College, Prince Consort Road, London SW7 2BP, UK

Abstract: A stratigraphical framework for the Turonian Chalk of southern England is described using lithostratigraphy, marker beds, cyclostratigraphy and biostratigraphy. For the Early and Mid-Turonian succession a basin-wide resolution of correlation of about a metre can be achieved. Sequences are identified from various criteria, which include bed-by-bed correlation along transects from platform to basin to provide a precise picture of sediment geometry in Turonian Chalk. Hardgrounds develop towards the platform margins and it is possible to identify some as lithified and commonly glauconitized sequence boundaries. For example, the Ogbourne Hardground, which represents a major Mid-Turonian sea-level fall, rests erosively on highstand chalks. Transgressive surfaces are also represented by hardgrounds; these onlap onto the basin margins and commonly show complex sedimentary histories and phosphatization. Transgressive events are marked by brief positive excursions in $\delta^{13}C$, detectable in basinal chalks. During the major Mid-Turonian sea-level lowstand, a thick 'shelf margin wedge' developed in the basin. A total of three complete and two partial sequences are identified in the Turonian of southern England. These compare well in number and biostratigraphical position with sequences described from a Turonian succession in Tunisia, which is an order of magnitude thicker.

The Turonian Age has long been identified as a time of major eustatic sea-level change and workers have generally agreed that the Early Turonian saw a major sea-level high, identified by Haq *et al.* 1988, fig. 15) as the highest sea-level stand of post-Palaeozoic time. In contrast, the Mid–Late Turonian sea-level fall registers the lowest sea levels known for the Late Cretaceous (Hancock & Kauffmann 1979). The latest Turonian shows widespread evidence of a further deepening episode, which continued into the Coniacian. Although this general pattern of eustatic change is widely agreed upon, the exact timing of deepenings and shallowings is debated. For example, Haq *et al.* (1988, fig. 14) record the Early Turonian maximum highstand as within the *Mammites nodosoides* Zone, whereas Hancock (1989) places it significantly higher, within the *Collignoniceras woollgari* Zone. One aim of this paper is to constrain the timing of eustatic changes precisely by using sequence stratigraphy.

The Turonian succession in the northern Anglo–Paris Basin consists of up to 100 m of nannofossil chalk with a variable coarser bioclastic component including inoceramid prisms, echinoderm debris, calcispheres and foraminifera. The chalks contain only a few per cent of clastic material, mostly clay, which is concentrated in thin marl beds. In deeper parts of the basin, chalk sedimentation continued with only minor interruptions through the Turonian, but on the northern margins of the basin, the Mid- and Upper Turonian are highly condensed and are represented by a group of hardgrounds called the Chalk Rock. Correlation of the Chalk Rock with basinal Turonian chalks should provide a key to the history of sea-level changes, but has hitherto proved elusive; a possible solution is provided here. Firstly, however, it is necessary to review and augment the stratigraphical framework for the Turonian Chalk of southern England. Localities mentioned in this paper, counties and the outcrop of the Upper Cretaceous are shown in Fig. 1.

Stratigraphical framework for the Turonian of southern England

Lithostratigraphy and marker bed stratigraphy

The past 15 years have seen a proliferation of lithostratigraphical schemes for the Chalk of southern England (Sussex, Mortimore 1983, 1986; Kent, Robinson 1986; Devon, Jarvis & Woodroof 1984, Jarvis & Tocher 1987) and the British Geological Survey now appears to be introducing new schemes for each remapped sheet (e.g. Westhead & Woods 1994). In 1987, Gale *et al.* proposed a compromise lithostratigraphical classification for the Chalk of southern England based on the historical precedent of nomenclature, but this was dismissed by Mortimore (1988) and largely ignored by other workers. The situation as it stands is fraught with confusion and requires rationalization. An attempt will be made to achieve this here for the Turonian Chalk.

From HESSELBO, S. P. & PARKINSON, D. N. (eds), 1996, *Sequence Stratigraphy in British Geology*, Geological Society Special Publication No. 103, pp. 177–195.

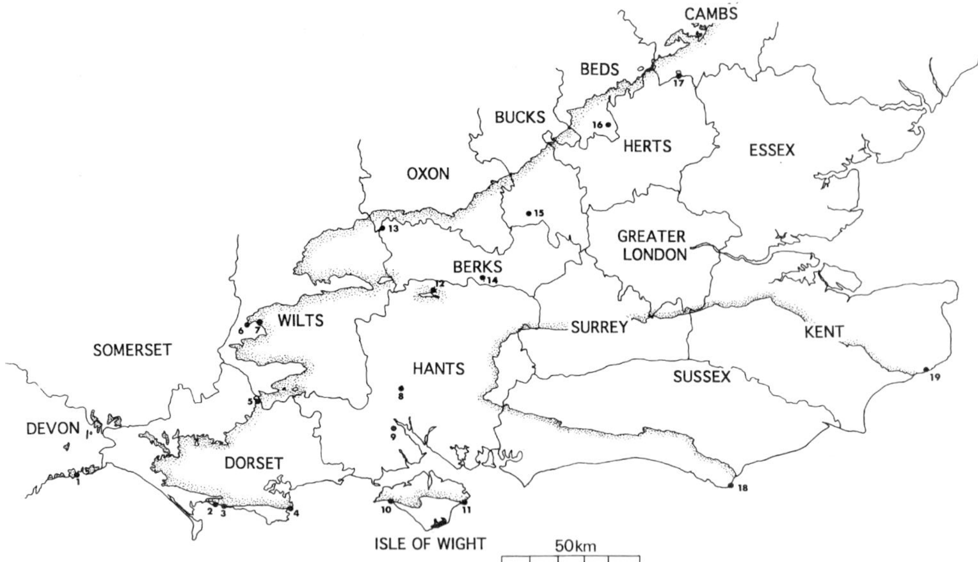

Fig. 1. Map of southern England showing counties and Upper Cretaceous outcrop (base stippled). Localities are: 1, cliffs immediately west of Beer harbour (Jarvis & Woodroof 1984); 2, White Nothe; 3, Lulworth Cove; 4, Ballard Cliff, Swanage; 5, Compton Down (Bromley & Gale 1982); 6, Cley Hill Warminster (Bromley & Gale 1982); 7, Beggar's Knoll, Westbury (Bromley & Gale 1982); 8, Twyford Down M3 cutting, Winchester; 9, Marchwood borehole; 10, Compton Bay; 11, Culver Cliff; 12, Burghclere, A34 cutting (Wray & Gale 1992); 13, Fognam Farm (Bromley & Gale 1982); 14, Faircross borehole; 15, Ewelme (Bromley & Gale 1982); 16, Luton Airport access road (Bromley & Gale 1982); 17, Beachy Head (Mortimore 1986); and 18, Aker's Steps and Langdon Stairs, Dover (Robinson 1986).

The succession at Beachy Head in Sussex, typical of the basinal development in southern England, is given in Fig. 2.

Ballard Cliff Member of the White Chalk Formation (= Melbourn Rock and base of Ranscombe Chalk, sensu *Mortimore 1983, 1986).* This new term includes beds from the top of the Plenus Marls up to the top of Meads Marl 6 of Mortimore (1986). The stratotype section is at Ballard Cliff, Dorset (SU 046 812, Fig. 3), where the member is well exposed and expanded. This basal member of the White Chalk Formation lies directly above the Plenus Marls and comprises 1–5 m of thinly bedded nodular chalk which contains abundant intraclasts, concentrated in marls and shallow scours. Numerous fine-textured green–grey marls are present, including both thin flaser partings and thicker (up to 10 cm) beds. Important marker beds include Meads Marls 1–6 and four beds containing the ammonite *Sciponoceras* in abundance. Correlation of this member is given in Fig. 3. In Devon, the member is represented by the highly condensed Haven Cliff Hardground and subjacent chalk (Jarvis & Woodroof 1984). The detailed stratigraphy of the Ballard Cliff Member has been well described in Sussex by Mortimore (1986, his fig. 4), who used the term Melbourn Rock Member for the lower part of this unit. The Melbourn Rock was coined by Penning & Jukes-Browne (1881) in the area of Melbourn, south of Cambridge, for rocky chalks and marl interbeds at the base of the Middle Chalk, and included the Plenus Marls as originally defined. Understanding and usage of the term Melbourn Rock has been enormously varied, and correlation between nodular units developed in Cambridgeshire and the coast of southern England has never been well established. For these reasons, I consider that use of the term is inappropriate on the southern English coast and recommend its abandonment.

Hollywell Member of the White Chalk Formation (emended from Holywell Beds of Mortimore 1983, 1986; = Connet's Hole Member of Jarvis & Woodroof 1984; partly equivalent to the Shakespeare Cliff Member of Robinson 1986). The Ballard Cliff Member is overlain by 10–30 m of nodular, intraclastic chalk which is conspicuously rhythmic on a half-metre to metre scale (Fig. 4). Beds are defined by thin flasered grey marls which are rough with sand-grade bioclasts, commonly inoceramid and rarely microcrinoid debris. Inoceramid coquinas are

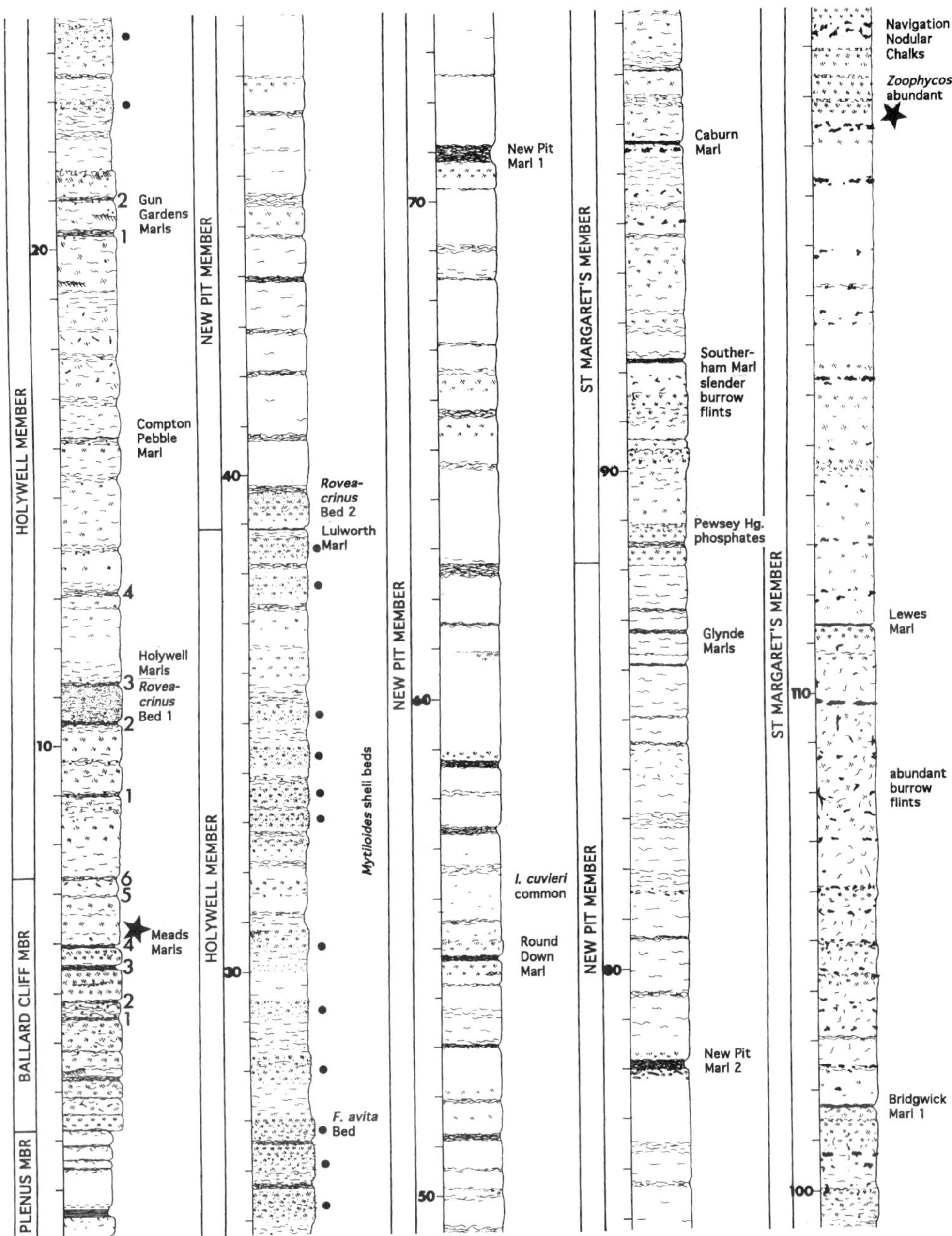

Fig. 2. Turonian succession at Beachy Head, Sussex. Asterisks represent base of Turonian and base of Coniacian. Measured by A. S. Gale, 1980–1991. Flints solid black, cross-hatching represents diagenetic hardening, marls represented by horizontal lines. Scale in metres.

present in the upper part of the member only. Intraclasts are common throughout and are concentrated in shallow scours. Matrix-supported debris flows containing abundant intraclasts and a flow-induced lamination occur more rarely. Nodules occur scattered throughout, but are concentrated in certain beds and hardgrounds are not developed in the more basinal successions. In Devon (Jarvis & Woodroof 1984) a thin development of the member is capped by the mature, mineralized Branscombe Hardground.

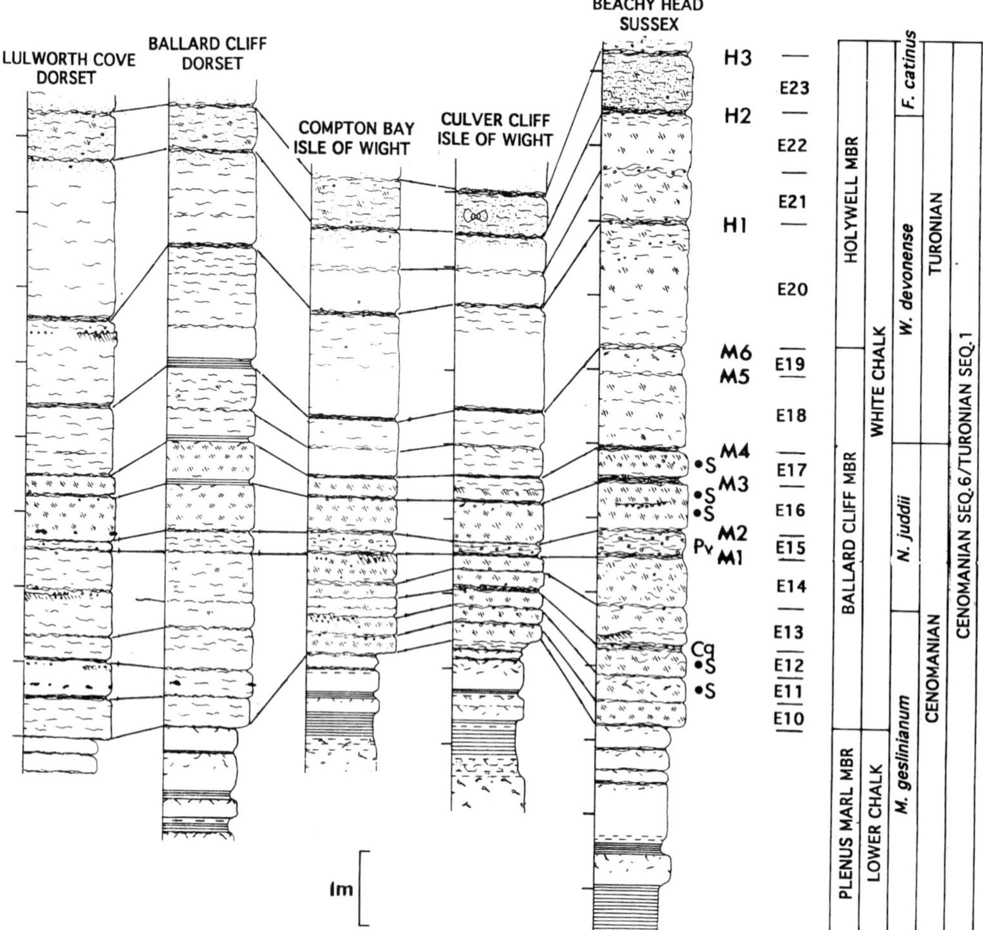

Fig. 3. Correlation of the Ballard Cliff Member from Dorset to Sussex. Note key lithological and faunal markers, including levels of abundance of ammonite *Sciponoceras* spp. (S), the asteroid *Crateraster quinqueloba* (Goldfuss) (Cq), the oyster *Pycnodonte vesicularis* (Pv), and Meads Marls 1–6 (M1–6) and Holywell Marls (H1–3). The M1–2 interval is marly and contains numerous white chalk intraclasts and oysters. The two overlying units are strongly nodular (E16–17). M6 characteristically contains white *Chondrites*. The H2–3 interval at the top of the figure is characterized by rock-forming abundance (up to 30%) of the microcrinoid *Roveacrinus communis*. E10–23 are rhythmic couplets interpreted as climatic cycles.

The Holywell Member can be correlated on a basin-wide scale (Fig. 4) by the use of marker beds and rhythmic couplets. Important marker beds include a bed containing abundant microcrinoid debris (bed 1) between Holywell Marls 2 and 3, and a thin, dark, flaser marl containing numerous intraclasts, called the Compton Pebble Marl. In the middle of the member is a horizon (10–15 cm thick) in which the shells of the inoceramid *Mytiloides* are encrusted (often densely) by the serpulid worm *Filograna avita* (J. Sowerby); this is uniquely found at this level, and was first noted by P. B. Woodroof 15 years ago. The upper part of the member contains 14 discrete shell beds containing abundant debris and valves of the inoceramid *Mytiloides mytiloides* (Mantell).

New Pit Member of the White Chalk Formation (emended from Mortimore 1986). The New Pit Member (Fig. 5) comprises soft coccolith chalks (sparse biomicrites) divided into conspicuous beds of even thickness (0.5–1.0 m) by thin flasered marls. Nodularity is developed only in thinner successions near the basin margins (e.g. Kent) and intraclasts are rare. Flints are sparse, but locally present, as in the basal bed in Dorset and Wiltshire where thin *Thalassinoides* flints are present. The New Pit Member differs from the underlying Holywell Member in the virtual absence of intraclasts and *Mytiloides* shell beds. The basal marker is a thin, dark marl, herein named the Lulworth Marl after Lulworth Cove in Dorset where it is well displayed. A bed-scale correlation of the lower part of the member is achieved by the use of rhythmic couplets and marker beds (Fig. 5). Important marker beds include the Round Down Marl (Robinson 1986), which can be traced to Dorset and Wiltshire, which immediately underlies several levels of abundant, small and fragmentary *Inoceramus cuvieri* (J. Sowerby). The thick, paired New Pit Marls (Mortimore 1983, 1986) are widespread markers in the thicker basinal successions, but are cut out by erosion to the west in south Hampshire, Wiltshire and Dorset. A closely spaced group of thin marls, the Glynde Marls (Fig. 6), include one dark 'plastic' marl called Glynde Marl 1. The upper limit of the New Pit Member is marked by the onset of nodularity, which is laterally variable in horizon.

St Margaret's Member of the White Chalk Formation (Robinson 1986; = Lewes Member of Mortimore 1983, 1986; = Pinhay Member of Jarvis & Tocher 1987). This unit (Figs 6 and 7) comprises nodular, flasered chalks including sparse to fossiliferous biomicrites, containing thin (<0.1 m) basinally traceable beds of marl and has been well described by Mortimore (1986). Flints are common in the upper part of the member. Marker beds have been described by Mortimore (1986) and include the Southerham Marl, which overlies a horizon of delicate *Thalassinoides* flints, the thin Caburn Marl, and the Bridgwick Marls with associated strong nodular flints. Nodular chalks above the Bridgwick Marls are richly fossiliferous and contain conspicuous silicified vertical, branched burrows, at the summit of which lies the Lewes Marl.

Chalk Rock Member of the White Chalk Formation (sensu Gale et al. *1987).* The Chalk Rock Member (Figs 5–7; Bromley & Gale 1982) is a thin group of mature hardgrounds and chalkstones which are laterally equivalent to the upper part of the New Pit Chalk and the lower part of the St Margaret's Chalk. They are highly condensed, and display a complex diachronism, which is described in detail in a following section (p. 187).

Assemblage zone biostratigraphy

Following Barrois (1876), the Turonian Chalk of southern England has been divided into three assemblage zones; those (in succession) of *Inoceramus labiatus* (or sometimes *Orbirhynchia cuvieri* is used as an index), *Terebratulina lata* (= *T. gracilis*) and *Holaster (Sternotaxis) planus*. These zones are easy enough to recognize, but offer only a very coarse stratigraphical resolution compared with that achieved by the use of marker beds and inoceramid bivalves. Additionally, taxonomic refinement has left what used to be understood as *Inoceramus labiatus* as numerous discrete species of the genus *Mytiloides*.

Ammonite biostratigraphy

Although ammonites are only common at two horizons within the Turonian of southern England, enough accurately located specimens are now known to establish a zonation. Important records are summarized in Fig. 8b. This is particularly important for the purpose of comparing sequences with other regions in which ammonites are the main zonal fossils, as in Tunisia (Robaszynski *et al.* 1990). The scheme presented here combines elements of a traditional three-fold division into Lower, Middle and Upper Turonian, with a zonation more widely applicable in Tethys.

Watinoceras devonense *Zone*. The base of the Turonian is defined by the first appearance of *W. devonense* (Wright & Kennedy) (Kennedy & Cobban 1991). In England, this species is extremely rare outside Devon, but in that region its appearance coincides with the disappearance of the Late Cenomanian species *Sciponoceras gracile* (Shumard) and *S. bohemicum anterior* (Wright & Kennedy). These species are common in Kent and Sussex and their highest occurrence can be established precisely at Beachy Head at 10 cm above Meads Marl 4. This level is therefore tentatively taken as the base of the Turonian Stage.

Fagesia catinus *Zone*. This zone is new herein and represents the total range zone of *F. catinus* (Mantell). In southern England, this species occurs frequently in a flasered bed containing abundant microcrinoids (*Roveacrinus communis*: Holywell Marls 2–3), where it is accompanied by *Lewesiceras peramplum* (Mantell). The specimen of *Pseudaspidoceras* cf. *footeanum* (Stoliczka) from Dover figured by Wright & Kennedy (1981, pl. 21 fig. 3) and collected by the author came from this bed. This zone is approximately equivalent to the *Pseudaspidoceras flexuosum* Zone and the *Vascoceras birchbyi* Zone in the Western Interior Basin in the USA (Kennedy & Cobban 1991).

Mammites nodosoides *Zone*. This zone is here defined as the total range of *M. nodosoides* (Schluter). In southern England this species first appears at the Compton Pebble Marl in Sussex, couplet E28, and ranges up to couplet E52[?]. Over

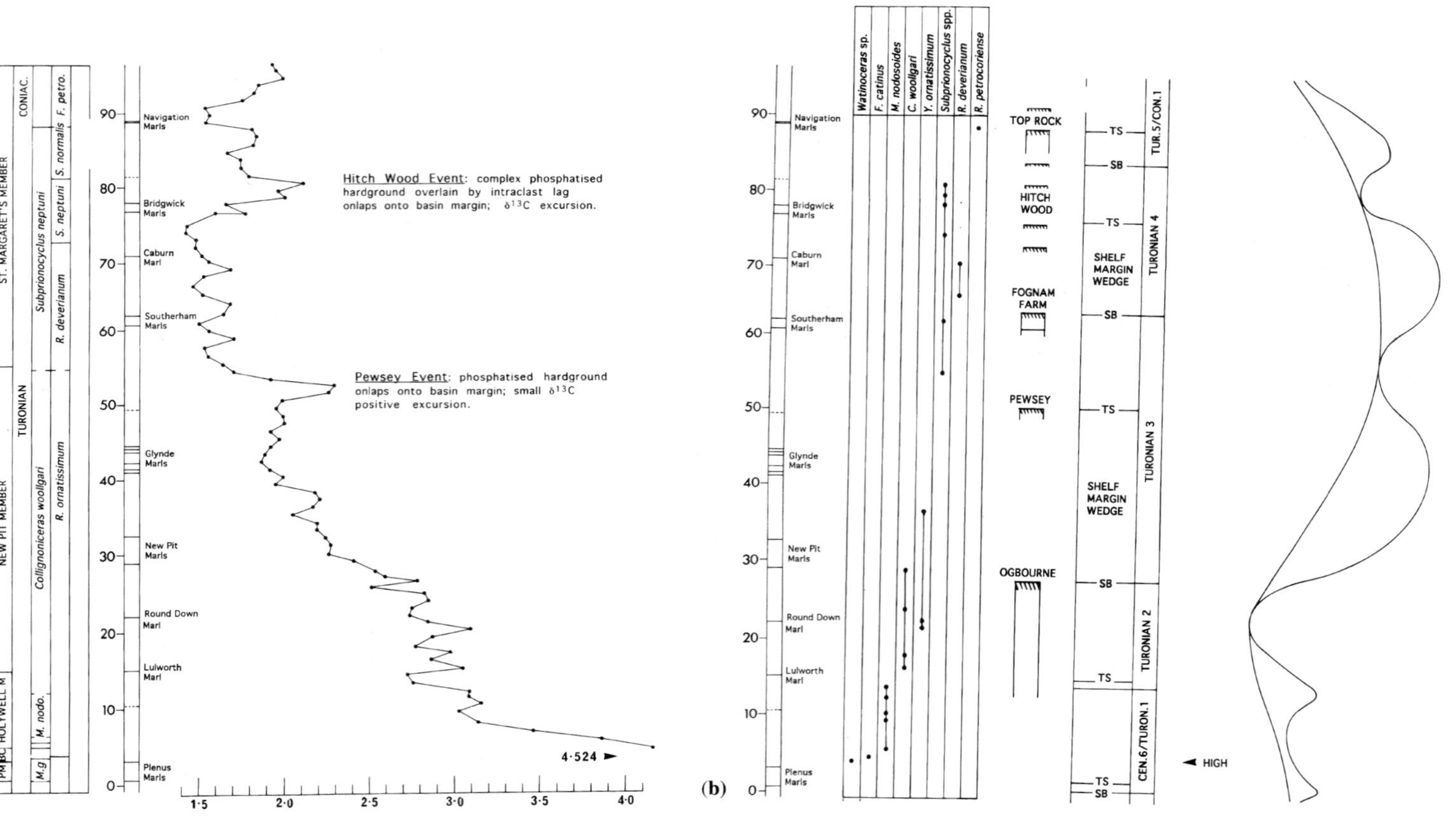

Fig. 8. (**a**) Simplified succession of Turonian Chalk at Dover showing main marker beds and $\delta^{13}C$ data replotted from Jenkyns *et al.* 1994. Scale in metres. Note overall correlation between low values and Turonian relative sea-level fall, and two minor peaks which correlate with onlapping surfaces represented by phosphatized hardgrounds on the adjacent platform. (**b**) Summary succession of Turonian Chalk stratigraphy in southern England; columns represent (left to right) skeletal succession at Dover; ammonite records compiled from various localities; position of Chalk Rock hardgrounds in relation to sequences (gaps between hardgrounds represent breaks); sequence interpretation; and putative sea-level curve for Turonian of southern England.

this range it co-occurs with *Morrowites wingi* (Morrow). *Lewesiceras peramplum* (Mantell) and *Metasigaloceras rusticum* (J. Sowerby) are also found in this assemblage (Wright & Kennedy 1981).

Collignoniceras woollgari *Zone*. The base of the *C. woollgari* Zone is defined by the first occurrence of the zonal species, a short distance above the Lulworth Marl (Figs 5 and 6), the top by the appearance of *Subprionocyclus neptuni*. The *Kamerunoceras turoniense* Subzone (not differentiated in UK) represents the partial range (above *M. nodosoides*) of *K. turoniense*. The *Romaniceras kallesi* Subzone (not differentiated in the UK) represents the total range of the nominal species. The *Romaniceras ornatissimum* Subzone (recognizable in the UK, but boundaries not precisely determined) represents the total range of the nominal species. Well-located specimens of the subzonal index are recorded from Fognam Farm in Berkshire (Wright & Kennedy 1981).

Subprionocyclus neptuni *Zone*. Recent publications have suggested that the range of *Romaniceras deverianum* lies entirely beneath that of *Subprionocyclus*, specifically *S. neptuni*, as shown convincingly in Uchaux, S. France by Devalque *et al.* (1982) and in the Aube by Amedro *et al.* (1982), where *C. woollgari* was recorded as co-occuring with *S. neptuni*.

In southern England, *Subprionocyclus* spp. have been collected *beneath* the range of *R. deverianum* at several localities. *S. hitchinensis* was collected by the author from immediately above Southerham Marl 1 at Dover (BGS Yd 6732) and phosphatized fragments of *Subprionocyclus*, one intermediate between *S. neptuni* and *S. branneri*, were found between the Pewsey Hardground and the Fognam Marl (= Southerham Marl at this locality) at Fognam Farm in Berkshire (identification by W. J. Kennedy). Accurately located specimens of *Romaniceras deverianum* are all higher, at the level of the Caburn (= Reed) Marl to a few metres beneath it (e.g. OUM K10482 from Luton and a specimen in Mortimore collection, the original of Wright & Kennedy 1981 pl. 16 fig. 2, from the Caburn Marl of Firle, Sussex). The diverse and locally abundant 'Chalk Rock' ammonite fauna (Wright 1979) of the *S. neptuni* Zone occurs in nodular chalks between the Bridgwick Marls and Lewes Marl in the basinal succession of southern England, and in the condensed fossil bed on the summit of the Hitch Wood Hardground where the Chalk Rock is developed (Bromley & Gale 1982). In the most northeasterly localities where the Hitch Wood fossil bed is present (Kensworth, Reed and Hitch Wood itself) *Subprionocyclus normalis* (Anderson) and *Baculites undulatus* d'Orbigny are present (Wright 1979). In the Munster Basin these species are confined to a *Subprionocyclus normalis* Zone, present above the *S. neptuni* fauna (Kaplan 1986). The presence of this same distinctive fauna in the UK is marked by the introduction of a *S. normalis* Subzone.

Because we can now demonstrate co-occurrence of *deverianum* Zone and *neptuni* Zone ammonites, the zonal nomenclature can be revised to incorporate both British and Gallic schemes. The base of the Upper Turonian is taken at the appearance of *S. neptuni*, the zone of which is subdivided into three subzones. The *Romaniceras deverianum* Subzone is an assemblage subzone, containing *Subprionocyclus*, *R. deverianum* and heteromorphs. The *Subprionocyclus neptuni* Subzone is characterized by the diverse assemblage of the Chalk Rock described by Wright, and includes many *Subprionocyclus*, etc. The *Subprionocyclus normalis* Subzone is characterized by *Baculites undulatus* and *Subprionocyclus normalis* and is found only by the youngest fauna of the fossil bed of the Hitch Wood Hardground at Kensworth (Bedfordshire), Hitch Wood and Reed (Hertfordshire). The subzone is present in the basinal successions of southern England, because records of *S. normalis* exist from Surrey (Wright 1979). The exact position of the subzone in the basinal succession is not known, however. The base of the Coniacian is taken at the entry of *Forresteria petrocoriense* in the Navigation Hardground (Gale & Woodroof 1981; see Birkelund *et al.* 1984).

Inoceramid biostratigraphy

Inoceramid bivalves offer one of the most refined and widely applicable means of correlation in the Upper Cretaceous; for example, Tröger (1989) was able to recognize 15 assemblage zones within the Turonian. However, much work still needs to be carried out to agree on uniform species concepts across the different traditions of study which have grown up. In the Turonian succession in southern England, inoceramids are common, but not always well preserved, and the assemblages outlined in the following are tentative.

Mytiloides columbianus – Mytiloides cf. hattini *assemblage*. The first *Mytiloides* appear within the *Watinoceras devonense* Zone, at approximately the level of Holywell Marl 1. The assemblage is diverse, and includes *M.* cf. *hattini* Elder, *M. columbianus* Heinz, *M.* gr. *mytiloides* and *Inoceramus* gr. *apicalis*.

Mytiloides mytiloides *assemblage*. Abundant *M. mytiloides*, accompanied by rarer *M. labiatus* and *Inoceramus* gr. *apicalis*, enter a short distance above Gun Gardens Marl 2.

Mytiloides hercynicus *assemblage*. Large, broad species of the *M. hercynicus* and *M. subhercynicus* group become frequent at the level of the Lulworth Marl.

Inoceramus cuvieri *assemblage*. Small, smooth *I. cuvieri* become common shortly beneath the Round Down Marl.

Inoceramus lamarcki *assemblage*. *I. cuvieri* is replaced by larger, rugate *I. lamarcki* between the New Pit and Glynde Marls. The boundary is hard to establish precisely.

Inoceramus securiformis *assemblage*. Characterized by *I. securiformis*, which enters between the Glynde and Southerham Marls, associated with species such as *I. costellatus*.

Mytiloides incertus *assemblage*. Characterized by abundant *M. incertus* Kennedy, which become common beneath the Lewes Marl or the equivalent level. The base of the Coniacian is close to the entry of *Cremnoceramus hannoverensis waltersdorfensis*.

Microfossil biostratigraphy

Hart *et al.* (1989) reviewed the distribution of foraminifera within the English Turonian and erected benthonic and planktonic zones based on the succession at Dover, Kent. Unfortunately, a detailed lithostratigraphy is not presented, which makes it impossible to identify zonal boundaries in terms of the succession given here. Burnett (pers. comm.) has developed a nannofossil zonation for the Turonian based on the Kent succession.

Cyclostratigraphy

Gale (1990, 1995) developed a cyclostratigraphy for the Cenomanian Stage of western Europe by correlation of decimetre-scale bedding couplets which are extensively developed in basinal hemipelagic chalks. He argued (Gale 1995) that these couplets are productivity cycles driven by the precession signal (mode at 21 ka) in the Milankovitch Band. Similar couplets, but with a reduced clay component, are present in the Lower and Middle Turonian of southern England, and can be very widely correlated by reference to marker beds and detailed biostratigraphy. These couplets are numbered in continuation with the Cenomanian scheme and are denoted (Figs 3–5) E 1–53 and F 1–36. The boundary between E 53 and F 1 is the base of the Lulworth Marl. Above F 36, couplets cannot be recognized on a basin-wide scale, probably on account of redeposition and condensation.

Carbon isotope stratigraphy

Jenkyns *et al.* (1994, fig. 3) have published a detailed carbon isotope curve from analysis of bulk chalk ($\delta^{13}C$ carbon) for the Chalk succession in east Kent, based on metre-spaced samples. The Turonian part of this curve is replotted here at a large scale (Fig. 8a). In general shape, the Turonian $\delta^{13}C$ curve has the form of an asymmetrical trough, falling from very high values near the Cenomanian–Turonian boundary to a low in the lower part of the St Margaret's Member, then starting to rise in the uppermost Turonian. In detail, a decrease in the rate of fall of $\delta^{13}C$ in the upper Holywell Member can be detected and two minor peaks are present within the trough; one between the Glynde and Southerham Marls, the other above the Bridgwick Marls. A nearly identical curve, containing all the same details, is present at Gubbio in Italy (Jenkyns *et al.* 1994, fig. 10), so the bulk carbonate $\delta^{13}C$ curve has considerable potential for long distance correlation of the Turonian. The association between transgressive events and $\delta^{13}C$ excursions is discussed in the following.

Criteria used to recognize sequences and infer sea-level changes from chalks

Nodular chalks and hardgrounds

Nodular chalks and hardgrounds are products of early diagenetic cementation which took place just beneath the seafloor and are generally associated with reduced sedimentation rates or hiatuses. Nodular chalks are made up of discrete carbonate concretions with diffuse boundaries, separated by soft burrow-fill chalk, but encroaching lithification progressively reduced the burrow diameter and a lithified chalk framework was produced (Kennedy & Garrison 1975; Bromley 1975). Seafloor exposure of this lithified chalk resulted in the formation of a true hardground which can be identified from the presence of boring and encrusting organisms, and commonly the replacement or coating of the surface by glauconite and phosphate. There is a close relationship between the length of seafloor exposure and the extent of lithification and mineralization of hardgrounds; mature hardgrounds are massively cemented to form chalkstone, possess intensely mineralized surfaces and represent major hiatuses of hundreds of thousands of years. Once formed, hardgrounds were modified by successive periods of sedimentation, lithification, and physical and biological erosion. Erosional stripping produced worn, planar surfaces and lithification of successive sediments onto the walls of burrows simplified complex surface topographies (Bromley 1975).

That hardgrounds represent very important events in chalk successions has long been appreciated (e.g. Hébert 1875). The fact that they are best developed and most mature in thin basin margin successions or on structural highs within basins led workers to associate their occurrence with relatively shallow water conditions (Hawkins 1942).

Hancock (1989) has made extensive use of both nodular chalks and hardgrounds in developing a sea-level curve from the Upper Cretaceous of Britain. He took each hardground as marking the end of a period of relative sea-level fall (bottom of the regressive trough) and placed a transgressive peak midway between two hardgrounds in white chalk. Hardgrounds are overlain by condensed successions representing a transgressive surge.

A difficulty with this model is that certain of the 'regressive' nodular chalks or hardgrounds formed at times of widespread sea-level rise inferred from other areas. For example, Hancock includes the latest Cenomanian and earliest Turonian interval in a regressive trough (1989, fig. 11) because nodular chalks locally containing hardgrounds formed widely at this time. However, in Sarthe, northwest France, the sea-level rise in the same interval was of such magnitude that outer shelf chalks, deposited in perhaps 150–200 m of water, rest directly on shallow marine sands, and even onlap onto basement (Juignet 1974; Robaszynski *et al.* 1996). The same sea-level rise is also associated with a major reduction in clastic input across Europe (the base of the White Chalk), presumably because clastic source areas were drowned. A possible solution is that hardgrounds do in fact represent both regressive and transgressive events. In sequence stratigraphical terminology, they developed both at sequence boundaries, at the maximum rate of sea-level fall, and as parts of transgressive systems tracts, when current winnowing was dominant. Thus the presence of regionally extensive hardgrounds is not evidence of relative sea-level fall *per se*.

Regionally developed erosion surfaces, which may be later lithified to form hardgrounds, formed during the maximum rate of sea-level fall and cut into the underlying chalk, either as local channels (exemplified by the sub-Totternhoe Stone erosion surface and the sub-Plenus erosion surface on the borders of the Anglo–Paris Basin) or as regional, planar surfaces downcutting into platform areas. Long exposure on the seafloor commonly resulted in massive lithification by marine cements of the superficial 30–40 cm of chalk and a glauconite coating formed during transient burial by shifting sediment. There is little simplification of the surface by the cementation of later sediments, as erosion was dominant over deposition. Hardgrounds of this type are found only on the basin margins (e.g. Ogbourne, Fognam Farm) and have a great preservation potential as the massive chalkstone resisted erosion.

Transgressive hardgrounds onlap onto platforms and were formed by currents which impinged on the seafloor during periods of rapid sea-level rise, thus preventing the deposition of coccoliths. They are commonly phosphatized and may show complex internal sediments formed by periods of intermittent sedimentation, lithification and erosion. In very condensed platform successions transgressive surfaces rest directly on or merge with sequence boundaries. Where they merge, the hardground mineralization characteristically shows glauconite coated with later phosphate; a good example of this is the Branscombe Hardground in Devon (Jarvis & Woodroof 1984).

Micro-, macro- and trace fossils

In a series of papers, Hart and co-workers (1980, 1990; Hart & Bailey 1979) have used ratios of planktonic:benthonic foraminifera and interpretations of planktonic foraminiferan morphology to elucidate sea-level changes from the Upper Cretaceous of the UK. The method is based on pickings of 500 specimens in the 250–500 μm size fraction (Hart 1990, fig. 1) and works on the general principle that a greater abundance of planktonic foraminifera indicate deeper water. Additional information is derived from the relative abundance of keeled planktonic species, interpreted as requiring deeper water for their life cycles. This approach can be criticized on several grounds. Firstly, the >250 μm size fraction only represents a part of the total planktonic assemblage which is dominated by small forms. Secondly, many factors other than depth control the distribution of larger planktonic species; many are dominantly low latitude in distribution and probably only became abundant in northern regions during very warm periods such as the Late Cenomanian and Early Turonian (Jenkyns *et al.* 1994). Hart's sea-level curve shows a highstand for all but the latest Turonian, which underwent a major relative sea-level fall, identified by a large reduction in the percentage of planktonic species at that time. However, the curve is not tied to any lithological succession and is therefore difficult to interpret.

Mortimore & Pomerol (1987, pp. 134-135, fig. 26) used the occurrence and shape varieties of the irregular echinoids *Conulus* and *Echinocorys* to identify transgressions and regressions from the Upper Cretaceous Chalks of the Anglo–Paris Basin, taken in conjunction with certain thin dark marls which are interpreted as 'anoxic events' and therefore associated with transgression. Their argument is that conical echinoids occur in shallower water and globose echinoids in deeper water. There is no empirical evidence to support these contentions. In the Anglo–Paris Basin *Conulus* increases in abundance and ranges westwards, irrespective of water depths, and so geography appears to be an important controlling factor.

In a subsequent paper, Mortimore & Pomerol (1991) used trace fossil distribution in the chalk to reconstruct a sea level curve. They assert that (p.228) 'a *Zoophycos–Bathichnus paramoudrae–Trichichnus* association is linked to transgression and *Thalassinoides suevicus* with *Spongeliomorpha–Ophiomorpha* are linked to regression or long stillstands.' The basis of this argument presumably follows Ekdale & Bromley (1984), who point out (p.330) the relative abundance of *Zoophycos* in deep sea chalks drilled in the Deep Sea Drilling Project compared with Cretaceous shelf chalks. However, there is no independent evidence that Cretaceous *Zoophycos* occurrence is in any way associated with transgression. In any case, the identification of certain of these trace fossils by Mortimore & Pomerol (1991) is questionable; for example, the supposed *Zoophycos* in the Plenus Marls and overlying Ballard Cliff Member may be more correctly referred to *Taenidium* (R. G. Bromley pers. comm.), and *Ophiomorpha* has never been recorded from the English Chalk by any other worker.

Sediment body geometry and basin margin onlap

The development of a high-resolution stratigraphy for the Turonian Chalk of southern England enables correlation on a bed scale for the lower part of the succession (Figs 4–6). These correlation diagrams can be used to identify sediment body geometries, onlap and offlap surfaces, similar to the way high-resolution seismic sections are used. In this work, particular attention has been paid to the transitional areas between basins and platforms, with the aim of identifying major erosional surfaces, basinal wedges and onlapping surfaces which extend onto platforms. In the Turonian this is particularly applicable to the Chalk Rock and its basinal correlatives. In chalks, sequence boundaries are only easily recognizable on platforms where erosion during the maximum rate of sea-level fall has led to the formation of disconformity surfaces, commonly but not invariably lithified to form hardgrounds. During lowstand, 'shelf margin wedges' of more marly chalk built out into the basin, adjacent to scoured platforms. Transgressive surfaces are also commonly represented by hardgrounds which onlap platform margins to rest directly on the preceding sequence boundary. Thus it is helpful to construct a basin to platform correlation to apply sequence stratigraphy successfully to chalk successions.

Carbon isotope stratigraphy

Jenkyns *et al.* (1994) discussed the relationship between the $\delta^{13}C$ curve and sea level in the English Chalk succession and suggested that the carbon-isotope curve can be used as a proxy for sea level. There is a broad correspondence between published sea-level curves (e.g. Hancock & Kaufmann 1979; Weimer 1984; Haq *et al.* 1988) and carbon-isotope curves for the Cenomanian and Turonian Stages, with an overall rise in both sea-level and $\delta^{13}C$ through the Cenomanian and a major fall through the Turonian (Gale *et al.* 1993; Jenkyns *et al.* 1994). A rise in sea level in the latest Turonian corresponds to a minor increase in $\delta^{13}C$. What is not clear, however, is the extent to which the carbon isotope profile reflects rates of relative sea-level rise rather than absolute sea-levels. Certainly, individual minor peaks in $\delta^{13}C$ correspond closely to onlapping transgressive surfaces in the Cenomanian of the Anglo–Paris Basin (Robaszynski *et al.* 1990; Jenkyns *et al.* 1994). The Turonian carbon isotope profile for the English Chalk given here (Fig. 8a is redrawn from data in Jenkyns *et al.* 1994). This shows two minor $\delta^{13}C$ peaks in the Upper Turonian, both of which correlate with onlap surfaces, and is taken as corrobatory evidence that these are transgressive surfaces.

Sequence stratigraphy of the Turonian, southern England

Ballard Cliff, Holywell and lower New Pit Members: sequences 1 and 2

The sequence stratigraphy of the Cenomanian of the Anglo–Paris Basin was described by Robaszynski *et al.* (1996). They placed the sequence boundary of their sixth and highest sequence at the base of the Plenus Marls Member (the sub-Plenus erosion surface of Jefferies 1962, 1963) and the transgressive surface of the same sequence at the base of bed 3 of the Plenus Marls. Robaszynski *et al.* (1996) interpreted the basal part of the 'Melbourn Rock' (Ballard Cliff Member herein), which is a coarse, calcisphere-rich chalk containing intraclasts, as the transgressive systems tract of the sixth sequence. The base of the Turonian Stage thus falls within the higher part of a transgressive systems tract. The maximum flooding surface above can be taken approximately at the base of the Holywell Member, where bed thickness increases two-fold, and nodularity concomitantly decreases across the northern Anglo–Paris Basin.

The Holywell Member can be correlated across southern England in very fine detail by using rhythmic couplets and faunal marker beds, but in thinner successions, such as in the North Downs and Chilterns, the rhythmicity is poorly developed. Although nodularity is common throughout, no regionally developed hardgrounds are present in the basinal succession and those which are present have formed in shallow scour channels up to a few metres across. Hardgrounds develop towards the basin margins, however, and help a tentative identification of systems tracts to be made. At Dover, a conspicuous iron-mineralized hardground is present in the upper part of the Holywell Member, 1.5 m above the *Filograna avita* marker bed, and a further hardground (the Limonitic Hardground of Robinson 1986) immediately overlies the top of the Holywell Member. The intervening beds are weakly nodular and distinctly marly and represent a shelf margin wedge. In Devon, the shelf margin wedge is present in thicker successions, such as at Whitecliff (Jarvis & Woodroof 1984, fig. 5), but thins westwards from here onto a local structural high in the Beer Head–Hooken area where the massively cemented Branscombe Hardground (Jarvis & Woodroof 1984) represents a coalesced sequence boundary and transgressive surface. The Branscombe Hardground, where well developed as at Beer Head (Jarvis & Woodroof 1984: p.206), has a glauconitized and phosphatized surface and a phosphatic veneer.

In the basin, the transgressive surface is represented by a bed containing abundant (winnowed?) echinoderm debris, immediately above the Lulworth Marl, at the base of the New Pit Member. The New Pit Member (Fig. 5) is strongly rhythmic on a metre scale, with thin flaser marls separating beds of fine white nannofossil chalk. Hancock (1989) identifies these chalks, which fall in the middle part of the *C. woollgari* Zone, as representing the deepest phase of the Turonian sea level highstand. The rhythms are widely traceable throughout the outcrop in southern England. There is little evidence of a maximum flooding surface, but it might be marked by a change in bedding style at the level of the Round Down Marl. The nodularity which immediately underlies the New Pit Marl 1 marks the overlying sequence boundary, but to understand this part of the succession we must first look at the Chalk Rock.

The Chalk Rock and its basinal correlatives; sequences 3 and 4

The Chalk Rock Member is a group of nodular chalks and chalkstones one to several metres thick which contains mature mineralized hardgrounds, which is most completely developed in Wiltshire and Berkshire (Jukes-Browne & Hill 1903; Bromley & Gale 1982). It is a highly condensed deposit which formed in relatively shallow water over structural highs (e.g. the London–Mendip Line which overlies shallow Palaeozoic basement) during the low sea levels of the Mid- and Late Turonian (Hancock 1989). In detail, the Chalk Rock displays complex diachronism, with the oldest hardgrounds occuring only in the west and the youngest extending furthest to the east (Bromley & Gale 1982; Gale *et al.* 1987). The picture is further complicated by the fact that formation of two of the Chalk Rock hardgrounds was preceded by erosion and they rest on different levels from locality to locality. The potential value of the Chalk Rock for understanding sea-level histories and sequence development in the Turonian is considerable, but has been hampered by the difficulties of correlation from the marginal areas to the expanded basinal succession present in Hampshire, Sussex, Surrey and Kent.

The Mid- and Late Turonian succession in the northern Anglo–Paris Basin has been reviewed earlier in this paper. The transitional area between the basinal succession and the Chalk Rock is commonly narrow and always poorly exposed (Bromley & Gale 1982). The summit of the highest hardground of the Chalk Rock, the Hitch Wood Hardground, contains a rich and distinctive fauna of ammonites, echinoids, brachiopods and inoceramid bivalves, which allow a precise correlation with fossiliferous nodular chalks present between the Bridgwick and Lewes Marls in Kent and Sussex (see Figs 6 and 7). Below this level, the calcitic fossils are long-ranging and little help in detailed correlation. For this reason, the equivalent of all but the highest part of the Chalk Rock within the basinal succession has been an enigma.

It appears likely that the marl seams present in the Chalk Rock succession are correlative with marls which are widely traceable in the basin, especially if some of these really are volcanic ash falls (Pacey 1984). Wray & Gale (1993) attempted to establish a correlation between the Fognam Marl of the Chalk Rock succession and the Glynde Marl of the basin using a trace element fingerprinting technique. Although this method initially looked promising, new data presented here cast doubt on its accuracy.

The correlation given here is based on two new pieces of evidence. Firstly, the precise position of the lowest Chalk Rock hardground (Ogbourne) has been established in the basinal succession by correlating the underlying beds in detail from Sussex through the Isle of Wight to Dorset, and on into the Chalk Rock heartlands of Wiltshire and Berkshire (Fig. 5). Secondly, the Pewsey Hardground, which immediately overlies the Ogbourne Hardground,

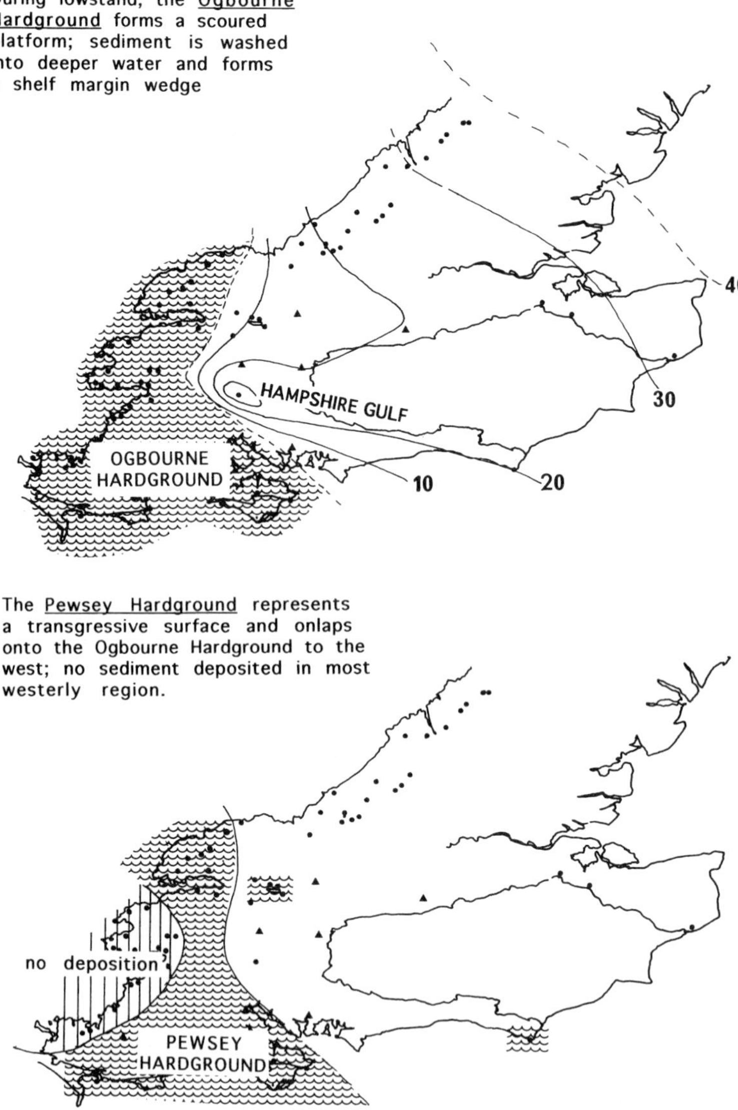

Fig. 9. (**a**) Map of Ogbourne Hardground. (**b**) Map of Pewsey Hardground and isopachs of shelf margin wedge in basin. The Ogbourne Hardground is interpreted as a sequence boundary (base of third sequence) which formed a storm-scoured platform to the west from which sediment was washed into the Hampshire Gulf and further eastwards to form a thick lowstand wedge. The Pewsey hardground formed during the subsequent transgression.

can now be identified in Sussex, the Isle of Wight, Dorset and at Burghclere in north Hampshire (Fig. 6) and this shows that the overlying Fognam Marl is equivalent to the Southerham Marl in the basin. These two correlations significantly alter our understanding of Turonian stratigraphy because they indicate the following three points:

(1) The base of the Chalk Rock rests in the *lower* part of what has traditionally been called the *Terebratulina lata* Zone, not in the upper part as has been presumed.

(2) The upper surface of the Ogbourne Hardground is a major non-sequence, missing 20–40 m of succession present in the basin, including the interval containing the New Pit and Glynde Marls. Thus published correlations between the standard

succession in Kent and Sussex and the Isle of Wight–Dorset region are incorrect (e.g. Mortimore 1987, fig. 2; Mortimore & Pomerol 1987, fig. 9). The new correlation opens the way to the sequence stratigraphical interpretation of the Chalk Rock given in the following section.

(3) The time interval represented by the Chalk Rock and its equivalent representation in the Weald Basin have been severely underestimated. At Cley Hill, near Warminster, Wiltshire (Fig. 6), the most condensed Chalk Rock succession known (Bromley & Gale 1982), the 10 cm of Hitch Wood Hardground resting on Ogbourne Hardground, is equivalent to about 60 m of chalk in Kent and Sussex – about half the total thickness of the Turonian Stage.

A major sea-level fall in the late *C. woollgari* Zone resulted in erosional bevelling of topographic highs and the consequent formation of a table-top erosional surface across Dorset, Wiltshire, Berkshire, the Isle of Wight and south Hampshire. This surface lithified to form the Ogbourne Hardground, which is a sequence boundary. The Ogbourne Hardground cuts down to progressively lower levels within the underlying chalk northwards into Wiltshire and Berkshire (Fig. 5); on the Isle of Wight and the Dorset coast it lithifies levels at couplets F 30–F 36, but in Berkshire and Wiltshire it lies at a level between F 21 and F 25. At Cley Hill, the Ogbourne Hardground probably rests even further down in the succession, but the precise horizon has not been demonstrated. The correlative hardgrounds to the Ogbourne in Normandy are the Tilleul Hardgrounds 1–2, which locally erode deeply into the Lower Turonian, as at Tancarville on the Seine where they lie a few metres above the base of the Turonian (Juignet, pers. comm.). The Ogbourne Hardground is a convoluted, glauconitized surface and the massive subjacent chalkstone contains characteristic patches of orange staining. In west Wiltshire, in the region of Mere, Warminster and Westbury, the Ogbourne has a planar morphology, probably reflecting longer exposure to erosive scour in this region.

The erosional agency which formed the Ogbourne Hardground was probably storm wave base, as Bromley (1967; see also Bromley in Bathhurst 1971) estimated the water depth of formation of the Chalk Rock at 50–70 m. The Ogbourne Platform (Fig. 9a) was thus a region of non-deposition, or temporary deposition, with periodic storms re-suspending ooze and washing it into the deeper waters of the basin. Where accommodation space was available in the Wessex Basin, a thick shelf-margin wedge developed.

The eastern limits of the Ogbourne Platform form a westerly-pointing V, herein called the Hampshire Gulf (Fig. 9a). The transition from platform to basin is poorly exposed, but appears to be fairly abrupt in south Hampshire at least. In the Marchwood borehole near Southampton, the Ogbourne Hardground can be identified in the resistivity signature, whereas in the Twyford Down M3 cutting (Fig. 6) only 17 km to the north, a thick wedge is already present. This boundary is probably controlled by a northwest–southeast fault downthrowing to the north, along the line shown by Stoneley (1992, his fig. 1).

The 'shelf margin wedge' comprises rather marly, well-bedded chalks, including the two groups of marls named New Pit and Glynde by Mortimore (1986). Their wedge geometry is very apparent in cross-section and from the isopachyte map (Fig. 9a), which shows the wedge expanding towards the southern part of the North Sea Basin and a small depocentre near Winchester. Although part of the sediment in the wedge must be re-deposited from the Ogbourne Platform, individual beds are widely traceable, and there is no evidence of the debris flows or turbidites characteristic of chalk redeposition in settings such as the Central Graben. A likely mechanism of deposition is frequent (probably annual winter storm) re-suspension of ooze on the platform, which subsequently settled out in deeper water. This sediment simply augmented the volume of chalk in the wedge and its composition (percentage of clay, etc.) would be identical to that of the indigenous sediment.

The Ogbourne Hardground is overlain directly by the Pewsey Hardground (Bromley & Gale 1982), except in north Dorset and west Wiltshire where the latter is onlapped by higher beds. The Pewsey Hardground (Fig. 9b) is a convoluted, weakly phosphatized surface (overlain by numerous small phosphatized intraclasts) and the underlying nodular chalk or chalkstone has a slightly pinkish tint. It is overlain in north Wiltshire and Berkshire by the conspicuous 5–10 cm thick Fognam Marl (Bromley & Gale 1982). The correlatives of the Fognam Marl and the Pewsey Hardground can now be identified confidently in the Wessex Basin as Southerham Marl 1, and the pinkish nodular chalk containing small phosphatic intraclasts which underlies it in sections such as Beachy Head, Sussex. Thus the Pewsey Hardground onlaps the lowstand wedge (New Pit and Glynde Marls) onto the platform to rest directly on the Ogbourne Hardground. The Pewsey Hardground is therefore identified as the transgressive surface which succeeds the thick 'shelf margin wedge' and can be reliably dated as earliest *S. neptuni* Zone from ammonites collected at Fognam Farm.

Above the level of the Pewsey Hardground the geographical distribution of the Chalk Rock hardgrounds shifts significantly to a more northerly and easterly position, and south Dorset and the Isle of

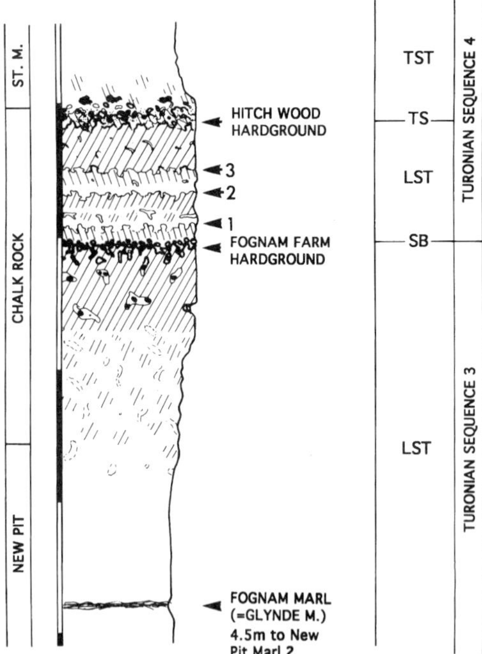

Fig. 10. Chalk Rock succession at Ewelme in Oxfordshire [SU 655 893]. The Chalk Rock here comprises five hardgrounds; FF, Fognam Farm; LHH, Leigh Hill Hardground; and HW, Hitch Wood. The massively lithified Fognam Farm Hardground (sequence boundary at base of fourth sequence) is here intensely glauconitized and overlain by a thick bed of glauconitized intraclasts; it rests a short distance above a marl (Wray & Gale 1994) which is equivalent to Glynde Marl 1 in the basinal succession. The Hitch Wood Hardground (condensed transgressive surface) has a complex phosphatized surface incorporating many sediment types.

Wight display a thin basinal type of succession above the Pewsey Hardground. This was perhaps caused by changing patterns of differential subsidence, such that storm wave base impinged on a shallower, more northeasterly area. A thin but typical succession is developed at Ewelme in Berkshire (Fig. 10).

The Pewsey Hardground is overlain by the Fognam Farm Hardground, which is one of the most distinctive hardgrounds in the Chalk Rock, on account of its strongly convoluted, intensely glauconitized surface, and an overlying bed (up to 10 cm thick) of dark green glauconitized intraclasts. The Fognam Farm Hardground is developed from Wiltshire across into Buckinghamshire (Fig. 11a) and represents the second and highest sequence boundary in the Chalk Rock. As is the case with the Ogbourne Hardground, pre-Fognam Farm Hardground erosion locally cuts down into the underlying chalk. This is best seen in the southern Chilterns, where the Fognam Farm Hardground can be shown to rest a short distance above the Glynde Marl (e.g. at Ewelme; Wray & Gale 1993; Fig. 10 herein). In the basin, the level equivalent to the Fognam Farm surface falls within nodular chalks between the Southerham and Caburn Marls. A thin 'shelf-margin wedge' developed in the Weald and north Channel Basins to the southeast of the platform formed by the Fognam Farm Hardground (Fig. 11a). This comprises chalks which are strongly nodular even in the thickest successions, and represents the lowest sea levels of the lowstand period of the Mid- and Late Turonian. In the Chalk Rock succession, minor hardgrounds are present above the Fognam Farm and below the Hitch Wood Hardground (Leigh Hill and Blount's Farm hardgrounds of Bromley & Gale 1982).

The Hitch Wood Hardground (Fig. 11b) is the highest in the Chalk Rock. It has a complex convoluted surface, often phosphatized, which is a product of repeated episodes of burial, lithification and scour and may incorporate up to 15 individual sediment types (Bromley 1975). It is the most extensive of the hardgrounds and onlaps onto the Ogbourne Hardground at Cley Hill in Wiltshire (Bromley & Gale 1982; Fig. 6 herein) and extends far east into Hertfordshire. In terms of basinal succession, the Hitch Wood represents the interval from beneath the Bridgwick Marls to the Lewes Marl (e.g. Fig. 7), a fact which can be accurately established from the distinctive fauna which surmounts the hardground in many localities. The nodular chalks of this interval commonly contain many small phosphatic pebbles in thinner successions such as at Dover.

The extensive phosphatization and onlapping relationship at the basin margins are here interpreted as evidence that the Hitch Wood Hardground is a condensed transgressive deposit formed during a Late Turonian sea-level rise.

The two minor but distinctive relative sea-level rises in the Upper Turonian *S. neptuni* Zone coincide with positive excursions in $\delta^{13}C$ (Fig. 8a). Increase in $\delta^{13}C$ correlates closely with transgressive surfaces in the Cenomanian, and a case can be made for using this ratio as a proxy for sea-level, probably because transgressions result in increased productivity on the shelves. The lowest of the Turonian peaks at Dover corresponds exactly with the level of the phosphatized, transgressive Pewsey Hardground on the platform and the ensemble is called the Pewsey Event. The higher excursion falls at the level of the similarly phosphatized and transgressive Hitch Wood Hardground and is therefore called the Hitch Wood Event.

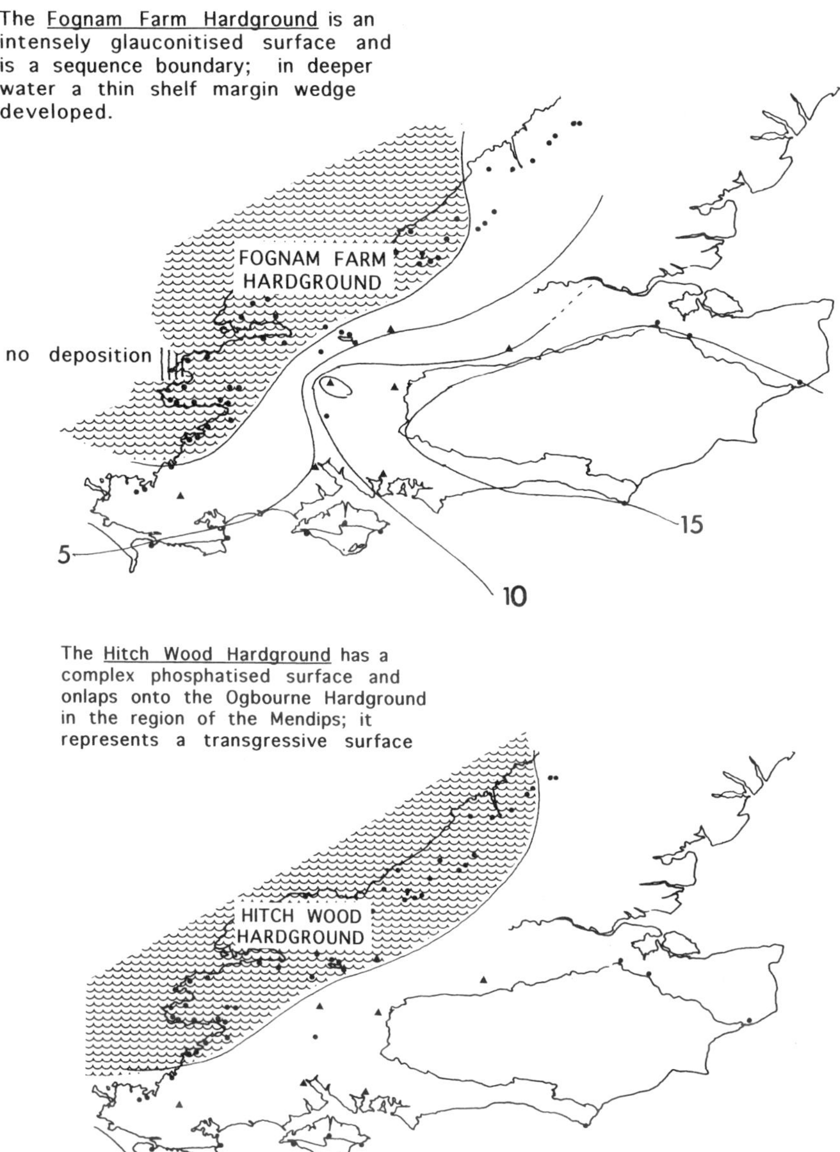

Fig. 11. Distribution of (**a**) Fognam Farm (sequence boundary) and (**b**) Hitch Wood (transgressive surface) hardgrounds and isopachs of intervening sediments. The Fognam Farm Hardground is a glauconitized sequence boundary (fourth-order sequence) forming a current-scoured platform, from which ooze was washed into a lowstand wedge in the Wessex Basin. The complex, phosphatized Hitch Wood Hardground represents the subsequent transgressive surface. Note the more northeasterly positions of these hardgrounds compared with the Ogbourne and Pewsey hardgrounds (Fig. 9).

Latest Turonian – Early Coniacian; sequence 5

In the northern Anglo–Paris Basin, the highest Turonian and earliest Coniacian chalks are represented by nodular beds containing abundant *Zoophycos*: the Navigation Hardgrounds (Mortimore 1986; really only nodular chalks in basinal areas), which are each overlain by a thin pair of Navigation Marls. The overlying Coniacian chalks contain a succession of nodular beds which thin and pass into hardgrounds towards the basin margin, and together with the Navigation Hardground locally converge to

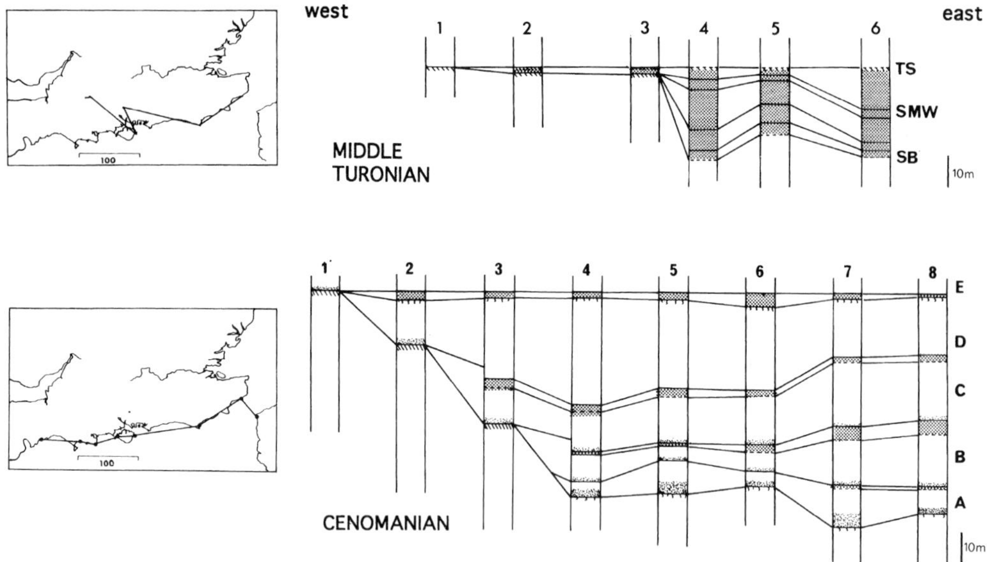

Fig. 12. Comparison of sequences in Cenomanian (lower panel) and Mid-Turonian (upper panel) in Anglo–Paris Basin showing relative development of lowstand wedges (shaded). Westerly onlapping sequences in Cenomanian are taken from the following localities: 1, Charton Cliff, Devon; 2, White Nothe, Dorset; 3, Ballard Cliff, Dorset; 4, Compton Bay, Isle of Wight; 5, Culver Cliff, Isle of Wight; 6, Beachy Head, Sussex; 7, Folkestone, Kent; 8, Escalles, Boulonnais, France. Base line of correlation at top of Plenus Marls. Bottom of sequence A at base of Cenomanian. Note thin lowstand development. Turonian localities: 1, Cley Hill, Wiltshire; 2, Beggars Knoll, Wiltshire; 3, Culver Cliff, Isle of Wight; 4, Twyford Down M3 cutting, Hampshire; 5, Beachy Head, Sussex; 6 Dover, Kent. Note thick lowstand deposit formed during Mid-Turonian major sea-level lowstand.

form the highly condensed Top Rock (Chatwin & Withers 1908; Bromley & Gale 1982), which is well developed locally in Dorset, and also in Bedfordshire and Hertfordshire (Fig. 7). The base of the Navigation Hardground, which is the lowest hardground in the Top Rock, is seen to cut down on the basin margins and platform, as at Reed and Luton, and is thus identified as a sequence boundary. The Cliffe, Hope Gap and Beeding Hardgrounds which form the upper part of the Top Rock are commonly phosphatized on the basin margins and represent flooding surfaces.

Discussion

The sequence stratigraphical interpretation of the Turonian of southern England allows us to pin-point exactly which intervals represent maximum sea-level stands. The highest sea levels within any sequence presumably fall at or a short distance above the maximum flooding surface. These are found in the *W. devonense*/*F. catinus* Zone (= lower *M. nodosoides* Zone of the old classification) in the lowest Turonian sequence and within the middle part of the *C. woollgari* Zone for the second sequence, both at or very close to maxima identified by Haq *et al.* (1988) and Hancock (1989). The fine, nannofossil chalks developed within the *C. woollgari* highstand (sequence 2 herein) were presumably deposited in deeper water than those of the *M. nodosoides* highstand (Fig. 8b).

Certain of the Turonian sequences described here are different in development to those defined using the same type of criteria for the Cenomanian Stage of the Anglo–Paris Basin (Robaszynski *et al.* 1996). The Cenomanian was a period of overall sea-level rise, and in consequence Cenomanian deposits progressive onlap onto platforms surrounding the basin (Fig. 12). Sea-level falls were relatively minor in extent and the basinal lowstand deposits are thin and poorly developed because the overall migration of sediment was towards the basin margins. In contrast, for much of the Mid- and Late Turonian sea levels were falling dramatically, and this is reflected in Turonian sequences 3 and 4. In these, thick ‘shelf margin wedges’ developed in deeper basins adjacent to current-scoured platforms on which highly condensed sequences are found (Fig. 12).

Juignet & Breton (1994) have described the sequence stratigraphy of the Turonian Chalk of the region of Fécamp, Normandie, France. This succession is comparable with that developed in southern England, and the sequence classification and sea-level curve of Juignet and Breton compare well with that produced in this work (Fig. 8b), with the exception that they recognize only one sequence beneath the Mid-Turonian Tilleul Hardgrounds 1–2 where I have two. The hardground they describe as Fagnet probably corresponds to the Branscombe Hardground in Devon, and Tilleul 1–2 with the two basal hardgrounds of the Chalk Rock (Ogbourne and Pewsey). The Senneville Hardgrounds 1–2 correlate broadly with the Fognam Farm/Hitch Wood Hardground of the Chalk Rock.

The biostratigraphy and sequence stratigraphy of a very expanded Turonian succession in central Tunisia has been described by Robaszynski *et al.* (1990), who show that the stage is represented by 810 m of marls and limestones, approximately an order of magnitude thicker than the Anglo–Paris Basin succession. From ammonite data it is possible to make a direct comparison of the sequences between the two areas. In both southern England and Tunisia, three complete sequences and two partial sequences are present in the Turonian (see Robaszynski *et al.* 1990, fig. 32). The lowest sequence boundary in Tunisia lies near the top of the range of the ammonite *Mammites nodosoides*, and the overlying transgressive surface (base of the *R. kallesi* Zone) are fairly close to equivalent surfaces in the Anglo–Paris Basin, as is probably the position of the overlying sequence boundary which must fall within the range of *Romaniceras ornatissimum*. Above this level, however, there are insufficient ammonites in common between the two sections to make any detailed comparison of the relative positions of sequence stratigraphical markers. However, the same number of sequences is developed in both areas and in view of the difference in thickness, tectonic setting and distance between the two areas, this must offer hope that a widely applicable sequence stratigraphy for the Turonian Stage may be recognized.

I am most grateful to R. Bromley for introducing me to the Chalk Rock. I also thank J. Hardenbol for inspiring my interest sequence stratigraphy. The manuscript was improved by comments from W. J. Kennedy, C. J. Wood, F. Robaszynski and C. C. Ebdon.

References

AMEDRO, F., COLLETE, C., PIETRESSON DE SAINT-AUBIN, J. & ROBASZYNSKI, F. 1982. Le Turonien Supérieur a *Romaniceras (Romaniceras) deverianum* de L'Aube (France). *Bulletin d'Information des Géologues du Bassin Paris*, 19, 29–37.

BARROIS, C. 1876. *Recherches sur le Terrain Crétacé Superieur de l'Angleterre et de Irlande*. de la Société géologique du Nord Mémoires.

BATHURST, R. G. 1971. *Carbonate Sediments and their Diagenesis*. Developments in Sedimentology, 12.

BIRKELUND, T., HANCOCK, J. M., HART, M. B., RAWSON, P. F., REMANE, J., ROBASZYNSKI, F. and SURLYK, F. 1984. Cretaceous Stage boundaries – proposals. *Bulletin of the Geological Society of Denmark*, **33**, 3–20.

BROMLEY, R. G. 1967. Some observations on the burrowing of thalassinidean Crustacea in chalk hardgrounds. *Quarterly Journal of the Geological Society, London*, **123**, 157–182.

—— 1975. Trace fossils at omission surfaces. *In*: FREY, R. W. (ed.) *The Study of Trace Fossils*, 399–428.

—— & GALE, A. S. 1982. The lithostratigraphy of the English Chalk Rock. *Cretaceous Research*, **3**, 273–306.

CHATWIN, C. P. & WITHERS, T. H. 1908. The zones of the chalk of the Thames Valley between Goring and Shiplake. *Proceedings of the Geologists' Association, London*, **20**, 390–421.

DEVALQUE, C., AMEDRO, F., PHILIP, J. & ROBASZYNSKI, F. 1982. Etat des correlations litho et biostratigraphiques dans le Turonien supérieur des massifs d'Uchaux et de la Ceze. Les ammonites et les rudistes. *Memoires du Museum National d'Histoire Naturelle Série C. Sciences de la Terre*, **XLIX**, 57–71.

EKDALE, A. A. & BROMLEY, R. G. 1984. Comparative ichnology of shelf-sea and deep-sea chalk. *Journal of Palaeontology*, **58**, 322–332.

GALE, A. S. 1990. A Milankovitch scale for Cenomanian time. *Terra Research*, **1**, 420–425.

—— 1995. Cyclostratigraphy and correlation of the Cenomanian of western Europe. *In*: HOUSE, M. R. & GALE, A. S. (eds) *Orbital Forcing Timescales and Cyclostratigraphy*. Geological Society, London, Special Publications, **85**, 177–197.

—— & WOODROOF, P. B. 1981. A Coniacian ammonite from the 'Top Rock' in the Chalk of Kent. *Geological Magazine*, **18**, 557–580.

——, JENKYNS, H. C., KENNEDY, W. J. & CORFIELD, R. M. 1993. Chemostratigraphy versus biostratigraphy: data from around the Cenomanian–Turonian boundary. *Journal of the Geological Society*, London, **150**, 29–32.

——, WOOD, C. J. & BROMLEY, R. G. 1987. Lithostratigraphy and marker bed correlation in the White Chalk (Cenomanian–Campanian) of southern England *Mesozoic Research*, **1**, 107–118.

HANCOCK, J. M. 1989. Sea-level changes in the British region during the Cretaceous. *Proceedings of the Geologists' Association, London*, **100**, 565–594.

—— & KAUFMANN, E. G. 1979. The great transgressions of the Late Cretaceous. *Journal of the Geological Society, London,* **136**, 175–186.

HAQ, B. U., HARDENBOL, J. & VAIL, P. R. 1988. Mesozoic and Cenozoic chronostratigraphy and eustatic cycles. *In*: WILGUS, C. K., HASTINGS, B. S., KENDALL, C. G. ST. C., POSAMENTIER, H. W., ROSS, C. A. & VAN WAGONER, J. C. (eds) *Sea-Level Changes: an Integrated Approach.* Society of Economic Paleontologists and Mineralogists, Special Publications, **42**, 71–108.

HART, M. B. 1980. The recognition of mid-Cretaceous

sea-level changes by means of Foraminifera. *Cretaceous Research*, **1**, 289–297.

—— 1990. Cretaceous sea level changes and global eustatic curves; evidence from SW England. *Proceedings of the Ussher Society*, **7**, 268–272.

—— & BAILEY, H. W. 1979. The distribution of planktonic Foraminiferida in the mid-Cretaceous of NW Europe. *Aspekte der Kreide Europas, IUGS Series A*, **6**, 527–542.

——, ——, CRITTENDEN, S., FLETCHER, B. N., PRICE, R. J. & SWIECICKI, A. 1989. *In*: JENKINS, D. G. & MURRAY, J. W. (eds) *Stratigraphical Atlas of Fossil Foraminifera*, 2nd edn. Ellis Horwood, Chichester, 273–371.

HAWKINS, H. L. 1942. Some episodes in the geological history of the south of England. *Quarterly Journal of the Geological Society, London*, **98**, xlix–lxx.

HÉBERT, E. 1875. Classification des terrains crétacés supérieurs. *Bulletin de la Société Géologique Français*, **3**, 595–599.

JARVIS, I. & TOCHER, B. A. 1987. Field meeting: the Cretaceous of SE Devon, 14–16th March, 1986. *Proceedings of the Geologists' Association, London*, **98**, 51–66.

—— & WOODROOF, P. B. 1984. Stratigraphy of the Cenomanian and basal Turonian (Upper Cretaceous) between Branscombe and Seaton, SE Devon, England. *Proceedings of the Geologists' Association, London*, **95**, 193–215.

JEFFERIES, R. P. S. 1962. The palaeoecology of the *Actinocamax plenus* subzone (lowest Turonian) in the Anglo–Paris Basin. *Palaeontology*, **4**, 609-647.

—— 1963. The stratigraphy of the *Actinocamax plenus* subzone (Turonian) in the Anglo-Paris Basin. *Proceedings of the Geologists' Association, London*, **74**, 1–34.

JENKYNS, H. C., GALE, A. S. & CORFIELD, R. M. 1994. Carbon- and oxygen-isotope stratigraphy of the English Chalk and the Italian Scaglia and its palaeoclimatic significance. *Geological Magazine*, **131**, 1–34.

JUIGNET, P. 1974. *La transgression crétacée sur la bordure orientale du Massif armoricain. Aptien, Albien, Cénomanien de Normandie et du Maine. La stratotype du Cénomanien*. Thesis, Université de Caen.

—— & BRETON, G. 1994. Stratigraphie, rhythmes sedimentaires, et eustatisme dans les craies Turoniennes de la region de Fécamp (Seine-Maritime, France). *Bulletin trimestrielle de la Societé Géologique et Amis Muséum du Havre*, **81**, 55–81.

JUKES-BROWNE, A. J. & HILL, W. 1903. *The Cretaceous Rocks of Britain. II. The Lower and Middle Chalk of England*. Geological Survey of the United Kingdom, Memoirs.

KAPLAN, U. 1986. Ammonite stratigraphy of the Turonian in Germany. *Newsletters on Stratigraphy*, **17**, 9–20.

KENNEDY, W. J. & GARRISON, R. E. 1975. Morphology and genesis of nodular chalks and hardgrounds in the Upper Cretaceous of southern England. *Sedimentology*, **22**, 311–386.

—— & COBBAN, W. A. 1991. Stratigraphy and interregional correlation of the Cenomanian–Turonian transition in the Western Interior of the United States near Pueblo, Colorado, a potential stratotype for the base of the Turonian. *Newsletters on Stratigraphy*, **24**, 1–33.

MORTIMORE, R. N. 1983. The stratigraphy and sedimentation of the Turonian–Campanian in the southern Province of England. *Zitteliana*, **10**, 27-41.

—— 1986. Stratigraphy of the Upper Cretaceous White Chalk of Sussex. *Proceedings of the Geologists' Association, London*, **97**, 97–139.

—— 1987. Upper Cretaceous Chalk in the North and South Downs, England; a correlation. *Proceedings of the Geologists' Association, London*, **98**, 77–86.

—— 1988. Upper Cretaceous White Chalk in the Anglo–Paris Basin: a discussion of lithological units. *Proceedings of the Geologists' Association, London*, **99**, 67–70.

—— & POMEROL, B. 1987. Correlation of the Upper Cretaceous White Chalk (Turonian to Campanian) in the Anglo–Paris Basin. *Proceedings of the Geologists' Association, London*, **98**, 97–143.

—— & —— 1991. Stratigraphy and eustatic implications of trace fossil events in the Upper Cretaceous Chalk of Northern Europe. *Palaios*, **6**, 216–231.

PACEY, N. R. 1984. Bentonites in the Chalk of central eastern England and their relation to the opening of the North Atlantic. *Earth and Planetary Science Letters*, **67**, 48–60.

PENNING, W. H. & JUKES-BROWNE, A. J. 1881. *Geology of the Neighbourhood of Cambridge*. Geological Survey of England and Wales, Memoirs.

ROBASZYNSKI, F., CARON, M. & 8 others 1990. A tentative integrated stratigraphy in the Turonian of Central Tunisia: formations, zones and sequential stratigraphy in the Kalaat Senan area. *Bulletin des Centres de Recherches Exploration et Production Elf Aquitaine*, **14**, 213–384.

——, JUIGNET, P., GALE, A. S., AMEDRO, F. & HARDENBOL, J. 1996. Sequence stratigraphy in the Cretaceous of the Anglo–Paris Basin, exemplified by the Cenomanian Stage. *In*: DE GRACIANSKY, P. C., HARDENBOL, J., JACQUIN, T., VAIL, P. R. & FARLEY, M. B. (eds) *Cenozoic and Mesozoic Sequence Stratigraphy of European Basins*. Society of Economic Palaeontologists and Mineralogists. Special Publication, in press.

ROBINSON, N. D. 1986. Lithostratigraphy of the Chalk Group, North Downs, Southeast England. *Proceedings of the Geologists' Association, London*, **97**, 141–170.

STONELEY, R. 1992. Review of the habitat of petroleum in the Wessex Basin: implications for exploration. *Proceedings of the Ussher Society*, **8**, 1–6.

TRÖGER, K.-A. 1989. Problems of Upper Cretaceous inoceramid biostratigraphy and paleobiogeography in Europe and Western Asia. *In*: WIEDMANN, J. (ed.) *Cretaceous of the Western Tethys*. Proceedings of the 3rd International Cretaceous Symposium, Tübingen 1987. Schweizerbart'sche, Stuttgart, 911–930.

WEIMER, R. J. 1984. Relation of unconformities, tectonics and sea-level changes, Cretaceous of Western Interior, USA. *In*: SCHLEE, J. S. (ed.) *Interregional Unconformities and Hydrocarbon Accumulation*. American Association of Petroleum Geologists Memoirs, **36**, 7–35.

WESTHEAD, R. K. & WOODS, M. A. 1994. Anomalous

Turonian–Campanian Chalk deposition in south Dorset; the influence of inherited pre-Albian structures. *Proceedings of the Geologists' Association, London*, **105**, 81–91.

WRAY, D. S. & GALE, A. S. 1993. Geochemical correlation of marl bands in Turonian Chalks of the Anglo–Paris Basin. *In*: HAILWOOD, E. A. & KIDD, R. B. (eds) *High Resolution Stratigraphy*. Geological Society, London, Special Publications, **70**, 211–226.

WRIGHT, C. W. 1979. The ammonites of the English Chalk Rock. *Bulletin of the British Museum (Natural History), Geology*, **31**, 281–332.

—— & KENNEDY, W. J. 1981. *The Ammonoidea of the Plenus Marls and Middle Chalk*. Monograph of the Palaeontographical Society.

Sequence stratigraphical analysis of late Ordovician and early Silurian depositional systems in the Welsh Basin: a critical assessment

N. H. WOODCOCK[1], A. J. BUTLER[1], J. R. DAVIES[2] & R. A. WATERS[3]

[1]*Department of Earth Sciences, University of Cambridge, Cambridge CB2 3EQ, UK*

[2]*British Geological Survey, Room G19, Sir George Stapledon Building, University of Wales, Penglais, Aberystwyth, Dyfed SY23 3DB, UK*

[3]*British Geological Survey, Kingsley Dunham Centre, Keyworth, Nottingham NG12 5GG, UK*

Abstract: Sequence stratigraphical concepts are applied to a 55 km long transect through the uppermost Ordovician and lower Silurian rocks of the Welsh Basin and the adjoining Midland Platform. The study focuses on sedimentary rocks deposited during the Llandovery epoch (about 439–430 Ma).

An early Llandovery slope apron of hemipelagite and laterally supplied mudstone turbidites shows the influence of eustatic sea-level changes. Two complete depositional sequences are recognized. Transgressions and highstands on the platform were accompanied by laminated hemipelagite deposition in the basin, recording anoxic bottom waters in a stratified watermass. The intervening regressions generated unconformities on the platform and produced bioturbated basinal sediments in a more oxic environment.

By contrast, the extent and timing of sandstone-turbidite and debrite systems in the basin are strongly affected by tectonic activity. Easterly derived late Llandovery facies relate directly to faulting and submarine mass wasting along the basin margin. Southerly derived late Llandovery to mid-Wenlock sandstone-lobe systems were the products of tectonic uplift in extrabasinal source areas and were partially confined within intrabasinal tilted fault blocks. Subsidence analysis confirms a basinal stretching event in late Llandovery (Telychian) time and shows that it also affected the adjacent Midland Platform. The depositional systems of this period constitute elements of a dual-sourced depositional sequence, within which eustatic effects are masked by the strong influence of relative base-level changes attendant to tectonism.

The architecture of the early Silurian Welsh Basin illustrates the complexities of applying sequence stratigraphical models where there is an interleaving of several depositional systems, each modulated by a different mix of eustatic, tectonic and input controls.

This paper results from an attempt to refine a preliminary sequence stratigraphical analysis of the Lower Palaeozoic Welsh Basin (Woodcock 1990) using data from new primary mapping by the British Geological Survey (Wilson *et al.* 1992; Waters *et al.* 1993; Davies and Waters 1994; Davies *et al.* in press; British Geological Survey 1993, 1994). In particular, these data constrain a stratigraphical transect from the basin centre to its eastern platform during the time of most rapid turbidite infilling in the latest Ordovician to early Silurian. However, the architecture of this transect does not fit a conventional sequence stratigraphical model involving changes in relative sea level. This paper examines the reasons for this mismatch.

The main depositional systems of the latest Ordovician to early Silurian basin are described, emphasizing the importance of both lateral and axial sediment supply paths. The genetic factors controlling the stratal geometry are diagnosed and it is argued that each system was influenced by a different balance of eustasy, intrabasinal tectonics and extrabasinal sediment supply. In such a basin, the architecture of any one transect will not fit a simple model of base level change, especially if much of the accommodation space remains unfilled by sediment.

Geological setting

For most of early Palaeozoic time, the Welsh Basin was an area of enhanced subsidence and sediment accumulation with respect to the Midland Platform further to the southeast and the Irish Sea Platform to its northwest (Fig. 1 inset). The basin lay on thinned continental crust of the Eastern Avalonia microcontinent. Its fill is dominated by turbidite sandstones and mudstones with important volcanic intercalations, mainly within mid-Ordovician sequences. These volcanic rocks record intra-arc and then back-arc extension above a subducting lithospheric slab of the Iapetus Ocean, which separated Eastern Avalonia from the Laurentian continent to the north (e.g. Kokelaar *et al.* 1984).

From HESSELBO, S. P. & PARKINSON, D. N. (eds), 1996, *Sequence Stratigraphy in British Geology*, Geological Society Special Publication No. 103, pp. 197–208.

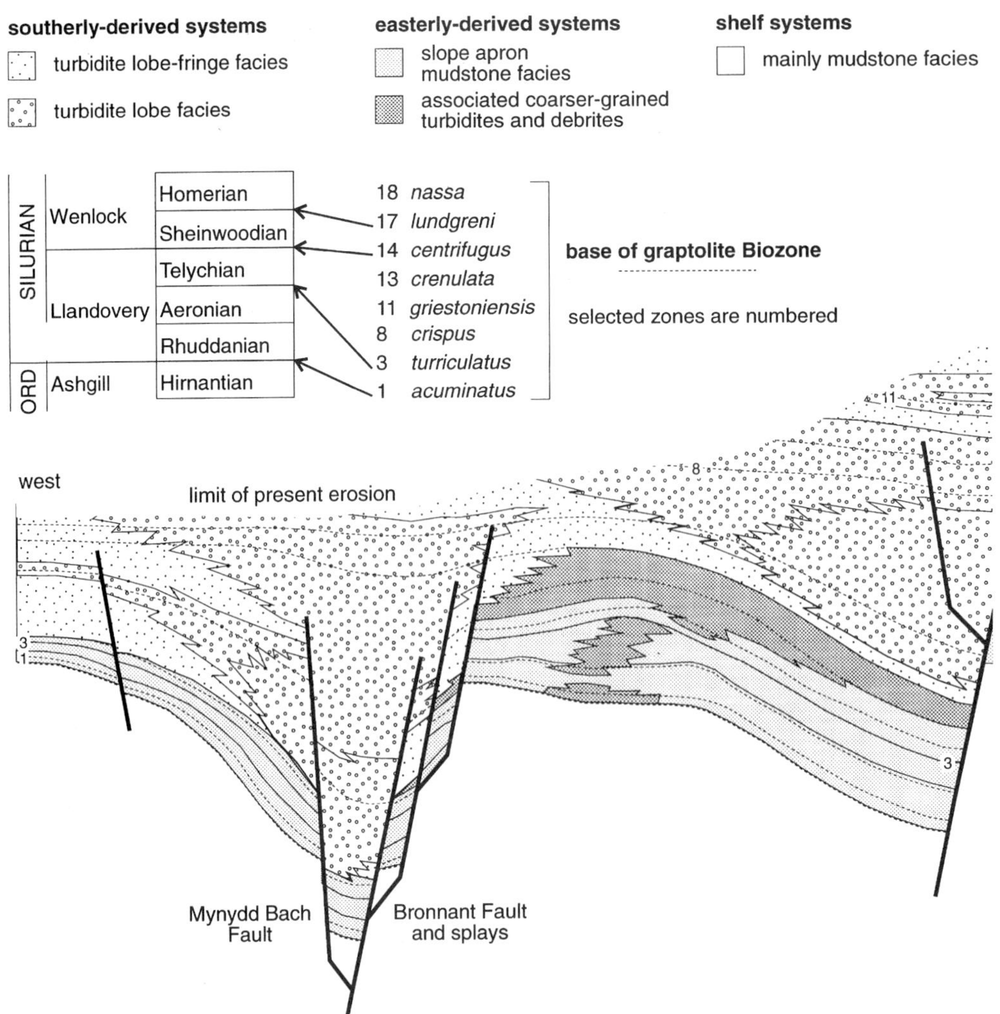

By late Ordovician (Ashgill) time, volcanic centres in the Welsh Basin had shut down as Eastern Avalonia approached and began to impinge on Laurentia. The uppermost Ordovician (Hirnantian) to lower Silurian (Llandovery to early Wenlock) rocks described in this paper were deposited in the developing convergence zone and probably include the first erosional debris from uplifted areas within it (Soper & Woodcock 1990). Marine conditions in the Welsh Basin gave way to non-marine sedimentation in latest Silurian (Pridoli) time. Sedimentation continued into the early Devonian, but the basin was progressively shortened and uplifted during the Acadian Orogeny, culminating in a major mid-Devonian unconformity.

The rocks described in this paper lie entirely within the Powys Supergroup (Woodcock 1990), the megasequence deposited after volcanic shut-down in the basin and before the Acadian orogenic climax. Reviews of the Powys Supergroup in Mid-Wales are provided by Woodcock & Bassett (1993) and Davies *et al.* (in press). Attention focuses here on the Llandovery Series, the stratigraphical interval that crops out most extensively across the mid-Welsh segment of the basin and its margin. The underlying upper Ordovician rocks crop out more locally in structural inliers, whereas the overlying Wenlock and Ludlow Series have been removed from above much of the basin, mostly by syn-Acadian erosion.

The stratigraphical transect to be discussed incorporates outcrop data from a band of ground stretching from the basin centre around Aberaeron eastwards to the basin margin near Llandrindod Wells (Fig. 1 inset). This band coincides with the Llanilar and Rhayader 1:50 000 geological sheets

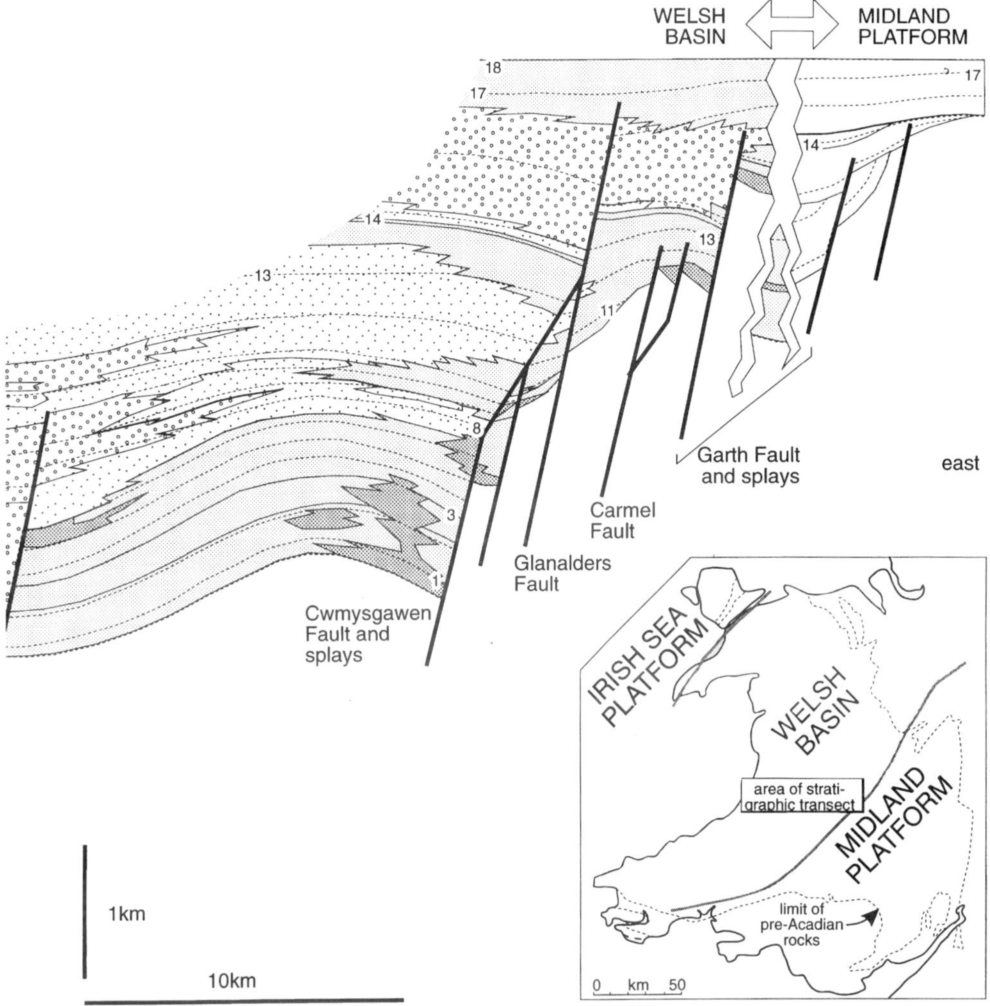

Fig. 1. Pre-Acadian stratigraphical cross-section of the latest Ashgill, Llandovery and Wenlock succession of the Welsh Basin along the transect shown on the inset map. The section is restored with the base of the *nassa* Biozone (late Wenlock) horizontal.

(British Geological Survey 1993, 1994). Outcrop quality is moderately good in rugged topography, including a number of valleys that provide long, continuous sections through the major turbidite systems.

Architecture of the platform to basin transect

Broad stratigraphical relationships are displayed on an east–west cross-section (Fig. 1) simplified from that in Davies *et al.* (in press). A version of the western half of this section has been discussed by Wilson *et al.* (1992) and of its eastern half by Waters *et al.* (1993). The section shows pre-Acadian facies and thickness relationships and is underpinned by detailed biostratigraphy (Zalasiewicz 1990; Loydell 1991; Davies *et al.* in press). Thicknesses have not been corrected, either for early compaction or for the Acadian shortening that was accommodated in cleavage formation and minor folding. The intensity of this deformation is broadly similar across the basinal section eastwards as far as the Garth Fault, and strain corrections would not grossly alter the across-strike thickness changes of individual units nor their geometrical relationships with adjoining units.

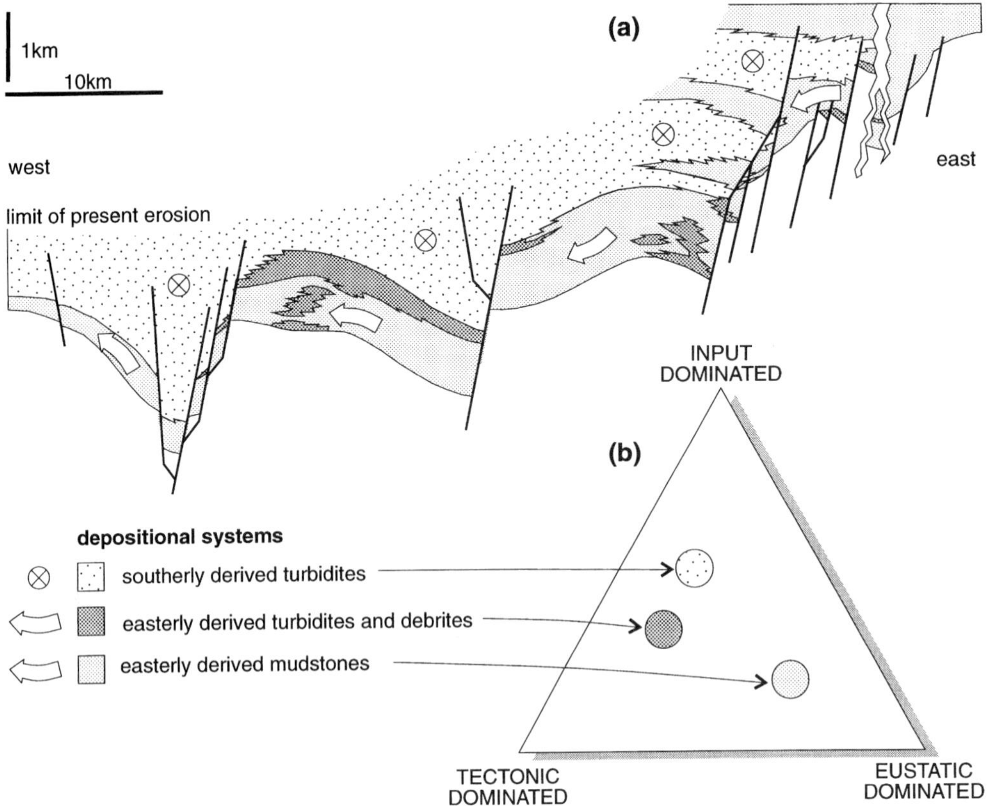

Fig. 2. Chronostratigraphical diagram (a) showing the inter-relationship of the three main depositional systems together with (b) their relative genetic controls plotted on the conceptual ternary diagram of Galloway (1991).

The thin solid lines on the cross-section (Fig. 1) are the boundaries of lithostratigraphical units. Selected units only are named and referred to in this paper, but detailed descriptions are provided by Davies *et al.* (in press). The bases to identified biozones are shown by pecked lines, with selected biozones named. The cross-section shows three main depositional components: shelf systems, easterly derived slope-apron systems and southerly derived sandstone-lobe systems.

The *shelf systems* comprise mostly distal shelf mudstones, deposited east of the Garth Fault and its splays. The Llandovery mudstone units are burrowed, recording oxic bottom waters, and the Wenlock units are laminated, recording anoxia. Transgressive sandstone overlies the two internal non-sequences below time lines 1 and 3. A regressive sandstone unit occurs in the latest Aeronian.

The *easterly derived slope-apron systems* are dominated by mudstone facies composed of interbedded mudstone turbidites and hemipelagites. The hemipelagites may either be burrowed (oxic) or laminated (anoxic). Lenticular coarser-grained clastic units locally punctuate the slope apron mudstones. These units vary from conglomeratic debrites and channelized turbidite sandstones and conglomerates to thin-bedded turbidite-lobe facies. The unifying feature of the slope-apron systems is their proved or inferred supply from the east or southeast, directly across the edge of the Midland Platform (Ball *et al.* 1992; Morton *et al.* 1992).

The *southerly derived sandstone-lobe systems* comprise thick sandstone-rich turbidite-lobe deposits passing into sandstone/mudstone lobe-fringe facies. In the lobes and, particularly, in the lobe fringe the resedimented turbidites may be interbedded with hemipelagic mudstones, either burrowed or laminated. In contrast with the slope-apron systems, the sandstone-lobe systems were sourced from the south or southwest, along-strike with respect to the eastern basin margin (e.g. Cave 1979; Smith 1987; Wilson *et al.* 1992).

The interdigitation of the southerly derived and easterly derived depositional systems means that the transect cannot be viewed simply as a record of linked platform-to-basin processes, shifting in

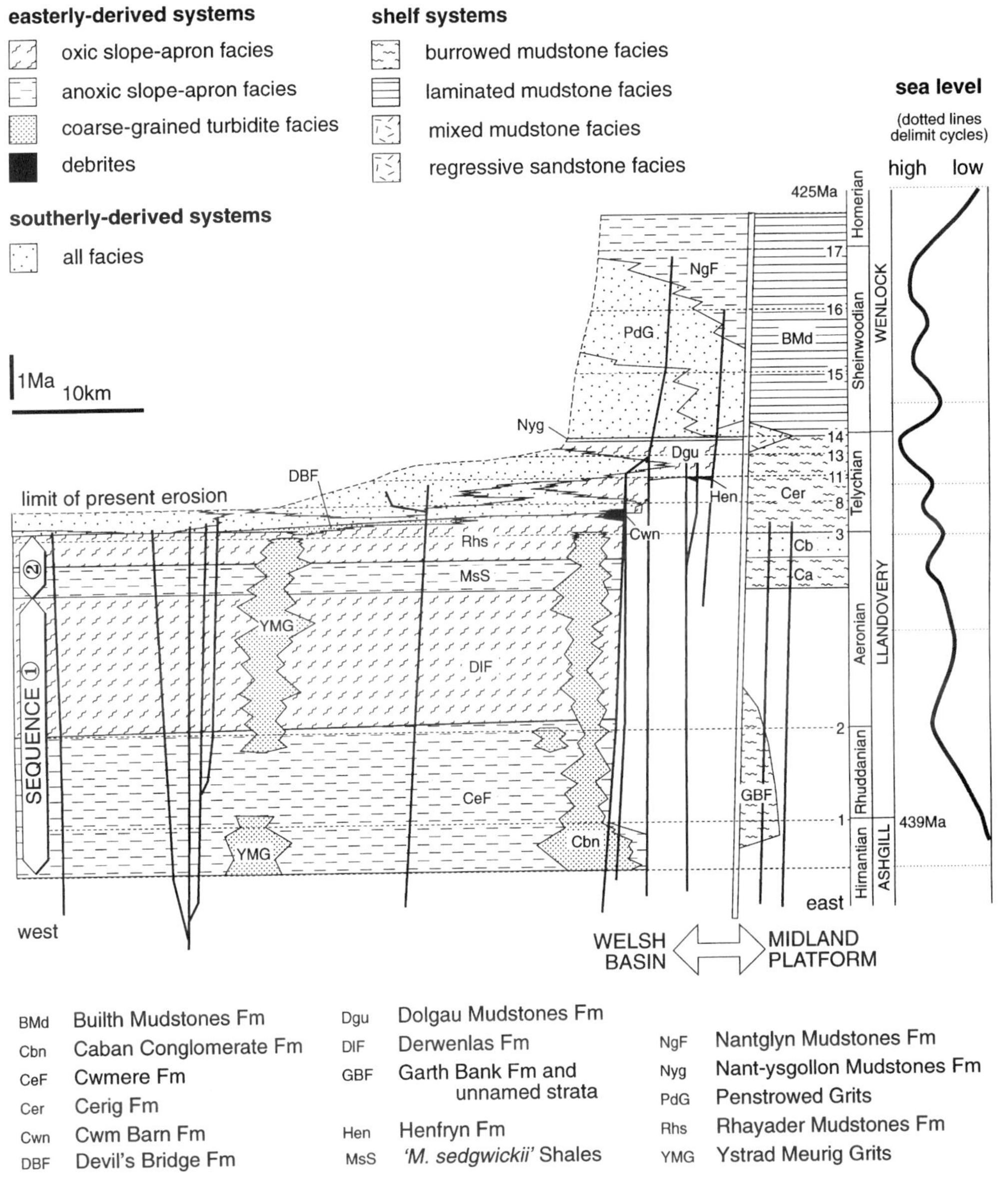

Fig. 3. Chronostratigraphical section detailing the shelf and slope-apron systems, correlated against the eustatic curve of Johnson *et al.* (1991*b*). Faults are shown cutting all the units that they affect, rather than to show their inferred duration of movement. Numbered time lines match those in the key to Fig. 1.

response to the same controlling factors. Contrasting geochemical and mineralogical characteristics of each system (Ball *et al.* 1992; Morton *et al.* 1992) suggest different provenance areas and the possibility of a different balance within them of eustatic influences and intra- or extra-basinal tectonic controls. The spatial relationships of these various systems are simplified in Fig. 2a. The systems are also plotted in Fig. 2b on a conceptual ternary diagram of Galloway (1991). This summarizes the inferred balance of controlling factors for each depositional system, which arises from the discussion in the next two sections of the paper.

Chronology and eustasy

Temporal variations in sedimentation patterns within the Welsh Basin transect are more easily assessed on chronostratigraphical diagrams (e.g. Fig. 3). The cross-section of Fig. 1 has been transformed so that the biostratigraphical time lines are horizontal and are spaced to match the radiometrically calibrated time-scale of Harland *et al.* (1990). Graptolite biozones have been taken of equal duration between stage boundaries and subzones of equal duration within biozone boundaries. Faults are shown on the chronostratigraphical sections as a guide to spatial location with respect to Fig. 1, rather than to indicate their assumed period of activity. The sea-level curve (Fig. 3), reproduced from Johnson *et al.* (1991*b*), was derived by correlating depth-related benthonic communities in carbonate-dominated cratonic shelf sequences on six early Palaeozoic continents: Laurentia, Baltica, Avalonia, Siberia, South China and Gondwana. Johnson *et al.* (1991*b*) deduce that this curve represents eustatic events, including five transgressive/regressive cycles (Figs 3 & 4).

Sequence stratigraphical analysis

Depositional sequences form in response to cycles of base-level movement. They are typically defined by unconformities in shelf successions and correlative conformities in basinal facies. Their initiating base-level movements can include either eustatic (absolute) changes in sea level, or regional (relative) changes related to basinal tectonism (subsidence or uplift). It follows, therefore, that the extent to which sequence boundaries and facies changes occur in step with postulated eustatic sea-level changes can be used as a measure of the relative importance of regional tectonism (Parkinson & Summerhayes 1985).

In view of their contrasting sources, depositional process and geometries, it is convenient, firstly, to establish sequence boundaries within the separately sourced systems of the basin-fill. In the older parts of the basin, sequences composed exclusively of easterly sourced shelf and slope-apron facies display features consistent with eustatic movements in base level. However, younger portions of the fill comprise sequences of mixed source that reflect the growing influence of relative base-level changes driven by regional tectonism. The vertical and lateral transitions between these sequences graphically illustrate the difficulties of defining sequence boundaries in deformed and uplifted ancient basins lacking seismic data.

Easterly supplied shelf and slope-apron mudstones

Chronostratigraphical analysis of the late Hirnantian to early Wenlock shelf and slope apron mudstone succession provides evidence for the first three eustatic cycles of Johnson *et al.* (1991*b*). These facies changes are used to define primary depositional sequences. The first such sequence records the late Hirnantian to early Aeronian post-glacial sea-level rise and the succeeding mid-Aeronian fall. On the shelf, a succession of bioturbated mudstone (Garth Bank Formation and unnamed strata of Williams & Wright 1981) with a basal transgressive sandstone is bounded by non-sequences. On the slope apron the eustatic sea-level rise is recorded by the laminated anoxic hemipelagite of the Cwmere Formation. Flooding of the platform is thought to have promoted an outflow of warm surface water carrying increased volumes of organic matter (e.g. Leggett 1980). In the resulting stratified basin watermass anoxic bottom waters were generated by oxidation of this organic matter. The subsequent relative sea-level fall allowed renewed mixing of this watermass, the re-establishment of a burrowing benthos and accumulation of the oxic hemipelagites of the Derwenlas Formation. The fall on the sea level curve is accompanied by a non-sequence on the platform spanning much of the Aeronian.

The second sequence, of short duration, formed in response to the widely recognized late Aeronian rise and fall in sea level. On the shelf, transgressive shelly sandstones pass upwards into fossiliferous burrow-mottled mudstones, which together constitute the Ca division of Andrew (1925). The succeeding granule-rich shoreface sandstones (Cb of Andrew) define the top of the sequence. In the basin, the transgressive rise in sea level introduced the anoxic slope-apron facies of the *M. sedgwickii* shales. The subsequent regression is represented by the overlying oxic facies of the Rhayader Mudstones.

On the shelf, the abrupt entrance of the burrow-mottled Cerig Formation above Cb records the transgressive rise in sea level at the base of a third sequence and equates with the early Telychian event of Johnson *et al.* (1991*b*). Although the horizon of this event can be identified biostratigraphically, in contrast with the earlier sequences, it has no lithological expression in the basin that is recognizable in the field. Evidence for the mid-Telychian regression and for the late Telychian to early Wenlock eustatic cycle of Johnson *et al.* (1991*b*) is absent from both shelf and slope-apron successions alike, suggesting that eustatic sea-level changes were no longer influencing deposition and, by inference, that regional subsidence had become the dominant control.

Coarse-grained turbidite and debrite facies of the slope apron

The coarse clastic rocks of the slope apron show little relationship with the relative sea-level changes

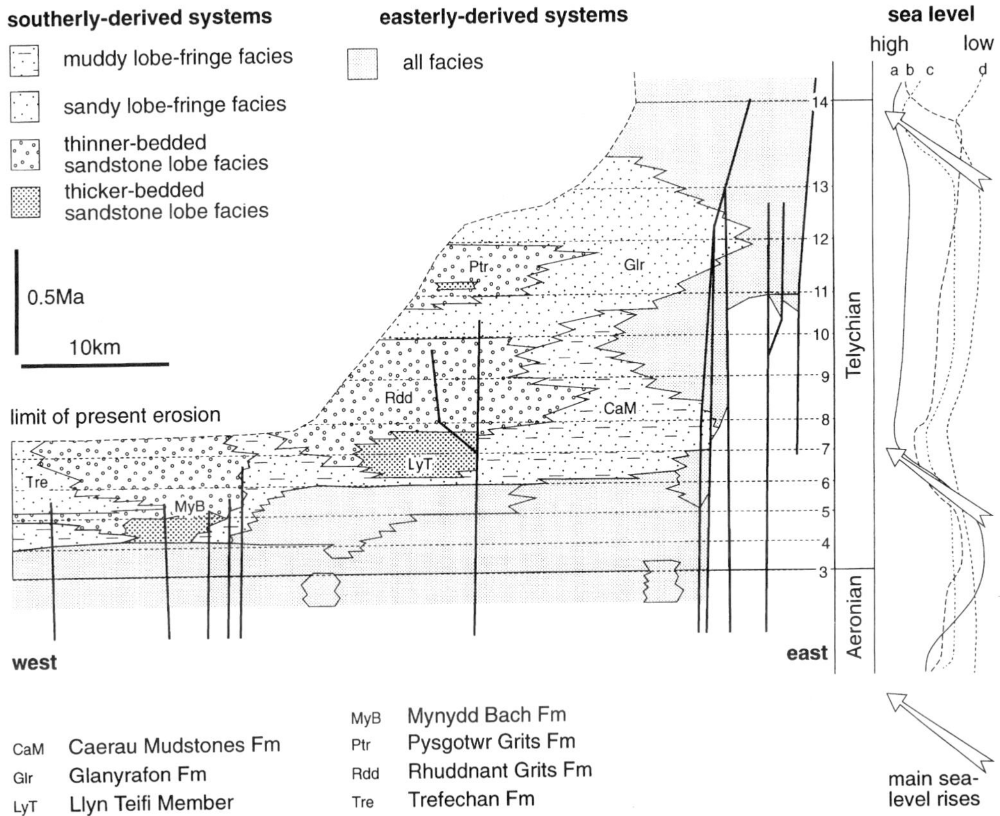

Fig. 4. Chronostratigraphical section of the Telychian (late Llandovery) southerly derived turbidite systems, compared with representative curves of relative sea level from Johnson *et al.* (1991*a*). Curves are from (a) Estonia, (b) Norway, (c) Iowa, USA and (d) Guizhou, South China.

inferred from their enveloping mudstone units. Small-scale, sandy turbidite lobes (Ystrad Meurig Grits Formation) were fed by an easterly sourced nested channel system (Caban Conglomerate Formation) throughout late Hirnatian to Aeronian time (Davies & Waters 1994). The channel system was initiated during the late Hirnantian post-glacial sea-level rise and contracted in size during the late Aeronian fall. Both these effects are out of phase with the expected eustatic influences on clastic supply to the basin (cf. Stow *et al.* 1985). A more likely control is tectonic uplift of the source areas, feeding sediment down an efficient shelf and slope-apron by-pass system.

Tectonic rather than eustatic control on Telychian debrites and turbidites in the proximal slope apron is well constrained. Two coarse-grained units (Henfryn and Cwm Barn formations), deposited during a period of uninterrupted sediment accumulation on the shelf (Fig. 3), were clearly sourced by local submarine mass wasting of normal fault scarps cutting the basin slope (Davies *et al.* in press). Displacement on these faults allowed erosion of their footwall blocks through earlier Llandovery units and into Ashgill strata, before they were eventually onlapped by slope-apron mudstones above a well mapped unconformity. Mass wasting occurred through mid-Telychian time (*utilis* to lower *griestoniensis* Biozones), suggesting a discrete short-lived episode of fault activity along the basin margin. The possibility of high-frequency eustatic events cannot be ruled out. However, an important tectonic event at this time is strongly indicated by the subsidence analysis described in a later section.

In summary, the coarse-grained resedimented facies associated with the slope apron do not accord with the eustatic sea-level changes of Johnson *et al.* (1991*b*). Instead, they probably record tectonic controls in sediment source areas or on the basin margin.

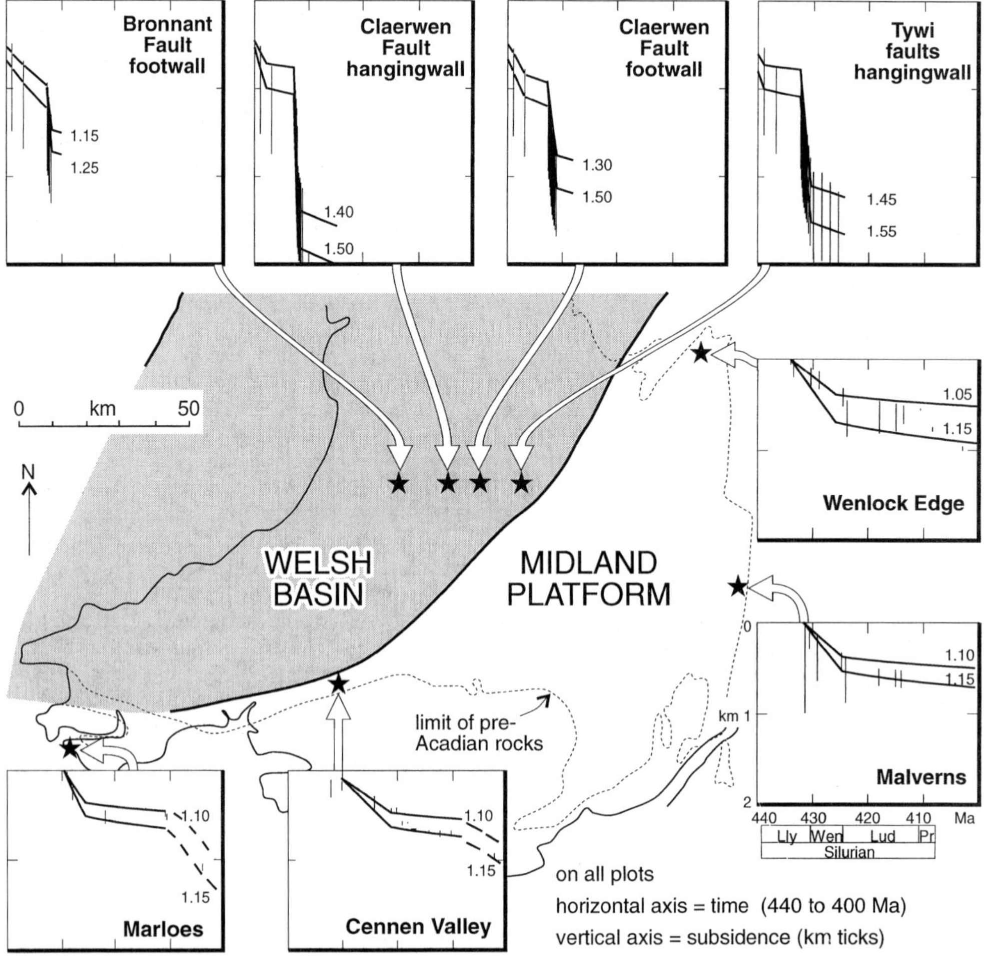

Fig. 5. Silurian subsidence curves for the basin transect and for selected points on the adjoining Midland Platform.

Southerly derived sandstone-lobe systems

Southerly derived sandstone-lobe systems entered the basin during Telychian and early to mid-Wenlock times (Fig. 3). Detailed studies of these systems (Dimberline 1987; Smith 1987; Wilson *et al.* 1992; Clayton 1992) have shown that they display little systematic vertical variation in bed thickness or grain size, inviting comparison with the aggradational systems of Macdonald (1986). Chronostratigraphical analysis of these systems and comparison with detailed sea level curves for the period, derived from Laurentia, Baltica and South China (Johnson *et al.* 1991*b*), confirm that eustatic effects exercised little influence on their development (Fig. 4). The sandstone lobes of the Mynydd Bach Formation aggraded during a period of rising global sea level. The Rhuddnant and Pysgotwr Grits were developed across fault blocks to the east, late in this rise and throughout the subsequent highstand and regression. The supply of southerly derived turbidites waned at the time of the late Telychian eustatic rise. Sandstone-lobe facies readvanced during early to mid-Wenlock times (Penstrowed Grits), but again do not display changes consistent with global sea-level movements (Fig. 3).

In not exhibiting facies responses attributable to eustatic events, it is clear that these southerly sourced systems formed in response to pronounced relative movements in base level associated with a marked increase in the grade and volume of sediment supply. Both were responses to regional tectonism. It is well established that the increase in sediment supply reflects tectonic rejuvenation of the source areas to the south (Cave 1979; Smith 1987; Clayton 1994*a*, *b*). Soper & Woodcock (1990) have suggested that this uplifted source area resulted from the newly developing collision between Eastern Avalonia and Laurentia. Tectonism in the sediment source areas was complemented in the basin by an increase in the rate of subsidence. In detail, the geographical and temporal distribution of the sandstone-lobe systems within the sequence was controlled by intrabasinal fault displacements (Wilson *et al.* 1992; Davies *et al.* in press). Clearly, these tectonic effects were able to counteract the effects of Telychian and early Wenlock eustasy, and led to a stratigraphical geometry which is analogous to the tectonic systems tract of Prosser (1993) (Fig. 2a).

Subsidence analysis as an indicator of tectonic influence

The importance of postulated tectonic controls on early Silurian depositional architecture has been tested by applying subsidence history analysis to representative successions determined at outcrop (Fig. 5). Data have been taken separately from the footwall and hanging-wall blocks of the basinal faults and comparative data sets have been compiled from representative shelf successions around the southeast margins of the basin. Each section has been backstripped to allow for compaction, deriving the original sedimentary thickness at the time of deposition. This sediment load is then replaced with a water load (including the original depths of water above the depositional surface) to isolate the tectonic components of subsidence. This backstripping process is a standard technique in basin analysis (e.g. Barton & Wood 1984). The parameters used are the same as those of Wooler *et al.* (1992). The error bars on the resulting curves mainly reflect the uncertainties in palaeobathymetric determination and hence are largest for basinal facies. The lithospheric stretching factors (β) have been determined by fitting theoretical subsidence curves to the water-loaded subsidence plots.

Best-fit curves to the backstripped data all show a marked increase in subsidence rate, beginning in early Telychian time. This subsidence event spans the time of major fault control on turbidite deposition in the basin and of fault-induced mass wasting and debrite deposition along the eastern basin margin. Significantly, it is also recognizable on the shelf, where it coincides with the decoupling of shelf deposition from strong eustatic influence. The smaller errors in water depth on the shelf define the subsidence well, but even the need for wide error bars for the basinal data cannot remove the need for enhanced Telychian subsidence. Estimated stretching (β) factors range from 1.1 on the shelf to around 1.5 in the basin.

As on the chronostratigraphical diagrams, the numerical time-scale is based on that of Harland *et al.* (1990), with the assumption of equal duration of graptolite biozones within stages and equal duration of subzones within biozone boundaries. Changing these assumptions does not smooth out the observed subsidence episode, although it may alter slightly the magnitude of β. No correction has been made for the tectonic thickening of units when backstripping. Again, this may lead to a different stretching value, but does not remove the event itself. The value of the stretching factor will be most in error in basinal sequences where cleavage is present and least in error in the platform sequences.

The main faults that take up differential subsidence, such as the Bronnant Fault, Claerwen Fault and those along the Tywi Lineament, have magnetic and gravity signatures that suggest their propagation from faults in pre-Silurian and probably Precambrian basement (McDonald *et al.* 1992). They therefore record crustal extension rather than superficial stratal extension. This extension must have occurred in late Llandovery time, because along the Bronnant and Claerwen Faults there is no localized slope facies to suggest that a pre-existing topography was being infilled. The subsidence data therefore reinforce the stratigraphical case for strong intrabasinal tectonic control on the system architecture of the southerly derived turbidites during late Llandovery time.

One unsolved regional problem is that crustal extension in Wales is postulated to be coeval with the uplift of a new source area to the southwest, presumably undergoing crustal shortening. One way of achieving this synchronous uplift and subsidence would be within a regional strike-slip regime. The spatial and temporal pattern of late Llandovery subsidence and uplift is broadly compatible with the suggestion of Soper & Woodcock (1990) that the western end of Eastern Avalonia was beginning its hard collision with Laurentia at this time. On this hypothesis, new source areas could be uplifted in a transpressive collision zone southwest of the Welsh Basin, and pre-existing faults that bounded and underlay the basin could have been reactivated in transtension.

Discussion

Lessons for Welsh Basin studies

Attempts to apply sequence stratigraphical thinking to the Welsh Basin have usefully focused attention on where the basin does not fit conventional models and on the reasons why. However, the possibility must also be considered that the misfit is an artifact of an inadequate or inappropriate observational database. Some of the relevant factors are as follows.

(1) The lack of any seismic control means that there is a reduced possibility of seeing transgressive or regressive signatures normally identified from stratal reflector geometries.

(2) In the sandstone-lobe systems, any such signatures should have been apparent in systematic grain-size and bed-thickness trends, none of which is found. In the slope-apron mudstone systems and the shelf systems, the non-recognition of transgressive or regressive stratal signatures is potentially more serious, because the scale of exposure in the field precludes the identification of any major regional discordances that may be present.

(3) The inverted, deformed and eroded nature of the basin allows a useful proportion of each depositional system to be studied in detail at outcrop, but creates uncertainties of extrapolation and interpolation where units are below or above the present erosion level. Projection of data along plunging folds resolves some uncertainties.

(4) Sequence stratigraphical analysis would be easier if the excellence of graptolite biostratigraphical resolution and correlation within the basin and slope apron was continued onto the adjoining shelf. There, graptolite faunas are generally very sparse, whereas the other biostratigraphical tools available (brachiopods and acritarchs) provide a comparatively coarse resolution, hampering event correlations with the slope and basin.

Lessons for sequence stratigraphical studies

Conversely, we suggest that the Welsh Basin study highlights a number of general problems in applying sequence stratigraphical methods to ancient sedimentary basins.

(1) The interpretation of basinal sequences is difficult if they cannot be followed continuously into shelf systems with which they are genetically linked.

(2) Uncertainties in interpreting stratal geometries are introduced in basinal systems by the presence of unfilled sediment accommodation space. This is the problem of distinguishing syn-sedimentary growth in space from the passive infilling of pre-formed space (e.g. Bertram & Milton 1989).

(3) There is a particular difficulty with basins filled by two or more coeval depositional systems, each of which may respond to a different balance of genetic controls. This situation arises commonly in tectonically controlled turbidite basins, where axial supply systems interdigitate with lateral systems.

(4) The distinction of the axial and lateral turbidite systems in the Welsh Basin has required outcrop-scale palaeoflow data (e.g. Smith 1987; Wilson *et al.* 1992; Clayton 1994*b*). Discrimination from their seismic character alone would be impossible, particularly in the absence of clear progradational signatures.

(5) The position of boundaries between depositional systems and, by inference, depositional sequences is scale-dependent. On the scale of analysis in this paper, they are deceptively sharp (Fig. 2a). On an outcrop scale the relationship is more complex. For instance, southerly derived turbidite beds can be interbedded with slope-apron mudstones on a vertical scale of centimetres to tens of metres and over lateral distances of kilometres. Some of the implications of this scale-dependence have been explored by Cartwright *et al.* (1993), particularly the consequences of thin hemipelagic mudstone units being below the limits of seismic resolution.

Synthesis

(1) The architecture of the studied transect across the Welsh Basin does not fit a simple sequence-stratigraphical model showing consistent basinward and platformward shifts in facies.

(2) Instead, on the scale of the units mappable in the field, there is a tendency for sedimentary systems to aggrade over a template of basement-linked faults across which there was syn-sedimentary displacement.

(3) A major reason for the complexity of the transect is that the different depositional systems intersected by it are not directly linked to each other in terms of source and depositional process (Fig. 2a).

(4) Late Hirnantian to early Telychian, easterly sourced, mudstone-dominated shelf and slope apron systems in the Welsh Basin accumulated during a period of tectonic quiesence. They display facies changes that are in step with the postulated eustatic changes in base level. Disconformities and/or abrupt facies changes in the shelf succession and their correlatable conformities in the basinal slope apron can be used to define eustatic depositional sequences. Two complete transgressive/regressive sequences are recognized: a late Hirnantian to mid-Aeronian sequence; and a late Aeronian sequence.

(5) Evidence for the transgressive base to a third eustatic sequence, of early Telychian age, is recognized in the eastern shelf succession, but has no expression in coeval slope-apron facies, suggesting that the factors influencing deposition in the deep water basin were decoupled from those affecting the shelf.

(6) Coarse-grained turbidite and debrite systems associated with the slope-apron display changes in geometry and facies that are out of step with those in the enveloping mudstone succession. These may relate instead to extrabasinal events affecting coarse

clastic source areas, with efficient shelf by-pass supply routes in operation or, as during the Telychian, to intrabasinal fault activity.

(7) Early Telychian to early Wenlock, southerly sourced sandstone lobe systems invaded the basin during a period of active tectonism. The entry, growth and migration of these systems are inconsistent with contemporary, eustatic base-level changes; effects related to the latter were evidently masked by those attendant to regional changes in base level. Uplift of terrigenoclastic source areas saw dramatic increases in the volume and grade of detritus supplied to the basin. This was complemented and accommodated by enhanced rates of basinal subsidence, notably in the hanging-wall blocks of intrabasinal faults. These sandstone-lobe systems constitute the deep water elements of a tectonic systems tract.

(8) The failure of both shelf and slope-apron successions to record a predicted late Telychian to early Wenlock eustatic sequence, combined with evidence of enhanced rates of local subsidence, suggest that deposition across the Welsh Basin as a whole (deep water, slope and shelf facies) became largely divorced from eustatic control at this time. Both easterly and southerly sourced systems now responded to the effects of regional subsidence and increased sediment supply. They can be viewed as elements of the same tectonic systems tract and components of a single, dual-sourced, early Telychian to mid-Wenlock depositional sequence.

(9) Thus the late Ashgill, Llandovery and early to mid-Wenlock succession of the Welsh Basin highlights the difficulties of applying sequence stratigraphical techniques to two-dimensional stratigraphical transects where these intersect systems that have different sources, and geometries that reflect contrasting extra- and intra-basinal controls. It illustrates the potential for separately sourced depositional systems, each responding to a different balance of absolute and relative base-level changes, to coexist within a single basin. Moreover, it demonstrates that as base-level movements related to regional tectonism displace eustasy as the dominate control on basinal sedimentation, a single source system, initiated as part of a eustatic sequence in one part of the basin, may evolve into a component of a dual-sourced, tectonic sequence which extends across the basin as whole. Under these circumstances sequence boundaries may become cryptic or artificial; simply artifacts of correlation and typically beyond the resolution of current field-based techniques.

We acknowledge the input from other members of the British Geological Survey Central Wales Project, notably C. Fletcher, D. Wilson and J. Zalasiewicz, in erecting the sedimentary architecture of the platform to basin transect. We thank N. White for the use of his subsidence analysis program. J.R.D. and R.A.W publish by permission of the Director, British Geological Survey (NERC).

References

ANDREW, G. 1925. The Llandovery rocks of Garth, Breconshire. *Quarterly Journal of the Geological Society of London*, **81**, 389–406.

BALL, T. K., DAVIES, J. R., WATERS, R. A. & ZALASIEWICZ, J. A. 1992. Geochemical discrimination of Silurian mudstones according to depositional process and provenance within the southern Welsh Basin. *Geological Magazine*, **129**, 567–572.

BARTON, P. & WOOD, R. 1984. Tectonic evolution of the North Sea basin: crustal stretching and subsidence. *Geophysical Journal of the Royal Astronomical Society*, **79**, 987–1022.

BERTRAM, G. T. & MILTON, N. J. 1989. Reconstructing basin evolution from sedimentary thickness: the importance of palaeobathymetric control, with reference to the North Sea. *Basin Research*, **1**, 247–257.

BRITISH GEOLOGICAL SURVEY 1993. *Rhayader, 1:50,000*. England & Wales Sheet 179. Solid Geology. British Geological Survey, Nottingham.

—— 1994. *Llanilar, 1:50,000*. England & Wales Sheet 178. Solid with Drift Geology. British Geological Survey, Nottingham.

CARTWRIGHT, J. A., HADDOCK, R. C. & PINHEIRO, L. M. 1993. The lateral extent of sequence boundaries. *In*: WILLIAMS, G. D. & DOBB, A. (eds) *Tectonics and Seismic Sequence Stratigraphy*. Geological Society, London, Special Publications, **71**, 15–34.

CAVE, R. 1979. Sedimentary environments of the basinal Llandovery of mid-Wales. *In*: HARRIS, A. L., HOLLAND, C. H. & LEAKE, B. E. (eds) *Caledonides of the British Isles: Reviewed*. Geological Society, London, Special Publications, **8**, 517–526,

CLAYTON, C. J. 1992. *The sedimentology of a confined turbidite system in the early Silurian Welsh Basin*. PhD Thesis, University of Cambridge.

—— 1994*a*. A rock volume accumulation curve for the late Ordovician–Silurian Welsh Basin. *Geological Magazine*, **131**, 539–544.

—— 1994*b*. Contrasting sediment gravity flow processes in the late Llandovery Rhuddnant Grits turbidite system, Welsh Basin. *Geological Journal*, **29**, 167–181.

DAVIES, J. R. & WATERS, R. A. 1994. The Caban Conglomerate and Ystrad Meurig Grits formations – nested channels and lobe-switching on the mud-dominated latest Ashgill to Llandovery slope-apron of the Welsh Basin, Wales, U.K. *In*: PICKERING, K. T., HISCOTT, R. N., KENYON, N. H., RICCI LUCCHI, R. & SMITH, R. D. A. (eds) *Atlas of Deep-Water Environments: Architectural Style in Turbidite Systems*. Chapman & Hall, London, 184–193.

——, FLETCHER, C. J. N., WATERS, R. A., WILSON, D., WOODHALL, D. G. & ZALASIEWICZ, J. A. *Geology of the Country Around Llanilar and Rhayader*. Memoir of the British Geological Survey, Sheets 178 and 179, England and Wales, in press.

DIMBERLINE, A. J. 1987. *The sedimentology and diagenesis of the Wenlock turbidite system*. PhD Thesis, University of Cambridge.

GALLOWAY, W. 1991. Genetic stratigraphic sequences in basin analysis I: architecture and genesis of flooding-surface bounded depositional units. *Bulletin of the American Association of Petroleum Geologists*, **73**, 125–142.

HARLAND, W. B., ARMSTRONG, R. L., COX, A. V., CRAIG, L. E., SMITH, A. G. & SMITH, D. G. 1990. *A geologic Time Scale 1989*. Cambridge University Press, Cambridge.

JOHNSON, M. E., BAARLI, B. G., NESTOR, H., RUBEL, M. & WORSLEY, D. 1991*a*. Eustatic sea-level patterns from the Lower Silurian (Llandovery Series) of southern Norway and Estonia. *Bulletin of the Geological Society of America*, **103**, 315–335.

——, KALJO, D. & RONG, J.-Y. 1991*b*. Silurian eustasy. *Special Papers in Palaeontology, London*, **44**, 145–163.

KOKELAAR, B. P., HOWELLS, M. F., BEVINS, R. E., ROACH, R. A. & DUNKLEY, P. N. 1984. The Ordovician marginal basin of Wales. *In*: KOKELAAR, B. P. & HOWELLS, M. F. (eds) *Marginal Basin Geology*. Geological Society, London, Special Publications, **16**, 245–269.

LEGGETT, J. K. 1980. British Lower Palaeozoic black shales and their palaeo-oceanographic significance. *Journal of the Geological Society, London*, **137**, 139–156.

LOYDELL, D. K. 1991. The biostratigraphy and formational relationships of the upper Aeronian and lower Telychian (Llandovery, Silurian) formations of western mid-Wales. *Geological Journal*, **26**, 209–244.

MACDONALD, D. I. M. 1986. Proximal to distal sedimentological variation in a linear turbidite trough: implications for the fan model. *Sedimentology*, **33**, 243–259.

MCDONALD, A. J. W., FLETCHER, C. J. N., CARRUTHERS, R. M., WILSON, D. & EVANS, R. B. 1992. Interpretation of the regional gravity and magnetic surveys of Wales, using shaded relief and Euler deconvolution techniques. *Geological Magazine*, **129**, 523–531.

MORTON, A. C., DAVIES, J. R. & WATERS, R. A. 1992. Heavy minerals as a guide to turbidite provenance in the Lower Palaeozoic Southern Welsh Basin: a pilot study. *Geological Magazine*, **129**, 573–580.

PARKINSON, D. N. & SUMMERHAYES, C. P. 1985. Synchronous global sequence boundaries. *Bulletin of the American Association of Petroleum Geologists*, **69**, 685–687.

PROSSER, S. 1993. Rift-related linked depositional systems and their seismic interpretation. *In*: WILLIAMS, G. D. & DOBB, A. (eds) *Tectonics and Seismic Sequence Stratigraphy*. Geological Society, London, Special Publications, **71,** 35–66.

SMITH, R. D. A. 1987. The *griestoniensis* Zone turbidite system, Welsh Basin. *In*: LEGGETT, J. K. & ZUFFA, G. G. (eds) *Marine Clastic Sedimentology*. Graham & Trotman, London, 190–211.

SOPER, N. J. & WOODCOCK, N. H. 1990. Silurian collision and sediment dispersal patterns in southern Britain. *Geological Magazine*, **127**, 527–542.

STOW, D. A. V., HOWELL, D. G. & NELSON, C. H. 1985. Sedimentary, tectonic and sea level controls. *In*: BOUMA, A. H., NORMARK, W. R. & BARNES, N. E. (eds) *Submarine Fans and Related Turbidite Systems*. Springer, New York, 15–22.

WATERS, R. A., ZALASIEWICZ, J. A. & CAVE, R. 1993. Llandovery basinal and slope sequences of the Rhayader district. *In*: WOODCOCK, N. H. & BASSETT, M. G. (eds) *Geological Excursions in Powys, Central Wales*. University of Wales Press, National Museum of Wales, Cardiff, 155–182.

WILLIAMS, A. & WRIGHT, A. D. 1981. The Ordovician–Silurian boundary in the Garth area of southwest Powys, Wales. *Geological Journal*, **16**, 1–39.

WILSON, D., DAVIES, J. R., WATERS, R. A. & ZALASIEWICZ, J. A. 1992. A fault-controlled depositional model for the Aberystwyth Grits turbidite system. *Geological Magazine*, **129**, 595–607.

WOODCOCK, N. H. 1990. Sequence stratigraphy of the Palaeozoic Welsh Basin. *Journal of the Geological Society, London*, **147**, 537–547.

—— & BASSETT, M. G. 1993. The geology of Powys: an introduction. *In*: WOODCOCK, N. H. & BASSETT, M. G. (eds) *Geological Excursions in Powys, Central Wales*. University of Wales Press, National Museum of Wales, Cardiff, 17–38.

WOOLER, D. A., SMITH, A. G. & WHITE, N. 1992. Measuring lithospheric stretching on Tethyan passive margins. *Journal of the Geological Society, London*, **149**, 517–532.

ZALASIEWICZ, J. A. 1990. Silurian graptolite biostratigraphy in the Welsh Basin. *Journal of the Geological Society*, **147**, 619–622.

ZIEGLER, A. M., COCKS, L. R. M. & MCKERROW, W. S. 1968. The Llandovery transgression of the Welsh Borderland. *Palaeontology*, 11, 736–782.

Tectonic controls on sequence development in the Palaeocene and earliest Eocene of southeast England: implications for North Sea stratigraphy

ROBERT W. O'B. KNOX

British Geological Survey, Keyworth, Nottingham NG12 5GG, UK

Abstract: Depositional sequences in the Palaeocene and early Eocene of southern England are grouped into three unconformity-bounded composite sequences, corresponding to the Ormesby Clay and Thanet Sand formations, the Lambeth Group (including the Woolwich and Reading formations) and the Thames Group (including the Harwich and London Clay formations).

A refined chronostratigraphical framework, coupled with a revised Palaeogene chronometry, reveals a clear relationship between long-term Palaeocene to early Eocene sea-level trends and radiometrically dated volcanic events in the North Atlantic Province. In gross terms, it is possible to relate the progressive Palaeocene uplift of NW Europe to the rise of the proto-Icelandic mantle plume beneath the continental crust of East Greenland, and the early Eocene subsidence of NW Europe to the transfer of the plume to the developing NE Atlantic mid-ocean ridge. However, both the sedimentary and volcanic record indicate that the uplift of NW Europe took place in three stages, expressed in the North Sea region as long-term regresssive–transgressive facies cycles. The first uplift phase was associated with the onset of Hebridean volcanic activity and with a short-lived influx of coarse clastic sediments into the North Sea Basin (basal Ekofisk Formation). No sediment associated with this cycle is preserved in southern England. The second uplift phase was associated with the main period of Hebridean activity and led to substantial uplift of Scottish source areas. It was followed by regional subsidence and transgression of the basin margins (Ormesby–Thanet composite sequence). The third uplift phase was associated with the onset of rifting and volcanism along the North Atlantic rift zone. Eastward tilting initially led to sedimentation in eastern and southern England (Lambeth composite sequence), but continued uplift of southern and western Britian eventually restricted sedimentation to the central and northern parts of the North Sea basin, with major progradation of the Dornoch delta complex taking place along the eastern margin of the Scottish land mass. This period marked the maximum isolation of the North Sea Basin, with reduced salinities being reflected in a near-absence of marine fauna and a highly restricted microflora.

The third uplift phase is believed to mark the culmination of plume-related tectonism in NW Europe and was associated with a resurgence of volcanic activity in the British Tertiary Igneous Province, including intrusion of the Cleveland–Blyth–Acklington dyke system along the uplifted Sole Pit inversion zone. A sharp reduction in intrabasinal activity, coupled with the initiation of a long period of regional subsidence (Thames composite sequence), is interpreted as representing thermal subsidence following the onset of seafloor spreading between Greenland and Scotland. Oceanic crust generation was initially subaerial and was associated with intense, largely basaltic, pyroclastic activity that affected the whole of NW Europe (Balder Formation and equivalent tuffs).

The cause of the episodic nature of the plume-related uplift is uncertain. It appears that the mechanics and configuration of plume emplacement were either intrinsically more complex than previously recognized, or were influenced by changes in the regional stress pattern within the NW European crust.

The early Palaeogene of southeast England was the subject of intensive stratigraphical study in the 19th Century (see Curry 1965), with major compilations carried out by Prestwich (1850, 1852, 1854) and Whitaker (1872). In recent years, the succession has been the subject of further intensive study, stimulated in part by international stratigraphy programmes (e.g. Knox *et al.* 1988; Ali 1994; Aubry 1994; Knox 1994*a*) and the need to place the hydrocarbon-bearing succession of the North Sea into a broader regional sequence stratigraphical context (e.g. Neal 1996). The North Sea successions themselves have been subjected to close sequence stratigraphical scrutiny in recent years, with particular attention being paid to assessing the relative importance of tectonism and eustasy in generating the sea level changes that led to the formation of some of Britain's most important hydrocarbon reservoirs (e.g. Stewart 1987; Milton *et al.* 1990; Armentrout *et al.* 1993; Den Hartog Jager *et al.* 1993; Neal 1996). Although all of these workers agree that the change from the pelagic sedimentation

From HESSELBO, S. P. & PARKINSON, D. N. (eds), 1996, *Sequence Stratigraphy in British Geology*, Geological Society Special Publication No. 103, pp. 209–230.

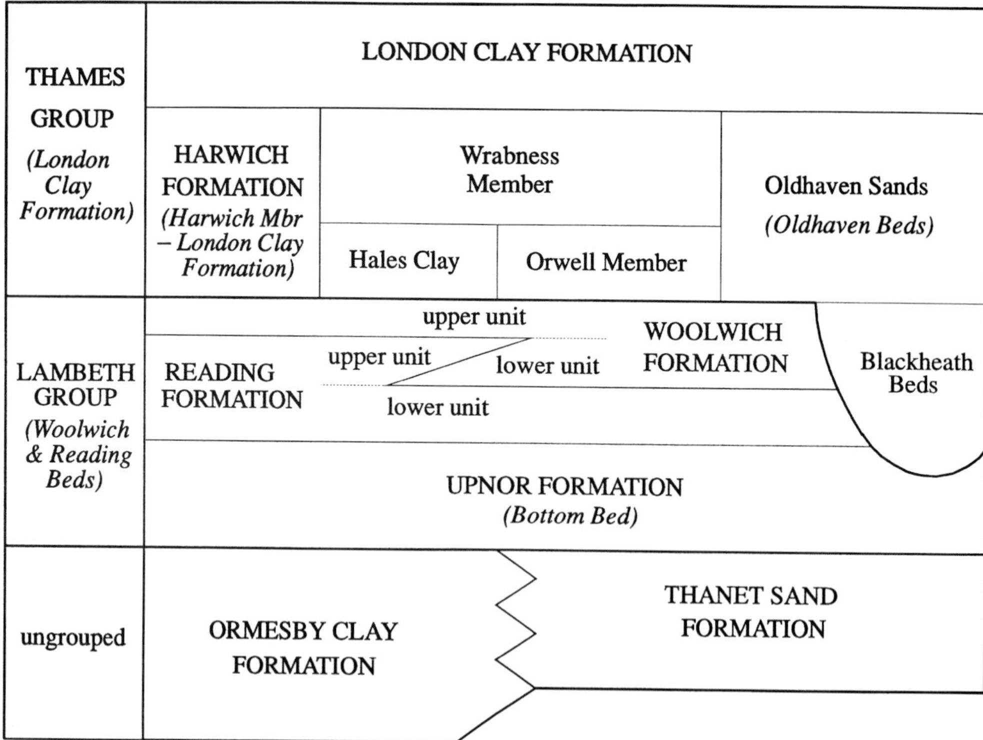

Fig. 1. Lithostratigraphic nomenclature for the Palaeocene and early Eocene of southern England (after Ellison *et al.* 1994; Jolley 1996).

of the Cretaceous and early Palaeocene Chalk Group to the clastic sedimentation of the late Palaeocene marks the onset of a major phase of uplift along the western margin of the North Sea Basin, as first proposed by Parker (1975), models for the generation of second- and third-order sequences differ considerably in their assessment of the role of tectonism (see Armentrout *et al.* 1993, p. 55).

This paper demonstrates how recently acquired, high-quality borehole data from SE England provide important information on the role of local and regional tectonism in the development of different orders of sequence developed within the North Sea region.

Regional context

Early Palaeogene strata are widely distributed in the North Sea and adjacent onshore areas, and include important hydrocarbon reservoirs in parts of the central and northern North Sea. These reservoirs are associated with lowstand fans (e.g. Armentrout *et al.* 1993; Den Hartog Jager *et al.* 1993; Neal 1996), developed within relatively complete successions of deep-water facies. By contrast, the succession of SE England, which is representative of the southern margin of the North Sea Basin as a whole, is thin and incomplete. Nevertheless, the record of sharp facies change and unconformity provided by this and other marginal successions provides a valuable complement to the basinal record and yields important information on the history of relative sea-level change.

Lithostratigraphical framework

The onshore lithostratigraphical nomenclature used in this paper is that of Ellison *et al.* (1994), supplemented by the proposals of Jolley (1996) for the subdivision of the Harwich Formation. Equivalence with earlier established nomenclature is indicated in Fig 1. The oldest formations are the Ormesby Clay Formation and Thanet Sand Formation, representing muddy outer shelf and sandy inner shelf sedimentation, respectively. The Lambeth Group consists of the Upnor Formation, a generally thin transgressive unit of shallow marine to lagoonal facies, overlain by an interdigitation of nearshore marine to lagoonal (Woolwich Formation) and

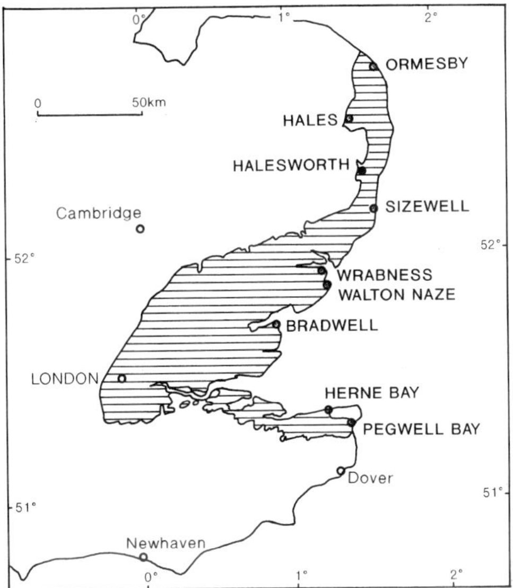

Fig. 2. Distribution of Palaeocene strata in the London Basin and East Anglia (shaded) and location of sections referred to in the text.

continental facies (Reading Formation). The Harwich Formation (including the Oldhaven Sands, Wrabness Member and Hales Clay) represents the re-establishment of shallow marine conditions throughout the area, with the base of the London Clay Formation marking a further deepening of the sea. London Clay sedimentation was terminated by a widespread sublittoral sand sedimentation of the Bagshot Beds. Reference to the offshore succession follows the terminology of Knox & Holloway (1992).

History of stratigraphical investigation in SE England

The succession in SE England was much studied even in the early part of the 19th century, with the first comprehensive account of the early Eocene (now Palaeocene and early Eocene) succession of the London Basin being provided by Prestwich (1850, 1852, 1854). He recognized three major units: the Thanet Sands; the Woolwich and Reading Beds; and the London Clay. Each of these units was shown to include a well-defined basement bed characterized by pebbly sands or pebble beds and abundant glauconite. These were referred to by Prestwich as an unnamed bed of green-coated flints (the Bullhead Bed of subsequent workers), the Woolwich Bottom Bed and the London Clay Basement Bed, respectively.

Although Prestwich and other early workers did much to elucidate the stratigraphy and facies distribution within the succession, a conceptual model of sea level change was not attempted until Stamp's publication in 1921 on the Palaeocene and Eocene of SE England. Stamp was influenced by the work of mainland European geologists such as Leriche (1905), who had divided the successions of the Paris Basin and Belgium into a series of unconformity-bounded transgressive–regressive units, which they termed stages. Stamp referred to these as cycles of sedimentation, which he explained in terms of successive uplift and subsidence of the depositional surface. For the irregular, erosional base of each cycle, presumed to have been generated in its final form during the marine transgression, he introduced the term ravinement.

In applying these concepts to the London Basin succession, Stamp (1921) identified two transgressive–regressive cycles, the first including the Thanet Sands and Woolwich and Reading Beds and the second including the London Clay and the sandy Bagshot Beds. This study marked an important step in the application of sedimentological–stratigraphical concepts to British stratigraphy. It was, however, flawed in stratigraphical detail by the failure to recognize the Woolwich and Reading Beds as constituting a separate cycle of sedimentation. This interpretation stemmed from a longstanding misidentification of the Woolwich–Thanet (now Lambeth–Thanet) boundary at Herne Bay in NE Kent (see Ward 1978, p. 5), which gave the impression that sedimentation was continuous across the boundary. It is now recognized that a significant hiatus exists at the base of the Woolwich and Reading Beds (Lambeth Group), as first proposed by Curry (1981). The true situation, of three cycles, had in fact been appreciated by Prestwich (1854) in his account of the London succession, in which (plate 1) he clearly illustrated ravinement surfaces at the base of each of the three major composite sequences identified in this paper.

Two main developments have taken place in recent years that have thrown significant new light on the nature and development of the early Palaeogene succession of SE England. The first is the acquisition of high-quality stratigraphical data from boreholes (see Fig. 2) in the London area (Ellison *et al.* 1996) and in East Anglia (Cox *et al.* 1985; Knox *et al.* 1990; Knox *et al.* 1994). The second is the development of a greatly improved chronostratigraphical framework, resulting from the application of dinoflagellate and calcareous nannofossil biostratigraphy coupled with magnetostratigraphy, and leading to a much better understanding of the location and duration of hiatuses within the succession (e.g. Aubry *et al.* 1986; Pomerol 1989; Knox 1990; Ali *et al.* 1993; Ali 1994; Aubry 1994; Knox 1994*a*; Ali & Jolley 1996; Fig. 3).

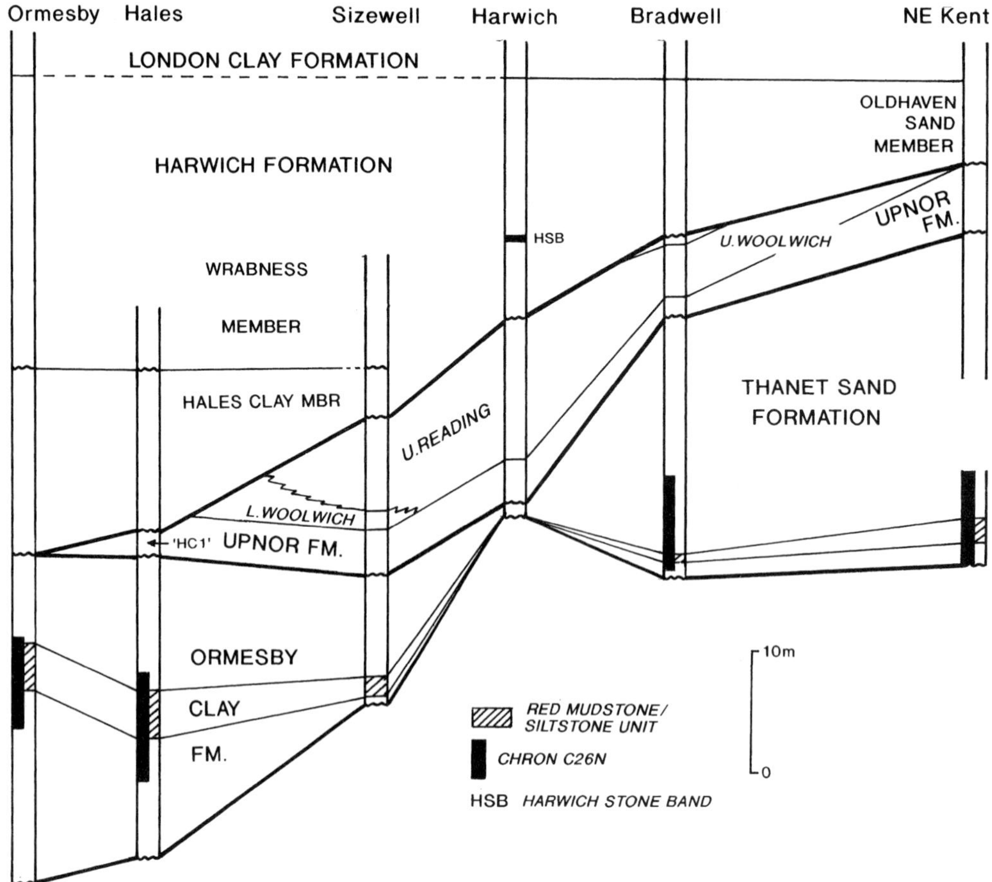

Fig. 3. Schematic N–S cross-section through the late Palaeocene and earliest Eocene of East Anglia and the London Basin illustrating (i) the contrasting thickness trends across the Sudbury–Felixstowe High and (ii) the base Thames Group unconformity, which is responsible for the total removal of the Lambeth Group in north Norfolk (Ormesby section). Details of biostratigraphical correlations between these sections are given in Jolley (1992) and Jolley (1996). Note that following Jolley (in press), what was originally designated the basal unit of the Hales Clay (unit 'HC1') is assigned to the Upnor Formation.

Sequence stratigraphical framework

Three major sequence boundaries can be recognized within the early Palaeogene of SE England, corresponding precisely to the boundaries of the primary lithostratigraphical divisions established by Prestwich (1852). These define three large-scale depositional sequences corresponding to: (1) the Ormesby Clay and Thanet Sand formations; (2) the Lambeth Group, including the Upnor, Woolwich and Reading formations; and (3) the Thames Group, including the Harwich and London Clay formations. These display strongly contrasting facies associations and subsidence patterns, reflecting deposition in markedly different environmental and tectonic settings (Fig. 3).

Each of these large-scale depositional sequences may be described as a third-order composite sequence (after Mitchum & Van Wagoner 1991), being made up of smaller scale (fourth-order) depositional sequences that are marked by less profound changes in facies association. Predictably, the composite sequences in the marginal succession of southern England are incomplete compared with those in basinal parts of the central North Sea and are dominated by deposits of the rising sea-level (transgressive) phase of each cycle.

The following account of sequence stratigraphy in SE England deals firstly with the gross stratigraphical relationships of the Ormesby–Thanet, Lambeth and Thames composite sequences. Detailed discussion of the constituent depositional

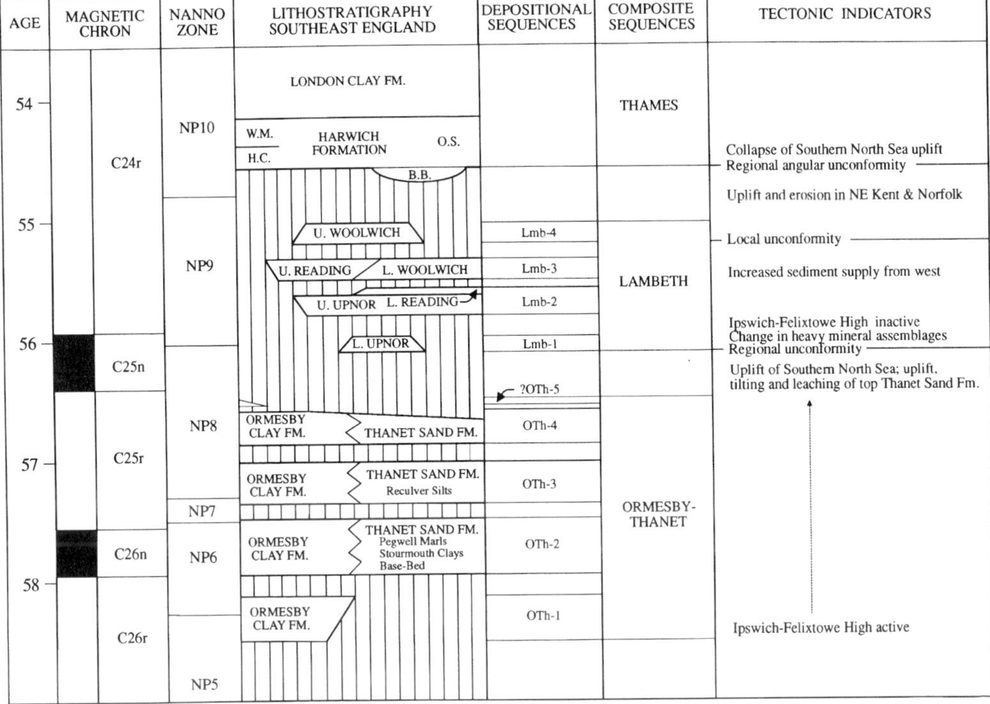

Fig. 4. Sequence stratigraphical summary of the early Palaeogene succession of SE England, with indicators of inferred tectonic events. Chronostratigraphy based on Ali (1994), Knox (1994*a*) and Ali & Jolley (1996). W.M., Wrabness Member; H.C., Hales Clay; O.S., Oldhaven Sands; and B.B., Blackheath Beds.

sequences is restricted to the Ormesby–Thanet and Lambeth composite sequences. Except where otherwise stated, the term 'sequence' is used in the following text in the sense of 'depositional sequence'. A summary of the proposed sequence stratigraphy is presented in Fig. 4.

Ormesby-Thanet composite sequence

Lithostratigraphy and facies

The Ormesby–Thanet composite sequence is represented by the mudstone-dominated Ormesby Clay Formation in East Anglia and the sand-dominated Thanet Sand Formation in the London Basin.

The Thanet Sand typically consists of a monotonous succession of very fine-grained glauconitic sandstones. It represents an extensive sublittoral sand sheet that extended eastwards into Belgium (Lower Landen Formation) and southwards into the Paris Basin. The Ormesby Clay represents outer shelf sedimentation and is in continuity with the Lista Formation of the southern North Sea (Lott & Knox 1994).

A representative section through the Thanet Sand is shown in Fig. 5.

Age

Both formations are of late Palaeocene age (NP6 to NP8, Chron 26r to Chron 25r). Data from Jolley (1992) indicate that the bulk of the two formations falls within the *Alisocysta margarita* Zone of Powell (1992). However, the occurrence of the dinoflagellate cyst *Palaeoperidinium pyrophorum* in the basal Ormesby Clay in Norfolk indicates assignment to an older zone, the *Palaeoperidinium pyrophorum* Zone (Ppy) of Powell (1992) (see Jolley 1992, fig. 9). A younger zone may also be represented at the top of the Thanet Sand in the type area of NE Kent, where both Heilmann-Clausen (1985) and Powell (1992) reported an absence of *A. margarita* in the upper part of the Reculver Silts, exposed at Herne Bay. This finding has since been confirmed by Powell *et al.* (1996), who reported the presence of *Apectodinium* species at the base of the Herne Bay section, on which basis they assign the strata to the *Apectodinium hyperacanthum* Zone. However, it should be noted that

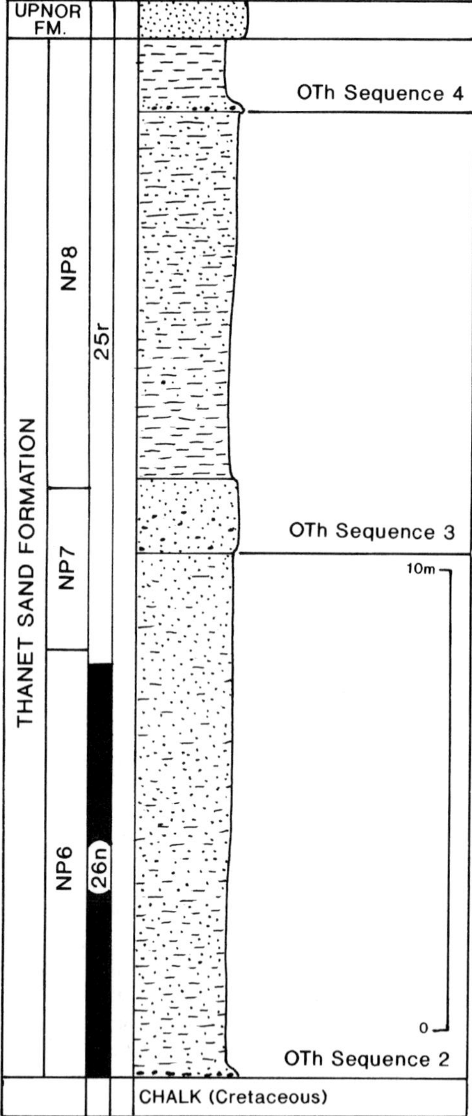

Fig. 5. Section through the Thanet Sand at Bradwell, Essex (see Knox *et al.* 1994).

this interpretation conflicts with data from the more complete, mudstone-dominated sections further to the north, where *A. margarita* persists to the top of the section (Jolley 1992). Resolution of the anomaly is beyond the scope of this paper, although it should be noted that *Apectodinium* species have been encountered in much older strata in the central North Sea (Maureen Formation: Thomas 1996), suggesting that the occurrence of *Apectodinium* may not in itself be a reliable biochronostratigraphic indicator. Pending clarification of the biozonal and/or lithostratigraphical assignment of the Reculver Silts at Herne Bay, the presumption is made in this study that the top of the Ormesby–Thanet composite sequence lies within the *A. margarita* Zone (see also Heilmann-Clausen 1994).

Lower boundary

Both formations rest unconformably on Cretaceous Chalk (see Curry 1965, fig. 2). The basal unconformity thus includes all of the early Palaeocene and part of the late Palaeocene.

Upper boundary

The Ormesby Clay and Thanet Sand are overlain unconformably by the basal deposits of the Lambeth composite sequence, with evidence for an intervening period of subaerial exposure (see Lambeth composite sequence: 'Lower boundary').

Constituent sequences

The detection of sequences within the Ormesby–Thanet composite sequence of SE England has been the subject of two recent studies. Knox *et al.* (1994) identified three depositional sequences within the Thanet Sand of the Bradwell 214 borehole (sequences BT1 to 3) (see Fig. 5). Each sequence is characterized by a coarse-grained sandy or pebbly, glauconitic basal lag deposit overlain by relatively fine-grained sandy or muddy deposits. The lower two sequences display upward-coarsening trends, whereas erosion at the base of the Upnor Formation has removed all but the basal portion of sequence BT3. These sequences are equivalent to sequences Th1–3 of Hardenbol (1994). Correlation with the succession of Norfolk (Hales and Ormesby boreholes, Fig. 3) is provided by a persistent unit of red–brown mudstone or siltstone, supported by magnetic polarity data. This indicates that the basal sediments of the Ormesby Clay pre-date those at the base of the Thanet Sand. They are here considered to represent the deposits of an earlier sequence, not represented within the Thanet Sand. To identify this and the overlying sequences as components of the Ormesby–Thanet composite sequence, they are here annotated as sequences OTh-1 to OTh-4. Sequences OTh-2 to OTh-4 equate with sequences Th-1 to Th-3 of Hardenbol (1994) and with sequences Tht-1 to Tht-3 of Powell *et al.* (1996).

The possibility of a fifth sequence (OTh-5?) is indicated by the presence at the top of the Ormesby Clay in the Hales Borehole of a sharp-based unit of dark brown grey, tuff-bearing, glauconitic mudstone (unit OC4 of Knox *et al.* 1990), suggestive of a change to more rapidly deposited, shallower water

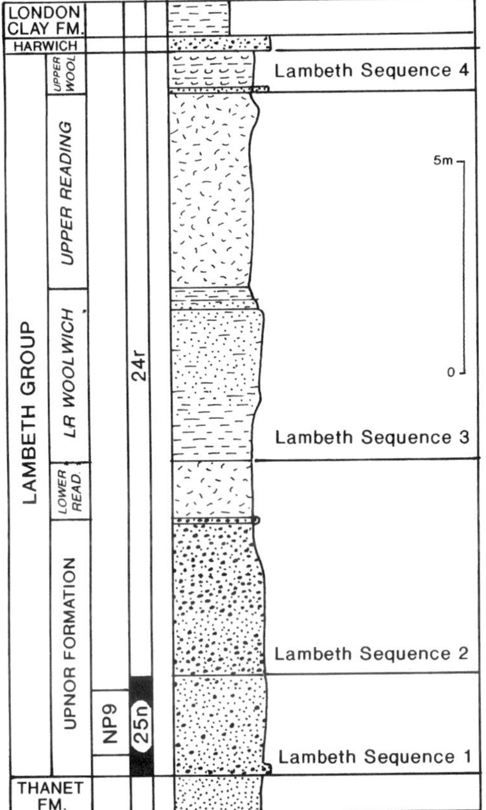

Fig. 6. Section through the Lambeth Group in borehole 404T, central London (stratigraphical data from Ellison *et al.* 1996).

facies. This unit is known only from this one locality, where it is preserved beneath the base-Harwich Formation unconformity. However, the upward change in facies is comparable with that observed in the Central North Sea at the transition from grey-green Lista Formation mudstones to grey Sele Formation mudstones with discrete tephra layers (e.g. 22/10a-4; see Knox & Holloway 1992, pp. 46, 47).

Interpretation

The base of the Ormesby–Thanet composite sequence is clearly unconformable with the underlying Cretaceous Chalk, although this cannot be attributed to a single event, but rather a coalescence of two or more unconformable sequence boundaries spanning the early Palaeocene and part of the late Palaeocene.

The three sequences defined within the Ormesby–Thanet composite sequence in East Anglia clearly represent successive phases in the transgression of the southern margin of the North Sea Basin following a major sea level fall early in late Palaeocene times. At the time of maximum transgression (sequence OTh-4) sand deposition appears to have been restricted to the nearshore marginal zone of the central and northern North Sea, with a larger spread in the shallow waters of the Anglo–Paris Basin (Fig. 7, left-hand panel). Little is known of the nature of the bounding surfaces between successive sequences, although palynological zonation indicates that they are broadly conformable (Jolley 1992; Powell *et al.* 1996).

The regressive phase of the Ormesby–Thanet composite sequence appears not to be represented in southern England, except perhaps for the uppermost unit of the Ormesby Clay in the Hales Borehole (Unit OC4 of Knox *et al.* 1990), which is questionally assigned to a fifth sequence (OTh-5?).

Lambeth composite sequence

Lithostratigraphy and facies

The Lambeth composite sequence equates with the Lambeth Group, consisting of the Upnor Formation and the overlying Woolwich and Reading formations (Fig. 6).

The Upnor Formation is essentially transgressive in character, consisting largely of marine glauconitic sands. However, recent studies of the succession below central London have shown that the most complete Upnor successions display a sharp two-fold facies division (Ellison *et al.* 1996). The lower division is represented by glauconitic, calcareous pebbly sandstones with a rich and diverse marine microfauna, palynoflora and nannoflora in the lower part. The upper division is represented by non-calcareous glauconitic sandstones and conglomerates with a relatively restricted microfauna and palynoflora. Two conglomerate units are commonly present within the upper division, with the lower unit consisting of black pebbles in a glauconitic matrix and the upper unit consisting of brown to cream coloured pebbles in a non-glauconitic sandy or muddy matrix. The presence of pebbles of silica-cemented conglomerate suggests that the upper conglomerate unit represents reworking of pebble beds that had previously undergone cementation under continental conditions. It thus seems likely that the colour contrast between the two conglomerate units represents a change from marine to continental conditions, and that the upper conglomerate unit is thus genetically related to the overlying Reading Beds (Ellison *et al.* 1996).

Lower boundary

The Lambeth Group overlies the Ormesby Clay in East Anglia and the Thanet Sand in central and eastern parts of the London Basin. Further west, the basal Upnor Formation oversteps the Thanet Sand to rest directly on the Cretaceous Chalk (see Hester 1965, fig. 7 and plate 10). A hiatus between the Lambeth Group and the Ormesby–Thanet composite sequence is revealed by the incompleteness of the magnetostratigraphical record (see Ali & Jolley, 1996). Angular discordance is not evident from lithostratigraphical relationships, but is revealed by the truncation of palynomorph associations (Jolley 1992).

Upper boundary

The Lambeth Group is overlain with distinct angular unconformity by the Harwich Formation, representing the base of the Thames composite sequence (see Fig. 4). As a result, the Woolwich Formation is locally absent in NE Kent, where the Harwich Formation (Oldhaven Sands) rests directly on the Upnor Formation. Similarly, the Reading Formation is locally absent in Norfolk, where the Harwich Formation (Hales Clay) rests directly on the Ormesby Clay (Fig. 3). In the east London area, the Lambeth Group is incised by fluvial–estuarine channels of the Blackheath Beds (see Fig. 4).

Constituent sequences

The most complete sections through the Lambeth Group are in the east London area, where four depositional sequences are recognized. The first sequence (Lmb-1) is represented by the pebbly sands of the lower part of the Upnor Formation ('lower Upnor unit'). The second (Lmb-2) comprises the pebbly and highly glauconitic sands of the upper part of the Upnor Formation ('upper Upnor unit') and the lower leaf of the Reading Formation ('lower Reading unit'), which represents paralic aggradation followed by intense pedogenesis. The third sequence (Lmb-3) comprises the lower leaf of the Woolwich Formation ('lower Woolwich unit') and the upper leaf of the Reading Formation ('upper Reading unit'). The lower Woolwich unit marks the westward extension of lagoonal environments over the 'Reading' land surface, at least as far as central London. The upper Reading unit represents the re-establishment of continental conditions, with pedogenic alteration of the underlying lower Woolwich unit. The base of the fourth sequence (Lmb-4) is taken at the base of the upper leaf of the Woolwich Formation ('upper Woolwich unit'), which represents renewed flooding of the 'Reading' land surface.

Outside the east London area, the succession is less complex and almost certainly less complete. Over western parts of the London Basin, the Upnor Formation is overlain by a single unit of Reading facies. In the west London area, at least, this single Reading unit must represent both the lower and upper units of the east London area. To the east of London, the lower leaf of the Reading Formation thins out, being represented in south Essex and NE Kent by pedogenic alteration and rootlet penetration of the top of the Upnor Formation ('Winterbourne Ironstone'). In these areas, the remainder of the succession is in Woolwich lagoonal facies, but to the north, at Bradwell, the Upnor Formation is overlain by the upper leaf of the Reading Formation, the top of which is incised by fluvial channels. A comparable succession is present in the Harwich area. Further north, at Sizewell, the succession is dominated by fluvial sands. These are underlain by grey, laminated mudstone facies of the Woolwich Formation and overlain by typical Reading Formation mottled clay. Finally, at Halesworth the bulk of the succession is in the Woolwich Formation, with facies typical of the Reading Formation occurring only as an alteration of the topmost few metres (A. N. Morigi, pers. comm. 1995). The Halesworth section is also notable for the absence of the Upnor Formation, the lower Woolwich unit overstepping the Upnor Formation to rest directly on the Ormesby Clay.

The upper Woolwich unit, corresponding to sequence Lmb-4, is known only from central and eastern parts of the London Basin. A very thin representative is present at Bradwell, in north Essex, where shelly, pebbly clays rest in most sections on the Reading Formation. However, analysis of an array of borehole sections at the Bradwell site reveals that the upper Woolwich unit is distinctly unconformable, with local removal of the Reading Formation over a local fault-related anticlinal ridge (unpublished data).

Interpretation

The base of the Lambeth composite sequence clearly marks a major sequence boundary, with evidence for erosion of the underlying Thanet Sand and Ormesby Clay. The presence of agglutinated foraminifera assemblages of '*Rhabdammina* biofacies' (see King 1981) in the fine-grained clays at the top of the Thanet Sand at Bradwell (C. King, pers. comm. 1992) indicates deposition under conditions of restricted bottom-water circulation, normally associated with water depths of 200 m or more (see Jones 1988). Although such a figure cannot be confirmed for the Bradwell section, the consistently fine-grained lithofacies indicates rapid lowering of the sediment surface below storm wave

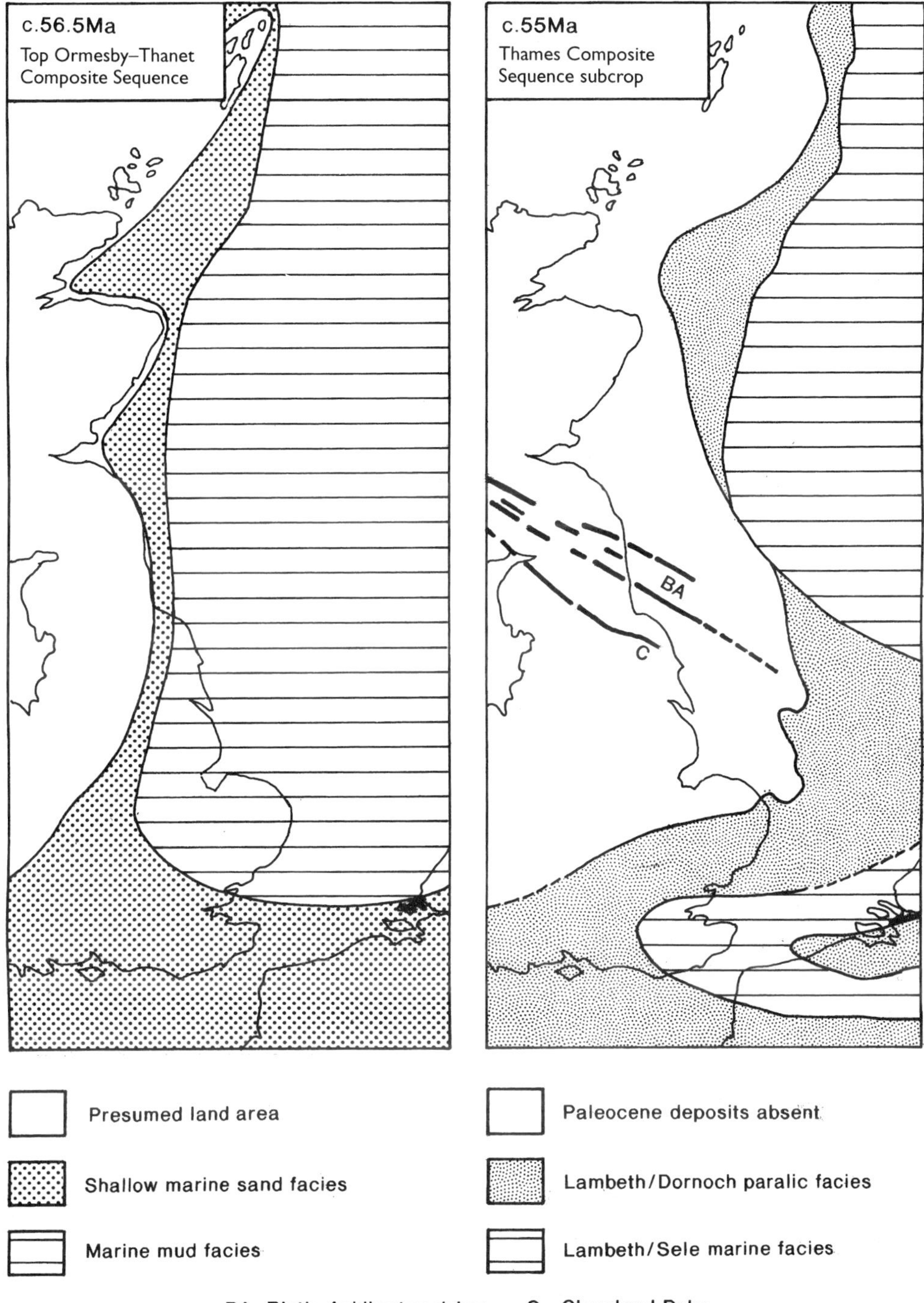

Fig. 7. Maps showing the contrast in palaeogeography and facies distribution between the highstand phase of the Ormesby–Thanet composite sequence (represented offshore by the uppermost Lista Formation) and the highstand phase of the Lambeth composite sequence (represented offshore by the Sele and Dornoch formations).

base, leading to water depths of at least several tens of metres, and possibly exceeding 100 m. The direct superposition of the restricted marine to continental deposits of the Lambeth Group is thus indicative of a substantial sea-level fall. Evidence for subaerial exposure and meteoric leaching before deposition of the Upnor Formation is provided by widespread decalcification of the uppermost Thanet Sand and by depletion of heavy mineral assemblages (Morton 1982).

In broad terms, the Upnor Formation may be regarded as representing the transgressive phase of the Lambeth composite sequence, with the Woolwich and Reading formations as representing maximum flooding, followed by aggradation during the highstand phase. Stratigraphical relationships indicate a more complex depositional history, however. The lower Upnor unit represents a temporary re-establishment of open marine conditions following the period of regression and subaerial exposure represented by the top surface of the Thanet Sand. The appearance of new nannoplankton assemblages, including the NP9 zonal marker *Discoaster multiradiatus* (Hine 1994) indicates renewed connection with the oceanic waters of the eastern Atlantic. This period of oceanic connection was short-lived, however, with the upper Upnor unit representing a change to restricted marine conditions throughout the Anglo–Paris Basin. It thus seems that the initial transgression, represented by the calcareous lower Upnor unit, was followed by a relative sea-level fall that led to the removal of the earlier deposits from most of the area and severance of the short-lived connection with the oceanic waters of the eastern Atlantic as a result of uplift to the southwest. Restricted marine or lagoonal environments, represented by coarsely glauconitic sands, appear to have become established in western parts of the area, whereas littoral to sublittoral environments, represented by relatively thick sands and local conglomerates, became established in the east. The establishment of continental conditions in sequence Lmb-1 (lower Reading unit) appears to represent further uplift to the west, severing links with the Atlantic ocean and leading to the establishment of highly restricted marine conditions throughout the North Sea Basin. This led to the development of bottom-water anoxia, as indicated by a change to delicately laminated mudstones within the basal part of the Sele Formation.

The onset of sedimentation of the Woolwich and Reading formations marked a substantial rise in relative sea level, leading to highstand aggradation throughout the region (Neal 1996). Continental environments were established along the southern, western and northern margins of the Anglo–Paris Basin. Lagoonal and marginal marine conditions were restricted to the central and eastern parts of the basin, extending eastwards into Belgium (see Fig. 7, right-hand panel). The continental facies of the Reading Formation extended northeastwards towards the southern margin of the Central Graben, where facies typical of the Lambeth Group pass laterally into the laminated mudstone facies of the Sele Formation. Dinoflagellate cyst assemblages within the lagoonal to marine facies of the Lambeth Group and equivalent formations in northern France and Belgium are characterized by an abundance of the genus *Apectodinium*.

Although no major change in the configuration of the Anglo–Paris Basin appears to have taken place during deposition of the Lambeth Group, there is evidence for changes in the pattern of marginal uplift. In particular, the transition from sequence Lmb-2 to sequence Lmb-3 (lower Reading unit to lower Woolwich unit) appears to have been marked by the relative uplift of source areas to the west, leading to an influx of clastic sediment that more or less kept pace with the sea-level rise. The widespread uplift that terminated Lambeth Group sedimentation indicates a further change in tectonic regime, with uplift centred on the southern North Sea area (Fig. 7, right-hand panel), causing removal of the Lambeth facies from northern East Anglia. The Lambeth composite sequence thus reflects the temporary development of a discrete structural basin bounded to the west and north by varying degrees of uplift.

Thames composite sequence

Lithostratigraphy and facies

The Thames composite sequence equates with the Thames Group, comprising the Harwich and London Clay formations. The Harwich Formation represents the re-establishment of marine conditions throughout the region, although the relatively low diversity of the fauna and marine palynoflora indicates continued restriction of the North Sea Basin. The Oldhaven Sand represents a sublittoral sand facies that passes northwards into the offshore muddy facies of the Hales Clay unit and Wrabness Member. The base of the London Clay represents a substantial deepening of the sea, with mud deposition extending over much of southern England. Further deepening of the London Clay sea eventually led to the re-establishment of full connection with oceanic waters, as indicated by an influx of planktonic foraminifera at the level of the 'planktonic datum' (see King 1981).

Lower boundary

The Thames composite sequence rests with marked unconformity on the underlying Woolwich and

Reading formations (Fig. 3). In the east London area, where the most complete Lambeth Group sections are preserved, the unconformity is locally marked by incision of the Woolwich Formation by the fluvial–estuarine conglomerates and sands of the Blackheath Beds (*sensu stricto*). Adjacent to these incisions, the Oldhaven Sand rests on a relatively complete Lambeth Group succession. To the east and north, however, the base of the Harwich Formation shows progressive down-cutting. To the east, the Oldhaven Sand cuts down through the Woolwich succession, eventually resting directly on the Upnor Formation (Shelford, Herne Bay: Fig. 2); to the north it cuts through the entire Lambeth Group succession, with the Hales Clay resting directly on the Ormesby Clay in Norfolk (Ormesby). Further north, in western parts of the southern North Sea subcrop, equivalents of the Hales Clay (Sele unit 3 of Knox & Holloway 1992) locally rest on Cretaceous chalk (e.g. offshore borehole 81/46A, Lott *et al.* 1983; see also Lott & Knox 1994, p. 12).

Constituent sequences

Detailed analysis of the Thames composite sequence is not attempted in this study. Sequence analysis of the Harwich Formation is hampered by abrupt lateral facies change from the mudstone-dominated sections in the north to the sandstone-dominated sections in the south, and by the absence of regionally distinctive lithostratigraphical markers. A detailed sequence biostratigraphical study has been carried out on the Harwich Formation by Jolley (1996), but work on the remainder of the Thames composite sequence has yet to be published.

In gross terms, the Harwich Formation represents the transgressive phase of the Thames composite sequence and the London Clay Formation the maximum flooding and highstand phases. The base of the Harwich formation, represented in the most distal sections by the base of the Hales Clay, marks a major sequence boundary.

Interpretation

The Harwich Formation represents the re-establishment of marine facies throughout southern England, following collapse of the southern North Sea uplift that had persisted throughout Lambeth Group sedimentation. In broad terms, this represents a return to the basin configuration of the Ormesby–Thanet composite sequence, with a south to north transition from nearshore to offshore facies. However, local structural features, such as the Ipswich–Felixstowe High, which had such a profound effect on the thickness of the Ormesby–Thanet composite sequence (Fig. 7), appear to have been more or less inactive. Jolley (1996) suggests that the Ipswich–Felixstowe High may even have undergone relative subsidence at this time.

Correlation with Central North Sea sequences

The most comprehensive sequence stratigraphical analysis of the onshore and offshore successions of the North Sea Basin is that of Neal (1996). The correlations presented in this study between the successions of southern England and those of the central North Sea (Fig. 8) broadly agree with those presented by Neal (1996), but differences exist in detail, both in stratigraphical terminology and correlation. Differences in terminology concern sequences around the Lista–Sele formational boundary. Neal (1996) recognizes two depositional sequences between the top of the lower Balmoral Sandstone unit and the top of the Forties Sandstone Member: an 'Upper Balmoral sequence' and an 'Upper Forties sequence'. In the present study, however, it is suggested that the interval includes three depositional sequences, represented by successive lowstand deposits of the Upper Balmoral Sandstone (in the lower part of Lista Unit 3 of Knox & Holloway (1992)), the Lower Forties Sandstone (within Sele unit 1a) and the Upper Forties Sandstone (within Sele unit 1b).

Ormesby–Thanet composite sequence

The occurrence of *Palaeodinium pyrophorum* at the base of the Ormesby Clay (Jolley 1992: 221), indicates correlation with the top of the *P. pyrophorum* Zone of Powell (1992). Comparison with the central North Sea succession would therefore indicate correlation of sequence OTh-1 with transgressive and early highstand deposits at the base of Lista unit 2 of Knox & Holloway (1992), above the Andrew Sandstone unit. Sequence OTh-2, the lower part of which is associated with superabundant *Areoligera gippingensis* (Powell *et al.* 1996), is interpreted as representing transgressive and early highstand deposits at the base of Lista unit 3, above the lower Balmoral Sandstone unit. This is supported by the widespread occurrence of red mudstones above the Lower Balmoral Sandstone in more distal basinal sections (e.g. c. 2655 m in well 22/10a-4, Knox & Holloway 1992, p. 85). These are believed to be correlative with the red mudstone that occurs near the base of sequence OTh-2 (see Fig. 3). Sequence OTh-3 is interpreted as representing the transgressive and highstand deposits in the upper part of Lista unit 3, above the main development of the Upper Balmoral Sandstone unit. Sequence Oth-4 has no obvious expression in the basinal North Sea sections.

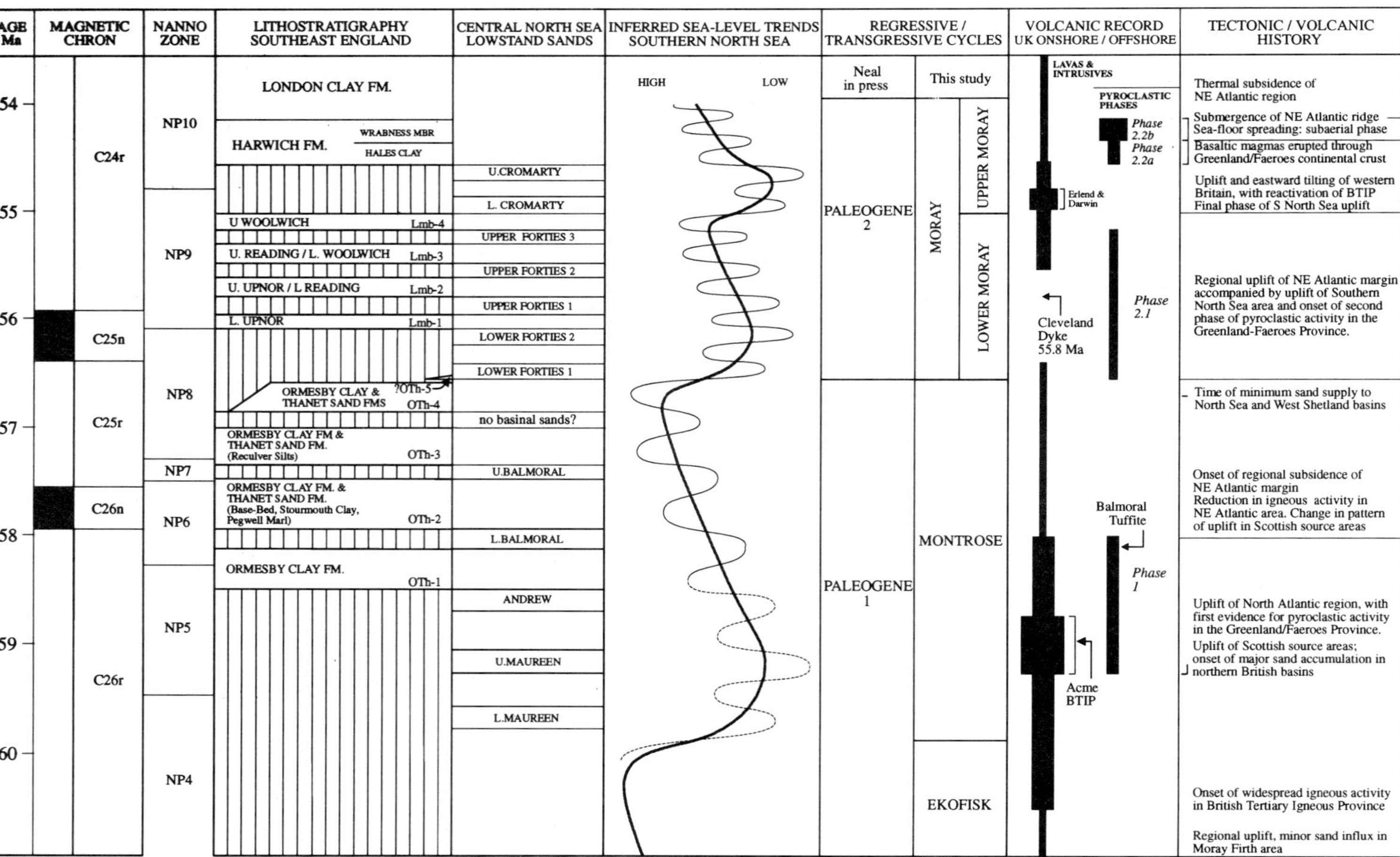

Fig. 8. Correlation of early Palaeogene stratigraphical and tectonic events in the NE Atlantic region. Biochronostratigraphical scale from Berggren *et al.* (1995). Curve of relative sea-level change shows general trends only, as the degree of sea-level change will have varied across the region. See text for data sources. BTIP, British Tertiary Igneous Province.

The correlation scheme proposed above differs from those of Neal (1996) and Powell (1996) in two key respects. The base of the Thanet Sand (Oth-2) is related to the Lower Balmoral highstand (not the Andrew highstand, as proposed by Neal 1966), and the entire Reculver Silts succession (Oth-3) is related to the Upper Balmoral highstand (not just the lower part, as proposed by Powell *et al.* 1996, see p. 213).

The Ormesby–Thanet composite sequence marks a period of long-term transgression within the North Sea Basin, constituting the younger, transgressive phase of what Neal (1996) terms a large-scale regressive/transgressive facies cycle (R/TF cycle). The regressive phase of the 'Palaeogene 1 R/TF cycle' was initiated in late Cretaceous times and culminated in the deposition of the Upper Maureen–Andrew Sandstone lowstand succession. Representatives of this latter phase are absent in southern England, where the Ormesby–Thanet composite sequence rests on an erosion surface that was by inference generated during the maximum lowstand of sea level.

Lambeth composite sequence

The correlation shown in Fig. 8 between the Lambeth sequences and those of the central North Sea should be regarded as provisional, pending detailed biostratigraphical comparisons. Correlation of the base of the Lambeth Group with the lower part of the Sele Formation of the Central North Sea is based on the occurrence of *Apectodinium* species in the lower part of the Upnor Formation (sequence Lmb-1) (see also Neal 1996).

The base of the Upnor Formation is correlated by Neal (1996) with the transgressive phase of his 'Upper Balmoral sequence' equivalent to the 'Lower Forties sequence' of this study. As discussed earlier, equivalents of the Upper Balmoral Sandstone are considered in this study to lie much lower in the sequence, within the Ormesby–Thanet sequence and within the *Alisocysta margarita* Zone. The base of the Upnor Formation is correlated in this study to a level within the lower division of the Forties Sandstone (Sele unit 1a) (see Knox & Holloway 1992: 55, well 22/8a-2), thus falling within the 'Lower Forties sequence' of O'Connor & Walker (1993).

The base of the Woolwich Formation (and the coeval Reading Formation) is correlated with the base of the 'Upper Forties sequence' of Neal (1996), as both are characterized by the first appearance of abundant *Apectodinium*. The occurrence of abundant *Apectodinium* throughout the Woolwich Formation indicates that its eroded top is no younger than the top of Sele unit 1 in basinal sections of the central North Sea, corresponding more or less to the top of the Forties Sandstone Member. However, the top of the Forties Sandstone is marked by a condensed succession and in the more expanded shelf succession of the Outer Moray Firth area, abundant *Apectodinium* extend about 50 m up into the Dornoch Mudstone.

Thames composite sequence

The basal unit of the Thames composite sequence in East Anglia, the Hales Clay, can be directly correlated with the upper, high-gamma section (unit 3) of the offshore Sele Formation through a combination of wireline-log and tephra correlation. Sele unit S3 overlies lowstand deposits of the upper Cromarty Sandstone in basinal sections of the central North Sea and marks the end of substantial progradation of the Dornoch Formation deltaic complex. The base of sequence Thm-1, represented by the base of the Hales Clay, thus represents the transgressive phase of the 'Upper Sele sequence' of Neal (1996). The Middle Sele sequence of Neal (1996), including the lower division of the Cromarty Sandstone, is not represented onshore.

The bulk of the Harwich Formation is equivalent to the lower, ash-rich part of the Balder Formation (unit B1 of Knox & Holloway 1992). The base of the London Clay Formation is broadly equivalent to the base of the ash-poor Balder unit 2. Precise correlation of the Wrabness Member with the complex sand systems of the Balder Formation in the northern North Sea is not attempted here and the lowstand phases shown in Fig. 8 are schematic.

Tectonic control on sequence development in the North Sea Basin

Evidence from southern Britain

A summary of evidence for tectonic control on sequence development in SE England is shown in Fig. 4.

Lithostratigraphical relationships reveal the presence of substantial angular unconformities at the base of the Ormesby–Thanet composite sequence and at the base of the Thames composite sequence. By contrast, angular unconformity at the base of the Lambeth composite sequence is less apparent, even though it represents a sea-level fall of over 100 m. In this instance, angular unconformity is revealed only by detailed biostratigraphical and sequence stratigraphical analysis (see Jolley 1992). Evidence for unconformity within an individual composite sequence is found only at the base of sequence Lmb-4 (base upper Woolwich unit). This unconformity has been positively identified at only one locality (Bradwell), but current studies on borehole sections in the east London area indicate that a widespread unconformity may exist at a similar stratigraphical level in that area (R. A. Ellison, pers. comm. 1995).

Unconformity cannot in itself be used as proof that a sequence boundary is of tectonic rather than eustatic origin, as an unconformable relationship could result from eustatically induced sea-level fall superimposed on an unrelated phase of tectonic flexuring. However, evidence from the southern England succession indicates that the base of each composite sequence marks a long-term change in subsidence pattern, palaeogeography and sediment supply. The most striking feature is the change in basin configuration that accompanied deposition of the Lambeth composite sequence, with widespread uplift taking place to the north of the London Basin, in an area previously occupied by the relatively deep-water facies of the Ormesby Clay. When viewed in the broader context of the UK southern North Sea Basin, the contrast in tectonic and palaeogeographical setting is even more marked (Fig. 7), with continental facies (Reading Formation equivalent) occurring around a broad uplift centred on the Sole Pit area. Although palynofacies data indicate that the Sole Pit structure formed a relatively positive area even during Ormesby Clay–Lista sedimentation (Jolley 1992, 1996), the uplift associated with Lambeth Group sedimentation was of greater geographical extent.

Mineralogical studies indicate that the changes in tectonic setting were associated with changes in sand provenance. Heavy mineral assemblages from sandstones in the Thanet Sand and Harwich Formation indicate Scottish provenance (Morton 1982), with transport having taken place by longshore drift down a coastline that ran approximately north–south through central or western England. Limited studies on sandstones in the Lambeth Group, however, indicate that a more local source of sand may have been tapped, at least during deposition of the Upnor Formation.

There is thus good evidence that the boundaries between successive composite sequences represent significant tectonic events that led to substantial changes in relative sea level and in facies distribution.

There is also evidence for long-term tectonically induced changes across sequence boundaries within the Lambeth composite sequence. These include evidence for increased source uplift and sediment supply at the base of sequence Lmb-3 (base Woolwich Formation) and a change in uplift pattern at the top of sequence Lmb-4. It is thus possible to identifiy a sequence grouping of lower order than the composite sequence identified in this study, reflecting less profound changes in regional tectonic setting.

Evidence for significant tectonic change at the level of individual sequences is also restricted to the Woolwich and Reading formations, in which the inferred change in a pronounced local, possibly regional, unconformity has been recorded at the base of sequence Lmb-4 (upper Woolwich unit).

As indicated earlier, none of the evidence summarized here can be taken as evidence for tectonism having been the driving mechanism behind the generation of individual (fourth-order) sequences. At the same time, the possibility cannot be ruled out, as it is evident that sedimentation of the late Palaeocene to early Eocene succession of NW Europe was strongly influenced by a complex tectonic setting, in which local structural elements reacted in different ways to crustal stresses associated with North Atlantic volcanism and plate tectonics. One interesting observation from this study is that the frequency of fourth-order sequence development appears to have been greater in the Lambeth composite sequence (c. 0.25 Ma) than in the Ormesby composite sequence (c. 0.5 Ma). Jolley (1996) suggests an average frequency of 0.38 Ma for sequences in the Wrabness Member of the Harwich Formation. These contrasts may simply be an artefact of the chronological calibration, but could also reflect a dominantly tectonic control, with the acceleration in the rate of sea-level variation representing increased tectonic activity before the onset of seafloor spreading. Alternatively, it could represent the superimposition of short-term tectonically induced events onto a steady (?eustatic) pattern of sea-level rise and fall. The upper Cromarty lowstand phase, for example, appears to coincide with a final phase of crustal stress adjustment before the onset of seafloor spreading and may possibly be entirely tectonic in origin. Removal of two such events from the Lambeth composite sequence would restore the 0.5 Ma average periodicity displayed by the Ormesby–Thanet composite sequence. However, only with a significantly finer chronological calibration would it be possible to detect superimposition of non-regular sea-level events in this way.

Implications for central North Sea sequence stratigraphy

The role of tectonism in controlling sea-level change in the early Palaeogene of the North Sea Basin has long been recognized, especially in relation to the onset of widespread clastic sedimentation in late Palaeocene times. Nadin & Kusznir (1996) estimated that the Outer Moray Firth area underwent uplift of 375–390 m in the mid-Palaeocene. Some workers have suggested that the North Sea Palaeocene sequences represent eustatic sea-level changes superimposed on a single tectonic cycle of uplift and subsidence (e.g. Milton *et al.* 1990, p. 345), whereas others have suggested a more complex tectonic history (e.g. Pomerol 1989; Galloway *et al.* 1993; Neal 1996). Neal (1996) considers the longer term changes represented by the major R/TF cycles to be of undoubted tectonic origin, noting that their magnitude and precise timing is not correlative throughout NW Europe. A tectonic signal is also often detectable at the level of individual sequences, but can be interpreted as indicating either tectonic control or tectonic overprint on eustatic control. This study confirms a tectonic control for the major R/TF cycles (equivalent to the composite sequences of this study) and provides evidence that some, at least, of the individual sequences may also reflect regional tectonic events.

Unfortunately, the Palaeogene succession in southern England lacks any representatives of the regressive phase of the 'Palaeogene 1' R/TF cycle of Neal (1996), who identifies a succession of sea-level falls within the regressive phase, commencing in the late Cretaceous, and culminating early in the late Palaeocene. The precise timing of the lowest sea-level stand is difficult to determine because of the difficulty in distinguishing between the regional relative sea-level signal and the effects of local tectonic activity. However, the widespread occurrence of basin-floor sandstones of the upper Maureen Formation (unit M2 of Knox & Holloway 1992), together with evidence for widespread erosion of adjacent shelf areas, suggests that this represents the lowest sea level in this particular regressive–transgressive cycle. In the present study, the culmination of the regressive phase is therefore equated with the mid-Maureen lowstand, corresponding to a regional change from calcareous to non-calcareous mudstone facies within the North Sea Basin (base Maureen unit 2 of Knox & Holloway 1992). Analysis of the regressive and lowstand phase of the 'Palaeogene 1 R/TF' cycle is beyond the scope of this paper, but brief reference is made to specific events in the context of the crustal evolution of the North Atlantic region (see below).

Correlation between the sequence stratigraphical scheme presented in this study and that of Neal (1996) demonstrates that the composite sequence defined in this study are closely related to the central North Sea R/TF cycles, though with representation being restricted to the transgressive phases. The tectonic events identified onshore from changes in the pattern of uplift and subsidence between composite sequence can therefore be applied directly to the central North Sea succession. Specific conclusions are as follows.

The initiation of the 'Palaeogene 2 R/TF' cycle, marked onshore by the base-Lambeth Group unconformity, led to widespread erosion of the youngest representatives of the Ormesby–Thanet composite sequence along the southern and western margins of the London Basin. This represents a substantial sea-level fall, probably of the order of 100–200 m, in parts of southern England. This figure may well have been exceeded in parts of the central North Sea area. Uplift to the north of the London Basin is indicated by the extension of shallow marine facies throughout the region, including areas that had previously been the site of relatively deep water (northern East Anglia). This change in configuration was probably responsible for the northeastward influx of sand into the southern Central Graben during deposition of the Forties Sandstone (see Morton 1987).

The onset of Woolwich Formation and Reading Formation sedimentation was accompanied by further changes in basin configuration, marked in particular by increased subsidence in the Hamphire Basin. A substantial rise in relative sea level is indicated by the aggradational nature of the Woolwich and Reading formations, but the association with increased terrigenous influx suggests that the sea-level rise resulted from the relative subsidence of southern England, accompanied by uplift to the west. Continued uplift to the north of the London Basin is indicated by the extension of continental facies of the Reading Formation over much of the southern North Sea area. The central and northern North Sea basins were the sites of continued deep-water sand sedimentation, with accumulation of the Upper Forties Sandstone and its equivalents. By contrast, major deltaic progradation was taking place to the west of Scotland (Mitchell *et al.* 1993).

A later phase of uplift within the 'Palaeogene 2 R/TF cycle' is reflected in the widespread erosion of Lambeth Group sediments in southern England. Erosion of the Woolwich Formation took place throughout the London Basin, with its total removal from NE Kent. Further north, in the northernmost parts of East Anglia and western parts of the southern North Sea area, erosion led to the total removal of the Lambeth Group. At the same time a major change took place within the Scottish source areas, with deltaic progradation switching from the western basins to the eastern basins (Dornoch Mudstone and Upper Dornoch Sandstone). This change in depositional style can be ascribed to eastward tilting of the Scottish land mass, causing erosion and eastward dispersal of older Palaeocene deposits from the Inner Moray Firth area (Milton *et al.* 1990) and relative sediment starvation in the West of Shetland basins. The culmination of this uplift phase is probably represented in deeper water facies by the upper unit of the Cromarty Sandstone Member.

A major change in basin configuration is identified within the 'Palaeogene 2 R/TF cycle' at the level of the base-Thames Group unconformity. This is interpreted as representing collapse of the southern North Sea uplift. It was associated with a marked reduction in the influence of local structural elements within the onshore area. Tectonic quiescence on a regional scale is indicated by the widespread distribution and 'layer-cake' geometry of the transgressive deposits of the Harwich Formation and its offshore equivalents. The thin, relatively high-gamma mudstones of Sele unit 3 (equivalent to the Hales Clay) can be identified in almost every basinal section in the North Sea region. In marginal areas, it can be identified throughout the northern North Sea area, where it transgressed over the Dornoch deltaic complex and was followed by deposition of the tuffaceous deposits of Balder unit 2. Only in the Outer Moray Firth area was the sedimentation rate initially sufficient to hold back the transgression represented by the Sele Formation, with paralic sedimentation being represented by the lignite-bearing Beauly Member. Analogy with southern England suggests that the sea-level rise that led to the termination of substantial progradation of the North Sea Dornoch Formation delta complex was of tectonic, rather than eustatic, origin. This is supported by the observation that the base of Beauly Member is locally unconformable (Milton *et al.* 1990, p. 340). The long-term reduction in sediment input associated with this transgression can therefore be ascribed not only to a regional rise in relative sea level, but also to a regional change in the pattern of uplift and subsidence. On a local scale, the Ipswich–Felixstowe High, active during Ormesby Clay sedimentation, appears to have undergone relative subsidence during deposition of the Harwich Formation (Jolley 1996).

Evidence from southern England indicates that although successive sequences within the regressive or transgressive sections of an R/TF cycle reflect a consistent regional tectonic style, there is evidence for minor tectonic adjustments across sequence boundaries. For example, the angular unconformity locally recorded at the base of the upper Woolwich unit (sequence Lmb-4) may also be reflected in the central and northern North Sea successions. The importance of such local tectonic activity on the source and distribution of potential reservoir sands in the Viking Graben has been demonstrated by Morton *et al.* (1993). Recognition of a regionally identifiable pattern in such tectonic events would therefore constitute an important predictive tool in hydrocarbon exploration.

The recognition of the Lambeth composite sequence as a distinct tectono-stratigraphical unit raises the possibility of subdividing the 'Palaeogene 2 R/TF cycle' of Neal (1996) into two shorter term 'subcycles'. The 'Palaeogene 1 R/TF cycle' may also be subdivided, with two separate long-term cycles being represented by (1) the Ekofisk Formation and (2) the Montrose Group (Maureen and Lista formations). On this basis, an alternative, lithostratigraphically based, R/TF cycle nomenclature is proposed, comprising the Ekofisk, Montrose and Moray R/TF cycles, with the latter being divided into a lower subcycle, including the Forties Sandstone, and an upper subcycle, including the Cromarty Sandstone Member and the Balder Formation (see Fig. 8).

North Atlantic tectonism

The tectonic events identified within the North Sea Basin are a reflection of wider changes in crustal evolution within the North Atlantic region (e.g. Parker 1975; Milton *et al.* 1990; Knox & Morton 1988; Armentrout *et al.* 1993; Den Hartog Jager 1993; Knott *et al.* 1993; Neal 1996). Two developments are widely recognized as having been the driving mechanisms behind these events. The first was the regional uplift of the entire North Atlantic area that started in the Late Cretaceous and which is regarded as reflecting the development of the proto-Icelandic mantle plume, centred beneath East Greenland but creating a mushroom-shaped head of hot mantle material that affected an area with a radius of 2000 km (White 1988). Evidence that the Late Cretaceous uplift was accompanied by igneous activity within the North Atlantic Igneous Province is provided by Late Cretaceous dating of basalts on Anton Dohrn Seamount (Jones *et al.* 1994) and Late Maastrichtian dating (71–69 Ma) of basalts on Rosemary Bank (Morton *et al.* 1995). The second development was the collapse of the North Atlantic uplift in the earliest Eocene, which is regarded as resulting from the onset of seafloor spreading between Greenland and Scotland, with the mantle plume becoming established beneath the spreading ridge (although without any causal link between the two). This long-term cycle of uplift and subsidence was termed the 'Palaeocene tectonosequence' by Galloway *et al.* (1993). The role of Alpine tectonics, resulting from the convergence of North Africa and Europe, is uncertain. England (1988: 388) has suggested that the stress field in NW England may have been generated by Alpine as well as Atlantic tectonics, whereas Dewey & Windley (1988: 29) have cited evidence for minimal convergence of the two continents from 65 to 51 Ma. The latter model does not, however, preclude some element of Alpine stress component in NW Europe. Indeed, a complex stress field with no one dominant component, coupled with the influence of basement structural lineaments (England 1988), might best account for the changing patterns of uplift and subsidence during the Palaeocene and for the apparently conflicting regional extensional stress directions inferred from the trend of the North Atlantic rift zone (NW–SE extension) and that of the dominant dyke trend in the British Tertiary Igneous Province (NE–SW extension).

Information from southern England provides clear evidence that the onset of seafloor spreading between Greenland and Scotland was preceded by a change in regional tectonic setting, lasting about 1.5 Ma and spanning the deposition of the Lambeth Group onshore and the Lower Moray R/TF cycle offshore. Regional uplift, coupled with a change in local uplift/subsidence pattern terminated a long phase of transgression that culminated in a period of greatly reduced sand input, reflected in the widespread distribution of mudstone facies at the top of the Lista Formation and its onshore equivalents. The mudstone at the top of the Lista Formation is also recognized in the West Shetland area. On this evidence, it appears that the concept of a single, long-term, plume-related uplift phase is an oversimplification. A comparable interruption to the long-term uplift trend is indicated by the widespread occurrence of chalk facies in the middle and upper part of the Ekofisk Formation. This is indicative of long-term transgression and relative tectonic stability following the sea-level fall that led to the temporary influx of coarse siliciclastic sediments in the central North Sea in earliest Palaeocene times. Following the earliest Palaeocene lowstand phase, further substantial sea-level fall did not take place for another 4 Ma, until the sandstones and reworked chalks of the Maureen Formation were deposited.

Three distinct phases of long-term uplift and subsidence may thus be recognized within the regressive phase of the long-term early Palaeogene tectono-sequence, separated by periods of reduced tectonic activity and widespread transgression. These define the Ekofisk, Montrose and Moray R/TF cycles of Fig. 8. Improved correlation of these uplift events with igneous activity throughout the North Atlantic region is possible as a result of new information on the chronology of the succession in southern England, together with the development of an improved Cenozoic time-scale (Berggren *et al.* 1995). Using this geochronometric framework, the three uplift phases may be broadly dated as follows (see Fig. 8). The first began in the Maastrichtian and culminated at about 65 Ma (base Ekofisk Formation). The second began at about 62–60 Ma (uppermost Ekofisk to Ekofisk–Maureen boundary) and culminated at about 59.5 Ma (base Maureen unit 2 of Knox & Holloway 1992). The third began at about 56.5 Ma (base Sele Formation) and culminated at about 55 Ma (base Sele Formaton unit S3 of Knox & Holloway 1992). This indicates that the long-term tectonic cycles are asymmetrical, with the uplift phases lasting about 1-1.5 Ma and the subsidence phases lasting for about 4 Ma.

The improved dating of the early Palaeogene R/TF cycles allows comparison with the periods of igneous activity identified in the various parts of the North Atlantic Igneous Province. Reviews of the temporal relationships of this igneous activity are provided by Noe-Nygaard (1974), by papers in Morton & Parson (1988) and, most recently, by Ritchie & Hitchen (1996), who have used the same time-scale as in this paper. These compilations reveal two main phases of igneous activity from 62.5 to c. 58 Ma and from c. 56 to 54 Ma, which clearly correspond closely to the regressive phases of the Montrose and Moray R/TF cycles.

The earlier igneous phase included the onset of activity in West Greenland and the main phase of activity in the Hebridean area. The culmination of Hebridean activity is placed at about 59 Ma (Mussett *et al.* 1988), corresponding approximately to the culmination of the mid-Palaeocene regression (base Maureen unit 2; 'upper Maureen' lowstand of Fig. 8). The second igneous phase included the onset of igneous activity on Rockall and on Hatton Bank (dipping reflectors). A temporary recurrence of activity in the British Tertiary Igneous Province also took place, but with the focus apparently having shifted northwards to the west of the Shetland Islands. Thus the Darwin and Erlend complexes were probably emplaced synchronously at 55.5–54.5 Ma (Ritchie & Hitchen 1996) during deposition of the Lambeth Group. Activity was not confined to the northern offshore area, however, as palynological data from Mull indicate lava eruption at about 55.5 Ma (D. W. Jolley pers. comm. 1995). Intrusion of the Cleveland Dyke (c. 55.8 Ma) also took place around this time and it is significant that this dyke, together with the subparallel Blyth–Acklington dyke system, runs along the axis of the uplift that was associated with the extension of continental facies of the Reading Formation into the southern North Sea area (see Fig. 7, right-hand panel). This late magmatic phase may have played a significant part in the devolatilization of Carboniferous coals in the southern North Sea area (see Dewey & Windley 1988, p. 29).

Although the second main phase of igneous activity (c. 55 Ma) in the British Tertiary Igneous Province appears to have been short-lived, substantial volcanism continued in East Greenland, on Hatton Bank (dipping reflector series) and on Rockall. This concentration of volcanic activity in the North Atlantic rift zone is interpreted as representing the onset of oceanic crust generation. It coincided with a marked reduction in stress within the NW European crust, reflected in a sharp decrease in intrabasinal tectonic activity.

The history of igneous activity outlined here is paralleled by the pyroclastic record in the North Sea. Knox & Morton (1988) identified two principal phases of pyroclastic activity. Phase 1 includes tephra layers that extend from the base of Maureen unit 2 (within the 'Maureen sequence'of Neal 1996) into the lower part of the Balmoral Sandstone unit (c. 59–58 Ma), thus corresponding to the culmination of the main phase of igneous activity in the British Tertiary Igneous Province. Phase 1 tephra layers also occur in the lower part (OTh-1) of the Ormesby Clay (Knox 1994*b*). Phase 2 is represented by tephra layers that occur in the Sele Formation, the Balder Formation and the Lower Eocene section of the Horda Formation (unit H1 of Knox & Holloway 1992).

Phase 2 was subdivided into four subphases (2a to 2d), but later studies have shown that the history of eruption was slightly different from that proposed by Knox & Morton (1988) and a different annotation is therefore adopted here. Phase 2.1 includes tephra layers within the Lower Moray R/TF subcycle, ranging from the base of the Sele Formation to the top of the Forties Sandstone (i.e. the top of Sele unit 1 of Knox & Holloway 1992), corresponding to the period of maximum regression within the 'Palaeogene 2 R/TF cycle'. Sporadic phase 2.1 tephra layers have been recorded onshore from the Upnor Formation.

After a period of quiescence (spanning most of Sele unit 2), pyroclastic activity resumed during deposition of Sele unit 3, followed by a marked increase in the frequency and magnitude of eruptions at the base of the Balder Formation. The same tephra record is present in the Harwich Formation onshore. This second period of activity, defined here as phase 2.2, spans the uplift and early subsidence phase of the Upper Moray R/TF subcycle. A further subdivision is recognized, with the earlier, sporadic tuff layers of mixed composition being assigned to phase 2.2a (Sele unit 3, Hales Clay) and the later, abundant tephras of tholeiitic composition being assigned to phase 2.2b (Balder unit 1, Wrabness Member). The sharp diminution of pyroclastic activity within the Balder Formation (top of Balder unit B1) may represent submergence of the source volcanoes as a result of thermal subsidence of the developing mid-ocean ridge (Knox & Morton 1988).

Combining all the available igneous, pyroclastic and tectonic data, it is possible to infer a more restricted age range for the igneous activity in the British Tertiary Igneous Province. Thus the observation of Mussett *et al.* (1988) that activity was concentrated around 59 Ma is supported by the coincidence with the period of maximum sea-level lowstand during the Montrose R/TF cycle. A more limited range for the late phase of activity in the British Tertiary Igneous Province may also be proposed, assuming coincidence with the maximum lowstand of the Upper Moray R/TF subcycle, which places the activity at about 54.7–55 Ma. This indicates that this phase of activity took place just before the phase 2.2 ash eruptions, which, as discussed earlier, are interpreted as marking the onset of seafloor spreading. The reactivation of the British Tertiary Igneous Province can thus be ascribed to a short-lived phase of crustal readjustment immediately before continental break-up.

Despite the apparent broad coincidence of North Sea pyroclastic sedimentation with phases of igneous activity in the British Tertiary Igneous Province, the geochemistry and mineralogy of the ashes is indicative of an origin in the East Greenland–Faeroes province (Morton & Knox 1990). Only the reworked Balmoral Tuffite (Andrew Tuff, Glamis Member of other authors) is regarded as being of Hebridean origin (Jolley & Morton 1992). An interesting feature in this respect is the relatively limited activity in the British Tertiary Igneous Province during eruption of the phase 2.1 tephras. It is possible that the direction of extensional stress at that time did not favour major activity within the Province, and it may be significant that this period was associated with intrusion of the Cleveland Dyke, well outside the normal range of Hebridean intrusion. It is also interesting to note that there appears to have been a period of minimal pyroclastic activity in the East Greenland–Faeroes area from c. 58 to c. 56.5 Ma, during which activity in the British Tertiary Igneous Province was also minimal. Two explanations can be offered for this situation. It may reflect a reduction in all igneous activity within the East Greenland–Faeroes province at that time, perhaps associated with the widespread regional subsidence that took place over the interval. This would require reinterpretation of the magnetic polarity data (e.g. Chron 25r not being represented). Alternatively, the scarcity of tephras may reflect, not a reduction in igneous activity as such, but a reduction in pyroclastic activity resulting from the drowning of the source volcanoes at a time of relatively high sea level. In either case, the coincidence between the three East Greenland–Faeroes pyroclastic phases and the Montrose, Lower Moray and Upper Moray uplift phases indicates a likely genetic link.

The revised timing of events in the North Atlantic region provides a further opportunity to consider the apparently anomalous orthogonal relationship between the dominant NW–SE trend of dykes in northern Britain and the NE–SW trend of the North Atlantic rift zone. England (1988) and Knox & Morton (1988) suggested that a NE–SW extensional stress pattern may have been replaced with a NW–SE extensional stress pattern immediately before the onset of seafloor spreading. As an alternative, Knott *et al.* (1993, p. 965) suggested that the dyke trend might represent the reactivation of NW–SE trending basement structures in response to superimposed E–W and NW–SE extensional stress fields. The present study indicates that the region was subjected to a series of changes in stress pattern, reflected in regional and local changes in tectonic style. This seems to favour a model of interacting regional and local stress regimes, as suggested by Knott *et al.* (1993). Another element in the equation is the possibility of subsidiary, ephemeral plumes developing far from the East Greenland centre. Such plumes could, for example, account for the concentrations of igneous activity in the West Greenland and Hebridean regions. Cope (1994) has proposed the development in the latest Maastrichtian of a plume centred on the eastern Irish Sea area, leading to an uplift of about 2 km in western England. Interaction between local stresses related to such a plume and regional stresses related to the NE Atlantic region could well have contributed to the complex tectonic history observed in this study.

Detailed analysis of the tectonic evolution of the North Atlantic province is beyond the scope of this paper, but it is clear that the combination of clear tectonic signals and relatively precise dating provided by succession in southern England can, through the intermediary of the North Sea succession, help to elucidate the history of crustal events to the west of Britain. A more detailed analysis of information relating to changing stress patterns may, for example, help to answer key questions concerning the nature and timing of mantle-plume emplacement and the extent to which plume activity or regional crustal stresses were primarily responsible for continental break-up.

Conclusions

Analysis of the Palaeogene succession in southern England has revealed that successive depositional composite sequences reflect local changes in tectonic style and palaeogeographical configuration. Correlation with regional tectonic and volcanic events confirms that the differentiation of the succession into composite sequences is the direct result of tectonic activity, rather than tectonic events superimposed on major eustatic events. Most of this tectonic activity can be related to regional uplift associated with the emplacement of the proto-Icelandic mantle plume beneath East Greenland (see White 1988), but correlation with North Sea successions reveals that the plume-related uplift of the NW European region took place in three stages. This episodicity in tectonic activity probably reflects changes in the regional stress pattern superimposed on a long-term trend of plume-induced uplift. However, the possibility that plume emplacement was itself complex and episodic cannot be ruled out.

Continental break-up occurred during the late Palaeocene to early Eocene deposition of the Moray R/TF cycle. It was preceded by a short-lived phase of igneous activity, accompanied by uplift and eastward tilting of the Scottish source area, leading to accelerated sedimentation along the western margin of the North Sea Basin, including the major progradation phase of the Dornoch Formation delta complex.

Less pronounced tectonic and palaeogeographical changes are associated with the boundaries of individual fourth-order depositional sequences in southern England. Evidence that these individual sequences are tectonically rather than eustatically driven is less compelling. Although proof of tectonic activity during sequence development is provided by local unconformity and facies changes across sequence boundaries, there is insufficient evidence to suggest that tectonism was the causal mechanism of sequence development. Stronger evidence for the occurrence of short-term tectonic events at this time is provided by changes in sediment provenance in the Viking Graben (Morton *et al.* 1993), but regional correlation of such events and their relationship to sequence boundaries is not yet well enough known to assess the role of tectonism as the driving mechanism behind the fourth-order depositional sequences.

The improved correlation presented in this paper between North Sea sequences and igneous events in the North Atlantic region would not have been possible without the generosity of M.-P. Aubry and W. Berggren in providing access to their unpublished chronometric calibration of the early Palaeogene nannoplankton zones. The paper has also benefited from reviews by C. Ebdon, D. Jolley and an anonymous referee. This paper is a contribution to IGCP Project 308, and is published with the approval of the Director, British Geological Survey (NERC).

References

ALI, J. R. 1994. Magnetostratigraphy of the Upper Paleocene and lowermost Eocene of SE England. *GFF*, **116**, 41–42.

—— & JOLLEY, D. W. 1996. Chronostratigraphic framework for the Paleocene/Eocene boundary deposits of southern England. *In*: Knox, R. W. O'B., CORFIELD, R. M. & DUNAY, R. E. (eds) *Correlation of the Early Paleogene in Northwest Europe*. Geological Society, London, Special Publications, **101**, 129–144.

——, KING, C. & HAILWOOD, E. A. 1993. Magnetostratigraphic calibration of early Eocene depositional sequences in the southern North Sea Basin. *In*: HAILWOOD, E. A. & KIDD, R. B. (eds) *High Resolution Stratigraphy*, Geological Society, London, Special Publications, **70**.

ARMENTROUT, J. M., MALECEK, S. J. & 6 others 1993. Logmotif analysis of Palaeogene depositional systems tracts, Central and Northern North Sea: defined by sequence stratigraphic analysis. *In*: PARKER, J. R.

(ed.) *Petroleum Geology of Northwest Europe: Proceedings of the 4th Conference*, Geological Society, London, 45–57.

AUBRY, M.-P. 1994. The Thanetian stage in NW Europe and its significance in terms of global events. *GFF*, **116**, 43–44.

——, HAILWOOD, E. A. & TOWNSEND, H. 1986. Magnetic and calcareous nannofossil stratigraphy of the lower Palaeogene formations of the Hampshire and London basins. *Journal of the Geological Society, London*, **143**, 729–735.

BERGGREN, W. A., KENT, D. V., SWISHER, C. C. III & AUBREY, M.-P. 1995. A revised Cenozoic geochronology and chronostratigraphy. *In*: BERGGREN, W. A., KENT, D. V. & HARDENBOL, J. (eds) *Geochronology, Time Scales and Stratigraphic Correlation: a Framework for an Historical Geology*. Society of Economic Paleontologists and Mineralogists, Special Volume, **54**, in press.

COPE, J. C. W. 1994. A latest Cretaceous hotspot and the southeasterly tilt of Britain. *Journal of the Geological Society, London*, **151**, 897–900.

COX, F., HAILWOOD, E. A., HARLAND, R., HUGHES, M. J., JOHNSTON, N. & KNOX, R. W. O'B. 1985. Palaeocene sedimentation and stratigraphy in Norfolk, England. *Newsletters on Stratigraphy*, **14**, 168–185.

CURRY, D. 1965. The Palaeogene beds of south-east England. *Proceedings of the Geologists' Association, London*, **76**, 151–173.

—— 1981. Thanetian. *In*: POMEROL, C. (ed.) *Stratotypes of Thanetian Stages. Bulletin d'Information des Géologues du Bassin de Paris, Mémoir Hors Série*, **2**, 255–265.

DEN HARTOG JAGER, D., GILES, M. R. & GRIFFITHS, G. R. 1993. Evolution of Paleogene submarine fans of the North Sea in space and time. *In*: PARKER, J. R. (ed.) *Petroleum Geology of Northwest Europe: Proceedings of the 4th Conference*. Geological Society, London, 59–71.

DEWEY, J. F. & WINDLEY, B. F. 1988. Palaeocene–Oligocene tectonics of NW Europe. *In*: MORTON, A. C. & PARSON, L. M. (eds) *Early Tertiary Volcanism and the Opening of the NE Atlantic*, Geological Society, London, Special Publications, **39**, 25–36.

ELLISON, R. A., ALI, J. R., HINE, N. M. & JOLLEY, D. W. 1996. Recognition of Chron 25n in the upper Paleocene Upnor Formation of the London Basin, UK. *In*: KNOX, R. W .O'B., CORFIELD, R. M. & DUNAY, R. E. (eds) *Correlation of the Early Paleogene in Northwest Europe*. Geological Society, London, Special Publications, **101**, 185–194.

——, KNOX, R. W. O'B., JOLLEY, D. W. & KING, C. 1994. A revision of the lithostratigraphical classification of the early Palaeogene strata of the London Basin and East Anglia. *Proceedings of the Geologists' Association, London*, **105**, 187–197.

ENGLAND, R. W. 1988. The early Tertiary stress regime in NW Britain: evidence from the patterns of volcanic activity. *In*: MORTON, A. C. & PARSON, L. M. (eds) *Early Tertiary Volcanism and the Opening of the NE Atlantic*. Geological Society, London, Special Publications, **39**, 381–389.

GALLOWAY, W. E., GARBER, J. L., LIU XIJIN & SLOAN, B. J. 1993. Sequence stratigraphic and depositional framework of the Cenozoic fill, Central and Northern North Sea Basin. *In*: PARKER, J. R. (ed.) *Petroleum Geology of Northwest Europe: Proceedings of the 4th Conference*. Geological Society, London, 33–43.

HARDENBOL, J. 1994. Sequence stratigraphic calibration of Palaeocene and Lower Eocene continental margin deposits in NW Europe and the US Gulf Coast with the oceanic chronostratigraphic record. *GFF*, **116**, 49–51.

HEILMANN-CLAUSEN, C. 1985. Dinoflagellate stratigraphy of the uppermost Danian to Ypresian in the Viborg 1 borehole, central Jylland, Denmark. *Danmarks Geologiske Undersøgelse, Serie A*, **7**, 1-29.

—— 1994. Review of Paleocene dinoflagellates from the North Sea region. *GFF*, **116**, 51–53.

HESTER, S. W. 1965. Stratigraphy and palaeogeography of the Woolwich and Reading Beds. *Bulletin of the Geological Survey of Great Britain*, **23**, 117–137.

HINE, N. M. 1994. Calcareous nannoplankton assemblages from the Thanet Formation in Bradwell Borehole, Essex, England. *GFF*, **116**, 54–55.

JOLLEY, D. W. 1992. Palynofloral association sequence stratigraphy of the Palaeocene Thanet Beds and equivalent sediments in eastern England. *Review of Palaeobotany and Palynology*, **74**, 207–237.

—— 1996. The earliest Eocene sediments of eastern England: an ultra high resolution palynological correlation. *In*: KNOX, R. W. O'B., CORFIELD, R. M. & DUNAY, R. E. (eds) *Correlation of the Early Paleogene in Northwest Europe*. Geological Society, London, Special Publications, **101**, 219–254.

—— & MORTON, A. C. 1992. Palynological and petrological characterisation of a North Sea Palaeocene volcaniclastic sequence. *Proceedings of the Geologists' Association*, **103**, 119–127.

JONES, E. J. W., SIDDALL, R., THIRLWALL, M. F., CHROSTON, P. N. & LLOYD, A. J. 1994. Anton Dohrn Seamount and the evolution of the Rockall Trough. *Oceanologica Acta*, **17**, 237–247.

JONES, G. D. 1988. A paleoecological model of Late Paleocene 'flysch-type' agglutinated foraminifera using the paleoslope transect approach, Viking Graben, North Sea. *In*: GRADSTEIN, F. M. & RÖGL, F. (eds) *Second Workshop on Agglutinated Foraminifera. Abhandlungen der Geologischen Bundesanstalt*, **41**, 143–153.

KING, C. 1981. *The Stratigraphy of the London Clay*. Tertiary Research, Special Publications, **6**.

KNOTT, S. D., BURCHELL, M. T., JOLLEY, E. J. & FRASER, A. J. 1993. Mesozoic to Cenozoic plate reconstructions of the North Atlantic and hydrocarbon plays of the Atlantic margins. *In*: PARKER, J. R. (ed.) *Petroleum Geology of Northwest Europe: Proceedings of the 4th Conference*. Geological Society, London, 953–974.

KNOX, R. W. O'B. 1990. Thanetian and early Ypresian chronostratigraphy in south east England. *Tertiary Research*, **11**, 57–64.

—— 1994*a*. The age and regional context of the Thanetian stratotype sections of SE England. The Paleocene Epoch–stratigraphy, global changes and events. *GFF*, **116**, 55–56.

—— 1994*b*. Stratigraphical distribution of tephra layers in the lower Paleogene of the southwestern margin of the North Sea Basin. *Bulletin de la Société Belge de Géologie*, **102**, 159–164.

—— & HOLLOWAY, S. 1992. 1. Paleogene of the Central and Northern North Sea. *In*: KNOX, R. W. O'B. & CORDEY, W. G. (eds) *Lithostratigraphic nomenclature of the UK North Sea*. British Geological Survey, Nottingham.

—— & MORTON, A. C. 1988. The record of early Tertiary N Atlantic volcanism in sediments of the North Sea Basin. *In*: MORTON, A. C. & PARSON, L. M. (eds) *Early Tertiary Volcanism and the Opening of the NE Atlantic*. Geological Society, London, Special Publications, **39**, 407–419.

——, BALSON, P., ELLISON, R. A. & HUMPHREYS, B. 1988. Lithostratigraphy: the London Basin and East Anglia. *In*: VINKEN, R., VON DANIELS, C. H., GRAMANN, F., KÖTHE, A., KNOX, R. W. O'B., KOCKEL, F., MEYER, K.-J. & WEISS, W. (eds) *The Northwest European Tertiary Basin. Results of the International Geological Correlation Programme Project No 124. Geologisches Jahrbuch, Reihe A*, **100**, 29–32.

——, HINE, N. M. & ALI, J. R. 1994. New information on the age of the Thanetian Stage in the type area of S.E. England. *Newsletters on Stratigraphy*, **30**, 45–60.

——, MORIGI, A. N., ALI, J. R., HAILWOOD, E. A. & HALLAM, J. R. 1990. Early Palaeogene stratigraphy of a cored borehole at Hales, Norfolk. *Proceedings of the Geologists' Association, London*, **101**, 145–151.

LERICHE, M. 1905. Observations sur la classification des assises paléocènes et éocènes du Basin de Paris. *Annales de la Société Géologique du Nord*, **34**, 383–392.

LOTT, G. K. & KNOX, R. W. O'B. 1994. 7. Post-Triassic of the Southern North Sea. *In*: KNOX, R. W. O'B. & CORDEY, W. G. (eds) *Lithostratigraphic Nomenclature of the UK North Sea*. British Geological Survey, Nottingham.

——, KNOX, R. W. O'B. & HARLAND, R. 1983. *The Stratigraphy of Palaeogene Sediments in a Cored Borehole off the Coast of North-east Yorkshire*. Institute of Geological Sciences Report 83/9.

MILTON, N. J., BERTRAM, G. T. & VANN, I. R. 1990. Early Palaeogene tectonics and sedimentation in the Central North Sea. *In*: HARDMAN, R. F. P. & BROOKS, J. (eds) *Tectonic Events Responsible for Britain's Oil and Gas Reserves*. Geological Society, London, Special Publications, **55**, 339–351.

MITCHELL, S. M., BEAMISH, G. W. G., WOOD, M. V., MALACEK, S. J., ARMENTROUT, J. A., DAMUTH, J. E. & OLSON, H. C. 1993. Paleogene sequence stratigraphic framework of the Faeroe Basin. *In*: PARKER, J. R. (ed.) *Petroleum Geology of Northwest Europe: Proceedings of the 4th Conference*. Geological Society, London, 1011–1023.

MITCHUM, R. M. JR. & VAN WAGONER, J. C. 1991. High-frequency sequences and their stacking patterns: sequence stratigraphic evidence of high-frequency eustatic cycles. *Sedimentary Geology*, **70**, 131–160.

MORTON, A. C. 1982. The provenance and diagenesis of Palaeocene sandstones of southeast England, as indicated by heavy minerals analysis. *Proceedings of the Geologists' Association, London*, **93**, 263–274.

—— 1987. Influence of provenance and diagenesis on detrital garnet suites in the Paleocene Forties Sandstone, Central North Sea. *Journal of Sedimentary Petrology*, **57**, 1027–1032.

—— & KNOX, R. W. O'B. 1990. Geochemistry of late Palaeocene and early Eocene tephras from the North Sea Basin. *Journal of the Geological Society, London*, **147**, 425–437.

—— & PARSON, L. M. (eds) 1988. *Early Tertiary Volcanism and the Opening of the NE Atlantic*. Geological Society, London, Special Publications, **39**.

——, HALLSWORTH, C. R. & WILKINSON, G. C. 1993. Stratigraphic evolution of sand provenance during Paleocene deposition in the Northern North Sea area. *In*: PARKER, J. R. (ed.) *Petroleum Geology of Northwest Europe: Proceedings of the 4th Conference*. Geological Society, London, 73–84.

——, HITCHEN, K., RITCHIE, J. D., HINE, N. M., WHITEHOUSE, M. & CARTER, S. 1995. Late Cretaceous basalts from Rosemary Bank, Northern Rockall Trough. *Journal of the Geological Society, London*, **152**, 947–552.

MUSSETT, A. E., DAGLEY, P. & SKELHORN, R. R. 1988. Time and duration of activity in the British Tertiary Igneous Province. *In*: MORTON, A. C. & PARSON, L. M. (eds) *Early Tertiary Volcanism and the Opening of the NE Atlantic*. Geological Society, London, Special Publications, **39**, 337–348.

NADIN, P. A. & KUSZNIR, N. J. 1996. Quantitative 2-D modelling of Early Paleogene post-rift stratigraphy in the Outer Moray Firth Basin, northern North Sea. *In*: KNOX, R. W .O'B., CORFIELD, R. M. & DUNAY, R. E. (eds) *Correlation of the Early Paleogene in Northwest Europe*. Geological Society, London, Special Publications, **101**, 43–62.

NEAL, J. 1996. A summary of Paleogene sequence stratigraphy in northwest Europe and the North Sea. *In*: KNOX, R. W. O'B., CORFIELD, R. M. & DUNAY, R. E. (eds) *Correlation of the Early Paleogene in Northwest Europe*. Geological Society, London, Special Publications, **101**, 15–42.

NOE-NYGAARD, A. 1974. Cenozoic to Recent volcanism in and around the North Atlantic Basin. *In*: NAIRN, A. E. M. & STEHLI, F. G. (eds) *The Ocean Basins and Margins*, Vol. 2. Plenum Press, New York, 391–443.

O'CONNER, S. J. & WALKER, D. 1993. Paleocene reservoirs of the Everest Trend. *In*: PARKER, J. R. (ed.) *Petroleum Geology of Northwest Europe: Proceedings of the 4th Conference*. Geological Society, London, 145–160.

PARKER, J. R. 1975. Lower Tertiary sand development in the central North Sea. *In*: WOODLAND, A. W. (ed.) *Petroleum and the Continental Shelf of North-west Europe*. Vol. 1, *Geology*. Applied Science, London, 447–453.

POMEROL, C. 1989. Stratigraphy of the Palaeogene: hiatuses and transitions. *Proceedings of the Geologists' Association, London*, **100**, 313–324.

POWELL, A. J. 1992. Dinoflagellate cysts of the Tertiary System. *In*: POWELL, A. J. (ed.) *A Stratigraphic Index of Dinoflagellate Cysts*. Chapman & Hall, London, 155–252.

——, BRINKHUIS, H. & BUJAK, J. *In*: KNOX, R. W. O'B., CORFIELD, R. M. & DUNAY, R. E. (eds) *Correlation of the Early Paleogene in Northwest Europe*. Geological Society, London, Special Publications, in press.

PRESTWICH, J. 1850. On the structure of the strata between the London Clay and the Chalk in the London and HampshireTertiary systems. Part I. *Quarterly Journal of the Geological Society of London*, **6,** 252–281.

—— 1852. On the structure of the strata between the London Clay and the Chalk in the London and Hampshire Tertiary systems. Part III. The Thanet Sands. *Quarterly Journal of the Geological Society of London*, **8**, 235–264.

—— 1854. On the structure of the strata between the London Clay and the Chalk in the London and Hampshire Tertiary systems. Part II. The Woolwich and Reading series. *Quarterly Journal of the Geological Society of London*, **10**, 75–170.

RITCHIE, J. D. & HITCHEN, K. 1996. Early Paleogene igneous activity on the northwest UK margin and its relationship within a stratigraphic framework for the North Atlantic Igneous Province. *In*: KNOX, R. W. O'B., CORFIELD, R. M. & DUNAY, R. E. (eds) *Correlation of the Early Paleogene in Northwest Europe*. Geological Society, London, Special Publications, **101**, 63–78.

STAMP, L. D. 1921. On cycles of sedimentation in the Eocene strata of the Anglo–Franco–Belgian Basin. *Geological Magazine*, **58**, 108–114; 146–157; 194–200.

STEWART, I. J. 1987. A revised stratigraphic interpretation of the Early Palaeogene of the central North Sea. *In*: BROOKS, J. & GLENNIE, K. (eds) *Petroleum Geology of North West Europe*. Graham & Trotman, London, 557–576.

THOMAS, J. E. 1996. Occurrences of *Apectodinium* in the Montrose Group (late Danian to early Thanetian) of the UK Central North Sea. *In*: KNOX, R. W. O'B., CORFIELD, R. M. & DUNAY, R. E. (eds) *Correlation of the Early Paleogene in Northwest Europe*. Geological Society, London, Special Publications, **101**, 115–120.

WARD, D. J. 1978. *The Lower London Tertiary (Palaeocene) Succession of Herne Bay, Kent*. Institute of Geological Sciences Report 78/10, 1–12.

WHITAKER, W. 1872. *The Geology of the London Basin. Part I. The Chalk and the Eocene Beds of the Southern and Western Tracts*. Geological Survey of England and Wales, Memoirs **4**.

WHITE, R. S. 1988. A hot-spot model for early Tertiary volcanism in the N Atlantic. *In*: MORTON, A. C. & PARSON. L. M. (eds). *Early Tertiary Volcanism and the Opening of the NE Atlantic*. Geological Society Special Publications, **39**, 3–13.

Gamma-ray spectrometry as a tool for stratigraphical interpretation: examples from the western European Lower Jurassic

D. N. PARKINSON

BP Research & Engineering Centre, Sunbury-on-Thames, Middlesex, UK
Present Address: Western Atlas Logging Services, 455 London Road, Isleworth, Middlesex TW7 5AB, UK

Abstract: Portable gamma-ray spectrometry provides an objective and quantitative method of studying sedimentary cyclicity in otherwise-cryptic mudrock successions. Spectral gamma-ray data are presented from the Lower Jurassic sections of the Yorkshire and Dorset Coasts (England), from Peniche (Portugal) and from inland exposures in southern Germany, including a recent road-cut. In Yorkshire, where proximal-distal relationships are readily demonstrated from sedimentological evidence, there is good correspondence between more proximal facies and elevated Th/K ratios. This relationship may be extended to the Dorset succession where the sequence-stratigraphical interpretation of Lower Lias mudrocks has been a source of some controversy. Th/K ratio data here suggest a distal, starved, Hettangian–earliest Sinemurian (Blue Lias) and a prograding Sinemurian interval (Black Ven Marls and Shales-with-'Beef'). Flooding and further progradation in the Early Pliensbachian (Belemnite Marls) is also suggested. These results provide support for common 'second order' stratigraphical forcing mechanisms between the Dorset and Yorkshire successions. Data from mainland Europe suggest that there are systematic regional variations in Th/K ratio, upon which local temporal variations are superimposed. These may reflect climatic or regional sediment transport effects.

Sequence stratigraphy, as defined by Van Wagoner *et al.* (1988), has two important components: a sedimentation model driven by relative sea-level change, and an assertion that many of the high-frequency sea-level changes which drive that model are global in extent. The Lower Jurassic of western Europe continues to play an important role in testing the assertion of globally-correlatable high-frequency sea-level changes for a number of reasons. Firstly, abundant, relatively non-provincial ammonites (Hallam 1975) provide the potential to correlate between different basins and different continents at this time. The northwest European Lower Jurassic ammonite biozonal scheme comprises some 52 subzones (Dean *et al.* 1961; Cope *et al.* 1980; Ivimey-Cook & Donovan 1983), allowing average time-resolution of perhaps 0.6 million years (Harland *et al.* 1990). Secondly, a long history of cycle-correlation studies in these rocks (see discussion in Arkell 1933) forms the basis for the widely-cited Jurassic sea-level curve of Hallam (1988). Thirdly, the study of the Lower Jurassic presents an opportunity to examine long-range correlation of sequences in the likely absence of glacio-eustatic sea-level change (Hallam 1985; but see Frakes & Francis 1988, for an opposing view).

Unfortunately, the corollary of excellent correlation based upon an abundant pelagic fauna, is that much of the western European Lower Jurassic is developed in distal carbonate and mudrock facies. In contrast to proximal siliciclastic settings, where the migration of facies belts may be monitored using well-understood sedimentological criteria, the sequence-stratigraphical interpretation of Lower Jurassic facies is often equivocal. It is principally for this reason that rather different sea-level curves can be interpreted for the same stratigraphical interval (compare Hallam 1988 and Haq *et al.* 1988). The objective of the work reported here is to investigate the potential of gamma-ray spectrometry to contribute to the sequence-stratigraphical interpretation of these rather cryptic successions.

Field gamma-ray spectrometry is an established technique in minerals exploration (Løvberg *et al.* 1971). Recent years have seen an extension of its application in sedimentological studies (e.g. Myers 1987; Hurst 1990). A scintillometer arrangement is used to produce an energy spectrum for the natural gamma-rays emitted from an outcrop. This spectrum is interpreted electronically to yield concentrations of the principal gamma-ray sources involved, namely thorium, potassium and uranium (Løvborg 1984).

In fine-grained rocks, high uranium concentrations are commonly associated with organic matter (e.g. Schmoker & Hester 1983). This is believed to be because uranium can be directly adsorbed onto organic matter, and also because it can be precipitated directly from seawater under the anoxic conditions associated with organic-matter preservation (Swanson 1960). Potassium is clearly a major component of rocks, and hence of soils.

From HESSELBO, S. P. & PARKINSON, D. N. (eds), 1996, *Sequence Stratigraphy in British Geology*, Geological Society Special Publication No. 103, pp. 231–255.

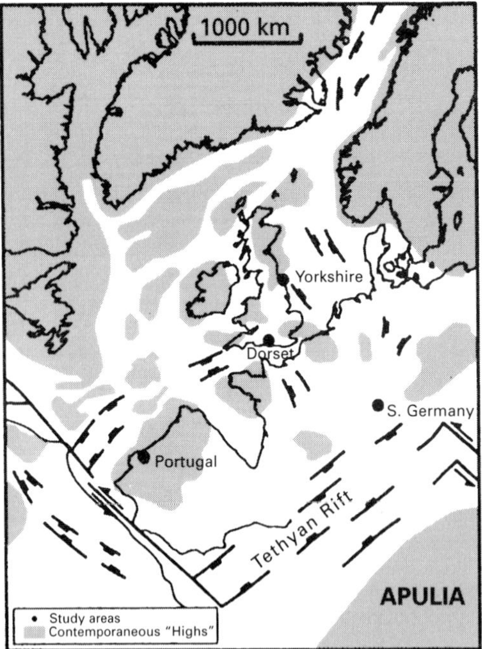

Fig. 1. Sketch paleogeography of the Early Jurassic of western Europe (after Lemoine 1983 and Ziegler 1988), showing the locations of studied sections.

Intense hydrolysis in hot and wet climates removes potassium in solution and leaves potassium-free, kaolinitic, soils (e.g. discussion in Summerfield 1991). Thorium, which is insoluble, is concentrated in those same soils, where it resides both in the heavy mineral component and adsorbed onto clays (Michel 1984). It is for these reasons that elevated Th/K ratios have been interpreted as indicators of hot and wet hinterland climate (e.g. Myers 1989). Studies of modern fine-grained systems, such as the mouths of the Niger (Porrenga 1966) and Amazon (Gibbs 1977), have also shown proximal-distal variations in clay mineralogy which are believed to reflect the differing hydraulic properties of clay grains and differences in flocculation behaviour. Th/K ratios have therefore also been suggested as proximal-distal indicators (Myers 1987).

In this study, spectral gamma-ray data are presented for the total accessible Lower Jurassic sections of Yorkshire and Dorset (England) and for sections in the Pliensbachian and Toarcian of Portugal and southern Germany (Fig. 1). Data from the Yorkshire section will be used to develop the thesis that Th/K can be used as a proximal-distal indicator in these rocks. This thesis will then be extended to the Dorset section, where sedimentological interpretation is more cryptic. Data are also presented for sections of the same age in Portugal and Germany. These sections provide further examples of systematic and sedimentologically useful variation in radionuclide concentration. They also suggest the possibility of systematic variations in Th/K ratio across Europe.

The use of radionuclide distribution in sedimentological studies is clearly dependent upon assumptions about the lack of post-depositional remobilisation of the measured elements (Cody 1971; Leythaeuser 1973; Clayton & Swetland 1978). The close correspondence between sedimentological and geochemical features shown in this paper is considered to provide empirical evidence for the primary nature of the geochemical signal. However, further work is needed to understand the sites and behaviour of Th, K and U in mudrocks.

Method

Data were obtained during 1991 & 1992 using an Exploranium GR-256 spectrometer (No. 1523, property of BP Exploration) equipped with a 76 × 76 mm sodium iodide detector (GPS-21, No. 1725). Readings were taken over periods of 3-6 minutes, depending on lithology. For these counting times, and given the radiochemical composition of rocks in this study (typically U = 3 ppm, K = 3%, Th = 15 ppm) the equations of Løvberg & Mose (1987) would predict an instrument precision of better than ±10% for all elements. In practice, departures of measurement geometry from a true plane (Fig. 2a &

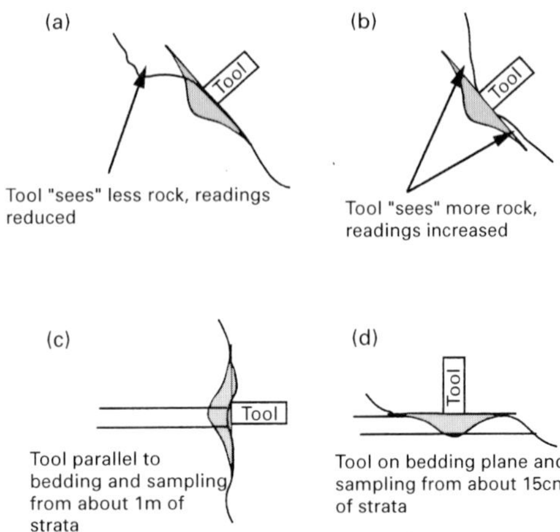

Fig. 2. The effects of sampling geometry upon field spectral gamma-ray measurements. The effective sample for a spectrometer of the type used in this study is shown in grey (after Løvborg *et al.* 1971). If the measurement surface is uneven the tool will 'see' more or less rock (*a* & *b*). Tool resolution is dependent upon its relationship to bedding (*c* & *d*)

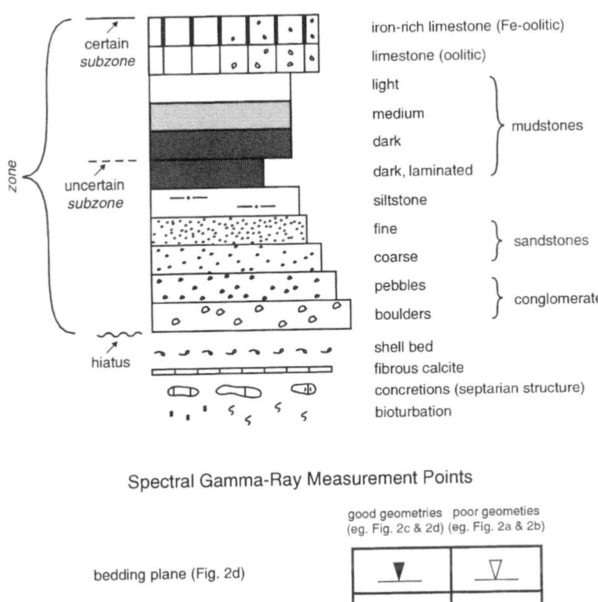

Fig. 3. Key to Figs 6–7, 10–12. Sedimentological key is that of Hesselbo & Jenkyns (1995).

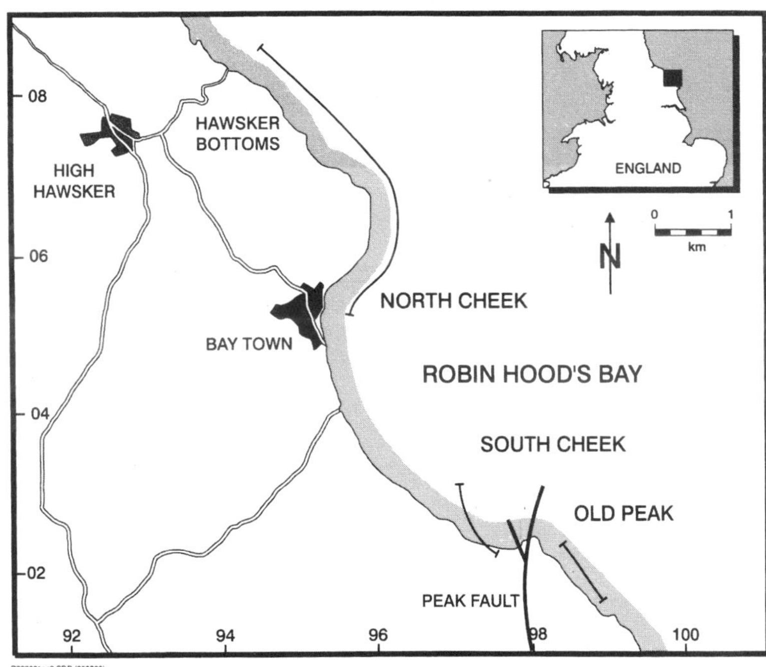

Fig. 4. Location map for the logged sections around Robin Hood's Bay on the Yorkshire Coast, NE England as shown in Fig. 5.

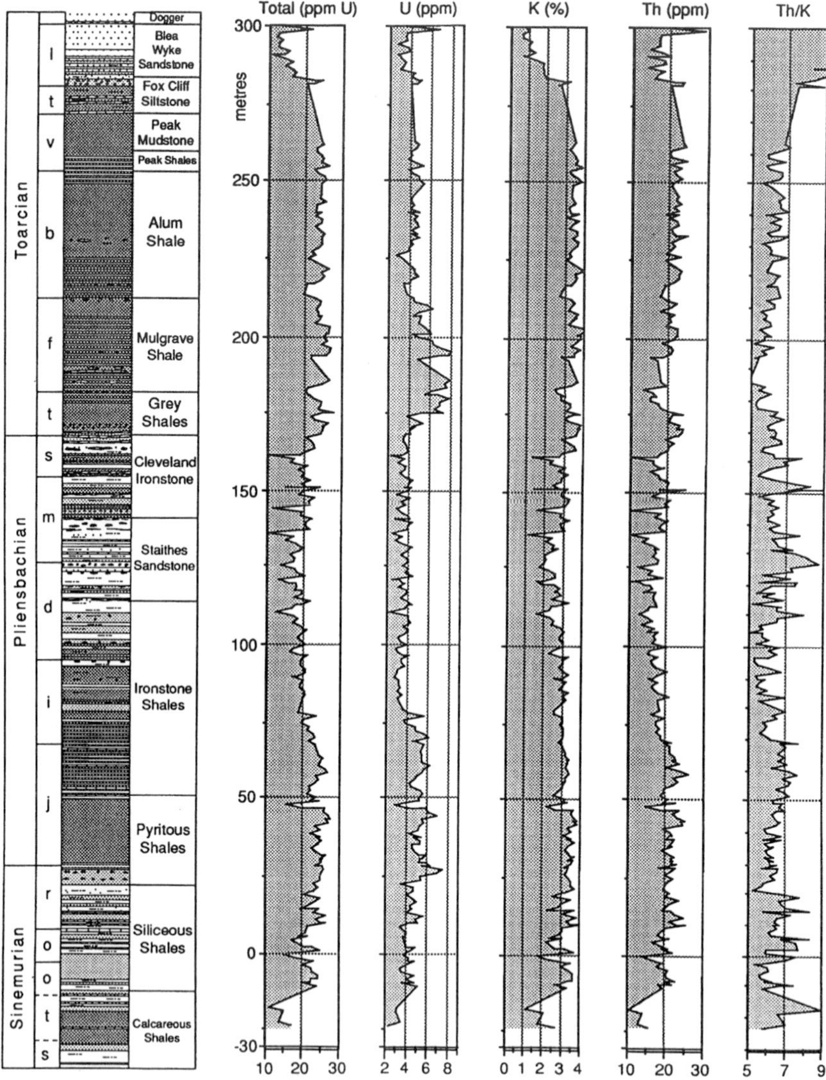

Fig. 5. A composite of spectral gamma-ray data from the Lower Jurassic of the Yorkshire coast around Robin Hood's Bay, see Figs 6–7 for details of some stratigraphical units. Ammonite zones have been abbreviated to their initial letters for clarity.

b) far outweigh instrument precision as a source of experimental error. Repeat readings along a bed and repeat logging of the same interval suggest errors nearer to ±7% for K, ±15% for Th and ±30% for U. These low accuracies should be remembered when interpreting the fine detail of the logs presented here. Ratios are unaffected by this error source.

The volume of rock from which gamma-rays are collected by a detector such as that used in this study, is approximately 1 m in diameter and 15 cm deep (Fig. 2, see also Løveborg *et al.* 1971). Maximum stratigraphical resolution is therefore achieved when the tool is used on a bedding plane (Fig. 2d). However, suitable bedding plane exposures are rare, particularly when all beds must be measured and not just those lithologies more resistant to weathering, which tend to form good platforms. Most measurements in this study were made perpendicular or sub-perpendicular to bedding at intervals of 0.5–1 m (Fig. 2c). They therefore represent a moving average of rock properties similar in character and somewhat coarser in resolution to that provided by borehole natural gamma-ray sondes (Anon 1992). Tool orientation relative to bedding is indicated on the detailed logs in this paper by the symbols shown in Fig. 3.

Further details of instrument calibration settings, together with data tabulations, will be found in Parkinson (1994). For a review of the theory and practice of spectral gamma-ray spectrometry in sedimentology the reader is referred to Myers (1987).

Lower Jurassic of Yorkshire, England

The Yorkshire data were acquired from the coastal exposures of the Lower Jurassic around Robin Hood's Bay (Fig. 4, see also Rawson & Wright 1992; Hesselbo & Jenkyns 1995). The lithostratigraphical nomenclature used here (Fig. 5) follows Powell (1984) except for the incorporation of amendments to the nomenclature of the Lower Toarcian proposed by Rawson & Wright (1992). The section is continuous from the Calcareous Shales in the core of the Robin Hood's Bay anticline (*semicostatum* Zone of the Sinemurian) to the Grey Shales at Hawsker Bottoms (*tenuicostatum* Zone of the Toarcian). Extension of the stratigraphy to the top of the Toarcian Stage requires a cross-fault correlation to the section south of the Peak Fault. This is an extensional feature across which some synsedimentary growth may be expected (Milsom & Rawson 1989). The cross-fault correlation used here is that of Howarth (1962). The total Lower Jurassic section is some 300 m thick.

The Yorkshire Lower Jurassic has already been the subject of a number of gamma-ray investigations. Myers (1987, 1989) logged the Cleveland Ironstone, Grey Shales and Mulgrave Shales at Staithes, some 20 km NW of the sections described here. Van Buchem (1990) logged the Calcareous Shales to Ironstone Shales at the localities described here and extended his log down into the Hettangian using the exposures at Redcar, some 40 km to the northwest. Both of these authors reported predominantly bedding-plane measurements. The objective of the present work was to cover as much as possible of the section with a consistent approach, using the same instrument and calibration settings applied to the other sections in this study and taking the opportunity to locate samples accurately against the new measured sections of Hesselbo & Jenkyns (1995). Figure 5 summarizes the data at a scale which enables the more significant trends to be recognized.

The Calcareous Shales, at the base of the sequence, are grey mudstones with occasional laterally extensive shell beds. These pass transitionally upwards into the Siliceous Shales, which are distinguished by the presence of fine-grained sandstones occurring both as scour-fills and more laterally persistent, decimetre-scale beds (Sellwood 1970). These scours have been interpreted by van Buchem & McCave (1989) to represent the impingement of storm wavebase. Some caution must be exercised in the interpretation of the lowermost spectral gamma-ray readings for this interval, as they were obtained from bedding planes in the foreshore and therefore may be biased towards more calcareous lithologies (see above). Taken at face value, however, the data suggest lower total counts in the Calcareous Shales than in the Siliceous Shales, a result of lower concentrations of Th, K and U. This observation corroborates that of van Buchem *et al.* (1992) and may be interpreted to indicate a lower proportion of terrigenous mud relative to carbonate in the Calcareous Shales. It is noteworthy that where, as here, carbonate is a significant component of the system, gamma-ray increases cannot be relied upon as an indicator of grain size and hence of proximality: the sandy and presumed higher-energy Siliceous Shales yield the higher gamma-ray signal.

The Siliceous Shales are overlain by the Pyritous Shales, a unit of dark grey mudstones with ferruginous concretions containing a sparse and restricted fauna which is interpreted to be indicative of dysaerobic, deep-water sedimentation (Sellwood 1972; van Buchem & McCave 1989). Total gamma-ray counts, Th and K increase from the Siliceous Shales into the Pyritous Shales (Fig. 5), again largely a reflection of terrigenous clay content. It is, however, only the uranium and Th/K ratio curves which clearly reflect the observed stratigraphy, with a sharp increase in uranium concentration and a sharp decrease in Th/K ratio passing upwards into the Pyritous Shales (Fig. 5). According to the ideas discussed in the introduction to this paper the increased uranium content might be interpreted as indicative of anoxia. The decreased Th/K ratio might be anticipated if these rocks are interpreted as having been deposited in a distal setting.

It should be noted at this point that the decreased Th/K ratio upwards into the Pyritous Shales does not appear to correspond with a decrease in kaolinite/illite. According to the clay-mineralogical data of van Buchem *et al.* (1992) the converse is true. This problem recurs when the Upper Toarcian part of the Yorkshire section is considered. The hypothesis in the literature that Th/K provides a proximal-distal indicator *originates* from a predicted simple relationship between Th/K and kaolinite/illite (e.g. Hassan *et al.* 1976). However, the data presented here support *only* a close empirical correspondence between Th/K ratio and sedimentologically-defined proximal-distal indications. They do not necessarily support a simple link with kaolinite/illite ratio.

The Ironstone Shales may be divided into two units: the lower part comprises decimetric interbeds of marl and mudstone with occasional horizons of sideritic concretions (van Buchem & McCave 1989; Fig. 6. The thinnest interbeds occur near the top of

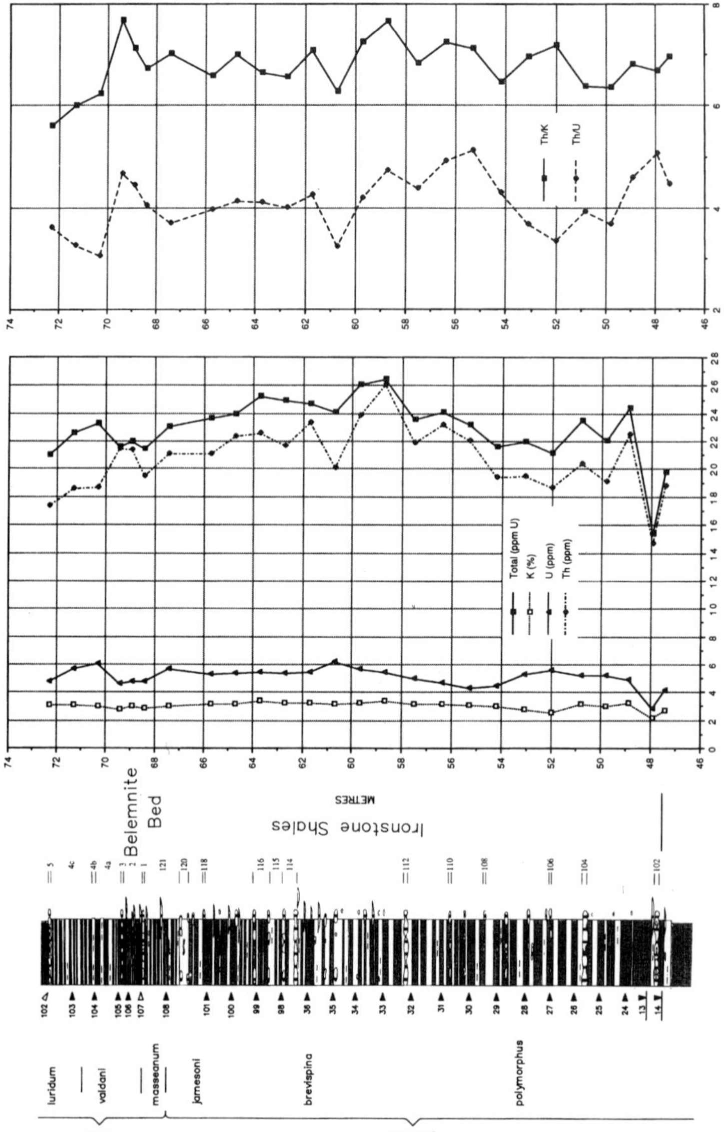

Fig. 6. Spectral gamma-ray data from the lower part of the Ironstone Shales, north side of Robin Hood's Bay, Yorkshire. Stratigraphical log from Hesselbo & Jenkyns (1995). For key see Fig. 3.

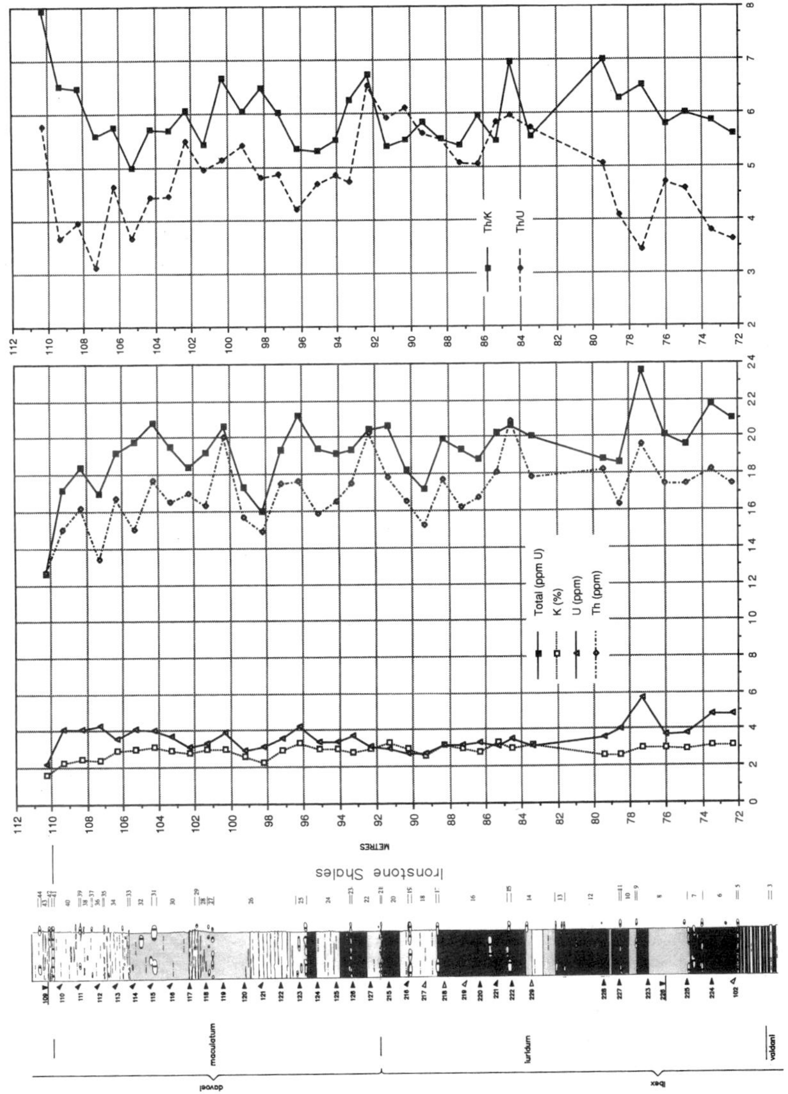

Fig. 7. Spectral gamma-ray data from the upper part of the Ironstone Shales, north of Robin Hood's Bay, Yorkshire. Stratigraphical log from Hesselbo & Jenkyns (1995). For key see Fig. 3.

the unit, and are associated with an increase in silt content and a belemnite concentration labelled the Belemnite Bed by Hesselbo & Jenkyns (1995). Th, K and U concentrations (and hence also total counts) decline at the base of the Ironstone Shales, reflecting the increased carbonate content of the unit. Concentrations then recover somewhat, to a peak at around the middle of the unit (Fig. 6). The Th/K ratio does not show a break at the base of the Ironstone Shales: rather, it is consistent within the lower Ironstone Shales themselves and forms the culmination of a cycle of increasing Th/K which has its base at the base of the Pyritous Shales and its top at the level of the Belemnite Bed (Fig. 6). Above the Belemnite Bed, Th/K declines sharply. In sequence-stratigraphical terms, the observed bed thinning, culminating in a belemnite concentration, may be interpreted as a result of decreasing rates of accommodation space creation, or as a result of decreasing sediment supply (starvation). Increasing Th/K ratios, if interpreted as indicative of progradation, might be taken to favour the former interpretation.

The upper part of the Ironstone Shales comprises relatively monotonous marls which Hesselbo & Jenkyns (1995) suggest coarsen upwards, passing transitionally into the overlying Staithes Sandstone. Total gamma-ray counts decline through the unit (Fig. 7), but with the exception of the uppermost few metres of the Ironstone Shales, this appears to be more a function of declining Th (and hence Th/K ratio) than a parallel decline in both elements, such as might be expected from a simple reduction in the ratio of clays to coarse siliciclastic sediments. Minimum Th/K ratio occurs at around bed 32, associated with a level of concretions and above which there is unequivocal coarsening, a marked decline in Th and K and an increase in Th/K. Spectral gamma-ray data, therefore, suggest continued transgression until within the *maculatum* Subzone of the *davoei* Zone and would suggest bed 32 as the 'maximum flooding surface'.

The Staithes Sandstone immediately north of Robin Hood's Bay, where it was logged for this study, comprises dominantly bioturbated argillaceous and silty sandstone with beds and lenses of cleaner, fine-grained sandstone (Howard 1985). Total gamma-ray counts are generally low, reflecting the high sand content of the unit, which occurs at the regressive maximum of an unambiguous 'second order' regressive-transgressive cycle bounded by transgressive maxima in the Pyritous Shales of the Early Pliensbachian and the Mulgrave Shale of the Early Toarcian (Fig. 5). There is, however, considerable variation in total count levels (Fig. 10), which appear to reflect both variations in sand/clay ratio and anomalously high Th/K ratios associated with horizons of ferruginous concretions. The latter may be a less extreme manifestation of the anomalously high Th/K ratios observed in ironstones within the Cleveland Ironstone Formation (Myers 1989; and below).

Considerable lateral facies variation exists within the Cleveland Ironstone (Howard 1985) and the ironstone horizons themselves are much more weakly developed at Hawsker Bottoms, where they were logged for this study, than at Staithes, where they were logged by Myers (1989). At Hawsker Bottoms, the Cleveland Ironstone comprises three well-defined coarsening-upwards cycles, each capped by a horizon of ferruginous concretions. Each concretionary horizon is marked by low total gamma-ray counts, but only the Pecten Seam exhibits a marked positive Th/K anomaly such as those described at the top of each cycle by Myers. Background Th/K appears to decline upwards through the unit, part of a trend of declining Th/K from peak values in lower part of the *margaritatus* Zone (within the Staithes Sandstone), to minimum values in the *falciferum* Zone (within the Mulgrave Shale, see below). Again, there is good correspondence between elevated Th/K ratio and sedimentological expectations of the most proximal facies.

The Cleveland Ironstone is overlain by the fine-grained rocks of the Whitby Mudstone Formation (Powell 1984; Knox 1984), which is defined to include the interval between and including the Grey Shales and the Fox Cliff Siltstone Members (Fig. 5). The Mulgrave Shale Member is a term introduced by Rawson & Wright (1992) to replace the Jet Rock Member of Powell (1984), which included the commonly used informal units 'Jet Rock', 'Bituminous Shales' and 'Ovatum Band'. Only the mudrocks of the Grey Shales and the lower part of the Mulgrave Shale may be observed in stratigraphical continuity with the Cleveland Ironstone at Hawsker Bottoms The Jet Rock may be distinguished by its distinctive concretions: both the Jet Rock and the overlying Bituminous Shales have a bituminous appearance and yield elevated Total Organic Carbon measurements (Küspert 1982; Raiswell & Berner 1985). The Jet Rock and Bituminous Shale interval has been interpreted to reflect oxygen-depleted bottom-water conditions (Raiswell & Berner 1985; Myers & Wignall 1987) and is the time-equivalent of distinctive organic-rich facies elsewhere (e.g. the Posidonienschiefer of Germany, see below), which may reflect a global Toarcian 'Oceanic Anoxic Event' (Jenkyns 1988).

As would be predicted from the increase in the proportion of terrigenous clay, the base of the Whitby Mudstone Formation (the Grey Shales) is marked by a sharp upwards increase in total gamma-ray counts. The organic-matter enrichment of the Mulgrave Shale is accompanied by a marked increase in uranium concentration and a decline in

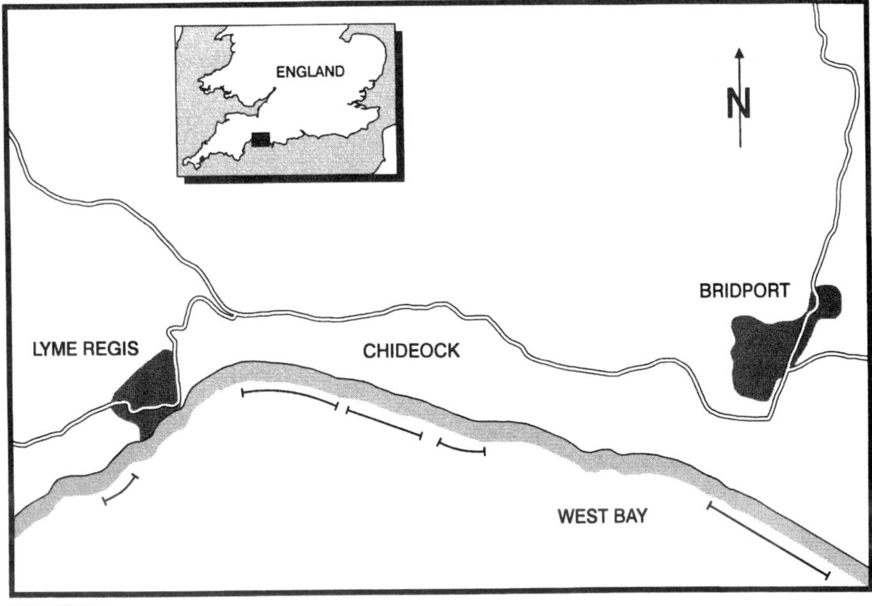

Fig. 8. Location map for the logged sections of the Dorset Coast, southern England as shown in Fig. 9. Data were recorded against the logs of Hesselbo & Jenkyns (1995).

Th/U ratio (Fig. 5). Interestingly, the continuing decline in Th/K ratio to minimum values in the Mulgrave Shale, discussed above, tends to counter the uranium increase so that the total-count signature of this stratigraphically important interval is suppressed (Fig. 5).

The upper part of the Mulgrave Shale, and overlying strata, were logged to the east of the Peak Fault (Figs 4 & 5). The upper units of the Whitby Mudstone Formation (the Alum Shale, Peak Mudstone and Fox Cliff Siltstone) are not distinctive in the field, but have been defined or re-defined on the basis of detailed grain-size and clay-mineralogical studies by Knox (1984). Knox's data demonstrate coarsening-upwards, both through this interval, and continuing through the overlying Blea Wyke Sandstone Formation. The upper part of the Blea Wyke Sandstone is strongly bioturbated, contains abundant belemnites and is compatible with a lower shoreface environment of deposition. At Blea Wyke there is a sharp contact with the overlying Middle Jurassic Dogger sandstones, marked by a shelly lag. On the western side of the Peak Fault the Dogger rests directly upon the Alum Shales. There is, therefore a significant structural unconformity at the base of the Dogger: the 'mid-Cimmerian' event, most recently discussed by Underhill & Partington (1993).

Uranium values decline before the top of the Mulgrave Shale (Fig. 5), but total count rates are maintained by an increase in thorium content. The Th/K ratios return to the levels of the Grey Shales by the middle of the Alum Shales (Figs 5 & 14). Limited spectral gamma-ray data recorded from the Blea Wyke Sandstone show much reduced, Th, K and U counts as would be anticipated given the near absence of clay. The Th/K ratios are much elevated, once again closely following the pattern of progradation and retrogradation which would be interpreted from the sedimentology. Knox (1984) presents percentages of chlorite, kaolinite and mica through the top Alum Shale to Dogger interval. Kaolinite percentages *decline* upwards, whereas the other two minerals increase upwards. As with the Pyritous Shales (see above), though we do not yet understand the sites of the radionuclides in the section, a simple correlation between Th and kaolinite does not appear to provide the answer.

Lower Jurassic of Dorset, England

The spectral gamma-ray data from Dorset were acquired from the classic Lower Jurassic coastal exposures between Seven Rock Point, west of Lyme Regis, and Burton Bradstock, east of Bridport (Fig. 8). Again, data were collected against the measured sections of Hesselbo & Jenkyns (1995), which should be consulted for details of lithostratigraphical, biostratigraphical and bed-numbering schemes. Details of access are provided in House

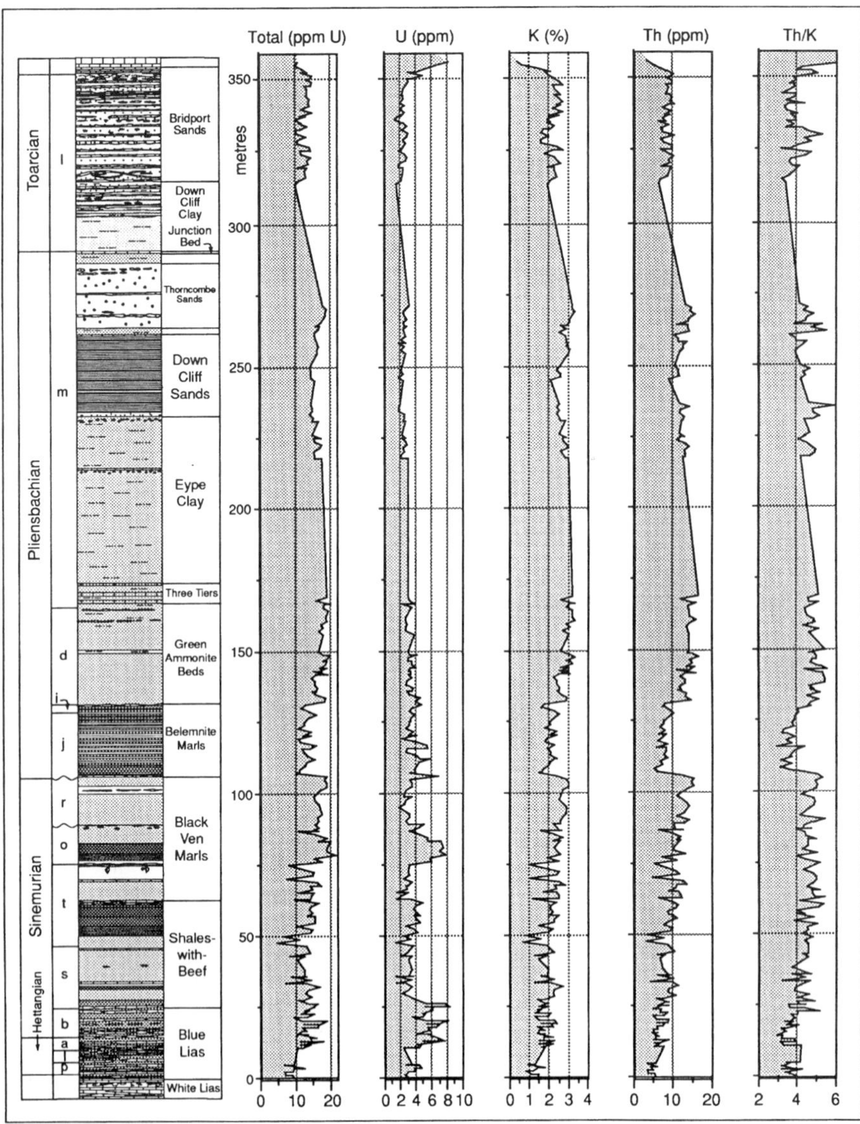

Fig. 9. A composite of spectral gamma-ray data from the Lower Jurassic of the Dorset, see Figs 10–12 for details of individual stratigraphical units. Ammonite zones have been abbreviated to their initial letters for clarity.

(1989). The section (Fig. 9) from the Triassic to the Middle Jurassic is continuously exposed, though the 350 vertical metres of log are acquired over 15 horizontal kilometres of coastal exposure, which includes numerous, dominantly east-west, faults across which Lower Jurassic growth is recorded (Jenkyns & Senior 1991; Hesselbo & Jenkyns 1995). The log does not, therefore, represent stratigraphical thickness at a single point.

The author is not aware of previous spectral gamma-ray studies of the Lower Jurassic in Dorset, though important work has been done on the Upper Jurassic Kimmeridge Clay (Wignall & Myers 1988). The Hettangian to middle Pliensbachian interval was continuously logged for this study and is the main focus of the discussion below. Some data were acquired for the Upper Pliensbachian and Toarcian but significant data gaps remain due to the inaccessibility of suitable exposures.

The lowermost lithostratigraphical unit of the Dorset coast Lower Jurassic is the Blue Lias (Lang 1924). It comprises some 26 m of limestone-marl

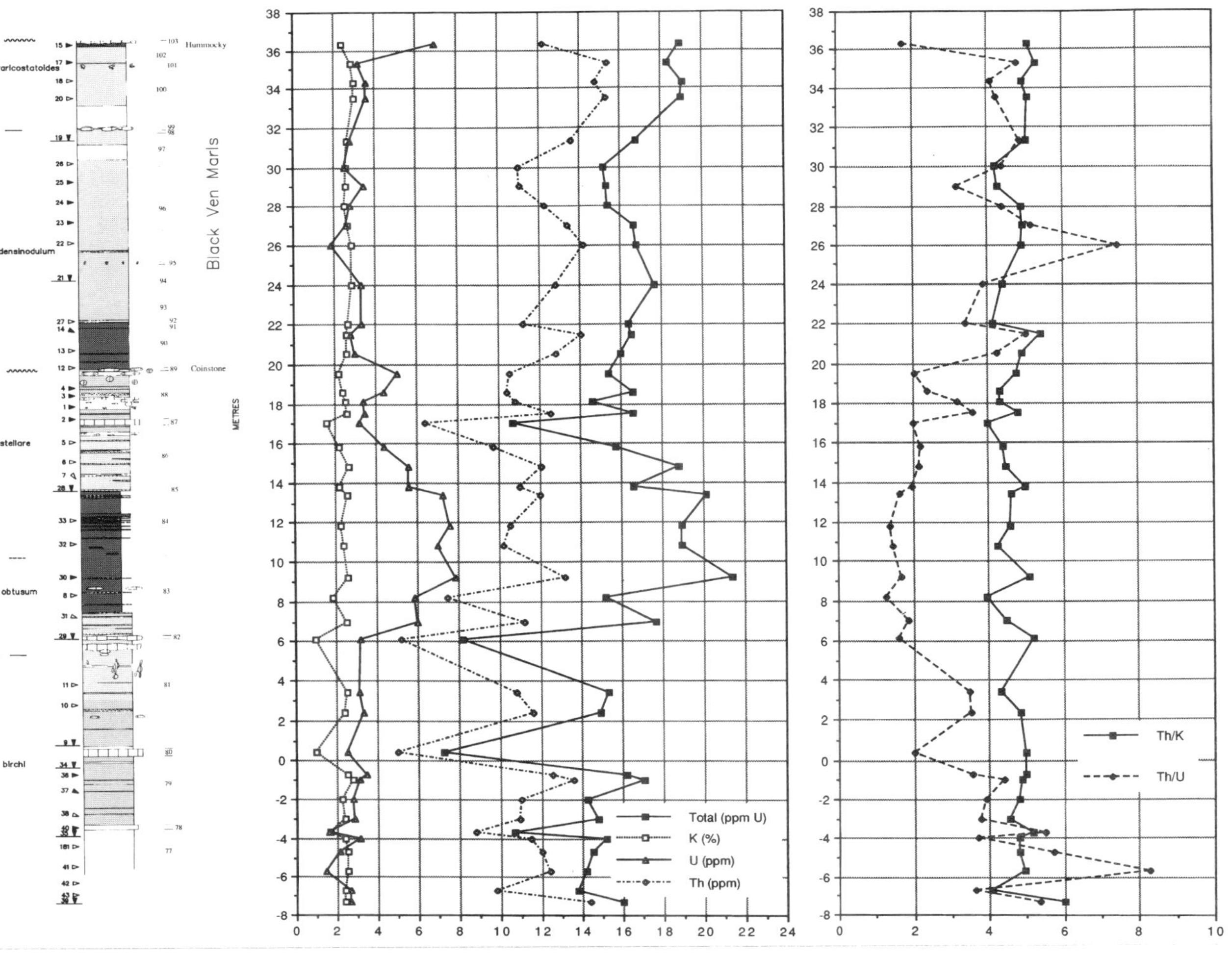

Fig. 10. Spectral gamma-ray data from the Black Ven Marls, foreshore east of Charmouth, Dorset. Stratigraphical log from Hesselbo & Jenkyns (1995). For key see Fig. 3.

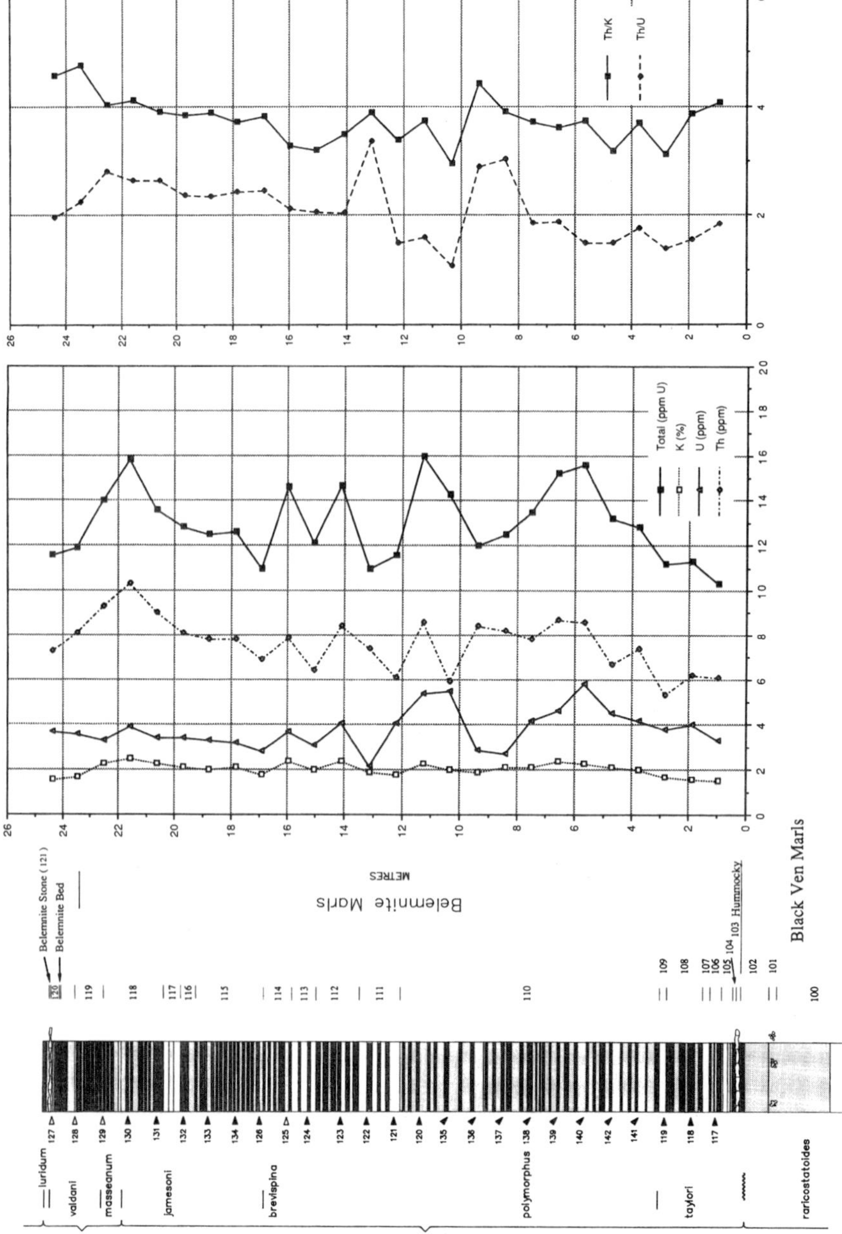

Fig. 11. Spectral gamma-ray data from the Belemnite Marls, foreshore between Stonebarrow and Seatown, Dorset. Stratigraphical log from Hesselbo & Jenkyns (1995). For key see Fig. 3.

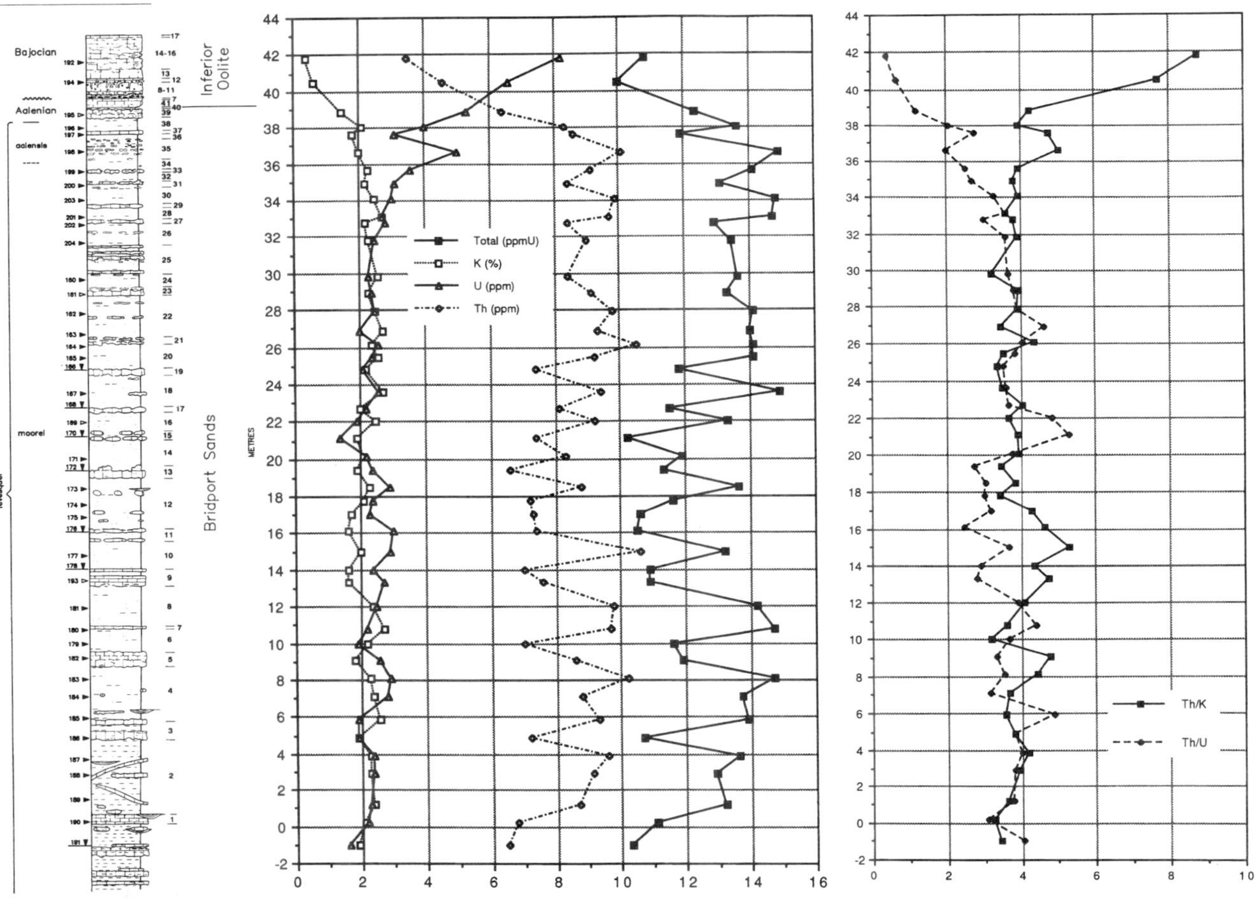

Fig. 12. Spectral gamma-ray data from the Bridport Sands and Inferior Oolite, West Bay to Burton Bradstock, Dorset (access to a complete section is complex: see House 1989, pp. 60–62 for details). Stratigraphical log from Hesselbo & Jenkyns (1995). For key see Fig. 3.

interbeds. In detail, five rock-types can be resolved: limestone, light and dark marl, laminated shale and laminated limestone: the limestones being formed by early diagenetic cementation of light marls and, occasionally, of laminated shales (Hallam 1960, 1964; Weedon 1986). Relatively shale-rich intervals occur in the *liasicus* and upper *bucklandi* Zones. A stratigraphical hiatus has been suggested within the intervening carbonate-rich interval (specifically within the *angulata* Zone) based on comparison with the section some 50 km to the north, in Somerset (Smith 1989).

Total gamma-ray count data in the Blue Lias (Fig. 9) are rather erratic, reflecting an interbed frequency greater than the sample interval. However, some interesting features can be determined. Firstly, there is a general increase in total gamma-ray counts upwards through the Blue Lias and into the overlying Shales-with-'Beef'. All elements increase upwards, which may be interpreted to reflect an increase in terrigenous clay content. However, Th/K ratios also increase upwards, suggesting a systematic change in the composition of terrigenous input. The Th/U ratio is uniformly low, suggesting anoxic conditions, commensurate with the Total Organic Carbon contents of 2–6% reported for the interval by Weedon (1986). Uranium content decreases sharply at the top of the Blue Lias, though this is not reflected in the total-count signal as the decrease in U counts is counterbalanced by an increase in Th and K concentrations. The top of the Blue Lias coincides with a significant incursion of benthic foraminiferal taxa, which may have been a response to improvements in bottom-water oxygenation (P. Copestake pers. comm.). This observation is compatible with the observed upwards decrease in U concentration.

The Shales-with-'Beef' (Lang *et al.* 1923) is a sequence of monotonous dark marls and dark laminated shales with occasional tabular and concretionary limestones. Thickening of ammonite subzones and a decrease in carbonate content compared with the Blue Lias may be interpreted in terms of upwards increased clastic supply at the boundary between the Blue Lias and the Shales-with-'Beef'.

The Th/K ratio remains stable at values of 4–5 within the Shales-with-'Beef' (perhaps increasing slightly upwards, Fig. 9) suggesting a rather constant clay composition. Low total counts are associated with carbonate-rich intervals. Perhaps the most interesting observation is that the shales of the *scipionianum* Subzone (at the base) have the highest Th/U ratio values and that uranium counts increase upwards through the unit to Birchi Tabular, where they fall off sharply (Fig. 9). That is, on spectral gamma-ray data the Shales-with-'Beef' are best defined using the uranium log, where they appear as a single increasing-upwards unit.

The Black Ven Marls (Lang & Spath 1926) are lithologically similar to the Shales-with-'Beef', with comparable accumulation rates as evinced by ammonite subzonal thickness (Fig. 10). They include two biostratigraphical omission surfaces: the Coinstone (*denotatus* Subzone and *oxynotum* Zone unrecorded) and the Hummocky (*aplanatum* and *macdonnelli* Subzones unrecorded). The Coinstone, in particular, has received much attention, being interpreted as a product of sea-level fall and associated winnowing by Hallam (1988) and a result of starvation associated with a sea-level highstand by Haq *et al.* (1988). Spectral gamma-ray data suggest that the Black Ven Marls represent the culmination of a cycle of increasing Th, K and Th/K which begins in the Blue Lias (Fig. 9). Within the Black Ven Marls there is little variation in Th/K suggesting essentially constant composition for the terrigenous clastic component (Fig. 10). A striking uranium peak is associated with the interval of organic-rich shales recorded within the lower part of the *obtusum* Zone by Hesselbo & Jenkyns (1995).

The Belemnite Marls (Lang *et al.* 1928) comprise decimetre interbeds of light and dark marls, similar in aspect to the lower, contemporaneous, part of the Ironstone Shales of Yorkshire. The light and dark marl alternations are grouped into variable length (191–705 cm) bundles (Weedon & Jenkyns 1990), with overall thinning towards the top of the unit in the mid-*ibex* Zone (Fig. 11). The unit is capped by two potentially important condensation surfaces: a concretionary carbonate at the top of the formation containing abundant belemnites and ammonites, known as the Belemnite Stone and, some 30 cm below, a belemnite concentration, the Belemnite Bed. These two beds closely approximate the top and base of the *luridum* Subzone, which expands to some 20 m in Yorkshire. In general, ammonite subzonal thicknesses in the Belemnite Marls suggest a return to the relatively low sediment accumulation rates of the Blue Lias. The analogy with Blue Lias conditions is supported by the spectral gamma-ray data. Total counts, and Th and K levels, are much reduced in the Belemnite Marls compared with underlying and overlying strata (Fig. 9), reflecting the increased carbonate content of the unit with a presumed decrease in the input of terrigenous clastics. Th/K ratios are also much reduced: the values of between 3 and 4 being comparable with those of the Blue Lias and interpreted here as reflecting a return to a distal setting (Fig. 9). The Belemnite Marls exhibit some uranium enrichment compared to immediately underlying and overlying strata, the highest values of U and the lowest Th/U being found near the base.

The relationship between the Belemnite Marls and the Green Ammonite Beds is closely analogous to that between the Blue Lias and Shales-with-

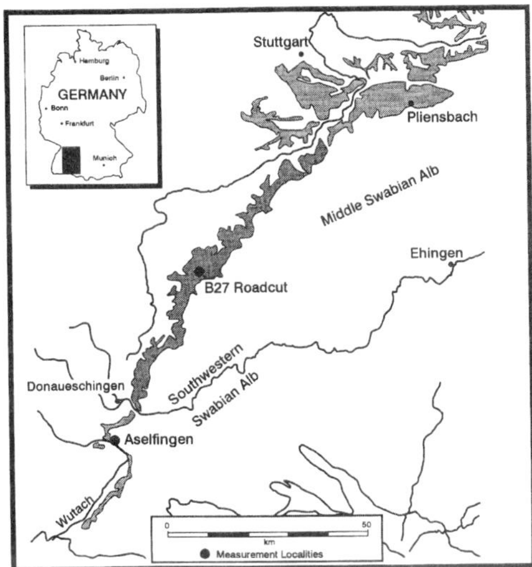

Fig. 13. Location map for the south German log (Fig. 14). The outcrop of the Lower Jurassic is stippled.

'Beef'. There is a sharp upwards change from relatively condensed decimetric light and dark marl alternations to relatively expanded, less calcareous deposits. This is accompanied by a sharp upwards increase in Th, K and Th/K ratio. These observations appear best explained by postulating an increase in the rate of clastic supply, with consequent progradation, i.e. interpreting the Belemnite Stone as a downlap surface in the sense of Van Wagoner *et al.* (1988).

Access to suitable exposures limited the acquisition of spectral gamma-ray data above the Green Ammonite Beds. The data obtained from the Down Cliff Sands interval are poorly constrained stratigraphically, but serve to confirm that similar values of Th, K and Th/K ratios prevail in the Middle Lias as in the Lower Lias (Fig. 9). This is of relevance to the discussion of pan-European trends (below). A well-constrained spectral gamma-ray data set was obtained for the Bridport Sands interval (Fig. 12). The most notable features of these data relate to the passage upwards in to the condensed limestone of the Inferior Oolite (Middle Jurassic), which is uranium-rich, Th and K (clay) poor, and marked by a very high Th/K ratio. Uranium enrichment extends down into the uppermost beds of the Bridport Sands. Interpretation of these observations must await more detailed geochemical studies.

Lower Jurassic, South Germany

The Lower Jurassic rocks of southern Germany outcrop along the line of the Swabian and Franconian Alps (Fig. 13). In the Swabian Alps, the Lower Jurassic varies in thickness between 50 and 150 m and comprises alternating intervals of marl and limestone-marl, similar to the lower part of the Dorset succession (Urlichs 1977). The system of Greek letters used to describe the stratigraphy dates from the work of Quenstedt (1843).

Two sections were examined in this study: the stream section at Aselfingen, some 4 km west of Blumberg, near the German-Swiss border, and a section within a new roadcutting for the B27 motorway between Balingen and Hechingen. The former section is well known biostratigraphically (Urlichs 1977), but much of the exposure is inadequate for spectral gamma-ray work. Biostratigraphical information on the latter section was supplied by Matthias Franz (Geological Survey of Baden-Würtemberg).

Total-count data for the Aselfingen section (Parkinson 1994) clearly differentiate the clay-rich 'Lias β' sediments from the carbonate-rich 'Lias γ' sediments. Declining Th and K upwards through the Lias β, with essentially constant Th/K, suggests an upwards-decrease in clay content. The 'Obliquabank', a distinctive phosphatic limestone which appears to separate biostratigraphical omission surfaces equivalent in age to the 'Coinstone' and 'Hummocky' of Dorset (Urlichs 1977) is anomalously enriched in Th and somewhat enriched in U.

The B27 roadcut enabled excellent fresh surfaces of Pliensbachian and lower Toarcian strata to be logged (Fig. 14). Total gamma-ray counts show two peaks: one centred upon the mid- to upper *margaritatus* Zone and one centred upon the *tenuicostatum* Zone. As with the Mulgrave Shale of Yorkshire, the strong uranium excursion associated with the organic-rich Posidonienschiefer, at the top of the section is suppressed on the total-count log by concommitant changes in Th, K and Th/K ratio. The concentrations of both Th and K are lower in the Posidonienschiefer than in underlying strata, reflecting a reduced content of terrigenous clay. The Th/K ratio is also somewhat reduced compared with underlying strata.

The peaks in total gamma-ray counts in the German section, as with the lower parts of the Dorset section, correspond to elevated concentrations of both Th and K, presumed to reflect relatively high proportions of terrigenous clay. Similarly to the Dorset section, there is systematic variation in Th/K ratio, which here is interpreted to suggest that the most Th-rich clay assemblages are associated with times of maximum terrigenous supply in the mid- to late *margaritatus* Zone.

Lower Jurassic, Portugal

The Lusitanian Basin is a roughly north–south-trending rift which extends both onshore and offshore western Portugal (Ziegler 1988; Wilson *et al.* 1989). A number of fragmentary exposures of

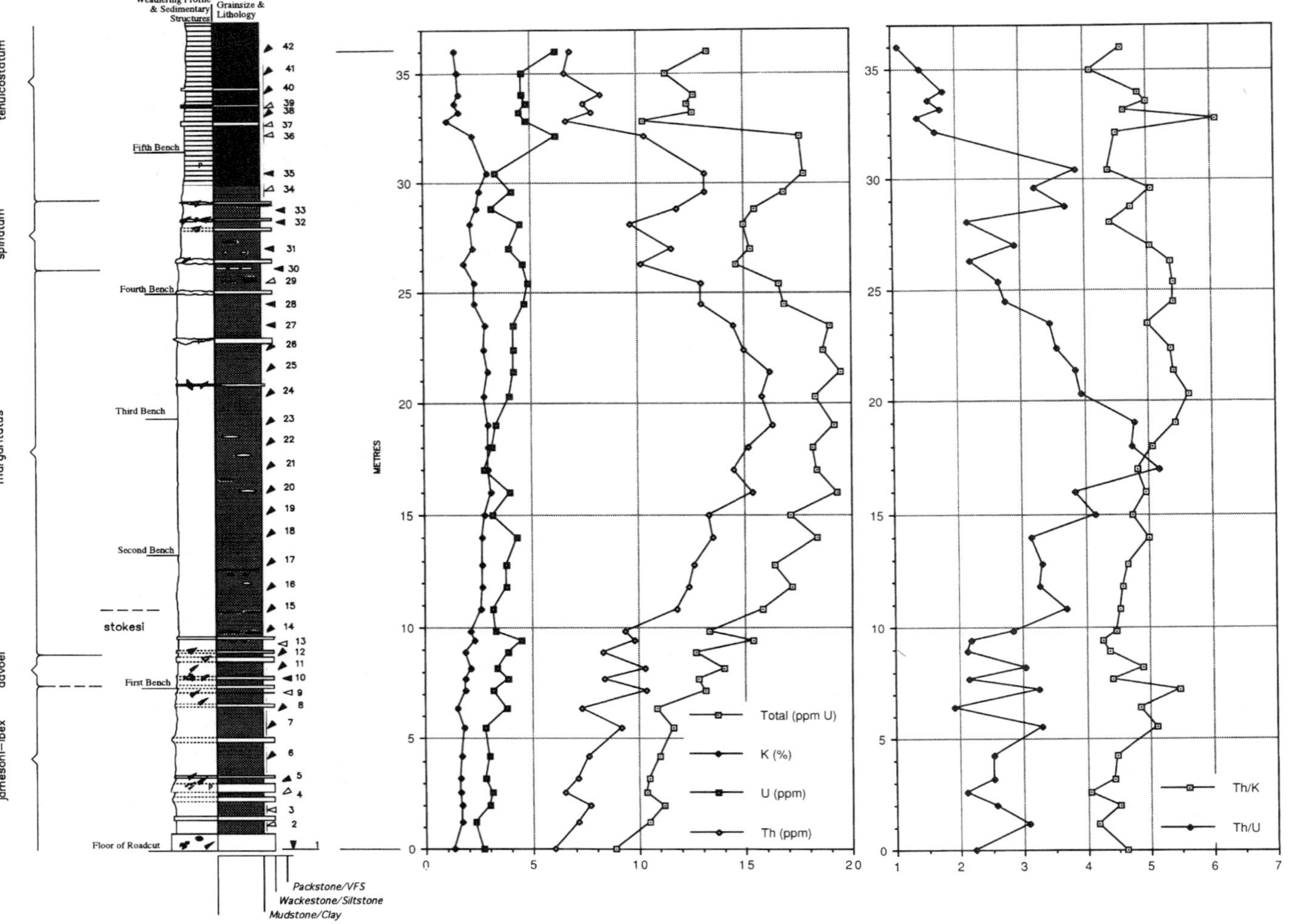

Fig. 14. Spectral gamma-ray data from the B27 roadcut between Balingen and Hechingen, south Germany.

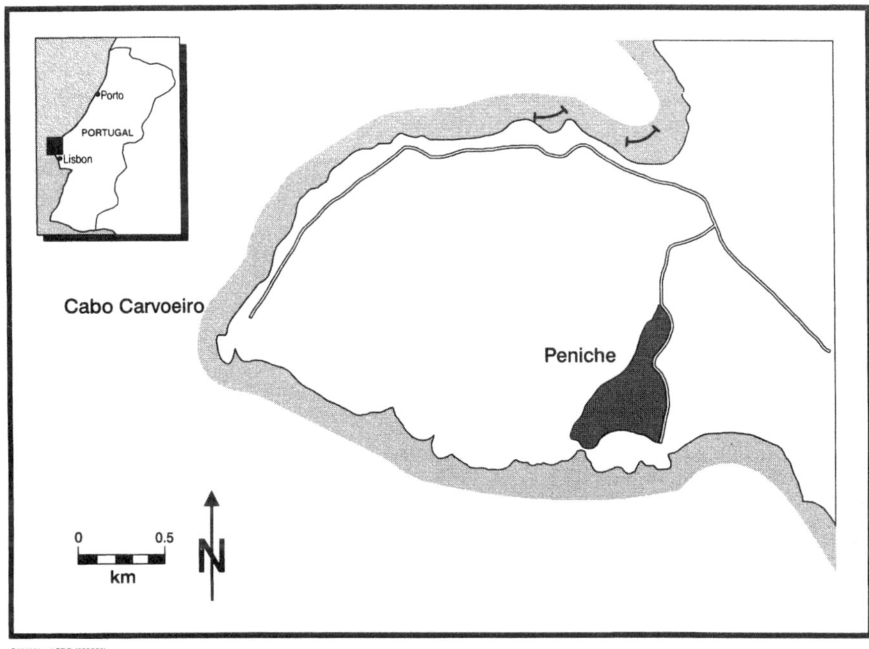

Fig. 15. Location map for the logs in the Lower Jurassic of Peniche, Portugal (Figs 16 & 17). Outcrops of the Lower Jurassic shown in inset are from Hallam (1971).

Lower Jurassic strata within this basin are available onshore, due largely to the effects of diapirism in Triassic salt. The most stratigraphically complete of these is at Peniche, some 70 km north of Lisbon (Fig. 15, see also Hallam 1971). The base of this section, of late Sinemurian age, is developed in shallow-water carbonate facies dominated by pelloidal and oolitic grainstones (Watkinson 1989). These deposits are abruptly overlain by some 200 m of Pliensbachian to mid-Toarcian outer-ramp facies, ranging in character from bioclastic packstones to marls. The more marly intervals within this section, particularly the Pliensbachian interval, contain an abundant ammonite fauna which has been studied by Mouterde (1955) and Phelps (1985). This biostratigraphical work enables unambiguous correlation with the western European scheme used elsewhere in this paper. The whole section is capped by Late Toarcian and younger submarine fan deposits (Wright & Wilson 1984). Spectral gamma-ray data were acquired through the two marly intervals within the Pliensbachian to mid-Toarcian outer-ramp sequence (Figs 16 & 17).

The prime interest of these data lies in their contribution to the overall pattern of western European Th/K ratios (see below). Two features of the data should be noted at this stage. Firstly, the variation in total-count data in the mid-Pliensbachian (Fig. 16), suggesting two cycles of upwards-increase in clay content, with the most clay-rich sediments being found near the top of the *masseanum* Subzone (*ibex* Zone) and the top of the *margaritatus* Zone, and the least clay-rich sediments being near to the top of the *valdani* Subzone (*ibex* Zone). There is some uncertainty in the validity of this observation, as *luridum* Subzone and *davoei* Zone observations were made in the foreshore, where there is a bias towards less-radioactive lithologies. However there is corroboration of the observation from a subtle minimum in Th/K ratios coincident with the total-count minimum.

The second interesting feature of these data is a peak in the total count data in the lower part of the *falciferum* Zone (Fig. 17). The common occurrence of organic-matter enrichment at this stratigraphical horizon has been discussed above. The total-count peak in Portugal is not, however, related to uranium enrichment, as might be expected. Rather, it reflects an increase in the proportion of terrigenous clay in the marls, demonstrated by a parallel increase in both Th and K. Sedimentologically, this influx is accompanied by a number of decimetre scale debris flows, with clasts up to cobble grade.

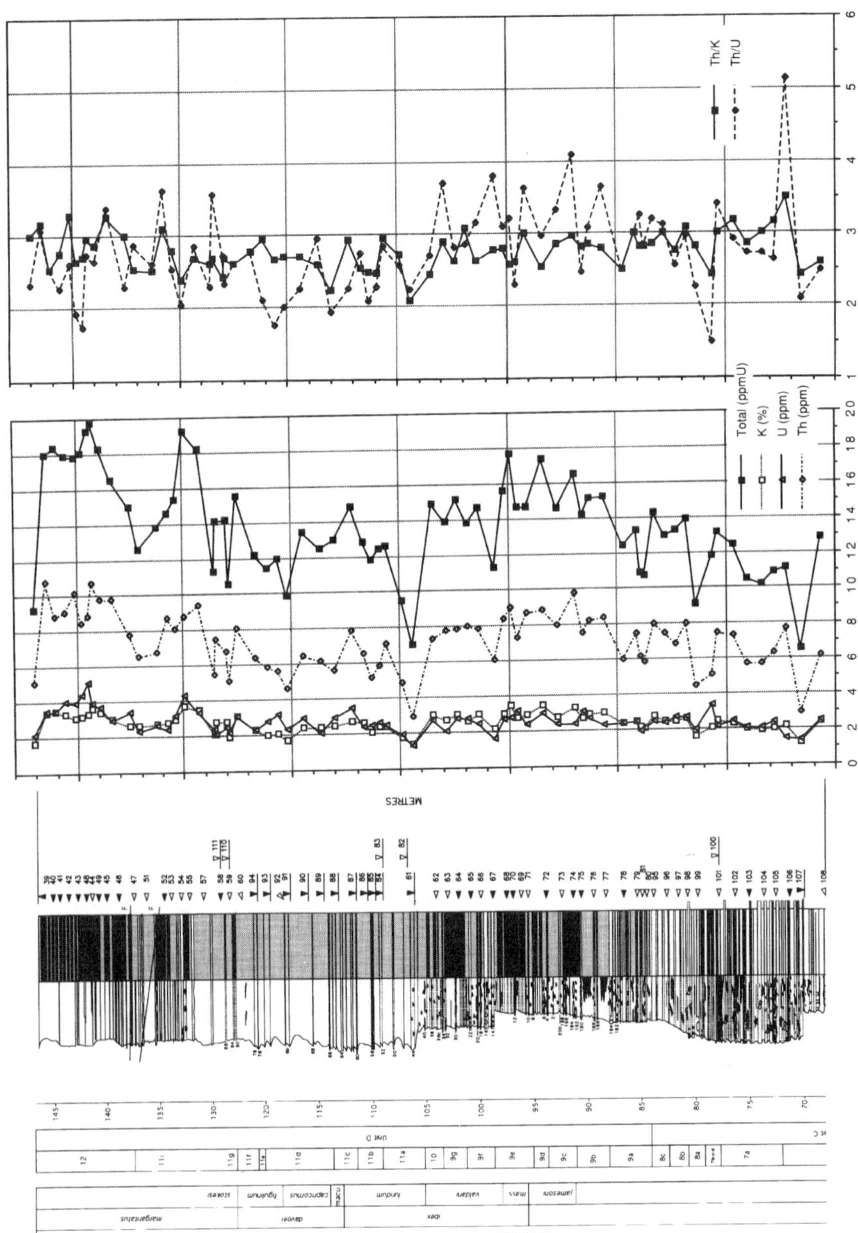

Fig. 16. Spectral gamma-ray data from the mid-Pliensbachian of Peniche, Portugal.

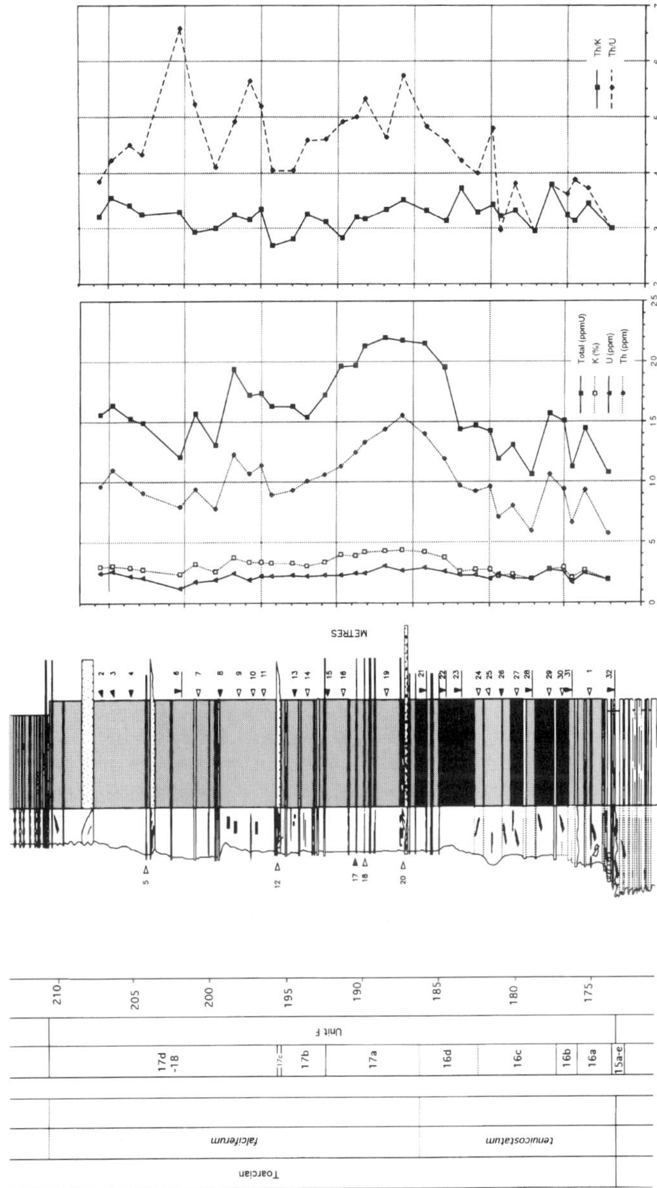

Fig. 17. Spectral gamma-ray data from the Lower Toarcian of Peniche, Portugal.

Discussion

Total-count data for Yorkshire define two gross sedimentary cycles ('second order' cycles in the sense of Vail *et al.* 1991). Gamma-ray minima (Fig. 5) are associated with the Calcareous Shales (Early Sinemurian) and with the Staithes and Blea Wyke Sandstones (Late Pliensbachian and Late Toarcian). Gamma-ray maxima are associated with the Pyritous Shales (Early Pliensbachian) and Mulgrave Shale (Early Toarcian). Such log-based cycles may be the primary data-source in a wireline log-based sequence-stratigraphical study (e.g. Steel 1993; Parkinson & Hines 1995) and it is of interest to understand the components of this cyclicity and how they relate to the stratigraphical units which are directly observable in the field. The first observation is that, to a first approximation, thorium and potassium co-vary (Fig. 18), a result common to many radiochemical studies of argillaceous rocks (Pliler & Adams 1962; Quinif *et al.* 1982; Myers 1987). This may be interpreted to suggest the presence of a clay mineral assemblage of relatively constant Th/K ratio, diluted by varying proportions of weakly radioactive carbonate and coarse siliciclastic material. That is, the Th and K contribution to the total counts is a measure of the amount of clay in the system, the most fundamental tenet of gamma-log interpretation (Rider 1991). The importance of carbonate in this analysis is demonstrated by the case of the Calcareous Shales. These are associated with a second order gamma-ray minimum, yet it is the overlying, more arenaceous, Siliceous Shales which on sedimentological criteria, probably mark second order maximum regression in the area. The Calcareous Shales represent miniumum argilaceous content, but not the coarsest-grained sediment.

The co-variance of Th and K is only approximate and in detail there are numerous exceptions. The high Th, low K outliers of Fig. 18 are generated by the ironstones of the Cleveland Ironstone and Staithes Sandstone, and by the Blea Wyke Sandstones. Variance within the 'normal' mudrock population is the result of systematic variation in Th/K ratio, discussed above and illustrated in Fig. 5. There appears to be a close correspondence between elevated Th/K and units interpreted as proximal on independent sedimentological grounds, thus providing empirical support for the use of the Th/K ratio as a proximal-distal indicator. Th/K ratios also draw out an additional sedimentary cycle in the Pyritous Shales and lower Ironstone Shales, which is not represented on the total-count log but is validly interpreted as a progradational cycle from field evidence (see above). Low Th/K ratios, resulting from low Th, are a particular feature of the bituminous facies of the Mulgrave Shale.

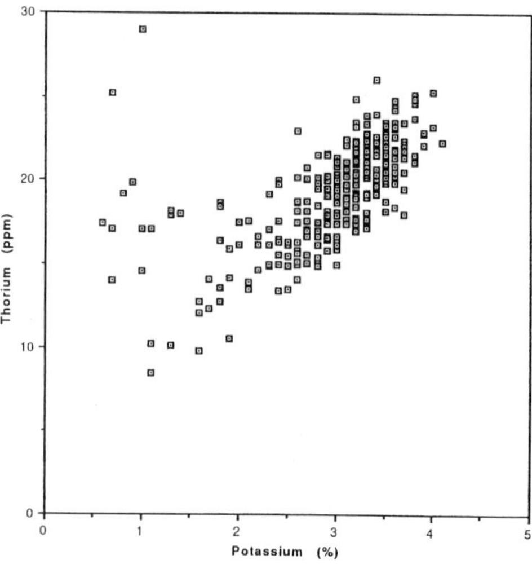

Fig. 18 Cross-plot of all spectral gamma-ray determinations of Th and K for the Yorkshire section (data of Fig. 5).

In summary, although the total-count data for Yorkshire broadly mirror the gross lithostratigraphy observed in the field, some important elements of that stratigraphy may be missed (lower Ironstone Shales regression), distorted (Siliceous Shales regression) or partially suppressed (Jet Rock) if only those total-count data are considered. It is clearly easier to recognize the key elements of the stratigraphy using the components of the spectral gamma-ray signal; U content and Th/K ratio appear to be particularly powerful in this regard.

In contrast to the Lower Jurassic section of Yorkshire, the Hettangian–Early Pliensbachian of Dorset is difficult to interpret in sequence-stratigraphical terms; largely as a result of the absence of coarse siliciclastic sediments from the depositional system. Total gamma-ray count data provide little assistance, clearly demarcating only the carbonate-rich sediments of the Belemnite Marls (Fig. 9). However, as discussed above, there is an unambiguous upward increase in Th/K from the middle Blue Lias to the top of the Black Ven Marls and, following the close correspondence between Th/K and proximal–distal facies in Yorkshire, it is suggested here that this may be the expression of a second-order progradational cycle. In this interpretation the Blue Lias is viewed as deposited under conditions of sediment starvation in which pelagic carbonate was an important component (Fig. 19a). The top of the Blue Lias is interpreted as a downlap surface across which the sediments of the Shales-with-'Beef' and Black Ven Marls would prograde, leading to the absence of further accommodation space by the mid to late Sinemurian. Relative sea-level rise in the latest Sinemurian–earliest Pliensbachian provides

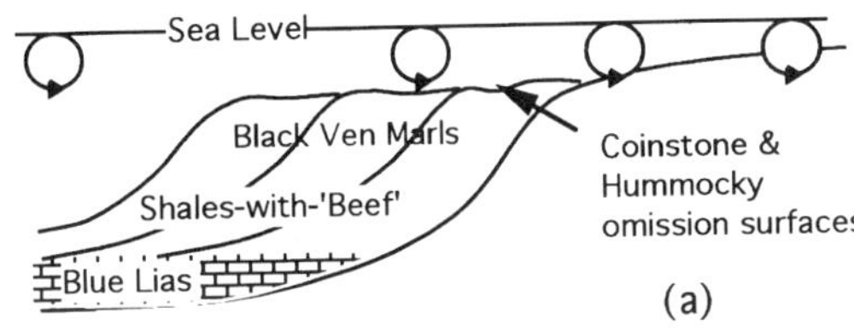

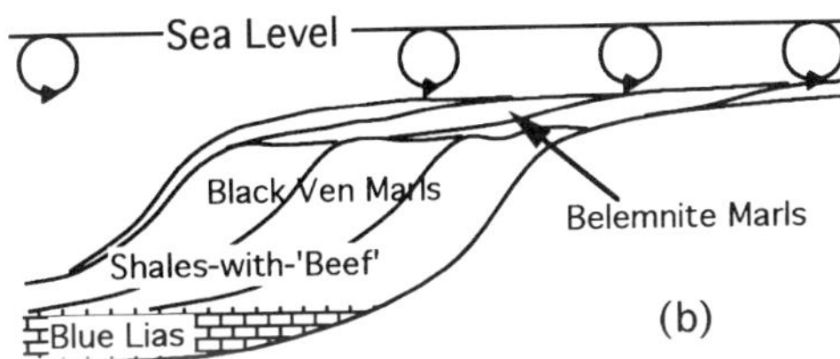

Fig. 19. Schematic model for progradational cycles in the Lower Jurassic of Dorset, subsequent to a postulated period of rapid accommodation space creation during deposition of the Blue Lias. (**a**) Sinemurian progradational cycle, (**b**) Early Pliensbachian progradationa cycle.

space for the Belemnite Marls (Fig. 19). Such an interpretation has the effect of reconciling the regressive-transgressive cycles of Dorset and Yorkshire; both sections showing Sinemurian progradation and earliest Pliensbachian flooding. If correct, it also has implications for the interpretation of the key horizons within the Dorset Sinemurian: the *turneri* and *obtusum* Zone 'hot shales' would occur during progradation, supporting models other than simple 'maximum flooding' for the optimum preservation of organic matter (Fleet *et al.* 1987); the Coinstone omission surface would similarly occur in the context of progradation, favouring the interpretation of Hallam (1988), who inferred stratigraphical omission due to lack of accommodation space, over the interpretation of Haq *et al.* (1988), who preferred to interpret starvation.

Continuing the comparison of the Dorset and Yorkshire stratigraphy into the Lower Pliensbachian, the Th/K ratio data for Yorkshire might describe progradation up to the mid-*valdani* Subzone (near the top of the lower Ironstone Shales, Fig. 5) and subsequent retrogradation up to the mid-*maculatum* Subzone (near the top of the Ironstone Shales). A precisely comparable situation occurs in Dorset, where the Th/K ratio maximum also occurs in the mid-valdani Subzone and the transgressive interval may be largely represented by the thin strata between the Belemnite Bed and Belemnite Stone (Fig. 11). A period of renewed accommodation space creation and subsequent progradation is envisaged. This reconciliation of the two basin stratigraphies is very similar to that argued on sedimentological grounds alone by Hesselbo & Jenkyns (1995). The subtle facies-belt migrations suggested within the mid-Pliensbachian of Portugal may be a manifestation of the same pattern, suggesting even more widespread synchroneity of stratigraphical events.

If Th/K ratio is a good indicator of proximal-distal relationships, then what of climatic control, as discussed in the introduction to this paper? Such control might be expected to be manifest on a pan-European scale, where there appears likely to have been significant latitudinal variation in climate (Chandler *et al.* 1992). To facilitate latitudinal comparison between logged sections of the same age, absolute ages have been assigned to the data using the following steps. Firstly, ages were assigned to subzonal boundaries by linear interpolation between the base Hettangian, base Sinemurian and mid-Bajocian ages of Harland *et al.* (1990), assuming subzones of equal length and assigning an arbitrary value of two subzones to any zones which are currently undivided into subzones. This is similar to the practice followed by Jones *et al.* (1994). The subzonal schemes adopted are those of Cope *et al.* (1980), as modified by Ivimey-Cook & Donovan (1983), and Callomon & Chandler (1990) for the required interval of the Aalenian and Bajocian. Secondly, ages were assigned to individual data points by linear interpolation between the subzonal boundaries shown on the figures of this paper.

A plot of gamma-ray data from all localities against time (Fig. 20) shows than whereas considerable variation occurs within an individual section,

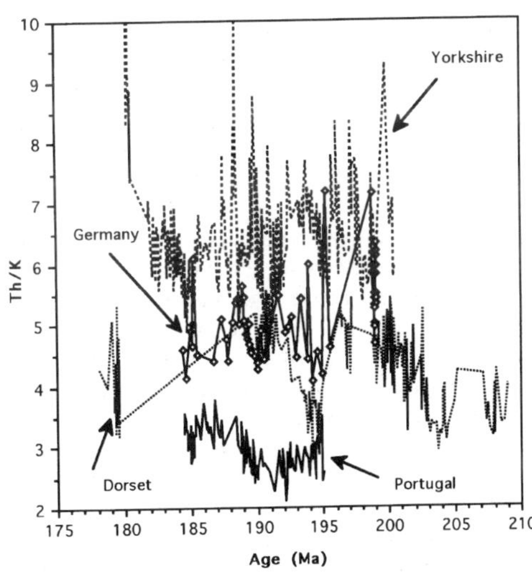

Fig. 20. Comparison of Th/K ratios for various sections studied, expressed against a common timescale (Harland *et al.* 1990).

there is much greater variation between sections. Specifically, the lower latitude carbonate-dominated section of Portugal has lower Th/K ratios than the less calcareous mudrock-dominated sections of Dorset and Germany, which in turn have lower Th/K ratios than the higher-latitude, more arrenaceous, section of Yorkshire. The Early Jurassic paleogeography of the European area is thought to have comprised a series of 'highs', associated with the ancient massifs, separated by fault controlled depocentres (Fig. 1). The extent to which these highs can actually be substantiated as sediment source areas is variable. However, all are relatively small and sediment supply is likely to have been dominated by the Fennoscandian and possibly the Laurentian Shields. Lower Jurassic deltaic deposits are extensively developed on Haltenbanken, offshore mid-Norway (Karlsson 1984). Whether the variation shown in Fig. 20 relates to the climate of individual local sediment source areas or is a result of sediment sorting in a Europe-wide suspended sediment transport system is a matter for speculation at present.

Conclusions

Outcrop gamma-ray spectrometry is a useful source of quantitative data for the stratigraphical interpretation of otherwise-cryptic mudrock successions. There appears to be a close empirical correspondence, in the Lower Jurassic of Yorkshire, between variations in Th/K ratio and proximal-distal relationships as interpreted from sedimentology: elevated Th/K being associated with the most proximal settings. Although the sites of Th and K in these rocks remains largely speculative, simple assumptions that Th/K rato is a proxy for kaolinite/illite ratio do not seem appropriate.

Use of Th/K ratio as a proximal-distal indicator can be extended to shed light upon the sequence-stratigraphical interpretation of mudrock and pelagic carbonate successions in the Early Pliensbachian of Yorkshire and the Hettangian–Early Pliensbachian of Dorset. Such interpretations lend quantitative support to an interpretation of synchronous second-order stratigraphical events between Dorset and Yorkshire. Further data from mudrock successions on mainland Europe provides a provisional insight into the systematic variations in Th/K ratios on a larger scale, which may relate either to local sediment provenance areas or to regional suspended-sediment transport systems.

The outcrop data presented here also provide some insights into the problems of interpreting wireline-log data sets, showing how important elements of that stratigraphy may be missed, distorted or partially suppressed if only the total gamma-ray log is available. Spectral gamma-ray data can do much to bring a log interpretation closer to the reality of the depositional units as observed at outcrop .

I am grateful to K. Myers for introducing me to the principles and practice of portable gamma-ray spectrometry; to P. Roberts and D. Talbot for allowing me to use spectrometer calibration facilities at the British Geological Survey, to C. Wilson for introducing me to the Portuguese section and to M. Franz for sending me his unpublished ammonite data on the B27 roadcut. H. Jenkyns and S. Hesselbo played a major role in allowing me to use their (then) unpublished logs of the Dorset and Yorkshire sections and providing many hours of stimulating discussion of Jurassic stratigraphy. Much of the work was undertaken in Oxford whilst on sabbatical leave from BP Exploration. I am grateful to BP for providing me with that opportunity and for permission to publish.

References

ANON 1992. *Introduction to Wireline Log Analysis.* Western Atlas International Inc., Houston, Texas.

ARKELL, W. J. 1933. The Jurassic System in Great Britain. Oxford University Press.

CALLOMON, J. H. & CHANDLER, R. B. 1990. A review of the ammonite horizons of the Aalenian-Lower Bajocian Stages in the Middle Jurassic of Southern England. *Memorie Descrittive della Carta Geologica d'Italia*, **40,** 85–112.

CHANDLER, M. A., RIND, D. & RETO, R. 1992. Pangean climate during the Early Jurassic: GCM simulations and the sedimentary record of paleoclimate. *Geological Society of America Bulletin*, **104**, 543–559.

CLAYTON, J. L. & SWETLAND, P. J. 1978. Subaerial weathering of sedimentary organic matter. *Geochimica et Cosmochimica Acta*, **42**, 305–312.

CODY, R. D. 1971. Adsorption and the reliability of trace elements as environmental indicators for shales. *Journal of Sedimentary Petrology*, **41**, 461–471.

COPE, J. C. W., GETTY, T. A., HOWARTH, M. K., MORTON, N. & TORRENS H. S. 1980. *A correlation of Jurassic Rocks in the British Isles, Part 1: Introduction and Lower Jurassic.* Geological Society, London, Special Reports, **14**.

DEAN, W. T., DONOVAN, D. T. & HOWARTH, M. K. 1961. The Liassic ammonite zones and subzones of the North-West European Province. *Bulletin of the British Museum (Natural History), Geology*, **4,** 435–505.

FLEET, A. J., CLAYTON, C. J., JENKYNS H. C. & PARKINSON, D. N. 1987. Liassic source rock deposition in Western Europe. *In*: BROOKS J. & GLENNIE K. (eds) *Petroleum Geology of Western Europe*. Graham & Trotman, London, 59–70.

FRAKES, L. A. & FRANCES, J. E. 1988. A guide to Phanerozoic cold polar climates from high-latitude ice rafting in the Cretaceous. *Nature*, **333**, 547–549.

GIBBS R. J. 1977. Clay mineral segregation in the marine environment. *Journal of Sedimentary Petrology*, **42**, 237–243.

HALLAM, A. 1960. A sedimentary and faunal study of the Blue Lias of Dorset and Glamorgan. *Philosophical*

Transactions of the Royal Society, London, **A243**, 1–44.

—— 1964. Origin of the limestone-shale rhythm in the Blue Lias of England: a composite theory. *Journal of Geology*, **72**, 157–169

—— 1971. Facies Analysis of the Lias in West Central Portugal. *Neues Jahrbuch für Geologie und Paläontologie*, **139**, 226–265.

—— 1975. *Jurassic Environments*. Cambridge University Press.

—— 1985. A review of Mesozoic climates. *Journal of the Geological Society, London*, **142**, 433–445.

—— 1988. A Reevaluation of Jurassic eustasy in the light of new data and the revised Exxon curve. *In*: WIIGUS, C. K., *et al.* (eds) *Sea-level changes: an integrated approach*. Special Publications of the Society of Economic Paleontologists and Mineralogists, **42**, 261–273.

HAQ, B. U., HARDENBOL, J. & VAIL, P. R. 1988. Mesozoic and Cenozoic chronostratigraphy and cycles of sea-level change. *In*: WIIGUS, C. K., *et al.* (eds) *Sea-Level Changes: an Integrated Approach*. Special Publication of the Society of Economic Paleontologists and Mineralogists, **42**, 71–108.

HARLAND, W. B., ARMSTRONG, R. L., COX, A. V., CRAIG, L. E., SMITH A. G. & SMITH, D. G. 1990. *A Geologic Time Scale 1989*. Cambridge University Press.

HASSAN, M., HOSSIN, A. & COMBAZ, A. 1976. Fundamentals of the differential gamma ray log interpretation technique. *Society of Professional Well Log Analysts, 17th Annual Symposium*, Paper H.

HESSELBO, S. P. & JENKYNS, H. C. 1995. A comparison of the Hettangian to Bajocian successions of Dorset and Yorkshire. *In*: TAYLOR, P. D. (ed.) *Field Geology of the British Jurassic*. Geological Society, London, 105–150.

HOUSE, M. R. 1989. *Geology of the Dorset Coast*. Geologists' Association, London, Field Guide.

HOWARD, A. S. 1985. Lithostratigraphy of the Staithes Sandstone and Cleveland Ironstone formations (Lower Jurassic) of north-east Yorkshire. *Proceedings of the Yorkshire Geological Society*, **45**, 261–275.

HOWARTH, M. K. 1962. The Jet Rock Series and the Alum Shale Series of the Yorkshire coast. *Proceedings of the Yorkshire Geological Society*, **33**, 381–422.

HURST, A. 1990. Natural gamma-ray spectrometry in hydrocarbon-bearing sandstones from the Norwegian Continental Shelf. *In*: HURST, A., LOVELL, M. A. & MORTON, A. C. (eds) *Geological Applications of Wireline Logs*. Geological Society, London, Special Publications, **48**, 211–222.

IVIMEY-COOK, H. C. & DONOVAN, D. T. 1983. Appendix 3: The fauna of the Lower Jurassic *In*: WHITTAKER, A. & GREEN, G. W. *Geology of the country around Weston-super-Mare*. Memoirs of the Geological Survey of Great Britain, Sheet 279 with parts of sheets 263 and 295, 126–130.

JENKYNS, H. C. 1988. The Early Toarcian (Jurassic) anoxic event: stratigraphic, sedimentary, and geochemical evidence. *American Journal of Science*, **288**, 101–151.

—— & SENIOR, J. R. 1991. Geological evidence for intra-Jurassic faulting in the Wessex Basin and its margins. *Journal of the Geological Society, London*, **148**, 245–260.

JONES, C. E., JENKYNS, H. C. & HESSELBO, S. P. 1994. Strontium isotopes in Early Jurassic seawater. *Geochimica et Cosmochimica Acta*, **58**, 1285–1301.

KARLSSON, W. 1984. Sedimentology and diagenesis of Jurassic sediments offshore mid-Norway. *In*: SPENCER, A. M., *et al.* (eds) *Petroleum Geology of the North European Margin*. Graham & Trotman, London, 389–396.

KNOX, R. W. O'B. 1984. Lithostratigraphy and depositional history of the late Toarcian sequence at Ravenscar, Yorkshire. *Proceedings of the Yorkshire Geological Society*, **45**, 99–108.

KÜSPERT, W. 1982. Environmental changes during oil-shale deposition as deduced from stable isotope ratios. *In*: EINSELE, G. & SEILACHER, A. (eds) *Cyclic and Event Stratification*. Springer-Verlag, Berlin, 482–501.

LANG, W. D. 1924. The Blue Lias of the Devon and Dorset Coasts. *Proceedings of the Geologists' Association, London*, **35**, 169–185.

—— & SPATH, L. F. 1926. The Black Marl of Black Ven and Stonebarrow, in the Lias of the Dorset coast. *Quarterly Journal of the Geological Society of London*, **82**, 144–187.

——, —— & RICHARDSON, W. A. 1923. Shales-with-'Beef', a sequence in the Lower Lias of the Dorset coast. *Quarterly Journal of the Geological Society of London*, **79**, 47–99.

——, ——, COX, L. R. & MUIR WOOD, H. M. 1928. The Belemnite Marls of Charmouth, a series in the Lias of the Dorset coast. *Quarterly Journal of the Geological Society of London*, **84**, 179–257.

LEMOINE, M. 1983. Rifting and early drifting: Mesozoic Central Atlantic and Ligurian Tethys. *In*: SHERIDAN, R. E. & GRADSTEIN, F. M. (eds) *Initial Reports of the Deep Sea Drilling Project*, **76**. US Government Printing Office, 885–895.

LØVBORG L. 1984. *The calibration of portable gamma-ray spectrometers – theory, problems, and facilities.* Risø National Laboratory Report, Roskilde, Denmark.

—— & MOSE E. 1987. Counting statistics in radioelement assaying with a portable spectrometer. *Geophysics*, **52**, 555–563.

——, WOLLENBERG, H., SØRENSEN, P. & HANSEN, J. 1971. Field determination of uranium and thorium by gamma-ray spectrometry, exemplified by measurements in the Ilímaussaq alkaline intrusion, South Greenland. *Economic Geology*, **66**, 368–384.

LEYTHAEUSER, D. 1973. Effects of weathering on organic matter in shales. *Geochimica et Cosmochimica Acta*, **37**, 113–120.

MICHEL, J. 1984. Redistribution of uranium and thorium series isotopes during involumetric weathering of granite. *Geochimica et Cosmochimica Acta*, **48**, 1249–1255.

MILSOM, J. & RAWSON, P. F. 1989. The Peak Trough - a major control on the geology of the North Yorkshire Coast. *Geological Magazine*, **126**, 699–705.

MOUTERDE, R. 1955. Le Lias de Peniche. *Comunicações dos Serviços Geológicos de Portugal*, **36**, 5–33.

MYERS, K. J. 1987. *Onshore-outcrop gamma-ray spectrometry as a tool in sedimentological studies*. PhD thesis, University of London.

—— 1989. The origin of the Lower Jurassic Cleveland Ironstone Formation of North-East England: evidence from portable gamma-ray spectrometry. *In*: YOUNG, T. P. & TAYLOR, W. E. G. (eds) *Phanerozoic Ironstones*. Geological Society, London, Special Publications, **46**, 221–228.

—— & WIGNALL, P. B. 1987. Understanding Jurassic organic-rich mudrocks - new concepts using gamma-ray spectrometry and palaeoecology: examples from the Kimmeridge Clay of Dorset and the Jet Rock of Yorkshire. *In*: LEGGET, J. K. & ZUFFA, G. G. (eds) *Marine Clastic Sedimentology – Concepts and Case Studies*. Graham & Trotman, London, 172–189.

PARKINSON, D. N. 1994. *The sequence stratigraphy of the Lower Jurassic of Western Europe*. DPhil thesis, University of Oxford.

—— & HINES, F. M. The Lower Jurassic of the North Viking Graben in the context of Western European Lower Jurassic Stratigraphy *In*: STEEL, R. J. *et al.* (ed.) *Sequence Stratigraphy on the northwest European Margin*. NPF Special Publications. Elsevier, Amsterdam, 97–107.

PHELPS, M. C. 1985. A refined ammonite biostratigraphy for the Middle and Upper Carixian (Ibex and Davoei Zones) in North-West Europe and stratigraphical details of the Carixian-Domerian boundary. *Geobios*, **18**, 321–362.

PLILER, R. & ADAMS, J. A. S. 1962. The distribution of thorium, uranium and potassium in the Mancos Shale. *Geochimica et Cosmochimica Acta*, **26**, 1115–1135.

PORRENGA, D. H. 1966. Clay minerals in recent sediments of the Niger delta. *In*: *Clays and Clay Mineralogy, 14th National Conference, New York*, Pergamon, Oxford, 221–223.

POWELL, J. H. 1984. Lithostratigraphical nomenclature of the Lias Group in the Yorkshire Basin. *Proceedings of the Yorkshire Geological Society*, **45**, 51–57.

QUENSTEDT, F. A. 1843. *Das Flözgebirge Würtembergs mit besonderer Rücksicht auf den Jura*. Laupp, Tübingen.

QUINIF, Y., CHARLET, J. M. & DUPUIS, C. 1982. Geochimie des radioelements: une nouvelle methode d'interpretation. *Annales de la Societe Geologique de Belgique*, **105**, 223–233.

RAISWELL, R. & BERNER, R. A. 1985. Pyrite formation in euxinic and semi-euxinic sediments. *American Journal of Science*, **285**, 710–724.

RAWSON, P. F. & WRIGHT, J. K. 1992. *The Yorkshire Coast*. Geologists' Association Guides **34**.

RIDER, M. H. 1991. *The geological interpretation of well logs*. Whittles Publishing, Caithness, Scotland.

SCHMOKER, J. W. & HESTER T. C. 1983. Organic carbon in Bakken Formation, United States portion of Williston Basin. *American Association of Petroleum Geologists Bulletin*, **67**, 2165–2174.

SELLWOOD, B. W. 1970. The relation of trace fossils to small scale sedimentary cycles in the British Lias *In*: CRIMES, T. P. & HARPER, J. C. (eds) *Trace Fossils*. Seal House Press, Liverpool, 489–504.

—— 1972. Regional environmental change across a Lower Jurassic stage boundary in Britain. *Palaeontology*, **15**, 125–157.

SMITH, D. G. 1989. Stratigraphic correlation of presumed Milankovich cycles in the Blue Lias (Hettangian to earliest Sinemurian), England. *Terra Nova*, **1**, 457–460.

STEEL, R. J. 1993. Triassic–Jurassic megasequence stratigraphy in the northern North Sea: rift to post-rift evolution *In*: PARKER, J. R., (ed.) *Petroleum Geology of Northwest Europe: Proceedings of the 4th Conference*. Geological Society, London, 299–315.

SUMMERFIELD, M. A. 1991. *Global Geomorphology*. Longman, London.

SWANSON, V. E. 1960. *Geology and Geochemistry of Uranium in Marine Black Shales, A Review*. United States Geological Survey, Professional Paper 356-C.

UNDERHILL, J. R. & PARTINGTON, M. A. 1993. Jurassic thermal doming and deflation in the North Sea: implications of the sequence stratigraphic evidence. *In*: PARKER, J. R., (ed.) *Petroleum Geology of Northwest Europe: Proceedings of the 4th Conference*. Geological Society, London, 337–345.

URLICHS, M. 1977. The Lower Jurassic in Southwestern Germany. *Stuttgarter Beiträge zur Naturkunde*, **B24**, 1–41.

VAIL, P. R., AUDEMARD, F. BOWMAN, S. A., EISNER, P. N. & PEREZ-CRUZ, C. 1991. The stratigraphic signatures of tectonics, eustasy and sedimentology – an overview *In*: EINSELE, G., RICKEN, W. & SEILACHER, A. (eds) *Cycles and Events in Stratigraphy*. Springer-Verlag, Berlin, 617–659.

VAN BUCHEM, F. S. P. 1990. *Sedimentology and diagenesis of Lower Lias mudstones in the Cleveland Basin, Yorkshire, UK*. PhD thesis, University of Cambridge.

—— & MCCAVE, I. N. 1989. Cyclic sedimentation patterns in Lower Lias mudstones of Yorkshire (GB). *Terra Nova*, **1**, 461–467.

——, MELNYK, D. H. & MCCAVE, I. N. 1992. Chemical cyclicity and correlation of Lower Lias mudstones using gamma ray logs, Yorkshire, UK. *Journal of the Geological Society, London*, **149**, 991–1002.

VAN WAGONER, J. C., POSAMENTIER, H. W., MITCHUM, R. M., VAIL, P. R., SARG, J. F., LOUTIT, T. S. & HARDENBOL, J. 1988. An overview of the fundamentals of sequence stratigraphy and key definitions *In*: WILGUS, ET AL. (eds) *Sea-level Changes: an integrated approach*. Special Publications of the Society of Economic Paleontologists and Mineralogists, **42**, 39–45.

WATKINSON, M. P. 1989. *Triassic to Middle Jurassic sequences from the Lusitanian Basin Portugal, and their equivalents in other North Atlantic margin basins*. PhD thesis, Open University, England.

WEEDON, G. P. 1986. Hemipelagic shelf sedimentation and climatic cycles: the basal Jurassic (Blue Lias) of South Britain. *Earth and Planetary Science Letters*, **76**, 21–335.

—— & JENKYNS, H. C. 1990. Regular and irregular climatic cycles in the Belemnite Marls (Pliensbachian, Lower Jurassic, Wessex Basin). *Journal of the Geological Society, London*, **147**, 915–918.

WIGNALL, P. B. & MYERS, K. J. 1988. Interpreting benthic oxygen levels in mudrocks: A new approach. *Geology*, **16**, 452–455.

WILSON, R. C. L., HISCOTT, R. N., WILLIS, M. G. & GRADSTEIN, F. M. 1989. The Lusitanian Basin of West-Central Portugal: Mesozoic and Tertiary Tectonic, Stratigraphic, and Subsidence History. *In*: TANKARD, A. J. & BALKWILL, H. R. (eds) *Extensional Tectonics and Stratigraphy of the North Atlantic Margins*. Memoirs of the American Association of Petroleum Geologists, **46**, 341–361.

WRIGHT, V. P. & WILSON, R. C. L. 1984. A carbonate submarine-fan sequence from the Jurassic of Portugal. *Journal of Sedimentary Petrology*, **54**, 394–412.

ZIEGLER, P. A. 1988. Evolution of the Arctic-North Atlantic and the Western Tethys. Memoir of the American Association of Petroleum Geologists, **43**.

A sequence stratigraphical approach to the understanding of basin history in orogenic Neoproterozoic successions: an example from the central Highlands of Scotland

B. W. GLOVER[1] & T. McKIE[2]

[1]*British Geological Survey, Kingsley Dunham Centre, Keyworth, Nottingham NG12 8GG, UK*

[2]*Badley, Ashton & Associates, Winceby House, Winceby, Horncastle, Lincolnshire LN9 6PB, UK*

Abstract: Sequence stratigraphical principles, developed largely in Phanerozoic basins, are applied here to a Neoproterozoic metasedimentary succession. The Grampian and Appin groups of the southern Monadhliath represent clastic and carbonate deposition within a multiphase rift basin. Subsequent polyphase deformation has resulted in considerable crustal shortening so that, within a comparatively small area, there is significant lateral variation in basinal settings. The succession is rationalized into a number of transgressive–regressive cycles. The northwestern area is dominated by shelfal metasedimentary rocks which exhibit a hierarchy of sequence boundaries distinguished by abrupt basinward shifts (southwards and southeastwards) in the proximality of marine facies, and onlap. The surfaces between the sequences provide the objective criteria for correlation into the basinal succession of the southeast. This sequence stratigraphical approach gives new insight into Neoproterozoic basin dynamics. Although rifting is confirmed as having a dominant role in creating accommodation for the deposition of at least two 'supersequences' thousands of metres thick, the basin depocentre is reinterpreted to have lain to the south and east throughout deposition of the Grampian and Appin groups. The upwards trend towards 'cleaner' sand from Grampian to Appin group deposition is suggested to reflect increasing sediment reworking by tides during a second-order transgression. The scale of major sequences means that patterns of deposition are likely to have been basin-wide or greater, so that it may be possible to identify similar transgressive–regressive stratal architectures in structurally isolated and disparate successions. This approach may help resolve the status of some suspect terranes.

The analysis of sedimentary basin history in orogenic (i.e. metamorphosed and deformed) successions is considerably hampered by a number of factors, including difficulties in correlation across complex structures, loss of sedimentary fabric and the lack of a high resolution biostratigraphical framework. This paper describes how the use of sequence stratigraphical concepts, in conjunction with detailed facies (rather than 'traditional' lithological) mapping, can be used to improve our ability to decipher basin history in complex terranes.

Sequence stratigraphy is a method of analysing the sedimentary history of marine or paralic basin-fills in terms of the depositional response to changes in accommodation space (the space available for sediment to fill in response to allocyclic forcing mechanisms, e.g. tectonic subsidence/uplift and eustasy). A consistent pattern in the depositional response to accommodation changes has been widely documented (cf. Van Wagoner *et al.* 1990). This, to a degree, is scale-independent, having been convincingly applied to a metre-scale delta prograding into a drainage channel (Posamentier *et al.* 1992) and to an entire Neogene succession (Vail *et al.* 1991). In broad terms, the sequence stratigraphical approach recognizes the landward and basinward migration of depositional systems in response to 'cyclical' accommodation changes. The approach emphasizes the existence of a variety of features attributable to relative sea-level falls: subaerial shelf exposure, incised valley-cutting, basinward shifts in coastal depositional systems and increased activity of mass flow processes (lowstand systems tract). During a subsequent rise in relative sea level, the shoreline depositional systems retreat landward, typically becoming increasingly depleted in sand, until during the maximum rate of relative sea-level rise a condensed section of shales is deposited (transgressive systems tract). As the rate of relative sea-level rise slows, sediment supply is increasingly capable of filling the accommodation space and coastal progradation ensues (highstand systems tract) until the relative sea level falls to create the next sequence boundary. Typically, a hierarchy of sequence stacking patterns occurs, such that sets of sequences can define lowstand, transgressive and highstand sequence sets in response to lower frequency relative sea-level changes (Mitchum & Van Wagoner 1991).

Within folded terranes, it should be possible in

From Hesselbo, S. P. & Parkinson, D. N. (eds), 1996, *Sequence Stratigraphy in British Geology*, Geological Society Special Publication No. 103, pp. 257–269.

N
0
5 km
fault
tectonic contact
TB
GD
CnL
Ap
A
GF
AG
BI
G
B
LS
DF
Corrieyairack Igneous Complex
CM
L
M
SBC
SV
BS
BL
C
LT
I
E
T
Iq
Gq
Location
axial planar trace of synform
axial planar trace of antiform
igneous intrusion
strata beneath tectonic contact

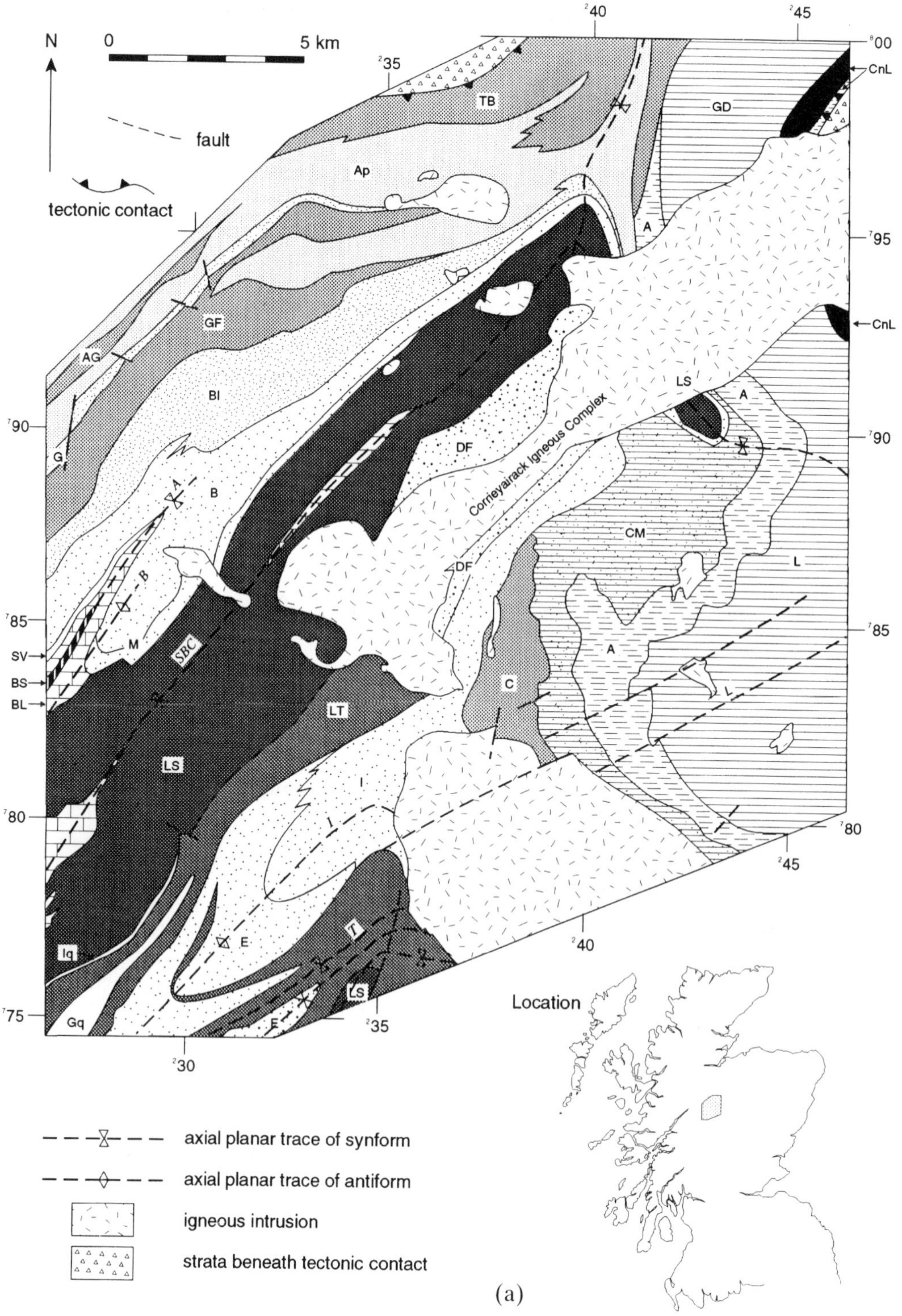

(a)

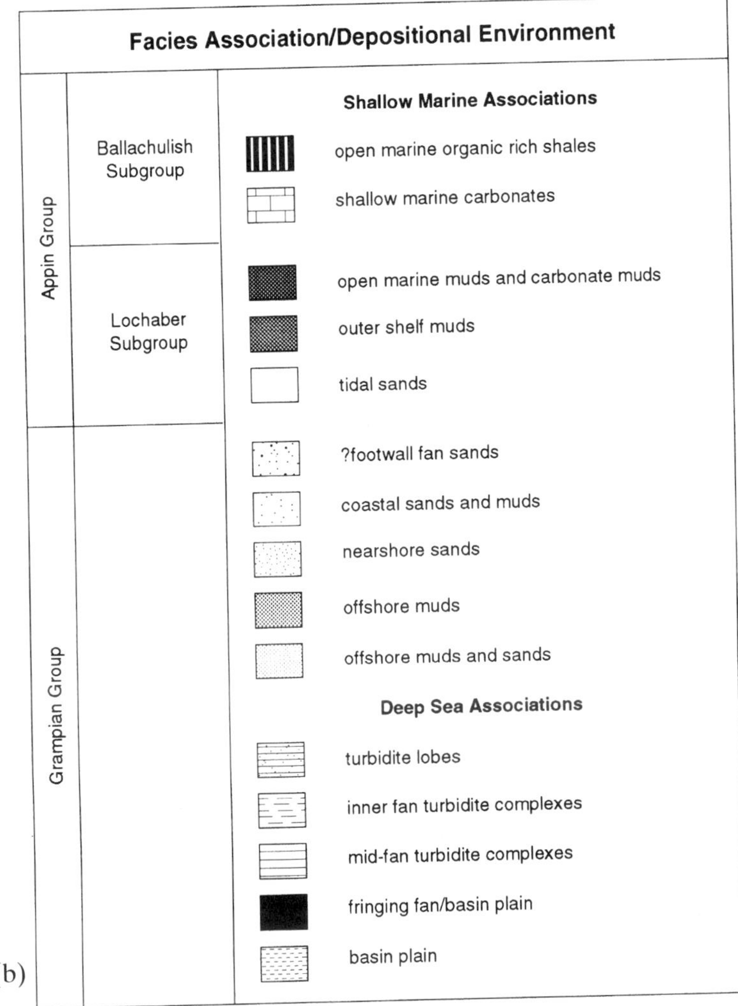

Fig. 1. (a) Geology of the southern Monadhliath based on British Geological Survey mapping of Sheet 63 West (Glen Roy). See Fig. 2 for key to facies associations within each formation. See Fig. 3b for the chronostratigraphical architecture. Key to abbreviations: Appin Group: BL, Ballachulish Limestone; BS, Ballachulish Slate; LS, Leven Schist; LT, Loch Treig Schist and Quartzite Formation; Iq, Innse Quartzite; Gq, Glen Coe Quartzite; and SV, Spean Viaduct Quartzite. Grampian Group ('shelfal succession' – shallow marine associations): DF, Dog Falls Formation; E, Eilde Flags Formation; M, March Burn Quartzite; B, Brunachan Psammite Formation; I, Inverlair Psammite Formation; C, Clachaig Formation; BI, Beinn Iaruinn Quartzite Formation; GF, Glen Fintaig Formation; G, Glen Gloy Quartzite Formation; Ap, Auchivarie Psammite Formation; AG, Allt Goibhre Semi Pelite Formation; and TB, Tarff Banded Formation. Grampian Group ('turbidite complex'– deep sea associations): CM, Creag Meagaidh Formation; A, Ardair Formation; L, Loch Laggan Psammite Formation (equivalent to) GD, Glen Doe Psammite Formation; and CnL, Coire nan Laogh Formation. Faults: LDF, Laggan Dam Fault. Folds: A, Appin Synform; B, Bohuntine Antiform; SBC, Stob Ban, Craig' a Chail Synform; L, Laggan Antiform; I, Inverlair Antiform; and T, Treig Syncline. (b) Key.

some instances to identify such hierarchical transgressive–regressive (T–R) patterns, particularly on fold limbs where the steeply dipping orientation of the bedding allows a geological map to function as a cross-section of the original stratigraphy (albeit deformed). In addition, basinward and landward shifts in facies and, in some instances, onlap/offlap patterns can be used to identify candidate sequence boundaries. To illustrate the application of these concepts an example is described from the Neoproterozoic Grampian and Appin groups of the southern Monadhliath Mountains in

Table 1. *Lithofacies descriptions for the Grampian and Appin groups of the study area*

Facies	Bouma	Lithology	Sedimentary structures	Bed thickness (m)	Depositional mechanism	Notes
B1.1	T_a, T_b	Muddy sandstone	Poorly formed normal grading	0.1–1.0	Rapid deposition of high-density turbidity current	Bed amalgamation common
B1.2	T_a, T_b	Pebbly, muddy sandstone	Massive, uncommon planar beds	0.1–0.5	Traction deposition, upper flow regime	Granules of altered alkali feldspar
C1.1	–	Muddy sandstone	Normal grading	0.1–1.0	High-density turbidity currents?	
C2.1	T_a, T_{ab}	Graded muddy sandstone	Normal grading, planar bedding	0.1–0.5	Medium-density turbidity currents	Some bed amalgamation
C2.2	$T_{a\text{–}e}$	Graded muddy sandstone	Normal grading, planar and ripple lamination	0.03–0.10	Dilute turbidity currents	
C2.3	$T_{a\text{–}e}$	Graded muddy sandstone	Normal grading, ripple lamination	0.01–0.10	Freezing of high-concentration traction carpet	
C2.5	T_{ae}, T_{be}	Muddy sandstone/ mudstone	Massive, uncommon planar bedding in sandstone	0.01–0.03	Dilute turbidity currents	Similar to C2.3 -distal equivalent
D2.1	T_b, T_{de}	Sandstone/ mudstone	Normal grading	0.01–0.03	Dilute turbidity currents	
D2.2	T_{cde}	Sandstone/ mudstone	Normal grading, planar lamination	0.01–0.10	Traction, suspension and dilute turbidity currents	Isolated ripples
D2.3	T_d, T_e	Sandstone/ mudstone	Planar lamination	0.01–1.00	Very dilute turbidity currents	May also be tectonically induced by shearing
E	–	Mudstone	Massive	0.01–1.00	Suspension	Recrystallization obliterated sedimentary structures
F2.1	–	Sandstone/ mudstone	Soft sedimentary folding, dewatering structures	0.1–0.5	Gravity sliding	
G	–	Sandstone and pebbly sandstone	Trough cross-bedding, ripple lamination, combined flow ripple lamination, planar bedding slumping, dewatering	5.0–30.0	In-channel traction deposition	Upwards-fining sequences, slumping at base
H	–	Sandstone	Cross-bedding, massive	5.0–30.0	Traction deposition	Parallel-sided, sheet-like
I	–	Sandstone	Planar and cross-bedding	0.1–5.0	In-channel traction deposition	
J	–	Sandstone	Planar and cross-bedding	0.1–5.0	Traction deposition	Parallel-sided, sheet like
K	–	Heterolithic sandstone and mudstone	Flaser bedding, planar lamination, ripple lamination, cross-bedding	0.1–5.0	Mixed traction and suspension possibly tidally influenced minor wave rippling	
L	–	Quartzite	Tabular and trough cross-bedding, ripple cross-lamination	0.1–10.0	Traction deposition possibly tidally influenced	
M	–	Calcareous mudstone	Planar lamination	0.10–10.0	Suspension	
N	–	Limestone	Planar and lenticular lamination	0.0–10.0	Biogenic and suspension	
O	–	Black mudstone	Massive	10s m	Organic accumulation and minor clastic input	Thin laminae of sand very uncommon

Table 2. *Summary of deep water lithofacies associations for the Grampian Group of the study area*

	Depositional setting					
Facies	Mid-fan	Inner fan	Mid-fan lobes	Mid-outer fan	Mid-outer fan	Basin plain
B1.1	√√Channel					
B1.2	√Channel		√Lobe			
C1.1		√Channel/ lobe				
C2.1		√√Channel/ lobe	√Lobe			
C2.2	√Inter- channel	√√Channel/ lobe	√√Lobe			
C2.3	√Inter- channel	√Inter- channel/lobe fringe	√Lobe fringe	√√Lobe fringe	√Lobe fringe	
C2.5	√Inter- channel	√Inter- channel/lobe fringe	√Lobe fringe	√√Lobe fringe	√Lobe fringe	
D2.1			√Lobe fringe		√√Lobe fringe	
D2.2					√√Lobe fringe	
D2.3					√√Lobe fringe	
E	√Inter- channel	√Inter- channel	√Lobe fringe	√Lobe fringe	√Lobe fringe	√√Basin plain
F2.1		√Channel and inter channel				
I						√?Contourites

Key to symbols: √, facies present: √√, facies common. Facies numbers refer to the deep water facies subdivision of Pickering *et al.* (1986).

the Central Highlands of Scotland. The pioneering work of Anderton (1971, 1979, 1982, 1985), Hickman (1975), Harris *et al.* (1978) and Litherland (1980) showed that it was possible to apply sedimentological principles and interpretations to this orogenic succession. It is demonstrated here how a sequence stratigraphical analysis provides new insights into the history of this metasedimentary basin.

Geological setting

The Grampian Group is the lowest subdivision of the Dalradian Supergroup (Harris *et al.* 1978). The base of the group is marked in places by a zone of high strain (Winchester & Glover 1988; Glover & Winchester 1989). The timing and origins of initial Grampian Group sedimentation is, as yet, unclear. Until recent years this group was interpreted as having formed during the early stages of Dalradian basin formation and, for the most part, was described as being largely of shallow marine and coastal origin (e.g. Hickman 1975; Thomas 1980). The work of the British Geological Survey (Glover *et al.* 1995; Key *et al.* in press) and Glover (1989; 1993) in the southern Monadhliath Mountains has demonstrated that the Grampian Group was deposited in a variety of shallow and deep marine and terrestrial depositional settings, probably in an ensialic basin. Syn-sedimentary faulting occurred at various times during deposition.

The overlying Appin Group is also assigned to the Dalradian Supergroup (Harris & Pitcher 1975; Harris *et al.* 1978). This was deposited largely within a marine environment (see below). Two subgroups are present in the study area; the Lochaber (oldest) and Ballachulish subgroups.

The Grampian and Appin groups have suffered polyphase deformation and metamorphism during Neoproterozoic and Ordovician times. This has resulted in significant horizontal shortening, so that within a comparatively small study area there is a considerable lateral variation in basinal settings.

Metamorphic grade across the study area is generally of mid-amphibolite facies (Phillips *et al.* 1994). The structure is dominated by a series of upright secondary NNE to NE trending folds (Fig. 1). In the south of the study area further complexity is provided by an earlier set of recumbent folds. The area is penetrated by several Caledonian igneous intrusions.

Sedimentary structures are found in abundance in

places, especially in fold hinges. These have allowed a detailed lithostratigraphical and sedimentological analysis of the area (Glover *et al.* 1995; Key *et al.* in press).

Lithofacies and lithofacies associations

Figure 1 shows the distribution of lithofacies associations in the Grampian and Appin groups of the study area. A wide variety of facies is exposed, ranging from shelfal deposits in the west to basinal deposits in the east. This type of map differs from the traditional style in that attention has been paid to the type of sedimentary structures that were present in each of the mapped units rather than just its lithology. Previous surveys of the Grampian and Appin groups would, therefore, not have differentiated clearly between turbiditic and cross-bedded sandstones, as the emphasis would have been structural rather than sedimentological; both 'facies' would probably have been mapped as 'psammites'. The approach outlined here, however, ultimately provides a greater degree of detail not only for basin analysis, but also for structural analysis (e.g. Phillips *et al.* 1994).

A summary of facies and facies associations for the Grampian Group is given in Tables 1–3 (after Glover *et al.* 1995). Four associations are recognized; two deep water associations and two shallow water associations (Fig. 2). The oldest strata recognized within the group comprise an originally muddy succession (Coire nan Laogh Formation) interpreted as having formed in a basin plain setting (Glover *et al.* 1995). In the east of the study area the overlying succession was dominated by turbidite sedimentation (Laggan, Ardair and Creag Meagaidh formations). Individual formations are recognized by differences in sand/mud content and differentiation (see below). In contrast, in the northwest, there is a succession of originally 'cleaner' sands and muds (psammites and quartzites, and semipelites). The former lithologies are frequently cross-bedded with less common combined-flow ripple cross-lamination. These indicate a shallow water origin (Table 1).

The overlying Appin Group comprises a muddy basal formation, the Loch Treig Schist and Quartzite Formation (LT; this and subsequent abbreviations in parentheses refer to Figs 1 and 3), within which four quartzite members are recognized (Figs 1 and 4). This formation forms a clastic tidal shelf facies association. The overlying Leven Schist Formation was also originally dominantly muddy. Carbonate-bearing rocks are present towards the top of the formation and these pass upwards into the Ballachulish Limestone. The Ballachulish Slate represents the youngest Appin Group rocks of the study area. A summary of the lithofacies and lithofacies associations for Appin Group formations is provided in Tables 1 and 2. A schematic representation of the Loch Treig Schist and Quartzite tidal shelf association is given in Fig. 2.

Sequence stratigraphy

This structurally complex area can be rationalized into a number of T–R cycles within the western shelfal succession, based on mappable facies associations. Within these cycles a hierarchy of sequences can be recognized by abrupt basinward shifts in the proximality of marine facies and in onlap. The recognition of these cycles and their bounding surfaces provide criteria for the correlation of the western shelfal succession with the eastern basinal succession across the Stob Ban – Craig a Chail synform (Fig. 1; SBC). The sequence stratigraphical interpretation of the study area is illustrated in Fig. 3.

Shelfal succession

The stratigraphy of the western limb of the SBC synform youngs to the southeast, and is best viewed by inverting Fig. 1. The base of the Tarff Banded Formation (TB) is a tectonic contact, which will not be dealt with in this paper (see Phillips *et al.* 1993), but the top of this shallow water, muddy facies succession is marked by a transition into nearshore sand and mud facies of the Auchivarie Psammite Formation (Ap). This transition represents a major basinward shift in facies (relative shallowing) and the contact therefore represents a candidate sequence boundary.

The stratigraphically overlying succession broadly comprises an interfingering of the Auchivarie Psammite (nearshore sand and mud facies) with the offshore mud facies of the Glen Fintaig Formation (GF). Within this interval, the offshore mud facies pinch-out to the northeast, whereas the more sand-prone Auchivarie Psammite, in the uppermost 'tongue', can be seen to pinch out to the southwest. The overlying Beinn Iaruinn Quartzite Formation (BI) defines a more pronounced, apparently southwestward extension of nearshore, sand-prone facies.

The proximality trends of the Allt Goibhre – Auchivarie Psammite – Glen Fintaig formations indicate an apparent increase in the proportion of offshore facies towards the southwest. The mapped progressive northeastward penetration of the muddy 'tongues' of the Glen Fintaig Formation, coupled with a converse pattern in the Auchivarie Psammite, suggests that this interval shows an overall retrogradational facies pattern (*sensu* Galloway 1989) and can be interpreted as a *long-term* transgressive

Table 3. *Summary of shallow water facies associations for the Grampian and Appin groups of the study area*

Facies	Depositional setting								
	Muddy off-shore shelf	Off-shore sandy shelf	Inner shelf, shallow water	Near-shore, coastal deposition possibly tidally influenced	Tidally dominated inner shelf	Off-shore sub-tidal	Off-shore	Open marine; low clastic input	?Marine, beneath wave base, anoxic
C2.3		√Storm beds							
C2.5		√Storm beds							
E	√√	√	√	√√	√	√√	√	√	
F2.1				√					
G				√Fluvial channels					
H	√Minor shallowing produces sandy shelf		√√	√√Marine sandstone bodies					
I	√Sandwaves	√Sandwaves	√√ Sandwaves	√√Sandwaves		√			
J				√√Minor fluvial channels					
K	√√	√√	√	√√					
L				√	√√	√	√		
M						√√	√	√	
N								√√	√
O									√√

Key to symbols: √, facies present: √√, facies common.

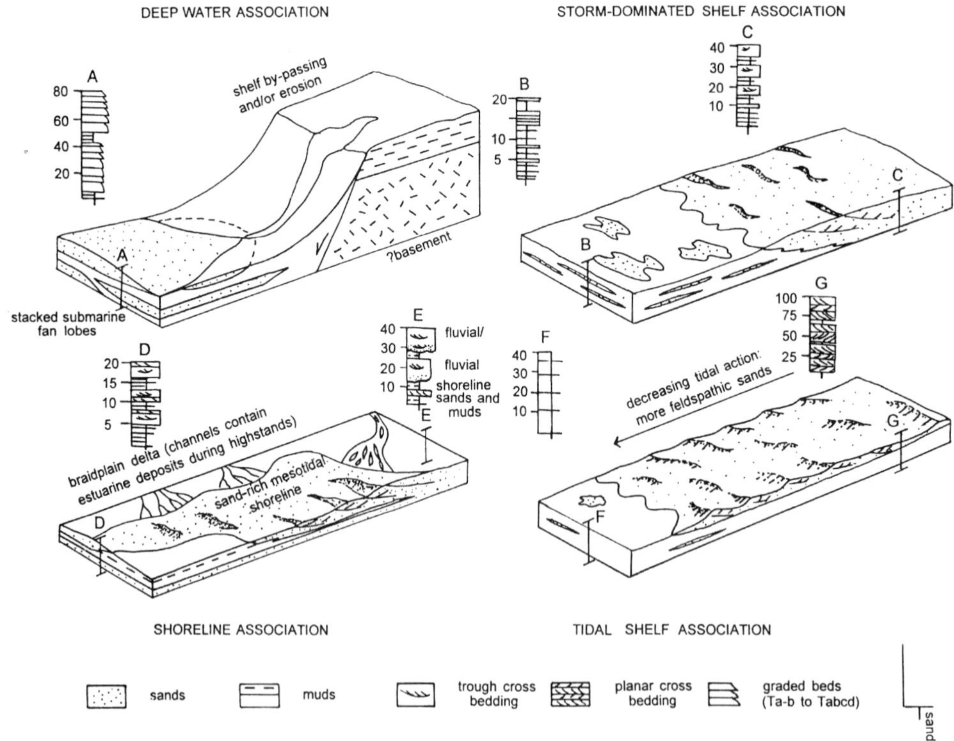

Fig. 2. Schematic summary of the main facies associations of the Grampian Group and the Loch Treig Schists and Quartzite Formation (LT) within the study area.

systems tract. The vertical scale of this succession (several kilometres) suggests that the long-term systems tract is an order of magnitude greater in thickness than transgressive systems tracts typical of 'third-order' depositional sequences (up to hundreds of metres, Van Wagoner *et al.* 1990) and that it may represent a transgressive sequence set. Evidence to support a sequence set interpretation is provided by the Glen Gloy Quartzite (G), which caps one Auchivarie Psammite tongue (Fig. 1). This quartzite represents an abrupt vertical shift to shallower water, nearshore facies and may also exhibit an eastwards or southeastwards basinward shift in onlap at its northeastern pinch-out.

That the coastline was not present to the northeast of the study areas is indicated by the presence of a muddy lateral equivalent to the Auchivarie Psammite in the north of the study area (cf. [4097], AP; Fig. 1) which may be partly equivalent to the Allt Goibhre Formation. Instead, the coastline must have lain to the north or northwest. This is also confirmed by the presence of deeper water facies in the southeast of the study area (see later).

The overlying, southwestwards extension of the Beinn Iaruinn Quartzite (BI) is interpreted as representing a change from overall transgressive to regressive facies which formed during a long-term highstand (or potentially highstand sequence set given the thickness of the succession).

Above the Beinn Iaruinn Quartzite, the Brunachan Psammite Formation (B) represents an abrupt shift to dominantly coastal facies. Its mappable correlatives on the eastern fold limb, the shallow water Inverlair Formation (I) and the Clachaig Formation (C), rest sharply on a succession comprising deeper water facies. This suggests that these formations collectively represent an abrupt basinward shift in facies and consequently long-term lowstand deposition. Their relationship with the underlying turbidite formations is discussed in the following section. Equal significance is attached to the Beinn Iaruinn Quartzite/Brunachan Psammite boundary and the Tarff Banded Formation/Auchivarie Psammite sequence boundary because these surfaces enclose a single, large-scale T–R cycle, within which higher frequency sequences are potentially enclosed.

The more basinally situated Inverlair Formation

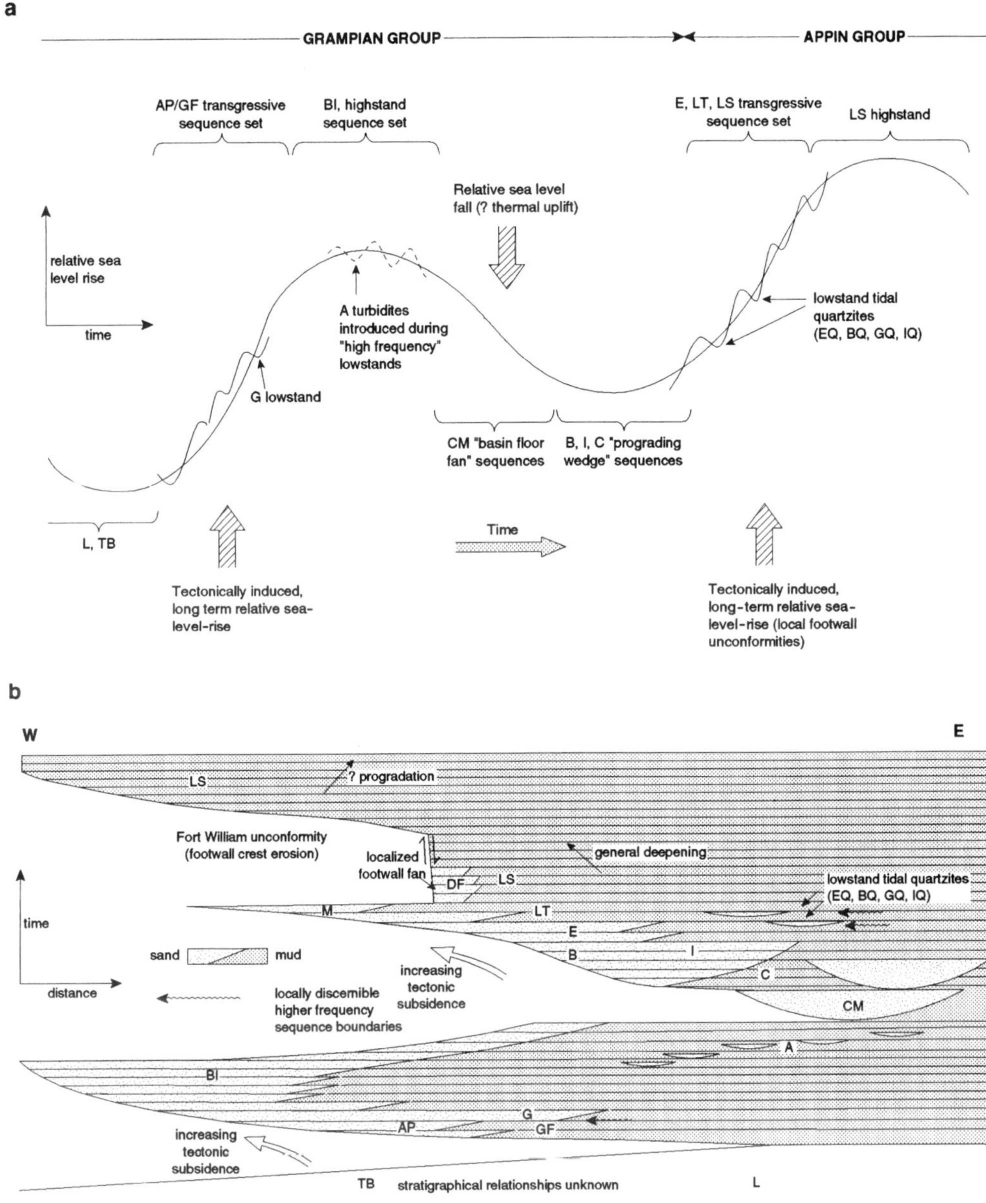

Fig. 3. Sequence stratigraphy of the Grampian Group and Loch Treig Schist and Quartzite Formation (LT) of the study area. (a). Schematic representation of relative sea level curve with the main depositional events indicated. (b). Simplified schematic second-order chronostratigraphical chart for the Grampian and Appin groups of the southwestern Monadhliath Mountains. Further local variations in stratigraphical architecture are shown in Fig. 1; these are due to erosion on footwall highs.

is overlain partly by the Eilde Flags (E), which shows an upward and southward increase in differentiation into quartzite and schistose partings (Hickman 1975). An increase in muddy partings towards the top of the Eilde Flags suggests an upwards increase in water depths (i.e. transgression), which is in keeping with its physical location overlying lowstand coastal deposits of the Inverlair Formation. The Eilde Flags pass upwards into the offshore mud facies of the Loch Treig Schist and

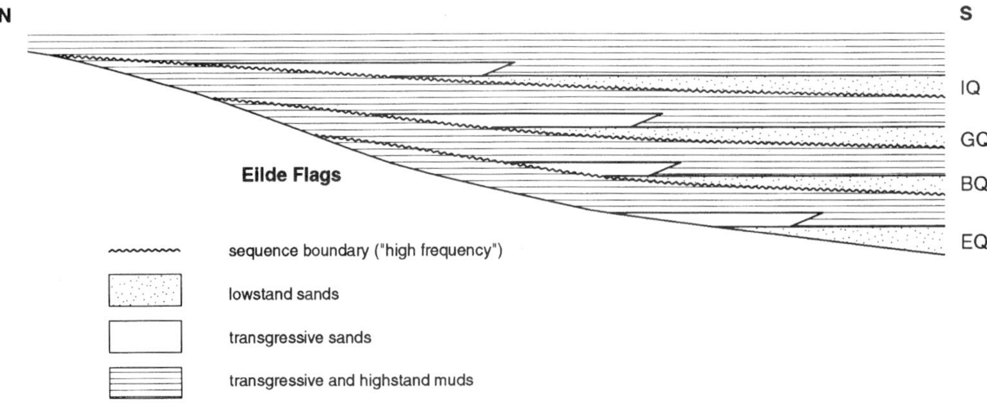

Fig. 4. Reinterpretation of the Loch Treig Schist and Quartzite Formation (LT) and the origin of the quartzite members.

Quartzite. In the southeast of the study area a tongue of the Loch Treig Schist, which wedges out to the north, is present within the coastal association of the Eilde Flags. This relationship suggests a complex interfingering of transgressive muds and regressive sands, analogous to the earlier, retrogradational Auchivarie Psammite – Glen Fintaig formations. [This has previously been interpreted as a structural phenomenon (e.g. Hickman 1975), which would be problematic because if it is assumed to be a fold closure, it would have opposite facing-sense to the fold set with which it is correlated (i.e. Treig Syncline and Kinlochleven Anticline).]

The Loch Treig Schist and Quartzite represents the deposition of offshore transgressive mud facies in the east of the area, as part of the overall deepening upward trend into the Leven Schist (LS), which ultimately blankets the shelfal succession in the west. This interval is punctuated by a number of quartzites of nearshore, tidal origin (Hickman 1975). These comprise, in ascending stratigraphical order: the Eilde, Binnein, Glen Coe and Innse quartzites, each of which progressively pinches-out farther northwards (cf. Figs 1 and 4). The quartzites are typically pebbly and become less feldspathic and finer grained towards their northern pinch-out. Between the Eilde and Binnein quartzites, the Loch Treig Schist shows an overall upward increase in the proportion of sand partings. Between the Binnein and Glen Coe quartzites, the Loch Treig Schist becomes increasingly pelitic upwards (Hickman 1975).

The abruptness of these incursions of shallow-water sands onto the offshore muddy facies of the Loch Treig Schist suggests that they were deposited following relative sea-level falls. As such, the base of each quartzite is likely to represent a 'high frequency' sequence boundary within the overall Appin Subgroup transgressive trend in this study area. The quartzites must largely represent lowstand deposition and may be analogous to the 'forced regressions' documented from Phanerozoic successions. The progressive northward pinch-out of the quartzites was taken by Hickman (1975) to indicate a palaeoshoreline to the south. However, we find this interpretation inconsistent with the stratigraphical and palaeogeographical context in two respects: all other proximality trends indicate a deeper palaeobathymetry to the south and east; and the successive northward pinch-out of each quartzite, if a southern palaeoshoreline is assumed, would imply long-term northward regression, when all other vertical trends point to long term transgression. Figure 4 illustrates an alternative, sequence stratigraphical interpretation for the quartzites as shoreline *detached* lowstand deposits, with the successive northward pinch-out of the quartzites now interpreted as long term, landward stepping, consistent with the overall transgressive trend. The northward decrease in grain size and increase in maturity of these quartzites is interpreted as a northward transition from pebbly lowstand deposits into onlapping, finer-grained transgressive deposits which experienced enhanced reworking.

In the western part of the study area interpretation of the sequence stratigraphy equivalent to the Eilde Flags to Leven Schist transition is complicated by the presence of a planar stratigraphical discontinuity, the Fort William Unconformity (Glover 1993). This is interpreted by Glover *et al.* (1995) as a footwall unconformity developed in response to active faulting during deposition of the Loch Treig Schist and Quartzite. Beneath the unconformity, however, the initial upwards trend towards cleaner sands (early part of the Loch Treig Schist and

Quartzite transgression) is represented by the March Burn (M) and Spean Viaduct quartzites (SV).

The overlying Leven Schist shows a muddying/cleaning-upward pattern (Hickman 1975), suggesting that the upper part of the T–R cycle (which began with the Eilde Flags) is represented within the Leven Schist as deep water equivalents to regressive nearshore facies nearer the basin margin. The overlying Ballachulish Limestone and Slate formations represent further changes in water depth and sediment supply. Interpretation of this part of the stratigraphy forms the basis of ongoing research.

Relationship with the eastern basinal succession

Within the study area the eastern, basinal succession comprises five formations (in ascending stratigraphical order): (1) the Coire nan Laogh Formation (CnL), which represents a muddy, basin floor facies; (2) the Laggan (and equivalent Glen Doe) Formation (L and GD), which are composed of turbiditic psammites with an overall muddying-upward trend; (3) the Ardair Formation (A), composed of thickly bedded, typically amalgamated turbidites; and (4) the Creag Meagaidh Formation (CM), which is a sand-rich turbidite succession, sharply overlain by the offshore muddy facies of the Clachaig Formation (C), which is in turn overlain by the Inverlair Formation.

From this overall succession a number of trends in the sand flux into the basin can be identified. The Laggan Formation/Glen Doe Formation section represents an initially rapid influx of turbidite deposits, which decreased in time, towards the top of the formation. The Ardair Formation again shows an increase in sandy turbidite lobes (albeit enveloped in relatively muddy strata; Table 2). The Creag Meagaidh Formation represents a further relatively abrupt increase in sand input. The overall pattern comprises a relatively rapid increase in turbidite input at the base of the Laggan section, which decayed upwards and was then reversed through the Ardair and Creag Meagaidh formations in a series of abrupt jumps in sand content. Assuming that the increased turbidite sand influx is broadly linked with lowstand deposition and basinward migration of the shoreline, then the overall pattern of basinal turbidite sand input can be linked to the T–R cycles observed on the western shelf succession. The sandy base of the Laggan Formation may record sediment by-pass across the western shelf. Unfortunately, this part of the stratigraphy is not exposed in the western shelfal area. The subsequent progressive muddying up succession of the Laggan turbidite complex may have been the result of increased sand storage on the shelf during a subsequent long-term transgression before a subsequent second-order fall in relative sea level. This second-order fall is interpreted as being represented by deposition of the Ardair Formation in the basinal succession. The increased abundance of turbidite sands through the Ardair and Creag Meagaidh formations may be linked to a shift to long-term lowstand deposition, which also led to the progradation over the Creag Meagaidh turbidite pile of the lowstand shelfal muds and sands of the Clachaig and Inverlair Formations (Fig. 3). This relationship is analogous (albeit on a much larger scale) to the lowstand fan and lowstand wedge geometries observed in third-order depositional sequences (Van Wagoner *et al.* 1988).

Within the area studied, two major sequence boundaries have been identified: the Tarff–Auchivarie boundary and the Beinn Iaruinn–Brunachan boundary. These enclose a major T–R cycle thousands of metres thick, within which higher frequency third-order sequences are locally resolvable. The scale of this major cyclicity is comparable with that of second-order supersequences (Vail *et al.* 1991), which are recognized to be largely driven by basin tectonics. Within this context, the onset of long-term transgressive events was the result of increased rates of accommodation creation (i.e. enhanced, tectonically driven subsidence). The Auchivarie Formation and Brunachan/Inverlair formations record two distinct episodes of increased basin subsidence in response to extensional events. The local presence of an unconformity (Fort William Unconformity) within the transgressive Appin Group stratigraphy may record erosion during the maximum rate of footwall uplift following the onset of accelerated subsidence in this basin. Such uplift may have been responsible for the deposition of localized footwall fans [e.g. the Dog Falls Psammites (DF); Figs 1a and 3b].

Conclusions

The sequence stratigraphical analysis of this succession has resulted in an improved understanding of the stratigraphical relationships between the shelfal and basinal facies and has highlighted a possible reinterpretation of the tidal quartzites within the Appin Group as lowstand and transgressively reworked sands. Russell & Allison (1985) have previously noted the superficial similarity between the mature, tidal quartz arenites of the Appin Group and the Lower Cambrian Eriboll Sandstone in northwest Scotland, and attributed the maturity of these deposits to intense alkaline weathering of the hinterland source. The new interpretation of these Appin Group quartzites is similar to that of the Cambrian Eriboll Sandstone

(McKie 1993), and both are suggested to be mineralogically mature through polycyclical lowstand and transgressive reworking in a tidal environment. Tidal activity may have been enhanced by basin widening during the second-order transgression.

This approach to mapping and interpreting metasedimentary rocks in structurally complex areas provides a valuable aid to stratigraphical correlations as it endeavours to identify broadly isochronous surfaces such as the Tarff–Auchivarie and Beinn Iaruinn–Brunachan Psammite boundaries. The scale of this cyclicity means that the trends are likely to have been basin-wide, so that if comparison is made with structurally isolated and disparate successions it may be possible to identify similar boundaries and cyclicity. This may then be of use in resolving the origins of some suspect terranes, especially when this approach is coupled with the more traditional methods of comparison, such as provenance studies and structural analysis.

The authors thank A. L. Harris, D. I. J. Mallick, S. Robertson, F. May, E. R. Phillips and N. H. Woodcock for their helpful comments and suggestions on an earlier version of the text. This paper is published with the permission of the Director of the British Geological Survey (NERC).

References

ANDERTON, R. 1971. Dalradian palaeocurrents from the Jura Quartzite. *Scottish Journal of Geology*, **7**, 175–178.

—— 1979. Slopes, submarine fans and syn-depositional faults: sedimentology of parts of the Middle and Upper Dalradian in the SW Highlands of Scotland. *In*: HARRIS, A. L., HOLLAND, C. H. & LEAKE, B. E. (eds) *The Caledonides of the British Isles – reviewed*. Geological Society, London, Special Publications, **8**, 483–488.

—— 1982. Dalradian deposition and the late Precambrian–Cambrian history of the North Atlantic region: a review of the early evolution of the Iapetus Ocean. *Journal of the Geological Society, London*, **139**, 421–431.

—— 1985. Sedimentation and tectonics in the Scottish Dalradian. *Scottish Journal of Geology*, **21**, 407–436.

GALLOWAY, W. E. 1989. Genetic stratigraphic sequences in basin analysis I: architecture and genesis of flooding-surface bounded depositional units. *Bulletin of the American Association of Petroleum Geologists*, **73**, 125–142.

GLOVER, B. W. 1989. *The sedimentology and basin evolution of the Grampian Group, Scotland*. PhD Thesis, University of Keele.

—— 1993. The sedimentology of the Neoproterozoic Grampian Group and the significance of the Fort William Slide between Spean Bridge and Rubha Cuil-cheanna, Inverness-shire. *Scottish Journal of Geology*, **29**, 29–43.

—— & WINCHESTER, J. A. 1989. The Grampian Group: a major Late Proterozoic clastic sequence in the Central Highlands of Scotland. *Journal of the Geological Society, London*, **146**, 85–96.

——, KEY, R. M., MAY, F., CLARK, G. C., PHILLIPS, E. R. & CHACKSFIELD, B. C. 1995. Neoproterozoic multiphase rifting; the Grampian and Appin groups of the southern Monadhliath Mountains of Scotland. *Journal of the Geological Society, London*, **152**, 391–406.

HARRIS, A. L. & PITCHER, W. S. 1975. The Dalradian Supergroup. *In*: HARRIS, A. L., SHACKELTON, R. M., WATSON J., DOWNIE, C., HARLAND, W. B. & MOORBATH, S. (eds) *A Correlation of Precambrian Rocks in the British Isles*. Geological Society, London, Special Reports, **6**, 52–75.

——, BALDWIN, C. T., BRADBURY, H. J., JOHNSON, H. D. & SMITH, R. A. 1978. Ensialic basin sedimentation: the Dalradian Supergroup. *In*: BOWES, D. R. & LEAKE, B. E. (eds) Crustal evolution in northwest Britain and adjacent regions. *Special Issue of the Journal of the Geological Society, London*, **10**, 115–138.

HICKMAN, A. H. 1975. The stratigraphy of the late Precambrian metasediments between Glen Roy and Lismore. *Scottish Journal of Geology*, **11**, 117–142.

KEY, R. M., MAY, F., CLARK, G. C., PEACOCK, J. D. & PHILLIPS, E. R. *Geology of the Area Around Glen Roy (Sheet 63W)*. Memoir of the British Geological Survey, in press.

LITHERLAND, M. 1980. The stratigraphy of the Dalradian rocks around Loch Creran, Argyll. *Scottish Journal of Geology*, **16**, 105–123.

MCKIE, T. 1993. Relative sea level changes and the development of a Cambrian transgression. *Geological Magazine*, **130**, 245–256.

MITCHUM, R. M. JR. & VAN WAGONER, J. C. 1991. High frequency sequences and their stacking patterns: sequence-stratigraphic evidence of high-frequency eustatic cycles. *Sedimentary Geology*, **70**, 131-160.

PHILLIPS, E. R., CLARK, C. G., KEY, R. M., MAY, F., GLOVER, B. W & CHACKSFIELD, B. C. 1994. Tectonothermal evolution of the Neoproterozoic Grampian and Appin groups, southwestern Monadhliath Mountains of Scotland. *Journal of the Geological Society, London*, **151**, 971–986.

——, —— & SMITH, D. I. 1993. Mineralogy, petrology and microfabric analysis of the Eilrig Shear Zone, Fort Augustus, Scotland. *Scottish Journal of Geology*, **29**, 143–158.

PICKERING, K. T., STOW, D. A. V., WATSON, M. P. & HISCOTT, R. N. 1986. Deep water facies, processes and models: a review and classification scheme for modern and ancient sediments. *Earth Science Reviews*, **23**, 75–174.

POSAMENTIER, H. W., ALLEN, G. P. & JAMES, D. P. 1992. High resolution sequence stratigraphy – the East Coulee Delta, Alberta. *Journal of Sedimentary Petrology*, **62**, 310–317.

RUSSELL, M. J. & ALLISON, I. 1985. Agalmatolite and the maturity of sandstones in the Appin and Argyle Groups and the Eriboll Sandstone. *Scottish Journal of Geology*, **21**, 113–122.

THOMAS, P. R. 1980. The stratigraphy and structure of the Moine rocks north of the Schiehallion Complex.

Journal of the Geological Society, London, **126**, 469–482.

VAIL, P. R., AUDEMARD, F., BOWMAN, S. A., EISNER, P. N. & PEREZ-CRUZ, C. 1991. The stratigraphic signatures of tectonics, eustasy and sedimentology – an overview. *In*: EINSELE, G., RICKEN, W. & SEILACHER, A. (eds) *Cycles and Events in Stratigraphy*. Springer Verlag, Berlin, 617–659.

VAN WAGONER, J. C., MITCHUM, R. M. JR., CAMPION, K. M. & RAHMANIAN, V. D. 1990. *Siliciclastic Sequence Stratigraphy in Well Logs, Cores and Outcrops: Concepts for High Resolution Correlations of Time and Facies*. American Association of Petroleum Geologists, Methods in Exploration Series, **7**, 55pp.

——, POSAMENTIER, H. W., MITCHUM, R. M. JR., VAIL, P. R., SARG, J. F., LOUTIT, T. S. & HARDENBOL, J. 1988. An overview of the fundamentals of sequence stratigraphy and key definitions. *In*: WILGUS, C. K., HASTINGS, B. C., KENDALL, C. G. St. C., POSAMENTIER, H. W., ROSS, C. A., VAN WAGONER, J. C. (eds) *Sea-level Changes: an Integrated Approach*. Society of Economic Paleontologists and Mineralogists, Special Publications, **42**, 39–46.

WINCHESTER, J. A. & GLOVER, B. W. 1988. The Grampian Group. *In*: WINCHESTER, J. A. (ed.) *Later Proterozoic Stratigraphy of the Northern Atlantic Regions*. Blackie, Glasgow, 146–161.

Index